THE ENDOCRINE HYPOTHALAMUS

AF756119

THE ENDOCRINE HYPOTHALAMUS

Edited by

S. L. Jeffcoate

Department of Endocrinology
Chelsea Hospital for Women
Chelsea, London, England

and

J. S. M. Hutchinson

Department of Developmental Biology
Aberdeen University
Aberdeen, Scotland

1978

ACADEMIC PRESS
London · New York · San Francisco
A Subsidiary of Harcourt Brace Jovanovich, Publishers

ACADEMIC PRESS INC. (LONDON) LTD.
24–28 Oval Road
London NW1

U.S. Edition published by
ACADEMIC PRESS INC.
111 Fifth Avenue
New York, New York 10003, U.S.A.

Copyright © 1978 by Academic Press Inc. (London) Ltd.

~~591.147.4~~
267012701

All Rights Reserved
No part of this book may be reproduced in any form
by photostat, microfilm, or any other means, without
written permission from the publishers

Library of Congress Catalog Card Number: 77-74372

ISBN: 0-12-382150-9

Printed in Great Britain by The Whitefriars Press Ltd.,
London and Tonbridge

Contributors

EVERETT, J. W., *Department of Anatomy, Duke University Medical Center, Durham, North Carolina*, 27710, *U.S.A.*

FRASER, H. M., *MRC Unit of Reproductive Biology, 2 Forrest Road, Edinburgh, Scotland*

HUTCHINSON, J. S. M., *Department of Developmental Biology, Marischal College, Aberdeen University, Aberdeen, Scotland*

JACKSON, I. M. D., *New England Medical Center Hospital*, 171 *Harrison Avenue, Boston, Massachusetts*, 92111, *U.S.A.*

JEFFCOATE, S. L., *Department of Endocrinology, Chelsea Hospital for Women, Chelsea, London S.W.*3, *England*

JONES, M. T., *The Sherrington School of Physiology, St. Thomas's Hospital Medical School, Lambeth Palace Road, London S.E.*1, *England*

KÖNIG, W., *Farbwerke Hoechst AG., Frankfurt-am-Main, West Germany*

LINCOLN, D. W., *Department of Anatomy, University of Bristol Medical School, University Walk, Bristol* 2, *England*

MARTIN, J. B., *New England Medical Center Hospital*, 171 *Harrison Avenue, Boston, Massachusetts*, 92111, *U.S.A.*

OLGIATI, V. R., *Department of Pharmacology, University of Milan School of Medicine*, 32 *Via Vanvitelli, Milan* 20129, *Italy*

PECILE, A., *Department of Pharmacology, University of Milan School of Medicine*, 32 *Via Vanvitelli, Milan* 20129, *Italy*

PICKERING, B. T., *Department of Pharmacology, The University, Bristol* 8, *England*

REICHLIN, S., *New England Medical Center Hospital*, 171 *Harrison Avenue, Boston, Massachusetts*, 92111, *U.S.A.*

SANDOW, J., *Farbwerke Hoechst AG., Frankfurt-am-Main, West Germany*

SHARP, P. J., *ARC Poultry Research Centre, West Mains Road, Edinburgh, Scotland*

TALEISNIK, S., *Instituto di Investigacion Medica, Mercedes y Martin Ferreyra, Cordoba, Argentina*

TINDAL, J. S., *National Institute for Research in Dairying, Shinfield, Nr. Reading, Berkshire, England*

Preface

The area of the brain known as the hypothalamus and the hormonal functions of the peptides it contains have been the subject of several books over the years including the publication of the Association for Research in Nervous and Mental Disease (1940), the monograph by G. W. Harris (1955), those edited by Haymaker, Anderson and Nauta (1969), and Martini, Motta and Fraschini (1970) and appropriate volumes published by the American Physiological Society in the Handbook of Physiology series. In addition aspects of the subject have been covered in the proceedings of conferences and in review series.

It seems appropriate at this time when information is increasing rapidly that there should be an attempt to delineate our knowledge of the endocrine hypothalamus (the term coined by Leveque in 1972) as a base-line for further developments. It is easy for a multi-author work to result in a disjointed book, and we have tried to achieve an approach as uniform and comprehensive as possible whilst reducing overlaps between chapters to a minimum.

The book has been designed as a source of knowledge and to be used as a reference book. Six general chapters on the history and anatomy of the hypothalamus, its control, the chemistry of its peptides and the methods that have been used to study its endocrine function are followed by chapters on the separate hypophysiotrophic functions. Cross-referencing allows specific instances to be used to illustrate general points made in the earlier chapters. Authors were requested to limit literature references and to cite reviews where applicable. The references are gathered together at the end of the book.

One aspect where uniformity is most important and particularly difficult is nomenclature. A guiding principle has been that a "releasing activity" (or release-inhibiting where appropriate) becomes a "releasing factor" during purification and a "releasing hormone" when chemically characterised; the adenohypophysial hormone involved is referred to in addition. Thus in Chapter 11, the corticotrophin releasing activity of the hypothalamus is called corticotrophin releasing factor (CRF) when semipurified; to write of corticotrophin releasing hormone would be premature. A second example, that of gonadotrophins (Chapter 8) is more complex but the principles are carefully followed to avoid ambiguity. Thus "gonadotrophin-releasing factor" (GnRF) is used to describe an extract that releases the gonadotrophins LH and FSH; LHRF and FSHRF are extracts shown to release LH and FSH respectively. LHRH is used to specifically refer to the decapeptide isolated and characterised by Schally and his colleagues—though it can also be described as a GnRF, LHRF and FSHRF.

This system of nomenclature is complicated by three considerations. Firstly, two international bodies (IUPAC/IUB and the World Health Organisation) have produced systems of nomenclature (see Chapter 5). Neither of these has been widely accepted. In the second place, the concept of specificity of hypophysiotrophic function has been eroded: thus LHRH releases both LH and FSH, thyrotrophin-releasing hormone releases both thyrotrophin and prolactin, somatostatin inhibits the release of both growth hormone and thyrotrophin (in addition to widespread non-hypophysiotrophic effects).

The distribution and functions of the "hypothalamic" peptides outside the hypothalamus constitute the third difficulty for their systematic nomenclature. Some have always maintained that these substances were not "hormones" in the Bayliss and Starling sense and their scepticism appears to have been justified by recent evidence for a wider neurophysiological role (e.g. as neurotransmitters) for these peptides. The term "cybernin" has been suggested by Guillemin to describe these carriers of chemical information. It has also been proposed that they are part of a "neurocrine system" alongside the endocrine system and a proposed paracrine system in the gut.

The importance of this field in medicine and physiology has been underlined by the award of the Nobel prize for 1977 to two leading investigators of the endocrine hypothalamus—Dr. R. Guillemin and Dr. A. V. Schally. It seems clear that the next few years will produce increased understanding of the wider physiological role of the peptides which have till now been thought of as the hormones of the endocrine hypothalamus.

February 1978.

S. L. J.
J. S. M. H.

Contents

CHAPTER 3

Control of the Endocrine Hypothalamus

J. S. M. HUTCHINSON

CHAPTER 4

The Biochemical Investigation of the Endocrine Hypothalamus

S. L. JEFFCOATE

CHAPTER 5

Chemistry of the Hypothalamic Hormones

J. SANDOW AND W. KÖNIG

CHAPTER 6

Neurohypophysial Hormones: Comparative Aspects

B. T. PICKERING

CHAPTER 7

Regulation of Thyroid-Stimulating Hormone (TSH) Secretion

S. REICHLIN, J. B. MARTIN AND I. M. D. JACKSON

CHAPTER 8

Control of Reproduction

P. J. SHARP AND H. M. FRASER

CHAPTER 9

Control of Prolactin Secretion

J. S. TINDAL

CHAPTER 10

Control of Growth Hormone Secretion

A. PECILE AND V. R. OLGIATI

CHAPTER 11

Control of Corticotrophin (ACTH) Secretion

M. T. JONES

CHAPTER 12

Control of Melanocyte-Stimulating Hormone (MSH) Secretion

S. TALEISNIK

CHAPTER 1

The Mammalian Hypothalamo–Hypophysial System

J. W. EVERETT

I. Introduction

The following comment was written several years ago in closing a review on the neural control of the adenohypophysis:

> " . . . most of the information outlined is descriptive of processes several steps removed from the fundamental ones. We still must speak in only general terms of . . . substances that mediate hypothalamic regulation of the adenohypophysis, of the identities and activities of nerve cells and pathways controlling discharge of these agents into the hypophysial portal circulation, and of the manner in which these neurones are influenced by sensory impulses and by substances which reach the brain through the general circulation" (Everett, 1964).

A few years later, at a conference devoted to the hypothalamic hormones, Geoffrey Harris (1971) remarked: "The main problem at the present time is to learn the molecular structure of one of the releasing factors." He prophetically suggested that once that was accomplished there would be rapid advances in many aspects of hypothalamo–hypophysial relationships. The chapters that follow here show the realisation of that prediction. By now, three releasing factors have been chemically identified and synthesised. By means of immunochemical techniques several laboratories are attempting, with considerable success, to locate the neurones and fibre tracts that manufacture and transmit these agents to their sites of discharge into the circulation. Radioimmunoassay techniques are being widely applied to investigate the dynamics of release of the adenohypophysial trophic hormones. Methods are rapidly being developed and used for the direct assay of the thyrotrophin-releasing hormone (TRH) and the luteinising hormone releasing hormone (LHRH). The specific uptake of steroid hormones in the hypophysis, the hypothalamus and related parts of the brain has been investigated at length. Sites of uptake have been localised by autoradiographic methods and the hypothalamic metabolism of the steroids is under study. No less important than the identification of TRH and LHRH has been the histochemical identification of monoamines by fluorescence techniques, disclosing the striking concentration of dopamine and noradrenaline in the median eminence. The functional relationships of the monoamines to releasing and inhibiting hormones controlling adenohypophysial secretion are under very active investigation. There are strong indications that the prolactin-inhibiting factor may actually be a catecholamine rather than a peptide. Closely related investigation by electron microscopy has been addressed to the median eminence, and attempts have been made to identify the several types of vesicles in nerve endings there, with neurochemicals known to be present. While this remains to be accomplished, one can expect that in the near future a combination of the wide variety of sophisticated techniques now available will fruitfully be brought to bear on this crucial question.

In a recent issue of *Endocrinology* (January, 1975) about one third of the articles relate to monoamines and releasing hormones in pituitary regulation, to localisation of neurones and fibre tracts containing LHRH, to feedback action of gonadal steroids, and to steroid binding in brain and hypophysis. An effort to place this modern scene in historical perspective will follow a short account of the descriptive terminology given the several hypophysial components and its embryological basis.

II. Hypophysial Ontogeny and Terminology

The hypophysis develops in the region immediately rostral to the notochord, arising from two ectodermal primordia: the saccus infundibularis, a ventral evagination from the diencephalon; and Rathke's pocket, a dorsal outgrowth of the buccal cavity (Fig. 1). The former differentiates from that portion of the neural tube immediately caudal to the chiasmatic plate, while the latter arises just rostral to the buccal membrane. The terminology applied to the fully developed complex is essentially that proposed by Rioch *et al.* (1940) (Table 1). The term "posterior lobe" is an operational expression based on the fact that the pars intermedia is firmly attached to, and in some species invades the neural lobe, hence remaining attached

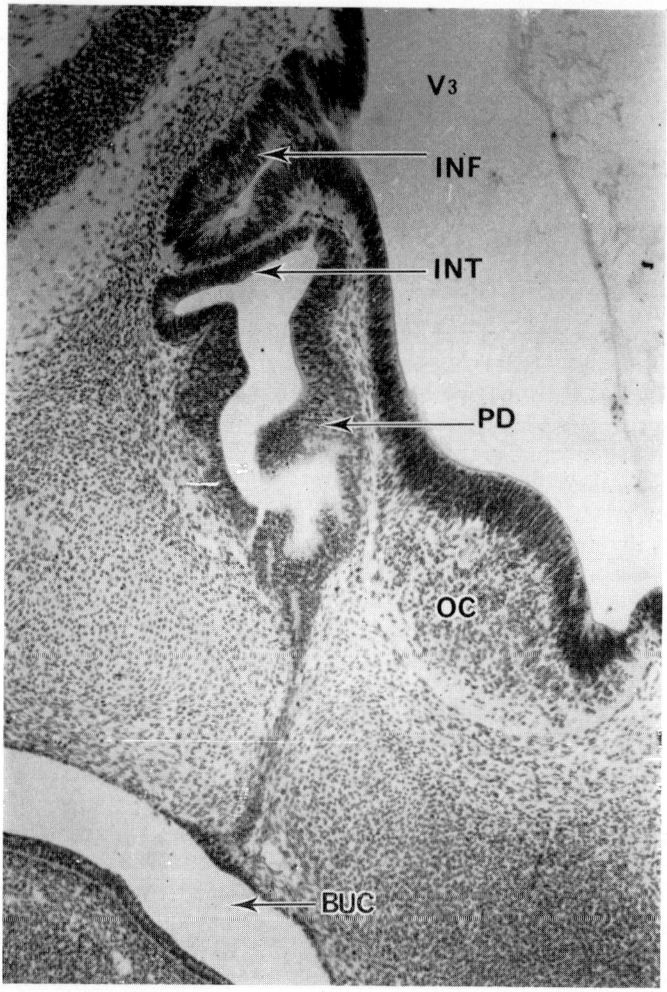

Fig. 1. The developing hypophysis in a 19 mm human embryo. Median sagittal section. 100×. Key: V_3, third ventricle; BUC, buccal cavity; INF, infundibulum; INT, pars intermedia; OC, optic chiasma; PD, pars distalis.

to it when the excised hypophysis cleaves at a persistent lumen of Rathke's pocket. There have been numerous discussions over the years about usage of the terms "infundibulum", "infundibular stem", and "median eminence". Thus, on the premise that the median eminence is not a part of the tuber cinereum, some authors have preferred to substitute "infundibulum" (see

Table 1. Subdivisions of the hypophysis

Adenohypophysis (Glandular lobe)	1. *Pars distalis*	Anterior lobe
	2. *Pars tuberalis*	
	3. *Pars intermedia*	Posterior lobe
Neurohypophysis	1. Lobus nervosus (*Neural lobe*, Infundibular process)	Posterior lobe
	2. *Infundibular stem*	Infundibulum
	3. *Median eminence*	Infundibulum

Haymaker *et al.*, 1969, pp. 219–220). However, while it is truly funnel-shaped in some species including our own, in others it is hardly so, as in the rat. On the same premise, the objection has been raised that the term "pars tuberalis" is inappropriate and ought to be replaced by "pars infundibularis". We shall be governed by common usage and simplicity, giving preference to those terms that are italicised in Table 1. In addition, the expression "hypophysial stalk" will mean the combination of the infundibular stem and the pars tuberalis.

Phylogenetically, the primitive neurohypophysis (lamprey) is a simple bulbous zone of the diencephalic floor, where neurosecretory fibres from the preoptic nucleus end along the external surface. The mammalian neurohypophysis, while beginning embryologically in much the same form, elongates commonly as a cylinder (*saccus infundibularis*) at the end of which extensive hypertrophy and folding takes place to form a more or less solid neural lobe. The proximal *median eminence* may remain thin-walled and retain primitive architecture, as in the rat. At the other extreme, it may be greatly thickened and structurally complex, as in man and most other higher primates. The *infundibular stem* varies greatly in length from species to species. The lumen is retained in a few cases, notably the cat and the dog, and may even continue into the neural lobe. The cells comprising the epithelium of the infundibular anlage differentiate almost exclusively as specialised neuroglia (pituicytes) and ependyma, although occasional neurones may form just beneath the ependyma.

Before the oral membrane ruptures, a nearby dorsal extension of the oral cavity extends along the base of the brain caudally toward the attachment of the notochord to the diencephalic floor. As connective tissue later moves between the brain and the oral epithelium this epithelial pocket remains close to the brain surface. The connection of the pocket with the oral epithelium narrows to a slender stalk, which eventually disappears except for vestigial nests of cells along the original path. After the rostral notochord

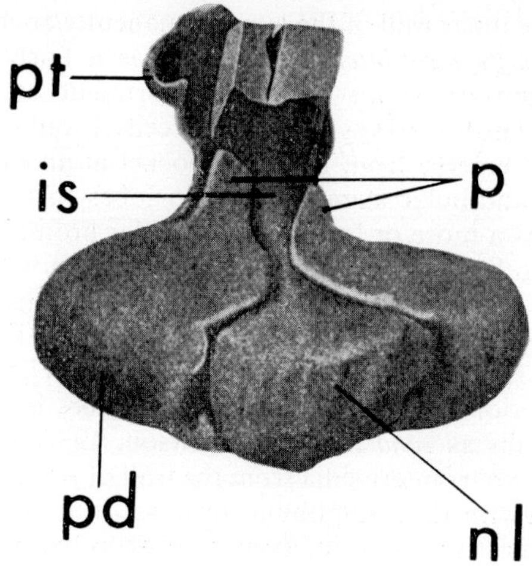

Fig. 2. Wax-plate reconstruction of the hypophysis from a 102 mm human foetus, viewed caudally and from above. 35 ×. (From Atwell, 1926, courtesy of the Wistar Institute Press). Key: *is*, infundibular stem; *nl*, neural lobe; *p*, process of the pars intermedia; *pd*, pars distalis; *pt*, pars tuberalis.

has regressed and the saccus infundibularis has formed, the deep end of Rathke's pocket remains adherent to this one derivative of the neural tube. The enlarging pocket becomes moulded about the infundibular process as a double-layered cup (Fig. 2). The outer wall of the cup rapidly thickens to form the *pars distalis*, encroaching on the adjacent cephalic veins so as to form a system of sinusoids amongst nests and columns of

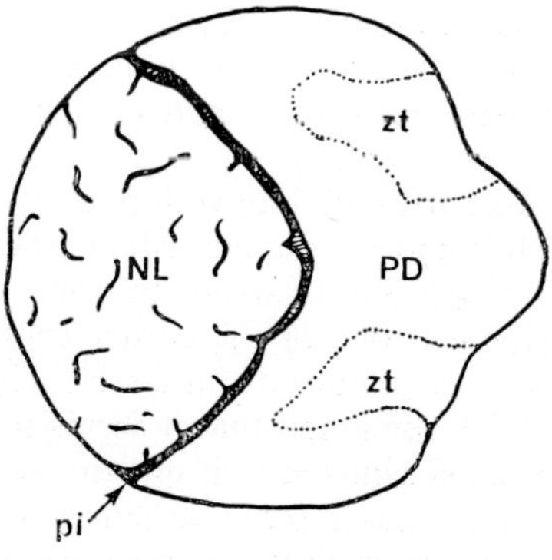

Fig. 3. Diagram of a horizontal section of the hypophysis of a rhesus monkey, especially to show the location of the "zona tuberalis" (see text). (From a photomicrograph by Dawson, 1948). Key: *nl*, neural lobe; *pd*, pars distalis; *pi*, pars intermedia; *zt*, zona tuberalis.

epithelial cells. The inner wall of the cup, permanently anchored to the neural lobe, differentiates as *pars intermedia*, either as a fairly uniform sheet of stratified epithelium or as an irregular assortment of follicles, compact epithelial lobules and/or arrays of isolated cells invading the neural lobe. Small lateral lobes emerge from Rathke's pocket at an early stage, opposite the narrowing infundibular stem. These two lobes form the *pars tuberalis*, moving together as a more or less complete collar around the stem and progressing upward to become approximately coextensive with the median eminence. Unlike the pars intermedia, the pars tuberalis remains separated from the neural tissue by a considerable amount of connective tissue, incorporating a vascular plexus from which the hypophysial portal circulation arises. The region of the anterior apposition of the lateral lobes is said to extend well into the pars distalis as *zona tuberalis* (Dawson, 1937, 1948) (Fig. 3).

Formation of the pars intermedia from the wall of Rathke's pocket depends on its close union with the infundibular process. In *Cetacea*, where a prominent wall of connective tissue intervenes between the infundibular process and pars distalis, neither the pars intermedia nor remnants of the original lumen of Rathke's pocket appear in the adult. Hanström (1966) describes similar deficiencies in the elephant, the armadillo, and a monotreme. The condition is characteristic of reptiles and birds. Experimental evidence of inductive action of the neural lobe was reported by Gaillard (1937), who found in tissue culture that epithelium derived from Rathke's pocket differentiated as recognisable pars intermedia only when combined with neural lobe tissue.

III. Historical Background

Fragments of information now recognised as fundamental to the science of neuroendocrinology can be traced back for many decades and, in some cases, for centuries. The effects of castration of man and domestic animals were known in ancient times. Seasonal influences on reproduction and behaviour must surely have been recognised prehistorically. Marshall (1922) notes reference by Aristotle and other classical authors to variation in the size of sea urchin gonads corresponding with phases of the moon, "common knowledge in the fish markets of the Mediterranean and elsewhere." According to Green (1966), Lister wrote in 1645 that swallows will lay additional eggs in a clutch if one of the earlier eggs is removed each day. Folley (1956), discussing a famous painting by Tintoretto entitled "The Origin of the Milky Way", remarked that it seems to have been inspired by an ancient myth. "It is as if Tintoretto, if not the ancients, intuitively knew that the suckling stimulus causes some response in the lactating breast." Haighton (1797) described in rabbits the phenomenon now recognised as reflex (provoked) ovulation. That report was made long before the recognition that bloodborne chemicals produced in one organ can influence another, that the pituitary is such an organ of internal secretion, and that it in turn depends to a great extent on central neural control.

Although neuroendocrinology is based historically in the endocrinology of reproduction, it owes much to lively clinical interests during the late 19th and early 20th centuries in intracranial tumours, diabetes, and defects of growth. Diabetic polyuria was variously ascribed to brain damage or to some defect of the pituitary gland. The former view appears to have stemmed from an observation by Claude Bernard (1878) that "piqûre" of the floor of the fourth ventricle results in glycosuria. Camus and Roussy (1913) demonstrated polyuria in dogs bearing surgical lesions of the hypothalamus without damage to the pituitary gland, thus giving strong support to protagonists of brain damage as the determining condition. In fact, they regarded the posterior pituitary as a "fragment nerveux atrophié". On the other hand, Cushing (1909, 1912) was a strong advocate of hypopituitarism as the cause, not only of diabetes insipidus, but also of asexual adiposity.

The correlation between acromegaly and hyperpituitarism can be traced to an 1887 review of case reports by Minkowski from patients having pituitary tumours (see Haymaker *et al.*, 1969). The correlation was wrongly interpreted to indicate impaired pituitary activity, on the assumption that a tumour necessarily represents loss of function. The concept of internal secretion had not yet emerged, and there was as yet no reason to differentiate the various parts of the pituitary functionally. As an appendage of the brain the neural lobe of the organ could even be viewed by Ramón y Cajal (1894) as a sense organ, presumably activated by nerve endings in the adjacent epithelium of the pars intermedia.

Berthold (1849) had shown that transplantation of testes in caponised cockerels resulted in normal male behaviour and the maintenance of normal appearance of the sex accessories. Thomas Addison (1855) had fully described the clinical syndrome of adrenal cortical deficiency, but without showing a cause-effect relationship. The techniques of ablation, transplantation and administration of organ extracts came into wide use as the 19th century drew to a close. Mering and Minkowski (1890) produced diabetes mellitus in dogs by pancreatectomy. Murray (1891) employed glycerine emulsions of sheep thyroids with some success in treating hypothyroidism. Extracts of the adrenal and pituitary glands were shown to have powerful vasopressor actions (Oliver and Schäfer, 1895). Finally, proof that chemical integration can occur without the intervention of the nervous system was produced by Bayliss and Starling (1902). They demonstrated that a substance (secretin), originating in the duodenal mucosa, stimulates pancreatic secretion after passage through the blood stream.

Meanwhile evidence was gathering that the sex accessories of vertebrates are governed by the gonads and that this can occur even when the gonads have been deprived of innervation by transplantation, as Berthold had first observed. Following the questionable report by Brown-Séquard (1889) of self-rejuvenation by injecting *aqueous* extracts of dog testis, there were many accounts of success or failure of such trials in both man and animals. Steinach (1894) succeeded in rats by transplanting testes either subcutaneously or into the abdominal cavity. Much later (1913) he noted the masculinisation of ovariectomised females by testicular grafts and the feminisation of castrated males by ovarian transplants. By that time it had become fully evident that the ovary is essential for maintenance of the uterus and mammary glands and for the cyclic phenomena of oestrus and menstruation; all these effects could be sustained by ovarian grafts, not only near the original site but in distant locations as well (Marshall and Jolly, 1907). However, the day of the pituitary gland as an agent controlling the ovary was several years in the future.

Suggestions of pituitary involvement in gonad function were apparent in the results of hypophysectomy experiments in dogs by Aschner (1909) and in Cushing's laboratory (Crowe *et al.*, 1910). Still earlier, Fichera (1905) noted pituitary enlargement in several species after gonadectomy. The presence within such pituitaries of abundant hypertrophied, vacuolated cells was described in rats by Biedl (1913), Schleidt (1914) and W. H. F. Addison (1917). Yet in 1922 the pituitary gland was not mentioned in either Long and Evans' monograph on "The Oestrous Cycle in the Rat" or Marshall's "The Physiology of Reproduction". The former authors were puzzled by the fact that pseudopregnancy could be induced by stimulating the cervix in rats whose ovaries had been transplanted and thus deprived of innervation. The possibility of a hormonal link was considered, but only through some agent liberated from the irritated cervix. Although Marshall discussed the coital induction of ovulation in rabbits, cats and ferrets, this was viewed as an effect of vascular engorgement of the ovary secondary to sexual excitement. Spontaneous ovulation was also pictured as the result of vascular engorgement, although with some reservation since that takes place during pro-oestrus instead of oestrus when ovulation normally occurs. Evans and Long, by the time their monograph appeared, however, had already reported (1921) stimulation of growth, oestrus and ovulation following the intraperitoneal administration of fresh adenohypophysial material. Analysis of trophic functions of the adenohypophysis advanced rapidly after the development of Philip Smith's (1926) parapharyngeal technique for the hypophysectomy of rats. Smith and Engle (1927) soon demonstrated the primary dependence of gonads, adrenal cortex and thyroid gland on adenohypophysial secretions. By that time it was clear that these activities were distinct from those of the neural lobe.

Recognition of functional reciprocity between the adenohypophysis and its target organs (see Moore and Price, 1932; and see Chapter 3) arose from the findings that gonadectomy results in not only a marked increase in weight of the hypophysis, but also a large increase in its gonadotrophin content (potency), together with the fact that administration of gonadal extracts suppresses gonadotrophin content and secretion (i.e. "negative feed-back" in modern parlance). The possibility that gonadal secretions might also have a stimulative action ("positive feedback") was raised by Engle (1931).

During the following two decades, as the various trophic secretions of the hypophysis and their actions on the target organs became increasingly well characterised, there was much interest in whether the pars distalis might be innervated and thus be under neural control. As mentioned earlier, there were already suggestions in the literature pointing in that direction. It seemed likely that coitally induced ovulation in rabbits involves a neural stimulus, especially after Fee and Parkes (1929, 1930) demonstrated that the hypophysis must remain in place for only an hour after coitus and that the effect could be prevented by vaginal anaesthesia. The classic experiment proving participation of the brain in the coital ovulation reflex was performed by Marshall and Verney (1936) who induced ovulation in rabbits by passing electric current through the head. The following year Harris (1937) and Haterius and Derbyshire (1937) obtained the same result by local stimulation of the hypothalamus. Westman and Jacobsohn (1937) demonstrated that postcoital ovulation would not occur if the pituitary stalk had been cut and a barrier of metal foil had been inserted between the hypothalamus and the hypophysis. The barrier was intended to prevent regeneration of nerve fibres assumed to be the final path of the stimulus. Although autonomic pathways had been considered possible routes, it now seemed clear that the stimulus took the more direct path via the stalk. Since Ramón y Cajal (1894) described nerve fibres passing from the neural lobe to endings in the pars intermedia, it had seemed reasonable to suppose that similar fibres could pass beyond to terminate in the pars distalis. However, searches for them failed to produce convincing results. Numerous claims for their existence were made, but were the results of vagaries of the silver techniques used and their tendency to stain reticular fibres as well as nerve fibres. A most thorough search by Rasmussen (1938) led to the firm conclusion that the only nerve fibres in the pars distalis are those innervating blood vessels.

Meanwhile, evidence of a sexual centre in the hypothalamus of the spontaneously ovulating rat was presented in a series of important studies by Hohlweg and associates (1932–1934). When the hypophysis had been transplanted to a site distant from the brain, gonadectomy failed to produce the usual large population of "castration cells", enlarged and vacuolated basophiles. It was also demonstrated that the administration of oestrogen to ordinary castrates prevents the castration cell reaction and that in the female rat at puberty there is an upward shift in threshold for the steroid. The administration of oestrogen to prepubertal females induced premature formation of corpora lutea.

Westman and Jacobsohn (1938a, b) confirmed this stimulative action of oestrogen and presented the additional information that the effect could be prevented by cutting the pituitary stalk and inserting a barrier. This was equivalent to transplanting the hypophysis to a distant site which in chronic preparations leads to target gland atrophy. From the modern vantage point, the acute effect was not the result of severing nerve fibres as then interpreted, but was caused by interrupting the major source of blood to the pars distalis, i.e. the long portal vessels. Nevertheless, it was apparent that the hypothalamo–hypophysial connection is necessary for the oestrogen-induced release of gonadotrophin.

A. HYPOPHYSIAL BLOOD VESSELS

Prominent vessels on the outer surface of the hypophysial stalk had been recognised in man for many years (see Romeis, 1940). Green (1966) quotes a comment by Willis (1634) suggesting a humoural influence of blood reaching the hypophysis via a "rete mirabile". The first description of the stalk vessels as a portal circulation is usually credited to Popa and Fielding (1930). Independent observations by Pietsch (1930) should also be noted, although he did not identify the prominent veins of the pars tuberalis as "portal" vessels. In both reports the direction of blood flow was assumed to be upward from the gland toward the brain. This was based partly on the observation by Popa and Fielding that, in living rabbits, if a clamp was placed on the hypophysial stalk the portal vessels below the clamp became distended. The explanation of that result seems to be that the rabbit, unlike all other familiar mammals, has a direct arterial blood supply to the pars distalis in addition to the principal source through the portal vessels.

The first report of "downward" flow in the portal vessels was by Houssay *et al.* (1935), not in a mammal, but in a South American toad. Blood flow toward the pars distalis was directly seen in the living animal. Furthermore, lesioning of the infundibulum caused the gland to become infarcted. Wislocki and King (1936) and Wislocki (1938) on histological grounds concluded that in the cat and the rhesus monkey the primary bed of capillaries is located in the median eminence and infundibular stem, whence the portal vessels transmit blood to the pars distalis sinusoids. That view was firmly substantiated later by direct observation of flow in several species of amphibia (Green, 1947), duck (Benoit and Assenmacher, 1955), rat (Green and Harris, 1947, 1949, and others), mouse (Worthington, 1955), dog and cat (Török, 1954, 1962), baboon and rhesus monkey (Daniel, 1966).

Figure 4 illustrates the circulatory pattern of the human hypophysis according to present views (Xuereb *et al.*, 1954). In principle, this pattern appears to be characteristic of mammals generally, with the exception of the rabbit. Virtually all blood reaching the pars distalis comes through either long or short portal vessels, the former draining primary capillaries of the median eminence and upper stem, the latter deriving from capillaries of the lower stem next to the neural lobe. The superior hypophysial artery feeds the median eminence, upper stem, and even a major portion of the lower

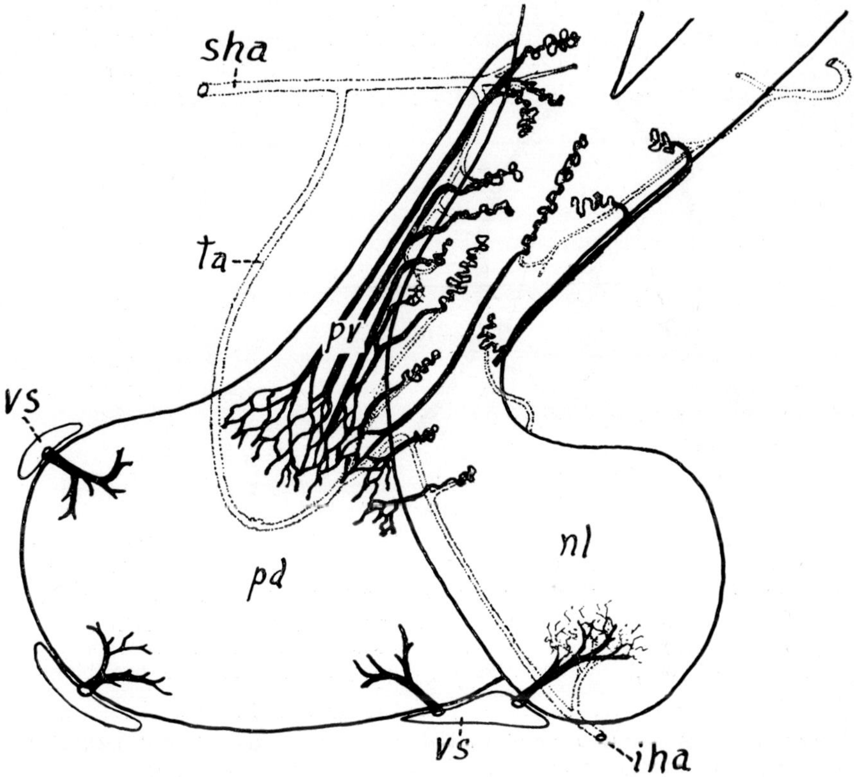

Fig. 4. Diagram of the blood vessels of the human hypophysis. (Re-drawn after Xuereb *et al.*, 1954). Key: *iha*, inferior hypophysial artery; *nl*, neural lobe; *pd*, pars distalis; *pv*, portal vessels; *sha*, superior hypophysial artery; *ta*, trabecular artery; *vs*, venous sinus.

stem through the trabecular artery. The inferior hypophysial artery, distributed mostly to the neural lobe, also supplies some blood to the pars distalis through the short portal vessels. This artery anastomoses with a branch of the trabecular artery. The long portal vessels traverse the pars tuberalis and empty into the anterior-superior region of the pars distalis referred to as the *zona tuberalis* (see Dawson, 1948), said to have an origin in common with the pars tuberalis from the lateral lobes of Rathke's pocket (Fig. 2). Some variation in the extent of anastomosis between the inferior hypophysial and trabecular arteries is suggested by the varied descriptions given by different authors (Landsmeer, 1963; Duvernoy, 1972).

The length and prominence of the long portal channels among different

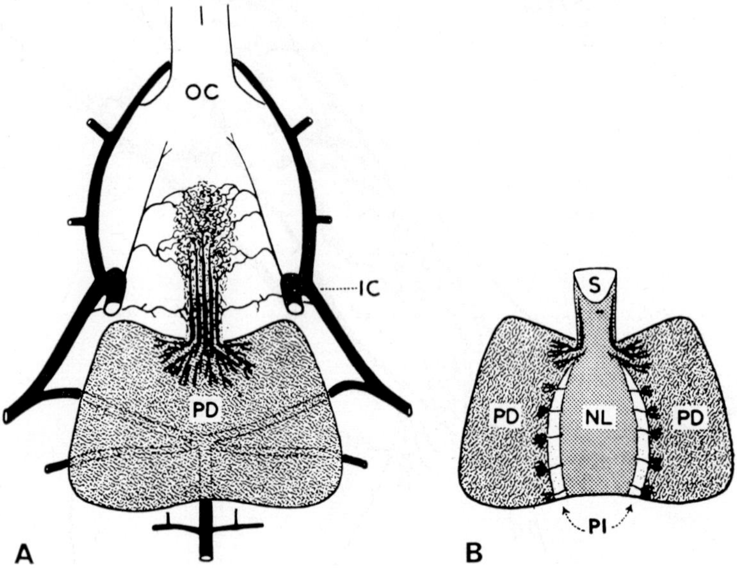

Fig. 5. Diagrams showing the blood supply of the rat pars distalis via long and short portal vessels. (From Adams *et al.*, 1963, by courtesy of the authors and the publisher of the *Quarterly Journal of Experimental Physiology*).

A. Ventral aspect. Superior hypophysial arteries pass as fine twigs from the internal carotids (IC) into the primary capillary bed in the median eminence and stem, whence the long portal vessels pass down the stalk into the pars distalis.

B. Dorsal aspect. Short portal vessels arising from capillaries in the neural lobe (NL) pass through the pars intermedia (PI) to enter the pars distalis. Some long portal vessels also appear on the dorsal aspect of the stalk (S).

species is governed by the length of the infundibular stem. Thus, in the cat and the dog, where the stem is very short and the pars distalis lies close to the median eminence, the long portal vessels are correspondingly restricted. Much of the pars distalis is supplied through abundant short vessels draining capillary loops of the adjacent infundibulum (Wislocki and King, 1936). In the rat, on the other hand, the principal blood supply to the pars distalis comes through the long portal vessels (Fig. 5). Short portal venules arise from capillary loops round the periphery of the neural lobe and pass through the pars intermedia to adjacent territory of the pars distalis.

Anatomically, there is a considerable territorial specificity among the various portal channels, with respect both to their origins from the primary capillary complex and to zones of distribution in the pars distalis. It has been proposed that this may determine the regional variations of cell types in the pars distalis (Daniel, 1966). Experimental evidence of this was obtained by Pasteels (1960) in the amphibian *Pleurodeles waltlii.* The hypophysis was removed and reimplanted in the original site, but inverted and rotated 180°. The regional cytology of the gland became reversed in accord with the new relationship to the regenerated portal vessel pattern. In the rat, sheep, goat and rhesus monkey, Adams *et al.* (1963) demonstrated that after partial transection of the hypophysial stalk the size and location of pars distalis infarcts varies with the number and position of the long portal vessels interrupted by the cut. A differing view has been given by Porter *et al.* (1970). Perfusion of the rat hypophysis with coloured fluids through microcannulation of single portal channels, indicated that distribution through the pars distalis from any given vessel is greatly varied and that sometimes the entire gland may fill. However, these variations may have represented a considerable imbalance of pressure between the vessel cannulated and the other channels. One may reasonably suppose that under physiological conditions of balanced pressure territorial specificity is maintained. It might be shifted, however, through surgical manipulation of the region, even from the mechanical effect of inserting an electrode into the tuberal hypothalamus.

Communication between capillaries of the median eminence and those of the adjacent tuberal hypothalamus have been recognised by several observers. Wislocki (1938) described them, but regarded them as of little consequence. Török (1954, 1962), on the other hand, directly observed in the dog and the cat, the movement of blood-borne dye from capillary loops into subependymal channels continuing into the hypothalamus. Examination by Duvernoy (1972) of the microcirculation of the hypothalamo–hypophysial system in 21 species of birds and mammals, including man, demonstrated the ubiquitous presence of anastomoses between vessels of the median eminence and hypothalamus, differing somewhat in detail from species to species. The functional significance of the anastomoses is not known, although some authors consider them to represent the anatomical basis for "short-loop feedback", whereby the hypothalamus may be influenced by neurohumours from the median eminence (Szentágothai *et al.*, 1972; see also Chapter 3).

Venous drainage from the pars distalis and from all of the neural lobe, except that limited part giving origin to short portal vessels, passes directly into the adjacent cavernous sinuses or their tributaries. The very restricted circulation of the pars intermedia drains with the neural lobe.

B. THE NEUROHUMOURAL HYPOTHESIS OF PARS DISTALIS REGULATION

The first suggestion that a humoural link intervenes between the central nervous system and the pars distalis, controlling its secretion, was made by Hinsey and Markee (1933) after failing to prevent postcoital ovulation in rabbits by sectioning the cervical sympathetic trunks. They proposed that some substance produced by the "posterior lobe" might pass "hormonally" to the pars distalis. After Wislocki and King (1936) presented convincing evidence that the direction of blood flow in the hypophysial portal vessels is downward, Friedgood (1970) suggested in 1936 that these vessels might serve as a humoural pathway. That possibility occurred to others as well (see Hinsey, 1937, and discussion that followed). However, for a long while the idea met resistance because of the frequent misunderstanding of the fact that the primary capillaries of the portal circulation lie within portions of the neurohypophysis, not the hypothalamus. Sceptics pointed out, for example, that in *Cetacea* a heavy connective tissue septum intervenes between the neural lobe and the pars distalis. However, it has been shown, in such cases, that the pars tuberalis with its incorporated portal vessels, nevertheless closely contacts the median eminence and infundibular stem; it is there that the primary capillaries are found.

Although in 1937 Harris dismissed humoural transmission as the likely means of neural activation of the pars distalis, he later became one of its champions. Green and Harris (1947) presented several strong arguments favouring it. Noting the variable results among different reports of the effects of cutting the hypophysial stalk, they proposed that this may have been due to regeneration of portal vessels. Harris (1950) substantiated that proposal by demonstrating that in stalk-sectioned rats reproductive functions were restored in a degree that was directly proportionate to the degree of portal vessel regeneration. The regenerative capacity of these vessels was also shown in the rhesus monkey (Harris and Johnson, 1950) and the rabbit (Jacobsohn, 1954). It is now recognised to be a general phenomenon and that to prevent renewed connection of the gland to the median eminence after stalk sectioning, an impermeable barrier must be inserted.

These observations appeared to explain the early findings by Greep (1936), who had obtained restoration of function in pars distalis autografts or homografts placed back in the original site after hypophysectomy. There was some uncertainty about that finding, however, since it would have been difficult to distinguish grafted tissue from possible fragments of the original gland left *in situ*. Harris and Jacobsohn (1952) resolved this problem by means of a dual operation. Rats were hypophysectomised by the parapharyngeal route, following which the pars distalis was implanted via the transtemporal route to a position immediately beneath the median eminence. These grafts, usually from the female subject's newborn offspring, restored normal reproductive function, often within a few days. The significance of this result was enhanced later by Nikitovitch-Winer and Everett (1958). The pars distalis was first autografted to the renal capsule where it lost most of its capacity to secrete FSH, LH, ACTH and TSH, but gained in ability to produce prolactin (Everett, 1954, 1956). After several weeks the graft was recovered and re-transplanted transtemporally to a position under the median eminence. In control preparations the graft was either left on the kidney or re-transplanted under the temporal lobe of the brain. Most of the rats having grafts revascularised from the median eminence began to have oestrous cycles; and after exposure to fertile males many became pregnant. There was also moderate restoration of ATCH and TSH secretion. Control grafts continued to secrete chiefly prolactin. Cytologic differences among the grafts in the several locations were equally significant. Whereas the control grafts, either on the kidney or under the temporal lobe, were comprised chiefly of prolactin-type acidophiles, grafts under the median eminence displayed many large gonadotrophic and thyrotrophic basophiles (Nikitovitch-Winer and Everett, 1959). Thus, it had become abundantly clear that some blood-borne agent(s) delivered from the median eminence to the pars distalis must be necessary for the manufacture and release of its various trophic secretions.

IV. Hypothalamic Nuclei and Pathways

The diencephalon in the embryonic vertebrate comprises four longitudinally oriented cell columns, each separated from the next by a distinct sulcus (Kuhlenbeck, 1954; see Haymaker *et al.*, 1969). These features are clearly evident in Fig. 6. The most ventral column, in early stages much the largest, is the hypothalamus, delimited dorsally by the hypothalamic sulcus. A portion of this primitive column caudal to the mamillary bodies eventually moves laterally and becomes the subthalamus. In early stages the preoptic hypothalamus is set off from the remainder by a sulcus that later deepens to form the preoptic recess. Although some authors in the past have regarded the preoptic region as partially or even completely a part of the telencephalon, Kuhlenbeck considers that it lies completely within the hypothalamus: that view appears to be well-founded and is adopted here.

The adult hypothalamus is characterised by a diffuse array of nerve cells, among which some are sufficiently aggregated to be recognised anatomically as nuclei. In general, the borders of these nuclei are ill-defined, showing only gradual transitions to relatively cell-poor zones. With the exception of the large cells of the supraoptic and paraventricular nuclei, distinguished by their neurosecretory features and the large amount of Nissl substance in the periphery of the cytoplasm, cytoarchitectural criteria are of limited value in the identification of nuclei. Relative positions of the nuclei with respect to outstanding landmarks are of primary importance. Figure 7 illustrates adaptations to interspecies variations in configuration of the diecephalon. In some instances where there is no marked aggregation of cells the expression

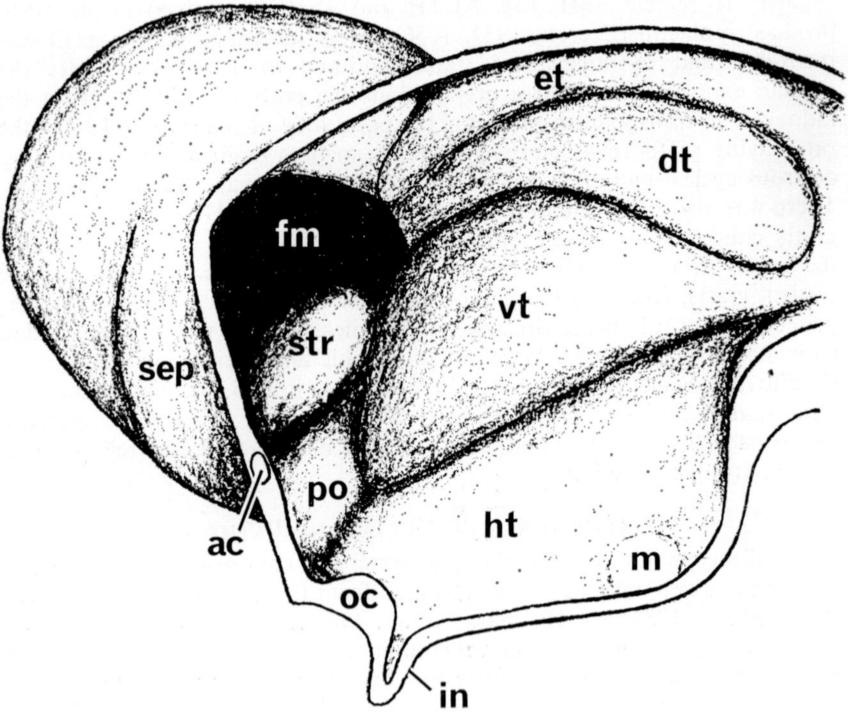

Fig. 6. Human diencephalon as in an 18 mm embryo. Median sagittal aspect. Re-drawn from a diagram prepared by H. Kuhlenbeck (Armed Forces Institute of Pathology, Neg. 92410. See Haymaker *et al.*, 1969, page 20). Key: ac, anterior commissure; dt, dorsal thalamus; et, epithalamus; fm, foramen of Monroe; ht, hypothalamus; in, infundibulum; m, mamillary body; oc, optic chiasma; po, preoptic hypothalamus; sep, septum; str, striatum; vt, ventral thalamus.

"area" is preferred. Thus, the *anterior hypothalamic area* is a population of cells dispersed between the *medial preoptic area* and the *suprachiasmatic nucleus* anteriorly and the *dorsomedial* and *ventromedial nuclei* posteriorly. Note that in the diagrams of both the rat and cat hypothalami, the dorsomedial nucleus is shown to continue without interruption caudally to become designated as *posterior periventricular area*. That in turn seems to blend directly with the *infundibular* (*arcuate*) *nucleus*.

For convenience the hypothalamus can be divided into three longitudinal zones (Crosby and Woodburne, 1940; Nauta and Haymaker, 1969), lateral, medial and periventricular. The lateral zone is that occupied by the medial forebrain bundle, the lateral preoptic area, the lateral hypothalamic area

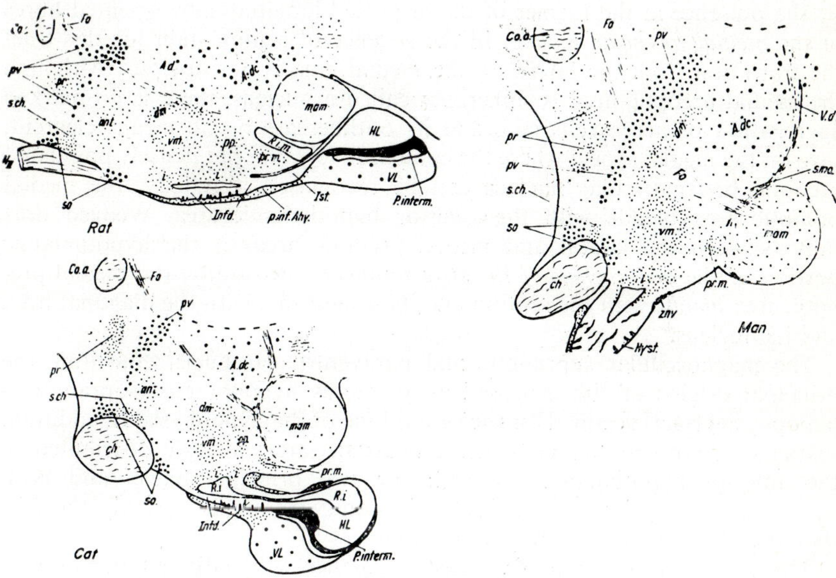

Fig. 7. Variations of positions, form and dimension of the hypothalamic nuclei in accord with species variation of configuration of the basal diencephalon. (From Diepen, 1962, by courtesy of the author and Springer-Verlag).

Hypothalamic nuclei or areas: A.d., dorsal hypothalamic area; *A.dc.*, dorsocaudal hypothalamic area; *ant*, anterior hypothalamic area; *dm*, dorsomedial nucleus; *i*, infundibular (arcuate) nucleus; *mam*, mamillary body; *pr.m.*, premamillary nucleus; *pp*, posterior periventricular area; *pr*, preoptic area; *pv*, paraventricular nucleus; *s.ch.*, suprachiasmatic nucleus; *sma*, supramamillary nucleus; *so*, supraoptic nucleus; *vm*, ventromedial nucleus.

Hypophysial components: HL, neural lobe; *Hyst*, hypophysial stalk; *Infd.*, infundibulum; *Ist*, infundibular stem; *p.inf.Ahy.*, pars infundibularis (tuberalis) of adenohypophysis; *P.interm.*, pars intermedia; *VL*, pars distalis; *znv*, neurovascular zone.

Other captions: ch, optic chiasma; *Co.a.*, anterior commissure; *fo*, fornix; N_{II}, optic nerve; *R.i.*, recessus infundibuli; *V.d'A.*, bundle of Vicq d'Azyr.

and the tuberomamillary nucleus. The medial zone includes the medial preoptic area, anterior hypothalamic area, paraventricular nucleus, dorsomedial nucleus, ventromedial nucleus, premamillary nucleus and the mamillary complex. The periventricular zone is distinguished from the medial zone as a comparatively homogeneous region of varied thickness intimately associated with the ventricular ependyma. The suprachiasmatic and arcuate nuclei are prominent subdivisions of this zone. The supraoptic nucleus extends from the medial zone basally along the optic tract ventral to the lateral zone.

In the preoptic region the lateral and medial areas can best be distinguished by the presence in the former of the dispersed longitudinally oriented fibres of the *medial forebrain bundle*. In the absence of a fibre stain for this tract the transition from the lateral to the medial area is not obvious. Although the medial preoptic area is sometimes called a nucleus, there is no marked aggregation of nerve cells here and its outlines are by no means clearcut, save on the medial aspect where the comparatively thin but densely populated preoptic periventricular nucleus can be recognised. Caudally, the medial preoptic area blends with the anterior hypothalamic area. Wedged dorsally between the lateral and medial preoptic areas is the hypothalamic portion of the *bed nucleus of the stria terminalis*. Rostrally, the medial preoptic area blends with the septum and, basomedially, with the diagonal tract and its nucleus.

The magnocellular supraoptic and paraventricular nuclei constitute the principal origins of the neurosecretory tractus hypophyseus (supraoptico-hypophysial tract) destined for the neural lobe of the hypophysis. In addition, scattered neurosecretory cells and cell groups are variously distributed in the anterior hypothalamus between the two principal nuclei and both rostral and caudal to the chiasma. In some mammals, notably the fox, they may even appear in the infundibulum.

The arcuate (infundibular) nucleus, together with adjacent portions of the anterior periventricular area, is thought to be the principal origin of the tubero-infundibular fibres that end in the median eminence and infundibular stem in close association with the primary capillaries of the hypophysial portal circulation. Collectively, this portion of the periventricular zone in the rat (Fig. 8) has become known as the "hypophysiotrophic area" (Halász *et al.*, 1962), the presumptive site of production of the hypothalamic hormones controlling the pars distalis. Theoretically some fibres of this system extend into the neural lobe to end about the primary capillaries of the short portal system.

Afferent neural pathways to hypothalamic neurones have been discussed in rich detail by Nauta and Haymaker (1969). Space allows only a summary here (see also Chapter 3). The medial forebrain bundle constitutes the principal longitudinal pathway and is regarded as the rostral continuation of the reticular formation of the midbrain with which it connects through the mamillary peduncle lateral to the mamillary body. The majority of its axons are short, with frequent synaptic relays through the cells of the lateral hypothalamic and lateral preoptic areas; relatively few fibres pass for long distances by this route. Rostrally it receives inputs from various basal

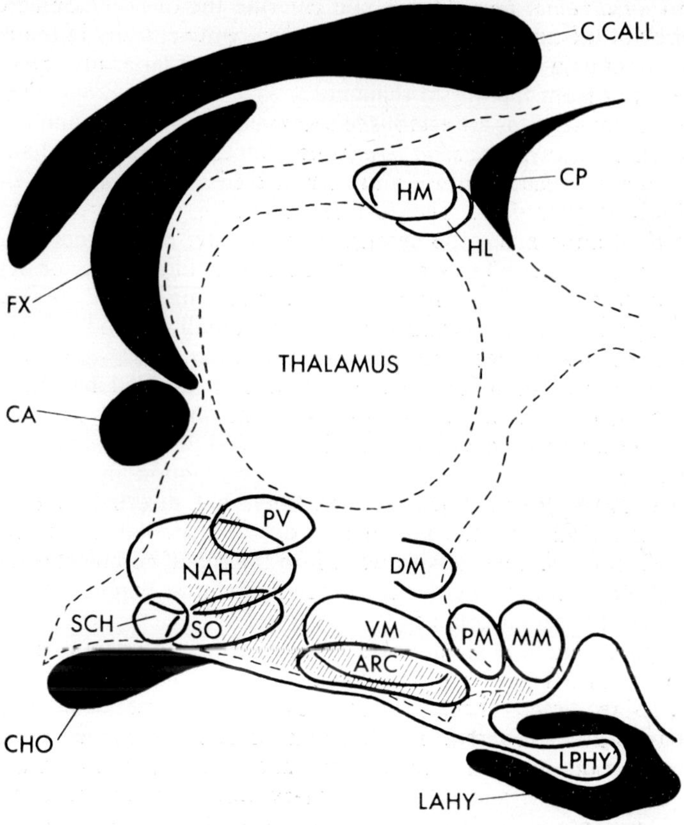

Fig. 8. Scheme of the rat diencephalon in median aspect, showing the hypophysiotrophic area (cross-hatching). ARC, arcuate nucleus; CA, anterior commissure; C CALL, corpus callosum; CHO, optic chiasma; CP, posterior commissure; DM, dorsomedial nucleus; FX, fornix; HL, lateral habenular nucleus; HM, medial habenular nucleus; LAHY, pars distalis of hypophysis; LPHY, neural lobe of hypophysis; MM, medial mamillary nucleus; NAH, anterior hypothalamic area; PM, premamillary nucleus; PV, paraventricular nucleus; SCH, suprachiasmatic nucleus; SO, supraoptic nucleus; VM, ventromedial nucleus. (From Halász *et al.*, 1965, by courtesy of the authors and the publisher of *Endocrinology*).

elements of the limbic forebrain, from the septum, the dorsal thalamus, and from the hippocampus by way of the precommissural fibres of the fornix. Conduction through the medial forebrain bundle passes both rostrally and caudally. Communication with the medial hypothalamus is established by short transverse fibres.

Another prominent connection with the midbrain is the continuation of the *dorsal longitudinal fasciculus*. Upon entering the diencephalon from the central grey of the aqueduct, the fasciculus fans out ventrally in the periventricular system as a vertical array of fine fibres. The latter also receives an extensive input from the dorsal thalamus.

Important connections are established between the amygdala and hypothalamus by two pathways, the *stria terminalis* from the corticomedial amygdala and the ventrally placed diffuse and complicated *sublenticular* system from the basolateral amygdala. The former gives rise to two principal subdivisions as it approaches the anterior commissure, one retrocommissural, distributing to the medial preoptic area and to part of the anterior hypothalamic area, the other supracommissural, sweeping round the anterior commissure rostrally and ventrally and then caudally to end principally in the cell-poor region about the ventromedial nucleus.

Hypothalamic input from the hippocampus is afforded by the massive *fornix system*, through both its pre- and postcommissural divisions. The latter, after entering the hypothalamus, just behind the anterior commissure, takes a straight course back to its major termination in the mamillary body. In passage, however, many of its fibres are diverted to the dorsal thalamus and thence to the lateral and dorsal hypothalamic nuclei. The precommissural fornix spreads forward into the septal complex, some of its fibres arching ventrally to enter the medial forebrain bundle, while others end widely in the septal complex.

Septo-hypothalamic fibres are said to travel chiefly by way of the medial forebrain bundle. There is physiological evidence, however, for a diffuse system passing directly backward into the medial preoptic and anterior hypothalamic areas (Everett, 1964). The septum is envisioned as a nodal point bringing the hippocampus, amygdala, and pyriform cortex into relation with the preoptic and other hypothalamic regions, as well as with the midbrain tegmentum (see Nauta and Haymaker, 1969).

Intrinsic neuronal connections in the medial and periventricular zones are not well understood. Aside from fibres of the stria terminalis anteriorly and the dorsal longitudinal fasciculus posteriorly it has been difficult to demonstrate longitudinal organisation. Szentágothai *et al.* (1972) comment that " . . . the medial hypothalamus ought to be considered a neuronal network of quasi random internal connexions, [in which] . . . excitation can spread from a given focus in any direction and can establish an infinite number of closed self-re-exciting chains." Hypothalamic neurones, with the exception of those in the magnocellular neurosecretory nuclei, are relatively small, with multiple long branching dentrites. Axodendritic synapses are frequent and far exceed axosomatic synapses in number. Characteristically, a given cell through its great numbers of dendritic spines is in synaptic relationship with many axons. Furthermore, a given axon not only ramifies freely but contacts dendrites of numbers of cells in passing.

V. Neurosecretion

The frequent presence of stainable "colloid" in the neural lobe and along the course of the tractus hypophyseus was the earliest visible evidence of a secretory function of this system. It was at first reasonable to consider that the material was formed by the epithelium of the pars intermedia (Cushing and Goetsch, 1910), stored in the neural lobe, and thence passed to the brain in some manner along the tract (Collin, 1928). That notion of upward movement was in line with other views depicting the hypophysis as subservient to the brain. Witness Ramón y Cajal's (1894) interpretation of the neural lobe as a sense organ and the Popa–Fielding (1930) hypothesis of centripetal flow of blood in the hypophysial portal vessels. As speculation gave way to accumulating facts, the basic error in all these conjectures became evident.

Cytological evidence suggesting that some neurones can secrete, was presented by Dahlgrén (1914) and Speidel (1919) who described cells in the spinal cord of a skate, containing inclusions that resembled secretory granules. Ernst Scharrer in 1928, described similar inclusions in the preoptic neurones of the killifish *Fundulus spp.* The inclusions were later recognised as consistent features of the large neurones of the supraoptic and paraventricular nuclei of mammals and the homologous magnocellular hypothalamic nuclei of lower vertebrates (see Scharrer and Scharrer, 1940). There were speculations about possible endocrine roles. Solid experimental evidence that the supraoptic nuclei have a secretory role was furnished by the classical studies of the Ranson school (Fisher *et al.*, 1938) demonstrating that ablation of the supraoptic nuclei, or transection of the neural stalk, resulted in atrophy of the neural lobe and diuresis. Both effects were quantitatively dependent on the extent of the lesions. Rasmussen (1938) proposed that the anti-diuretic hormone is liberated from the terminals of the abundant fine nerve fibres in the neural lobe. Green (1947) described specialised nerve endings of such fibres in the neural lobe of the frog, noting their radiating patterns about the capillaries. Another development essential to present-day views of neurosecretion, was the demonstration by Weiss (1944) and Weiss and Hiscoe (1948) of axoplasmic streaming in nerve fibres, consistently in the direction away from the perikaryon. Finally, came the serendipitous observation by Bargmann (1949) that fibres of the tractus hypophyseus, their cells of origin and their terminals in the neural lobe are selectively stainable by Gomori's chrome-alum haematoxylin (CAH) technique. This break-through quickly led to the view that neurosecretory material is synthesised in the large cells of the supraoptic and paraventricular nuclei, passed along their axons to the neural lobe for storage and eventual release from the axon terminals near the capillaries (Bargmann and Scharrer, 1951). It was shown that the stainable material disappears from the neural lobe in dehydrated animals.

Stalk section or other lesions interrupting the tract resulted in the accumulation of the neurosecretory substance proximal to the point of interruption (Hild and Zetler, 1951). Neural lobe hormone activity was demonstrated in all parts of the system stainable by CAH, including the two nuclei.

Although as seen under the light microscope the neurosecretory material was first thought to ensheath the unmyelinated fibres of the tractus hypophyseus and terminals in the neural lobe, this concept was dispelled by studies with the electron microscope. Palay (1957) summarised ultrastructural features of the magnocellular neurosecretory apparatus in the rat, essentially as now recognised. In the first place, the secretory material clearly remains within the axons and terminals in the form of membrane bounded granules. Other features of the fine, unmyelinated fibres of the tractus hypophyseus, are their content of abundant microtubules, filaments and occasional mitochondria. The axons are imbedded in pituicytes much as unmyelinated fibres of the peripheral nervous system rest in the periphery of Schwann cells. After the tract enters the neural lobe the axons are no longer enclosed in this way and pituicyte processes intermingle with them in the manner of astrocytes. The axons divide into numerous twigs whose swollen tips lie in palisade arrangement around capillaries. The terminals contain many mitochondria, many small electron lucent ("synaptic") vesicles 25 to 50 nm diam. and an abundance of comparatively large membrane-bounded electron dense granules having diameters of 100 to 150 nm or larger. Palay noted that the distribution of the large vesicles corresponds exactly with that of neurosecretory material stained by CAH, i.e. in the perikarya, tract and neural lobe. The terminals rest against a continuous basal lamina, interspersed by occasional foot processes of pituicytes. A narrow connective tissue space containing collagen fibres, occasional fibroblasts and other cells, intervenes between this and the basal laminae of capillaries. The capillaries are fenestrated.

The cell bodies of the magnocellular nuclei are often tightly clustered and tend to be closely apposed to capillaries, which are especially numerous in the supraoptic nucleus. Even the more scattered cells lying between the principal nuclei and along the brain floor toward the infundibulum, tend to be strung out along blood vessels. Unlike the capillaries of the neurohypophysis, these are not fenestrated.

The neurosecretory perikarya differ from those of other hypothalamic neurones not only in their large size, but in several other respects. Nissl substance is localised peripherally, leaving a more lightly stained perinuclear cytoplasm. The cell nuclei tend to be eccentrically placed and display prominent infoldings of the nuclear membrane. Two morphological types have been recognised on the basis of staining characteristics and ultrastructure (see, for example, Zambrano and DeRobertis, 1966). Type 1 cells, darkly staining with Toluidine Blue and related dyes, appear to be in a resting state.

As described by Clattenberg (1973), in the rabbit supraoptic and paraventricular nuclei, these cells are featured by an accumulation of elementary neurosecretory granules at the periphery of the soma, only occasional clusters of flattened cisternae of rough endoplasmic reticulum (RER) and poorly developed Golgi apparatus. Mitochondria and lysosomes are sparse. Type 2 cells, lighter staining, appear to be synthetically active. An elaborate RER round the periphery of the cytoplasm comprises dilated cisternae containing a finely granular substance. Intercisternal polysomes are abundant. The perinuclear cytoplasm contains well-developed Golgi profiles, many mitochondria, lysosome-like elements and electron lucent vesicles. Neurosecretory dense core vesicles, although occasionally seen budding from the Golgi apparatus, are comparatively few. Once formed, they seem to move into the axon rather promptly. Their scarcity in these active cells is a common characteristic among familiar mammals, with the dog as the outstanding exception (Sloper and Bateson, 1965; Zambrano and DeRobertis, 1967). The dog is also exceptional in that many of its supraoptic perikarya are highly vacuolated. In dehydration studies, accelerated activity of supraoptic cells is expressed by still further depletion of secretory granules from the perikarya and proximal portions of the axons. The RER cisterns become more dilated and there is a further increase in numbers of free ribosomes. Lysosomes become enlarged. After greatly prolonged dehydration, neurosecretory granules reappear in the perikaryon.

There is an obvious distinction between the dimensions of the elementary granules visualised by the electron microscope (usually < 200 nm) and dimensions that can be resolved by the light microscope. It appears that in some instances the inclusions formerly thought to be secretory granules, may in fact have been lipofuscins or aggregates of the elementary granules.

Chemically speaking, the neurosecretory material of the magnocellular system is a complex of proteins (neurophysins) and the octapeptide neural lobe hormones, vasopressin and oxytocin. There is strong evidence for separate neurophysins binding the two hormones, Np–I for oxytocin and Np–II for vasopressin (see Hayward, 1975). All these substances are rich in cystine, a fact accounting for the selective staining of the perikarya, axons and terminals. Cytochemical staining techniques substantially specific for cystine-containing polypeptides, are as effective as CAH and aldehyde fuchsin (see Sloper, 1966). These methods include performic acid oxidation followed by a basic dye such as Alcian Blue. Fluorescent basic dyes have also been used in this way. Another useful procedure involves reduction of cystine to cysteine by sodium thioglycollate; this is followed by ferric ferricyanide which the cysteine then reduces to form Prussian Blue.

The recent application of immunological techniques affords a more direct means of localising and tracing the neurophysins and hence, presumably, the neural lobe hormones. In the rat (Watkins and Evans, 1972) and in the calf and the rhesus monkey (Zimmermann *et al.*, 1973) neurophysin has thus been disclosed throughout the supraoptic and paraventricular nuclei, their axons and terminals. Significantly, it was absent from the pituicytes, which at one time were proposed to be the source of the neural lobe secretions. Immunological methods have also demonstrated neurophysins in systemic blood. By the use of antibodies prepared for Np–I and Np–II, respectively, the former was found to increase during oxytocin related events and the latter during vasopressin related events (Robinson *et al.*, 1971). A third neurophysin has been recognised in neural lobe extracts of the pig, but its significance is not yet apparent (Uttenthal and Hope, 1970).

Although the neurophysins exist together with neural lobe hormones within the elementary granules (Ginsburg and Ireland, 1966), this is not to say that all dense core vesicles in the neural lobe contain these materials. There is evidence for more than one population, 120 to 200 nm and 50 to 120 nm diam., respectively (Baumgarten *et al.*, 1972). The latter resemble in size those present in catecholamine-containing fibres elsewhere. Their significance in the neural lobe is not yet clear.

The mechanism of discharge of neurosecretion, like that of other peptide secretions, is thought to involve exocytosis (Douglas, 1973). According to this view the increased number of electron lucent vesicles found in association with increased secretion are considered to represent the recycling of membrane. The process of exocytosis is activated by impulses passed down the tractus hypophyseus, presumably over the neurosecretory fibres themselves. The possible role of pituicytes in the secretory process remains a mystery.

The dynamics of neurosecretion have been examined experimentally in several ways (see Chapter 2). The effects of stalk section or even of partial constriction of the stalk have given important information. Once the specificity of CAH staining was discovered, it was quickly found that interruption of the tractus hypophyseus results in loss of CAH staining as the fibres degenerate distal to the cut (see Sloper, 1966). There is a gradual accumulation of neurosecretion in the tract proximal to the cut and even in the median eminence, where a degree of reorganisation takes place to form a partially effective miniature neural lobe. Raisman's (1973) account of this reorganisation in hypophysectomised rats describes, first a phase of engorgement during which the nerve terminals, axones and perikarya become filled with neurosecretory vesicles. Axonal sprouts, in gradually increasing number, push outward into the external zone of the median eminence and even beyond it into the perivascular space. They lose their glial sheaths and come to rest directly on the pial basal lamina, thus establishing new neurohaemal contacts. During the recovery phase that begins in about two weeks, capillaries of the mantle plexus proliferate profusely; this is accompanied by a great increase of neural outgrowths, to the extent that the original tubero-infundibular fibres become widely dispersed. The new terminals now contain an abundance of large, mostly electron lucent neurosecretory vesicles and synaptic vesicles. Structural reconstitution is accompanied by recovery of antidiuretic function.

As noted earlier, dehydration of the animal has profound effects on the neurosecretory apparatus, especially when prolonged. After several days of water deprivation or salt loading, there is a marked depletion of stainable neurosecretory material and the cytological picture is one of greatly increased activity. A valuable method of following the time course of synthesis of the cystine-bearing polypeptides and their movement to the neural lobe is based on the administration of [^{35}S] cysteine. According to Heap *et al.* (1975), after such treatment of rats radioactivity first appears in the neural lobe in about 2 hr, increasing linearly until about 9 hr and then more slowly to a maximum at about 2 days. Synthesis itself is said to occupy about 1·5 hr (see Hayward, 1975). Consequently, the most rapid transport is accomplished in about 0·5 hr. Rates of transport have been classed as fast, medium and slow, respectively.

VI. Median Eminence, Infundibular Stem, and the Parvicellular Neurosecretory System

The median eminence and its continuation into the infundibular stem are of special interest for the present volume, containing as they do the primary capillary bed of the hypophysial portal circulation. Here are the neurohaemal contacts whereby the portal blood receives the hypothalamic hormones destined to regulate pars distalis secretion. Earlier, reference was made to the fact that axons of the tubero-infundibular tract ending here arise from cells of the arcuate (infundibular) nucleus and adjacent portions of the periventricular hypothalamus constituting the "hypophysiotrophic area". Additional origins from cells in other sites are in question. Unlike fibres of the magnocellular system, fibres of this "parvicellular" system until very recently have been difficult to trace to their exact origins, because of the lack of selective means of staining the neurohormones which they are presumed to transmit. Under normal circumstances few, if any, fibres terminating in the median eminence or stem are stainable by CAH or aldehyde fuchsin. Szentágothai *et al.* (1972) by means of silver staining have traced the tubero-infundibular fibres to the arcuate nucleus and in some cases to the periventricular nucleus. In agreement with Spatz (1951), they insist on the distinct separation of the tubero-infundibular and supraoptico-hypophysial tracts. Selective lesioning of the arcuate nucleus results in orthograde degeneration of nerve terminals in the rat median eminence on both the same and the opposite sides (Raisman, 1972; Rêthelyi and Halász, 1970). Dopaminergic fibres selectively identifiable by the Falck–Hillarp formaldehyde fluorescence method have been clearly traceable not only to the arcuate nucleus (see Hökfelt and Fuxe, 1972), but to adjacent portions of the periventricular nucleus (Björklund *et al.*, 1973; Björklund and Nobin, 1973). With the recent advent of immunocytological techniques for localisation of LHRH and TRH, the outlook brightens for following peptidergic fibres to their origins.

The architecture of the mammalian median eminence and stem is fundamentally rather simple and often suggests the primitive organisation of the entire neurohypophysis of certain fish (Kobayashi and Matsui, 1969). This is generally true of small mammals like the rat and rabbit, as well as a number of larger animals that possess a deep infundibular recess, notably the carnivores. On the other hand, in some cases the primitive configuration is elaborately modified through hypertrophy of the infundibular wall and invasion by complex loops of the primary portal capillaries.

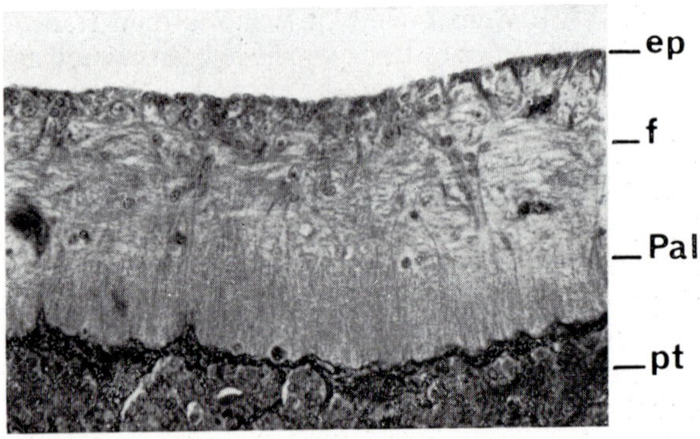

Fig. 9. Median eminence of a cat, showing the several distinct zones: ep, ependyma; f, the middle fibrous zone transmitting the tractus hypophyseus; Pal, the palisade zone, containing the long basal processes of ependymal cells and glia as well as the axons and terminals of the tubero-infundibular tract. Capillaries of the primary portal plexus are not distinct. Pars tuberalis (pt) appears below. Occasional neuronal nuclei beneath the ependyma probably represent extensions of the arcuate nucleus. 200×.

This is characteristic of the higher primates. Figure 9 illustrates the uncomplicated median eminence of the domestic cat. Its three principal regions are the ependymal (inner) zone, the fibrous (middle) zone transmitting the tractus hypophyseus, and the neurovascular (external) palisade zone in which terminal fibres of the tubero-infundibular tract intermingle with the vertically oriented basal processes of the specialised ependymal cells (tanycytes). A rich mantle plexus of primary capillaries of the portal system occupies the narrow interface or "contact zone" between the palisade zone and pars tuberalis. Numerous capillary loops project from this plexus into the median eminence for varying distances, some as far as the fibrous zone.

Occasional communications have been described between these deeper loops and a subependymal plexus (Duvernoy, 1972). Like capillaries in the neural lobe, all of those in the median eminence and stem are fenestrated and thus differ from capillaries of the overlying hypothalamus and other

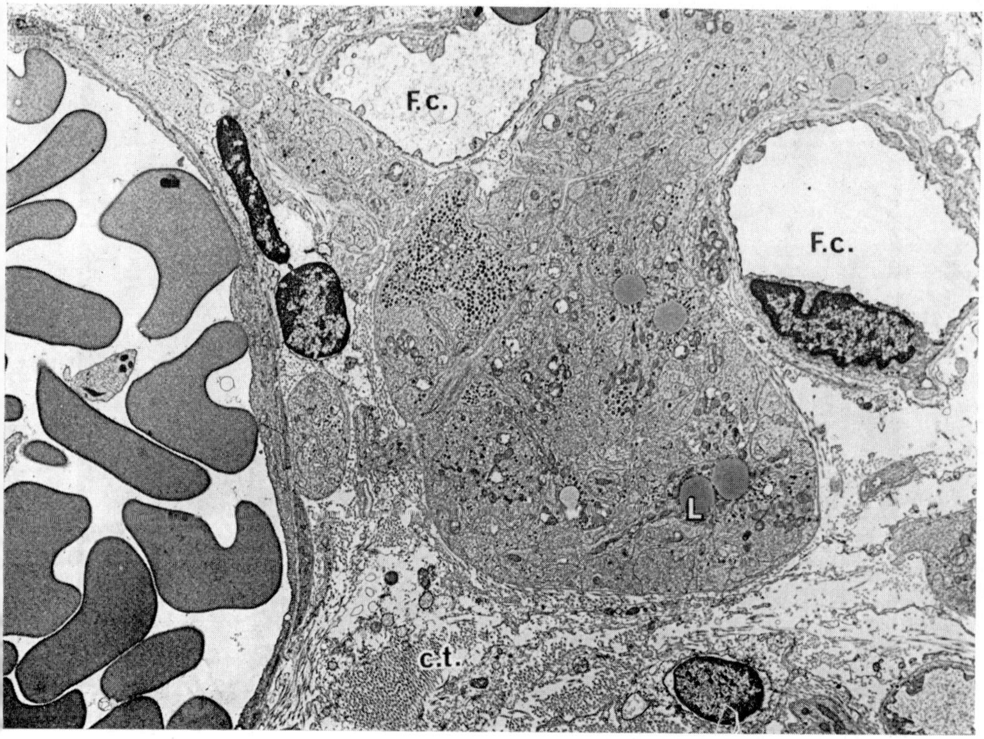

Fig. 10. Electron micrograph of the contact zone of the rat median eminence, slightly rostral to the tuberoinfundibular recess of the third ventricle. A large cluster of axon and ependymal terminals lies free in the perivascular connective tissue (ct) surrounding fenestrated capillaries (Fc). Some axon endings have many small dense-core vesicles, others few or none at all. Ependymal processes have large, lightly stained inclusions with characteristics of lipid droplets (L). The widebore vessel with many erythrocytes is a post-capillary portal venule. 4,560×. (By courtesy of Dr W. H. Fletcher.)

parts of the central nervous system. The endothelium is covered by a continuous basal lamina. Collagenous fibres in variable numbers intervene between this and the basal lamina of the neural tissue (Figs 10 and 11).

The neural basal lamina is continuous over the flattened outer surface of the median eminence and turns inward to line the channels occupied by the capillary loops and accompanying connective tissue. Nerve terminals, together with the many foot processes of tanycytes and glial cells, rest in close array along the full extent of the basal lamina.

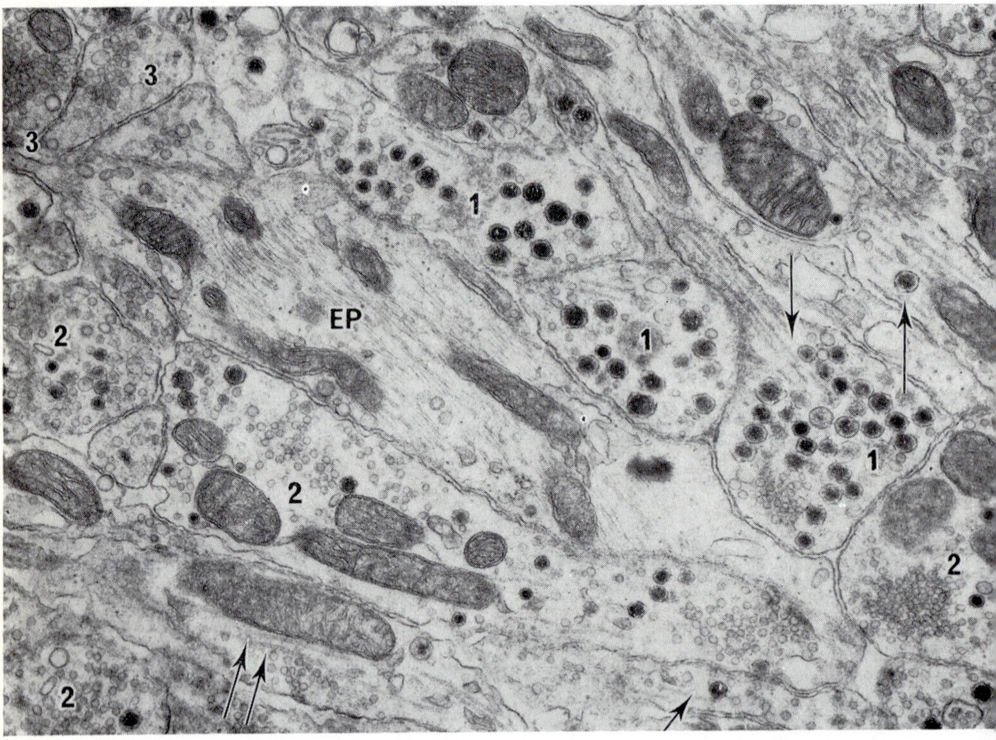

Fig. 11. Electron micrograph of the rat median eminence below the lateral recess of the third ventricle. The trunk of an ependymal cell (EP) lies in the middle of the field, identified by the abundant microfilaments. In some axon profiles (1), the dense-core granules and smaller electron-lucent "synaptic" vesicles occur in nearly equal numbers. In others (2), the vesicles predominate but a few granules are visible, while in a third variety (3) there are only the "synaptic" vesicles. It is uncertain whether these variations represent differences in proximity to the end of the fibre, variations in neurone type, or fluctuations in physiological state. Microtubules, often in close relation to dense-core granules (arrows) and to "synaptic" vesicles (double arrows) may provide motive force in the secretory processes. 25,250×. (By courtesy of Dr W. H. Fletcher.)

Electron microscopic studies of the median eminence (Figs. 10–12) have disclosed the presence of abundant vesicles of differing size and content within the various nerve fibres and their terminals. As described by various

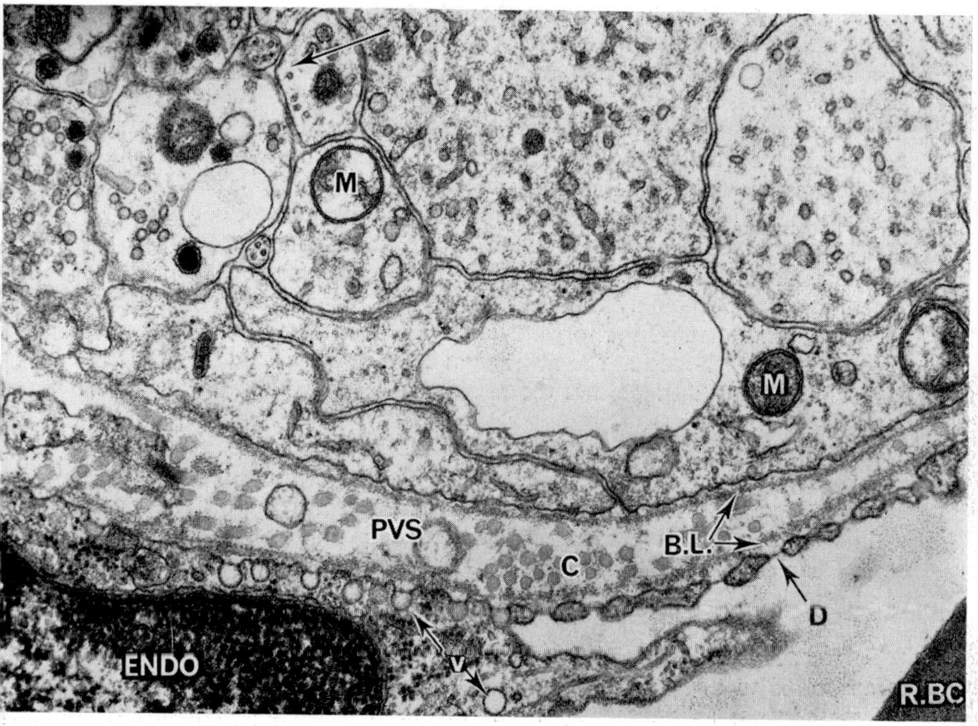

Fig. 12. Electron micrograph of the contact zone of the rat median eminence showing details of the neurovascular relationship. Axon processes with dense-core granules and "synaptic" vesicles terminate near a fenestrated capillary (an erythrocyte, RBC, lies in the lumen). Microtubules appear in cross-section at the arrow; compare with Fig. 11. The perivascular space (PVS), delimited from nervous tissue above and from the capillary endothelium (ENDO), respectively, by the two continuous basal laminae (BL), characteristically contains numbers of collagen fibres (C). Identities of the other objects in the space are uncertain.

There are several conceivable routes by which substances may pass from the neural tissue into the portal capillaries once they have passed through the neural basal lamina: (a) through narrow 10 nm hiatuses between endothelial cells, which lack occluding junctions; (b) by diffusion through the vascular basal lamina and thence through the tenuous diaphragms (D) that cross the fenestrations in the endothelial wall; (c) diffusion through the vascular basal lamina and movement via transport vesicles through the endothelium (V). There also exists a theoretically possible extravascular route through the perivascular space all the way to the pars distalis. At best this would offer slow diffusion. 30,500×. (Courtesy of Dr W. H. Fletcher.)

authors in mammals, birds, amphibia, and fish (see Ishii, 1972), these may be classified in general terms as small, medium and large. The smallest are consistently electron lucent, with diameters of 50 nm or less, and thus resemble synaptic vesicles. The vesicles of medium size, commonly possessing electron dense cores, have diameters usually ranging from 80–120 nm. The largest vesicles, also predominantly having dense cores, commonly range in size up to around 200 nm. These large vesicles are thus like the granular vesicles of the tractus hypophyseus and its terminals in the neural lobe and, in the median eminence, are largely confined to the fibrous layer. The nerve profiles in sections of the palisade layer, on the other hand, contain aggregates of the other two types. Both "synaptic" vesicles and dense core vesicles may exist together in the same fibre, or a given profile may be filled with one or the other. Commonly the synaptic vesicles are most abundant in the perivascular zones. Other organelles to be seen are neurotubules and occasional mitochondria. The absence of both ribosomes and endoplasmic reticulum clearly distinguishes the nerve profiles from those of ependymal and glial cell processes.

The ependymal cells ("tanycytes") of the median eminence and other neurohaemal organs (e.g. the organum vasculosum of the lamina terminalis) have attracted much attention during the last several years because of certain features suggesting functions other than simple mechanical support (see Chapter 3). One current hypothesis, strongly championed by Knigge and associates (Knigge and Scott, 1970), holds that hypothalamic hormones destined for the pars distalis, are first released into the third ventricle whence they are absorbed by ependymal cells of the median eminence and stem and transported to the primary portal capillaries through the long basal processes of these specialized cells. The attention which this hypothesis has received of late warrants general consideration of ependymal ontogeny and the relationships of ependymal cells to other varieties of macroglia, the astrocytes and oligodendroglia.

As the epithelium of the embryonic neural tube proliferates and becomes multilayered, some of the epithelial cells (spongioblasts) remain attached to the external limiting membrane by basal processes that become progressively elongated and branched as development proceeds. In some regions, such as the ventral sulcus of the spinal cord, the retina and the lateroventral walls of the third ventricle, this relationship persists in the adult. Spongioblasts that move away from the ventricular lumen give rise to astrocytes and oligodendroglia. Some of the astrocyte processes remain attached to the pial surface like the longest basal processes of ependymal cells. Conversely, processes both of astroglia and of the elongated ependymal cells contact blood vessels and neurones.

The word "tanycyte" was coined by Horstmann (1954) as a strictly descriptive term for ependymal cells having the stretched-out basal processes.

Millhouse (1971), in studies of tanycytes in the third ventricle of the rat and mouse by the rapid Golgi (bichromate-silver) method, describes the typical cell as characteristically possessing three portions: soma, neck, and tail. The soma rests in the ependymal layer *per se*, the neck projects into the periventricular layer for some distance and the tail, if present, extends distantly to end as a small bulbous enlargement on the pial surface or on a blood vessel. The soma and especially the neck have numerous fine lateral processes. Although these cells are to be found in small groups throughout the ventricle wall, they are concentrated ventrally. In the median eminence the slender tail is missing, whereas in the region of the arcuate nucleus the long tails arch laterally and ventrally through the nucleus to reach the pia near the lateral border of the median eminence. The floor of the third ventricle in the infundibulum and for some distance rostrally toward the optic chiasma and caudally toward the mamillary body, is formed entirely by tanycytes: it is these which are of special interest for present purposes.

The ultrastructure of infundibular tanycytes has been described by several authors (see Holmes and Ball, 1974; and Monroe and Paull, 1974). These cells differ significantly in several respects from the ciliated ependyma lining the more dorsal wall of the third ventricle. Infundibular tanycytes lack cilia entirely and instead, their apical surfaces variously possess microvilli that suggest absorptive function or bulbous protrusions containing a flocculent material suggestive of apocrine secretion into the ventricle. Whereas the ciliated ependymal cells are joined by gap junctions, the infundibular tanycytes are joined near their apices by pentolaminar tight junctions (Brightman and Reese, 1969), which prevent the intercellular movement of large molecules to and from the ventricles. The soma, and especially its processes, contain a well developed Golgi apparatus, small to moderate amounts of rough endoplasmic reticulum, polysomes, microtubules, bundles of microfilaments, lysosomes and occasional lipid droplets. The relative amounts of these organelles and inclusions vary from cell to cell. The foot processes characteristically contain aggregates of large pleomorphic inclusions. These have been interpreted as possible secretory products. Frequent "synaptoid" junctions between tubero-infundibular axones and the tanycyte processes appear throughout the palisade zone. Interestingly, there are always small aggregates of synaptic vesicles in the nerve profiles directly opposite these junctions.

Several recent studies, in the rat and the mouse, have been addressed to the chronology of development of the median eminence and maturation of the neurohaemal apparatus in the palisade zone (see Monroe and Paull, 1974). At 15 days of gestation in the rat, the prospective infundibulum is still merely a stratified epithelium several cells in thickness, resting on a basal lamina below which are occasional capillaries. A few growing axons soon infiltrate the basal layer of the epithelium. Processes of the epithelial

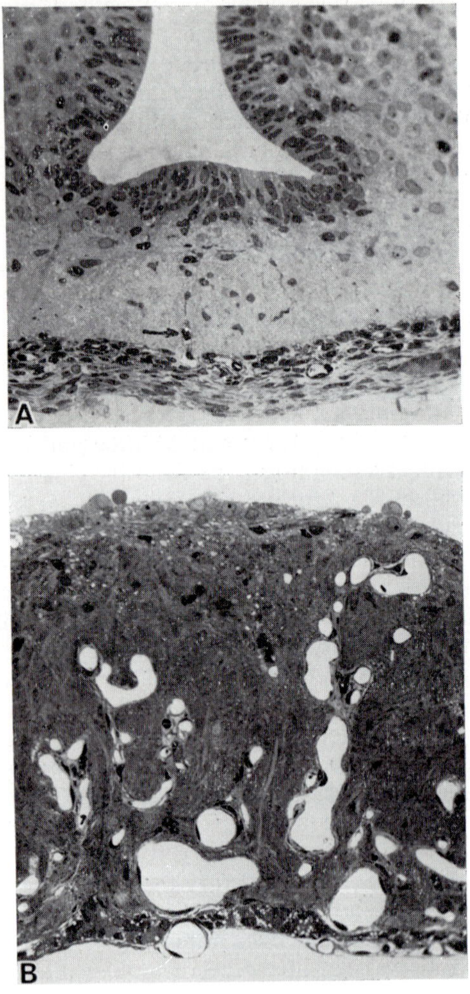

Fig. 13. The rat median eminence. Photomicrographs of frontal, 1 μm sections. (From Monroe and Paull, 1974, by courtesy of the authors and the publishers.) 200×.

A. From a 5-day neonate, showing the mantle plexus of capillaries along the ventral surface and a rare capillary loop containing a red blood cell (→). Loop formation has barely begun at this age.

B. From an adult rat, showing a strip of the mid-portion of the median eminence with its elaborate capillary loops. Fixed by perfusion.

cells, however, remain attached to each other ventrally to form for a while a continuous covering for the basal lamina. The ingrowth of axons rapidly accelerates, so that by day 18 the ventral half of the median eminence is occupied by multitudes of axons among elongated processes of the epithelium. The expanded feet of these processes continue to form a continuous ventral layer. The epithelial cells have now begun to display the cytological features of adult tanycytes. As the palisade zone thus becomes well defined, there is a pronounced increase in the numbers of capillaries at its outer surface. During the last 3 days before birth, axon terminals intrude in increasing numbers among the ependymal feet to establish contact with the basal lamina, with the result that at birth there are many more terminals than ependymal feet. The terminals contain at that time abundant synaptic vesicles but only occasional vesicles having dense cores. Although neurohaemal contact has thus been established, further changes are in prospect (compare Figs. 13A and 13B).

The 5-day postnatal median eminence remains immature in several respects. The vasculature is still mostly superficial, although a few capillary loops have begun to move inward. Dense core vesicles remain few and largely confined to parts of the axons near the terminals. Synaptoid contacts of axones upon ependymal processes are extremely rare. During the second postnatal week the capillary loops develop extensively and thus bring about a marked increase in the neurohaemal contact area, making room for many more axon terminals and ependymal end feet. Maturation also involves a marked increase in the number of dense core vesicles, especially at a distance from the terminals, concentrated in dilated axonal profiles. The chronology of maturational changes in the ultrastructure of the tubero-infundibular neurones after the fifth postnatal day, however, has yet to be described. Inasmuch as the median eminence vasculature is well developed before birth in the rabbit and the guinea-pig (Campbell, 1966; Donovan and Peddie, 1973), it seems possible that correspondingly the cytological features may also be largely complete by the time of birth in species having longer intrauterine life than the rat and mouse. Clearly, much more work needs to be done in this area.

CHAPTER 2

Investigation of Hypothalamic Function: Anatomical and Physiological Studies

D. W. LINCOLN

I. Introduction

The hypothalamus forms only a small proportion of the brain (4–5% in man); yet its essential contributions to cranial organisation far exceed its size. It is positioned deep within the cranial cavity and its neurones are small, undistinguished, and complex in their connections. These features preclude any easy approach to the study of the hypothalamus, but far from halting research, investigations continue at an evermore enthusiastic pace—motivated by the fact that within this small area resides the control for many functions which represent the basis of life itself. Amongst these are the determination of biological time, behaviour, homoeostasis, consciousness, sexuality, autonomic events and endocrine phenomena; but this very diversity generates a problem. Scientists of any one school rarely have the interest or breadth of knowledge to consider concurrent developments in areas other than their own, and these shortcomings often restrict the value of a single line of investigation. The methods whereby the hypothalamus may be studied are equally diverse and multidisciplinary, and involve the work of many fields of research, including endocrine biochemistry, neurophysiology and clinical medicine. In this chapter an attempt will be made to delineate the various anatomical and physiological approaches used to study the hypothalamus, to analyse their particular advantages and disadvantages, and thus to formulate a synthesis of the contribution each has made to our knowledge of hypothalamic function.

II. Surgical Manipulations of the Hypothalamus

A. Ablation Reconsidered

Ablation can involve anything from the gross removal of large parts of the brain to the carefully controlled destruction of minute areas (lesions). The removal of the hypothalamus, if it were surgically possible, would completely destroy the co-ordination of the endocrine system and many other functions; though life of a form would continue through the actions of the vital centres in the brain stem which determine respiration and cardiac performance. The localised lesioning of the hypothalamus is an entirely practical technique, and it is an experimental approach which played a formative role in the early days of neuroendocrinology.

Electrocoagulation of the brain, the procedure used to create most experimental lesions, is performed by the passing of an electric current through an indwelling electrode, insulated except in the region of the brain to be destroyed. The circuit it then completed by an indifferent electrode connected to another part of the animal's anatomy. The size of the lesion that results depends on several factors, such as current, duration and polarity of the stimulus and the electrode size and composition. Twenty millicoulombs of current passed through a fine steel or platinum electrode will coagulate a sphere of brain tissue about 0·5 mm in diameter. Radiofrequency currents can also be used for this purpose, and in some respects these have advantages, since they minimise the thermal and electrolytic problems inherent in the more traditional lesioning procedures. Some "lesions", made through steel electrodes, are noted for their stimulatory actions on the brain through the electrolytic deposition of metal ions. These will be discussed in Section IV.

The interpretation of results from experiments with lesions presents a difficult problem; and one that is exaggerated in the context of the hypothalamus by the fact that there is no single point where one neuroendocrine system can be destroyed in isolation from the others. A lesion is characterised by an area of total destruction surrounded by a halo within which there is a varying amount of tissue damage. If, as suggested in Fig. 1, the zone of trauma extends to twice the diameter of the area of destruction, then the volume of tissue that is traumatised is seven times greater than that which is destroyed, and the proportions increase dramatically as the zone of trauma enlarges—a point rarely appreciated. A neuroendocrine response to a given

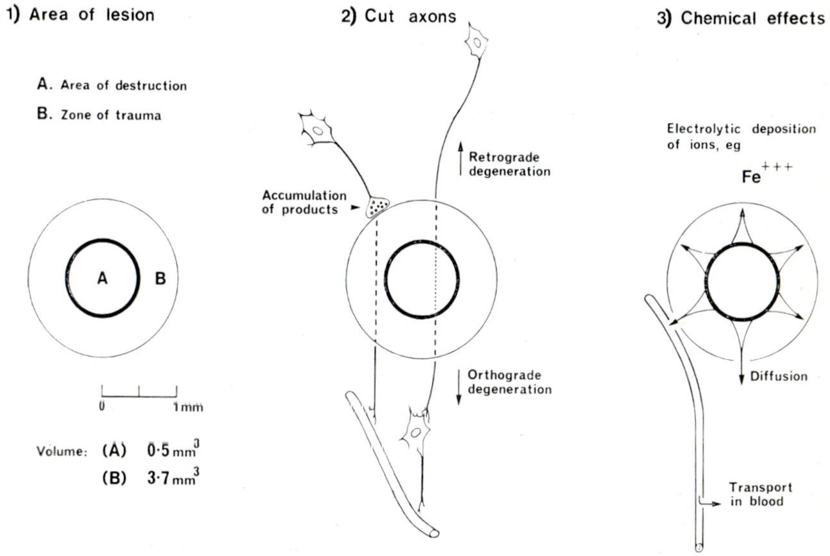

Fig. 1. Electrolytic lesions of the hypothalamus: problems of interpretation.

lesion may also change with time, for nervous tissue displays considerable plasticity with regard to the reorganisation of its synaptic connections. Alternatively, the neuroendocrine response might parallel the phenomenon of spinal shock where the injury to the system is of such severity that the whole neural apparatus is completely disorganised. Such damage is not necessarily permanent and, if these observations are applied to the hypothalamus, some neuroendocrine functions might be expected to return after a few hours or days. Another possibility is that the observed results might relate to the destruction of a fibre tract or a part thereof (Fig. 1), e.g. lesions of the anterior hypothalamus or septum might destroy efferent projections of neurones in the amygdala or hippocampus. Mechanisms of control within the nervous system often involve a delicate equilibrium between opposing excitatory and inhibitory influences, and frequently in experiments with lesions it has proved very difficult to decipher whether a lesion has interfered with an excitatory or inhibitory control process or both. An example often quoted to illustrate the removal of inhibition is the induction of precocious puberty by lesions of the anterior hypothalamus; but with reference to the foregoing observations, it is not surprising that similar results are obtained with lesions of the stria terminalis and amygdala. Ultimately, therefore, little can be concluded about the functions of a structure through the use of lesions alone.

The "reversible" lesion, based on the principle that nerve conduction ceases when nervous tissue is either heated or cooled by a few degrees, has considerable potential as an experimental tool for the evaluation of both afferent and efferent pathways within the brain. Temperature changes in the tissue may be caused either by the pumping of fluids through fine steel probes or by electrical methods. The two principal problems are the miniaturisation of the probes to a size where they can be used, without causing too much brain damage, on animals smaller than sheep, and the development of methods for the measurement of the temperature to which the brain tissue has been effectively heated. Neither problem is beyond the scope of current technology.

Chemical methods may also be used to cause the temporary or selective ablation of nervous tissue. Local anaesthetics and other drugs can be applied to eliminate (or even potentiate) the function of a particular nervous structure temporarily, though as with the electrical methods already discussed they lack selectivity. There are, however, methods which permit the selective destruction of neuronal populations. One example is the destruction of dopaminergic nerve terminals by 6–OH dopamine. Unfortunately this approach suffers from the disadvantage that effects will be experienced wherever dopaminergic terminals are found throughout the brain.

B. DEAFFERENTATION, ISLANDS AND PITUITARY STALK SECTION

Deafferentation has been employed with two objectives in mind. One school, based on the use of the "Halász knife", has examined hypothalamic connections through the section of afferent inputs and by the separation of one part of the hypothalamus from another. The other school, organised on the use of the "Cross hypothalamic island", has examined the question of hormone feedback in an island of hypothalamic tissue devoid of all neural connections, but retaining an intact vascular supply.

In the first approach a fine curved blade is lowered through the brain to predetermined stereotaxic co-ordinates, and there a curved incision is made by the rotation of the blade through a given arc (Halász and Pupp, 1965). This procedure can be used in chronic studies as the surgery can be performed under an anaesthetic such as ether, sodium pentobarbitone or tribromethanol, and the animal allowed to recover. Considerable vascular damage cannot be avoided and if the arc of section is too great, it may result in necrosis of the encircled tissue. Several useful neuroendocrine observations have emanated from the use of these knife cuts. Incisions isolating the basal tuberal region of the hypothalamus from the rest of the brain have shown convincingly that the basal region contains the inherent ability to maintain the tonic release of some pituitary hormones. Secondly, cuts slightly caudal to the suprachiasmatic nuclei in the rat, separating the rostral parts of the hypothalamus from the median eminence, have been found to eliminate the ovulatory release of LHRH.

The "Cross island" is formed by lowering a sharp-edged cylinder of 4–5 mm diameter through the brain to a position where it totally encloses the hypothalamus and sections all nervous inputs to the structure. The surgery is performed under a very short-acting anaesthetic, such as sodium methohexitone. Once the cylinder is in position, the unwanted area of the forebrain is removed by aspiration and thus the need for further anaesthetic is eliminated (Cross and Kitay, 1967). The end product is an island of unanaesthetised hypothalamic tissue, devoid of neural connections but retaining a reasonable vascular supply from the cerebral arterial circle below. Both electrophysiological and electron microscopical studies have shown the neurones within these islands to remain functionally viable for periods in excess of 24 hours. One particularly interesting fact to emerge from the use of the island preparation, was the observation that the firing rate of neurones within the island depended on the hormonal status of the animal. Firing rates were higher in islands prepared from rats on the day of proestrus than in islands prepared from rats at other stages of the oestrous cycle, although the differences in "mean" activity were never more than a few spikes/sec (Cross and Dyer, 1971a). A more detailed interpretation of these changes in firing rate is given in Section VI.

The transection of the hypophysial stalk creates a more definable result than the surgical manipulations mentioned hitherto: it transects the portal blood vessels as they pass from the median eminence to the adenohypophysis and the axons of the supraoptic and paraventricular neurones as they pass into the neurohypophysis. However, the portal blood vessels are noted for their ability to regenerate, and neuroendocrine functions may resume at some time after the transection of the stalk unless a permanent barrier is placed across the plane of section. Similarly, grafts of pituitary tissue to the region of the median eminence are readily revascularised. The ablation of the neurosecretory axons that enter the neurohypophysis is not without its complications. While the section of these axons in the region of the stalk will result in a peripheral degeneration of the fibres and depletion of the neurohypophysial stores of hormone, a miniature neurohypophysis may form on the proximal side due to an accumulation of materials transported by the neurosecretory fibres (Fig. 1). This pseudoneurohypophysis, if not fully functional, could provide for a basal output of the neurohypophysial hormones. Nevertheless, pituitary stalk sectioning, combined with hypophysectomy and the grafting of the pituitary to other parts of the body, played an important part in the establishment of many of the basic principles of neuroendocrinology. Transplants of pituitary tissue, to sites distant from the portal blood vessels, were found to release large amounts of prolactin, whilst the release of other hormones was reduced (Everett, 1956). This, among other observations, established the theory that the hypothalamus produced an inhibiting factor which normally suppressed the release of prolactin—the so-called, prolactin inhibiting factor (PIF) (Fig. 2). On the other hand, pituitary tissue grafted into the region of the median eminence supported "normal" endocrine functions when vascularised by the portal vessels (Harris and Jacobsohn, 1952). These studies also established that sexual differentiation was of hypothalamic and not of pituitary origin, since the transplantation of an adult male pituitary into the region of the median eminence of an adult hypophysectomised female rat maintained the cyclical reproductive functions of the latter.

III. Investigation of Hypothalamic Structure

A. Neglected Facts

The absence of well-organised compartments within the hypothalamus to which observations can be related, forces many investigators to create their own antomical image of the hypothalamus. Unfortunately, there are a number of anatomical facts that tend to be neglected in their reconstructions. The hypothalamic neurone is often depicted as an isolated element devoid of connections (see Figs 1, 2, 8, 9), but this is far from the truth. Ultrastructural evidence indicates that a single hypothalamic neurone may receive a synaptic input from each of several thousand other neurones; and, in turn, the named cell may make synaptic contact with thousands of other cells—such is the complexity of the wiring diagram. On the other hand, the hypothalamus is not so large that the number of neurones cannot be fairly accurately enumerated. The rat, for example, has been reported to have about 4,000 paraventricular and 14,000 supraoptic neurones (Bandaranayake, 1971), but even these figures are now thought to be over-estimates on the grounds that the author did not make the necessary adjustments for section thickness (Swaab *et al.*, 1975). In the hypothalamus as a whole there may be fewer than 200,000 neurones, and the majority of these will be in areas with no well-defined neuroendocrine function, such as the lateral and posterior regions. Consequently, the hypothalamus should be considered as a structure of relatively few elements but extremely rich in interconnections.

Hypothalamic neurones are of two distinct types: interneurones and endocrine neurones. The interneurones, which could be further classified according to function, abstract and process information, and they probably outnumber the endocrine neurones which they eventually control. The endocrine neurones are the more interesting, at this stage in the evaluation of hypothalamic function, for they represent the motor component of the neuroendocrine system and synthesise hypothalamic peptide hormones for release into the vascular system. Unfortunately for the analyst, however, endocrine and interneurones appear to be intermingled in all but the magnocellular nuclei. But even the cells of these nuclei, with their distinctive neurosecretory inclusions, have electrical properties which are virtually identical to those of neurones in many other parts of the mammalian brain.

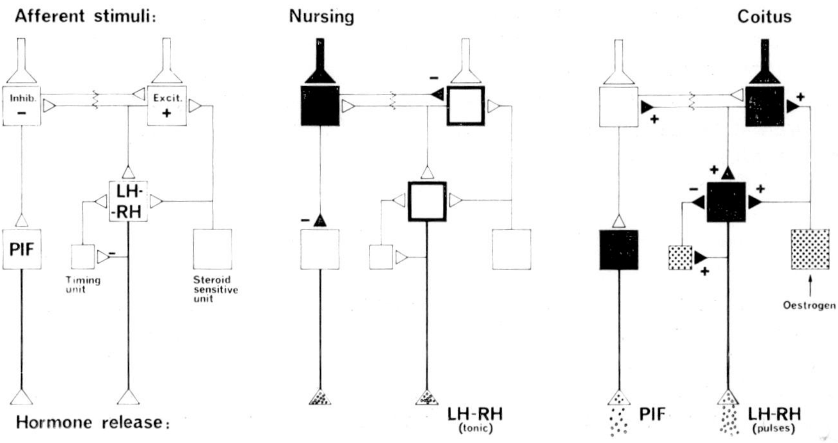

Fig. 2. A theoretic model of a six-component hypothalamus, consisting of two endocrine neurones, a steroid detector, a timing unit and two interneurones. The operation of these units, where coitus promotes the release of pituitary LH and suckling the release of prolactin, is shown in the righthand parts of the figure. An electrically active unit and synapse is shown in solid or hatched form; a partially active unit is illustrated by a thickened border. Synaptic effects are shown as + (= excitation) and − (= inhibition).

The hypothalamus is a bilateral structure, a feature often neglected, and there is no prominent tract of commissural fibres connecting the two halves. One might justifiably ask whether the two halves of the hypothalamus function separately, like the two cerebral hemispheres of a split-brain preparation. Does the right side dominate the left, or *vice versa*, and, if they function in unison, what controls their simultaneous activation? These questions cannot be adequately answered, but they serve to emphasise that in some studies, particularly those involving lesions, it is important to consider whether procedures have been applied bilaterally or unilaterally.

Duplication of function extends beyond the simple consideration of whether each individual has one or two hypothalami. With the evolutional progression from simple to more advanced nervous systems, the number of neurones producing each hypothalamic hormone would appear to have increased, whilst the number of individual hormones has remained constant. Such absolute neuronal specificity of cell to hormone is more theoretical than established fact, but it follows the widely accepted rule that a neurone produces only one substance, whether peptide hormone or neurotransmitter. Someday, endocrine neurones may be discovered that release, for example, LHRH at their main axonal terminals, whilst releasing acetylcholine at the terminals of recurrent collateral branches. Alternatively, an overlap in function may be discovered with cells producing more than one hormone. The evidence gathered so far does not appear to point in this direction. Immunofluorescence microscopy and electrophysiological recording indicate that there are discrete classes of oxytocin- and vasopressinergic neurones within both the supraoptic and the paraventricular nuclei; and the oxytocinergic neurones at least appear to be linked in an electrical syncytium that allows all four (two left and two right) neurosecretory nuclei to function in unison. There is virtually no information on other neuroendocrine systems.

B. STUDIES OF CELLULAR ARCHITECTURE

Developments in the biochemical and electrophysiological study of the brain have in recent years outpaced our knowledge of the structural anatomy of the areas under investigation, and our knowledge of hypothalamic structure is now, without question, limiting many other experimental approaches. From studies of the cerebral cortex, the thalamus, the gracile and cuneate nuclei and the cerebellum, it has become increasingly apparent that the precise layout of the nervous tissue relates very closely to physiological function, and similar structural parallels could apply to the organization of the hypothalamus. To appreciate the full significance of electrophysiological data information is required on the identity and connections of the hypothalamic neurones from which the recordings are taken. Finally, precise information is required on the cellular localisation of all the hypothalamic hormones.

The Golgi methods for the staining of brain tissue, introduced in 1873, remain the single most important group of techniques for the study of the structural organisation of the brain (see Chapter 1). The technique, with numerous variants, involves the repeated immersion of chromate-hardened nervous tissue in dilute silver nitrate, and this results in a progressive accumulation of silver on some of the neurones, but the technique is both capricious and time consuming. About 10% of the neurones are stained in a manner which allows the whole structural geometry of the cell to be examined; the remaining 90% are not impregnated and thus are not revealed. The basis of this all-or-none staining is not understood, and one can only speculate about the functions of the cells which the Golgi methods stain within the hypothalamus. The problem is well illustrated by the fact that the supraoptic and paraventricular neurones, which are easy to locate by standard histology, are quite refractory to Golgi staining. These limitations may soon be overcome, however, as techniques are now being developed which permit the staining of *selected* neurones to a degree of clarity hitherto only achieved by the Golgi methods.

These new techniques involve the extrusion of dyes and other "labels" from micropipettes, by the passage of small amounts of electric current (iontophoresis). Three variations of this method are now being used to study the hypothalamus and its connections. The crudest of these, but one particularly useful for the tracing of nerve pathways, is *axonal iontophoresis*. In this process a nerve is sectioned and the label is borne along the axons in either an orthograde or a retrograde direction from the cut nerve stumps, by passing a suitable current (Isles and Mulloney, 1971). The second method is *regional iontophoresis*, where a small quantity of a marker is ejected from a glass pipette (10–20 μ tip) into a selected region of the brain; the surrounding cells take up the dye through their damaged processes or by pinocytosis (Mason, personal communication). The third and most precise of these methods is that of *intracellular microiontophoresis*. An individual neurone is first identified by extra- or intracellular recording of electrical activity, then impaled with the recording micropipette (1 μ top), and the dye injected intracellularly (Pitman, *et al*., 1972). Such studies have already shown, in areas other than the hypothalamus, that it is possible to record, and then to label and reconstruct the geometry of a single cell. Soon, maybe, similar studies will be performed on hypothalamic neurones.

The procion dyes (yellow and brown), horseradish peroxidase, ^{3}H-labelled amino acids and a number of heavy metals have all been used to label cells by one or more of the above methods, but no one of them is entirely satisfactory. The ideal material should be ionizable, and the molecular weight and reactivity should be such that it stays within the cell into which it has been infused; at the same time it should be quickly and evenly distributed throughout the cell and its processes. Finally, the label should be visible by light and electron microscopy. Horseradish peroxidase cannot be used effectively from micropipettes and, though readily taken up by cells after regional iontophroesis, its distribution within the cell remains uneven. Tritiated amino acids have a tendency to leak from the cells into which they are placed, and the definition provided by light and electron autoradiographs is not adequate for the study of fine neuronal branching patterns. Procion yellow, a textile dye, is useful in some situations but it is not electron-dense and the requirement for fluorescence microscopy limits the resolution once again. Procion brown, has the advantage of being electron dense, but is not readily visible in the light microscope.

Cobalt appears to be a label with considerable potential. Cobaltous ions may be passed into nervous tissue by each of the methods mentioned above, and they may be visualised by precipitation as cobaltous sulphide—an intensely black material. The cells so labelled may be studied in whole amounts of cleared brain (providing a 3-dimensional view) and by standard light and electron microscopy. With invertebrate neurosecretory and non-neurosecretory cells the cobaltous chloride-ammonium sulphide technique, as it has been named, has provided a structural picture at least as good, if not better than that of a Golgi preparation (Mason and Nishioka, 1974). The first application of this technique in the study of the hypothalamus has been to visualise the much disputed retinohypothalamic projection, through the application of cobalt to the sectioned optic nerve. Cobalt ions move along the nerve with considerable speed, faster than many suggested rates of axonal flow. Furthermore, the ions appear to move just as effectively in the brain tissue of an animal that has been recently killed, but how these ions move remains a mystery. The end result is clearly visible, and fibres from the optic nerve have been seen to enter the hypothalamus from the posterior part of the optic chiasma and terminate in the posterior portion of the suprachiasmatic nucleus (Mason and Lincoln, 1976).

Electron microscopy has provided a great deal of information about the cellular organisation of the hypothalamus, particularly with regard to the synaptology of the cells and the production and release of hormone-containing granules, but standard electron microscopy has been limited largely to the provision of descriptive information. The quantification of ultra-structural information, as now practised in a few laboratories, adds a further dimension to these investigations and permits a more functional analysis of hypothalamic events. Quantitative studies provided the first evidence to suggest that sexual differentiation of the hypothalamus had a morphological basis (see Chapter 8). This work, by Raisman and Field (1971b), involved a painstaking evaluation of the numbers, types and positions of synapses on hypothalamic cells, and revealed a sex difference in the upper parts of the anterior hypothalamus. There is unfortunately no other information to link this area with reproductive functions; lesions of this area do not block the cyclical ovulation of the female rat. Within the context of the hypothalamo-neurohypophysial system, counts of the neurosecretory granules of different forms and position have provided a new understanding of the maturation, movement and release of neurosecretory products, and added a new feature to the analysis of hormone dynamics. Morris (1976a) has shown that the 2 μg of oxytocin and vasopressin in the neurohypophysis of the rat are contained within $1{\cdot}44 \times 10^{10}$ granules (i.e. 8,400 hormone molecules/granule), and these he has shown, in a further paper (1976b), can be compartmentalised with 31% of granules in axonal endings which contain microvesicles and 55% in axonal swellings devoid of vesicles. This approach has been coupled with EM autoradiography (Heap *et al.*, 1975), and the fate of granules has been traced with time, providing a morphological basis for the observation that new synthetised hormone may be released in preference to that produced earlier and passed into the dilated "storage" compartments. Such quantitation is also invaluable in the evaluation of previous work and in the formulation of further experimentation. For example, given the number of neurosecretory granules, the daily turnover and maximum release rates of the hormones and the time taken for granule release, calculations can be made as to the probability with which an exocytocic profile might be found in an electron micrograph of the neurohypophysis. The chance of observing a granule in the process of release, even under conditions of maximal physiological stimulation, is found to verge on the impossible (Cross *et al.*, 1975). Surprisingly, several authors claim to have observed exocytosis of granules; perhaps fixation halts the mechanism of release in mid-stream. However, this paucity of visual evidence should not be taken as proof that exocytosis does not occur.

C. LOCATION OF HORMONE PRODUCING NEURONES

Much effort has been expended in the search for areas of hormone production within the hypothalamus. One approach, which appears naive in retrospect, was to assay (by bioassay or radioimmunoassay) small portions of hypothalamic tissue in the hope of discovering regions rich in one particular hormone. There were, however, several insurmountable limitations to this experimental approach. There is no structural organisation within the hypothalamus, with the exception of the paraventricular and supraoptic nuclei, to suggest that cells are aggregated according to hormone production. Further, the pro-hormones present in the perikarya of the cells may be biologically inactive by comparison with the final hormone molecules which are detected by radioimmunoassay and which are released from the nerve terminals some distance away.

A somewhat earlier approach, and the approach which eventually pioneered the techniques currently used, was the development of histochemical methods specific for the staining of a hormone, its precursor, or some feature of the hormone producing cell. The concept of "neurosecretion", as applied in the context of neurohypophysial physiology, was founded on a series of histochemical staining procedures for the localisation of products concerned with the neurohypophysis (Bargmann, 1949; Sloper, 1966). One of the most successful staining methods, was acid permanganate chrome alum haematoxylin. This is thought to stain cystine-rich structures, and thereby reveal cells rich in oxytocin or vasopressin. There remains, however, much uncertainty as to precisely what is stained.

Developments in the field of immunohistochemistry overcame the poor specificity inherent in such histochemical methods as mentioned above.

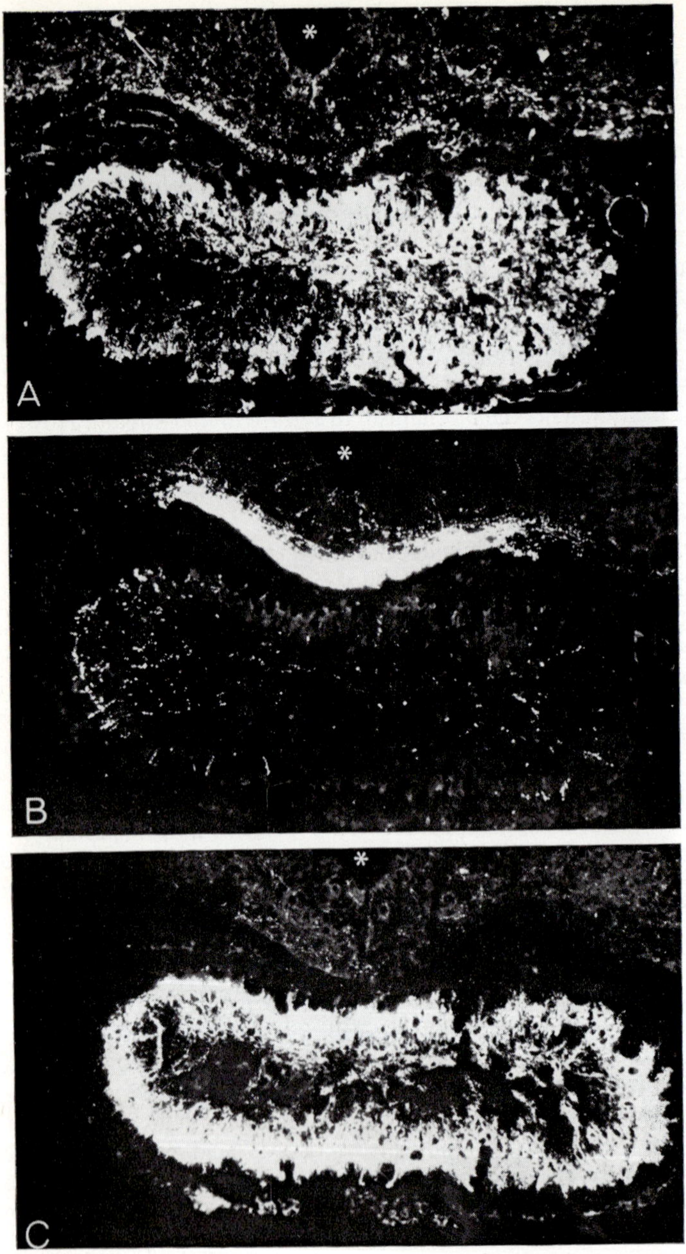

Fig. 3. Immunofluorescence micrographs of consecutive sections of the posterior part of the median eminence and the stalk after incubation with antibodies to (A) tyrosine hydroxylase (TH); (B) LHRH; (C) somatostatin. The ventral surface is covered by moderately dense TH, very dense LHRH and sparse somatostatin positive plexuses. In the stalk, dense TH and somatostatin and sparse LHRH positive plexuses are seen. Asterisks indicate the third ventricle. The arrow points to a TH-positive cell. × 140. (From Hokfelt *et al.*, 1975f).

The principle is simple and largely based on immunological developments for the measurement of hormones in blood (see Chapter 4). Typically, thin slices of fresh brain are prepared and exposed for some time to an antibody raised against one of the hypothalamic hormones—the antibody being selectively retained by those tissues containing antigenic material. The first problem, and one which still gives rise to some elements of doubt, concerns the production of good antibodies to the small molecular weight hormones of the hypothalamus. The second concerns the development of suitable methods for the visualisation of the antibody-antigen complex. This latter problem was overcome by the production of a second antibody against the first; and fluorescein was added to this second antibody as a label. Thus, with the aid of fluorescent microscopy, it became possible to visualise the antigenic sites within the tissue. Claims have been made for the localisation, both within the hypothalamus and elsewhere, of cells rich in oxytocin, vasopressin, neurophysins, LHRH, TRH, somatostatin and enzymes involved in amine metabolism (Fig. 3) (see Watkins and Evans, 1972; Barry *et al.*, 1973b; Burlet *et al.*, 1973; Hökfelt *et al.*, 1975a, f; Swaab and Pool, 1975), and this list will continue to increase. Some caution must still be exercised, for there are several questions concerning these procedures that have not been adequately answered. Does the antibody gain access to all the hormone containing cells? Does the hormone exist in the perikarya of the cells in the form that has been extracted, synthesised and used for antibody production; or is it present as a prohormone which is immunologically unreactive? In many studies, colchicine or methanol (by intraventricular adminstration) have been used to increase the number of cells displaying immunological reactivity. Presumably these materials, by blocking axonal transport, cause an accumulation of "mature" hormone within the perikarya of the cells. Even then, many observers seem surprised by the low numbers of cells which react to each antibody; histological sections of the rostral hypothalamus of the guinea-pig show the number of LHRH cells to be numbered in tens rather than thousands. However, these small numbers are to be expected, when the cells are classified first into endocrine neurones and interneurones, and then subdivided according to the numerous functions that the hypothalamus controls.

IV. Stimulation of the Hypothalamus

"Stimulation" includes a number of procedures designed to evoke the physiological activation of the hypothalamus. These stimuli are most diverse, but broadly they can be placed into three categories: natural stimuli (e.g. coitus and suckling), chemical treatments (e.g. implants and infusions of hormones), and electrical stimuli.

A. APPLICATION OF NATURAL STIMULI

The natural exteroceptive and proprioceptive stimuli that are involved in hypothalamic activation are difficult to define or quantify, and none can be expressed in terms of the activation of a single afferent pathway. Furthermore, many of these stimuli fail to elicit their natural response when applied under the experimental constraints of factors such as stress and anaesthesia. These complexities and failures have made it common practice to seek experimental stimuli which mimic, in part or in full, the natural events. Milking machines have been used to replace the suckling of the young, the electric toothbrush has been used as a vaginal vibrator to replace coitus and odours of laboratory chemicals have been used as olfactory cues. Interoceptive stimuli have been handled in much the same way; experimental changes have been made in blood volume, osmolarity, flow and temperature to activate components of the process of homoeostasis. The degree of stimulation, achieved by these manipulations, is most debatable, and the interpretation of the results is confused by this uncertainty. The natural stimuli are complex and "rich" in terms of their neural expression, and, though some experimental stimuli may work in the conscious and unoperated animal, the probability of these artificial stimuli working in an animal after the trauma of surgery and restraint is very much decreased. The analysis of neuroendocrine reflex arcs is further complicated by the fact that they often involve higher nervous centres. The stimuli of coitus and suckling, which under the appropriate circumstances will cause the release of LHRH and oxytocin respectively, elicit widespread changes in brain activity. At the same time, these two neuroendocrine responses can be blocked by inhibitory signals emanating from the cerebral cortex.

Our measurement of neuroendocrine responses, particularly to natural stimuli, are not for the most part sufficiently accurate to allow a full interpretation of the neural events which mediate these processes. Changes in electrical activity are measured in seconds, or even milliseconds. Measurements of hormones in peripheral blood on the other hand, indicate changes which at best are measured on a minute by minute basis. Peripheral blood values are also of limited use in the assessment of the "precise" patterns with which hypothalamic hormones are released. The adenohypophysis acts as a transducer in the pathway of hormone release, and this function may be far from direct. LHRH causes the release of LH, but the response is dependent on the previous exposure of the gland to LHRH and the feedback effects of oestrogens (Aiyer *et al.*, 1974a). Thus, a change in peripheral blood LH does not necessarily indicate a concomitant change in the release of hypothalamic LHRH (see also Chapter 4).

There is presently only one situation, within the limits imposed for the recording of the electrical activity of single hypothalamic neurones, i.e. anaesthesia and stereotaxic restraint, where the requirements of stimulus

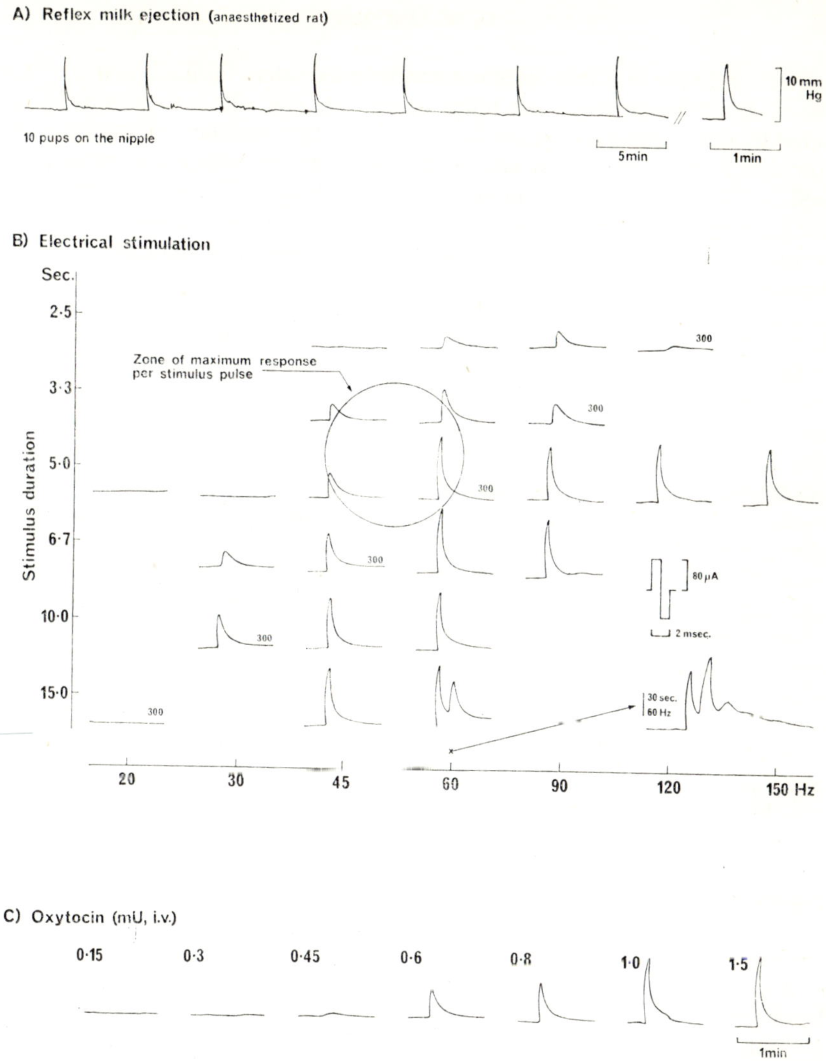

Fig. 4. Recordings of intramammary pressure from a lactating rat anaesthetised with urethane. A. Suckling. Uniform pressure peaks recur every 5–7 min, and each represents a milk ejection during which the pups obtain milk. B. Electrical stimulation of the neurohypophysis. Stimulation was applied every 3 min, and the frequency and duration were varied at random. The stimulation matrix which has been presented provides for the delivery of a constant number of pulses when read diagonally. The zone of maximum response was determined by dividing the rise in pressure (a measure of an effective milk ejection) by the number of stimulus pulses applied. C. Simulation of the milk ejections shown in A and B by bolus injections of oxytocin into the great saphenous vein. The latencies to milk ejection following electrical stimulation and injection of oxytocin were 12–16 sec and 10–14 sec, respectively.

control and response measurement can be adequately fulfilled, and this is in the study of the milk-ejection reflex of the rat. The afferent stimulus is provided by the young suckling on the nipples of the anaesthetised mother. This stimulus is both prolonged and continuous—as the young are prepared to feed for hours, but the hormone release is pulsatile (Fig. 4) (Lincoln *et al.*, 1973). This efferent response, i.e. the release of oxytocin, can be assessed continuously from the pressure changes in the main teat duct of one of the unsuckled mammary glands.

B. ADMINISTRATION OF HORMONES AND OTHER CHEMICALS

Three levels of hormone feedback to the brain have been proposed: the long short and ultrashort loops representing respectively the feedback of steroid, adenohypophysial and hypothalamic hormones (see also Chapter 3). Two effects can be furnished by these actions on the brain. The hormones may operate directly on the hypothalamus to bring about changes which eventually modify their own release (i.e. positive or negative feedback). Additionally, these hormones may modify other neural functions. Gonadal steroids indirectly control their own release by actions on both the hypothalamus and adenohypophysis, but they also determine the behavioural expressions of sexual receptivity and aggression, interact with learning processes and sleep, and modify the neural response to many drugs. Such a wide spectrum of neural actions is also indicated by the finding that radioactively labelled oestrogens accumulate in many extrahypothalamic nuclei (Pfaff and Kleiner, 1973). Such an accumulation of a steroid within a nucleus of the brain provides no evidence that the hormone is actually active within that structure, but such evidence can be used as an indicator of where in the vast expanses of the brain implants of hormones or recording electrodes should be positioned. A number of neural actions, other than those of self-regulating feedback, have also been claimed for both the pituitary and hypothalamic hormones. Prolactin controls maternal behaviour, and as such its effects are expressed through the nervous system. Vasopressin has been implicated in learning processes (van Wimersma Greidanus *et al.*, 1975); whilst LHRH has been shown to promote sexual receptivity in animals suitably primed with ovarian steroids (Moss and McCann, 1973). There are several methods by which these actions have been explored.

(1) *Implantation*

One of the most widely applied techniques involves the chronic implantation into the brain of fine steel capillaries filled at the tip with a hormone preparation. The hormone diffuses from the exposed surface in minute amounts and creates effects which, it is hoped, are localised in the surrounding brain tissue. Some transport of the hormone by the local vasculature may occur and, indeed, some of the confusion attached to the effects of implants in the region of the median eminence may have been caused by the implanted hormones passing in the portal blood to act on the cells of the adenohypophysis; even so, such implants have provided some most useful information. Oestrogens, for example, promote sexual receptivity when implanted into the anterior hypothalamus of the ovariectomised animal (Harris, *et al.*, 1958); proof—perhaps—that the anterior hypothalamus contains a centre for the control of sexual expression.

An intrahypothalamic injection, when performed on a scale larger than by microiontophoresis, resembles a crude implant and as such merits no further attention; but such a result was accidentally achieved during some of the early studies on ovulation in the rat. The application of an anodal DC current to the anterior hypothalamus and preoptic area through a steel electrode creates a lesion. At the same time iron is deposited in the tissue by electrolytic action (Fig. 1), and this iron acts as a stimulatory agent to promote ovulation, presumably through the release of LHRH. This procedure is often referred to as electrochemical stimulation, and it is the most effective stimulus (other than exogenous LHRH or LH) known for the induction of ovulation in the rat in which the spontaneous ovulation has been blocked by the administration of a light anaesthetic dose of sodium pentobarbitone (40 mg/Kg, i.p.) early on the afternoon of proestrus. The electrochemical stimulus is many times more effective in terms of coulombs of current applied than pulsed electrical stimulation, and for this reason there is much uncertainty as to whether the irritative deposition of iron activates the normal mechanism for the release of LHRH. The electrochemical stimulus is not without its actions on other neural functions; there are changes in behaviour such as increased running activity and mating, the sympathetic nervous system is activated and the electroencephalogram is changed in anaesthetised animals from slow-wave sleep to a condition of arousal (Lincoln *et al.*, 1972). None of these changes in brain activity are observed with a "clean" hypothalamic lesion, i.e. a lesion of equal size made by the passing of the same amount of current through an electrode of platinum.

(2) *Intravenous injection*

Implants of hormones are only of limited value in acute studies. The obvious route of hormone administration for these investigations is that of intravenous injection or infusion, but this simple procedure harbours many potential pitfalls. More often than not, investigators fail to seek or obtain proof that the injected hormones actually reach the areas of the brain under study, much less determine whether or not the concentrations there are physiological. Steroid hormones, when injected in a highly concentrated bolus—in an alcoholic solution for example, may precipitate when diluted by the blood; and presumably much of this will collect within the first capillary bed encountered in circulation, and with an i.v. injection this will be the lungs. Alternatively, the blood-brain barrier may restrict the access of the hormone to the brain tissue.

(3) *Intraventricular injection*

Intraventricular injections bypass the blood-brain barrier and depend on the injected materials passing through the ependymal cells lining the ventricular walls. Not much is known about the functions of the ependymal cells or of their ability to transfer materials between the cerebrospinal fluid and the brain tissue. The cells have several notable anatomical features (see Chapter 1); their ventricular surface is covered with cilia, microvilli and cytoplasmic projections and from their basal surface projections interdigitate between the superficial neurones; there is no basement membrane. Experimental evidence with intraventricular injections of putative neurotransmitters and steroid hormones suggest that these materials readily gain access to the brain. The investigator should appreciate, however, that the natural flow of the cerebrospinal fluid is from the lateral ventricles, through the third ventricle and aqueduct of the midbrain to the fourth ventricle and from there out into the subarachnoid space in the region of the cisterna magna. Ideally, therefore, materials for hypothalamic action should be injected into the lateral ventricles. Injections into the third ventricle may result in considerable damage to the hypothalamus, for this ventricle is no more than a slit-like cavity. Many studies, whilst apparently successful, have defied the law of cerebrospinal fluid movement and used injections into the cisterna magna; one can only speculate about the route by which these injected materials reach the hypothalamus.

C. ELECTRICAL STIMULATION

Marshall and Verney (1936) provided the first direct evidence to implicate the nervous system in the control of endocrine phenomena when they induced ovulation through the application of crude electrical stimuli to the head and lumbar regions of the rabbit. The same results were soon produced by Harris (1937), and with more sophisticated methods he was able to localise the effective site of stimulation to the basal hypothalamus. In some ways Harris was fortunate, for in stimulating the hypothalamus he was limited to the use of 50 Hz mains frequency: this, it was learned much later, was the optimal frequency for the stimulation of the unmyelinated axons of hypothalamic neurones. Had 0·1 msec pulses of higher or lower frequency been used, which are optimal for the stimulation of large myelinated fibres and easily produced by all modern stimulators, ovulation would not have been observed.

The present methods of hypothalamic stimulation involve the stereotaxic placement of bipolar electrodes and the passage of precisely matched biphasic square wave pulses. The electrodes are commonly of concentric structure with about 0·5 mm separating the two terminals. This electrode assembly minimises the spread of current, and the selection of the biphasic waveform reduces the polarization and electrolytic effects which result from the use of monopolar stimulation. The maximum current applied to the hypothalamus should not exceed about 800 μA, peak to peak, when applied in continuous pulse trains, otherwise the spread of current will lead to convulsions, seizures and ocular effects. The latter can on occasion be useful in determining the placement of the electrode and the amount of current spread. Asymmetrical eye movements, due to a preferential stimulation of one of the oculomotor nerves as it passes adjacent to the hypothalamus, would indicate a displacement of the electrode from the midline; whilst exophthalmos and changes in pupil size would provide information on the effect of stimulation on the central activation of the autonomic nervous system. Pulse duration and pulse shape are still somewhat arbitrary considerations; a 2–4 msec pulse is usually quite effective and limits the stimulus artifact in electrical recordings to an acceptable duration.

The most extensive evidence on the relationship between electrical stimulation of the hypothalamus and hormone release, relates to the stimulation of the neurosecretory fibres entering the neurohypophysis. Here, oxytocin (and vasopressin) release is related to both the frequency and duration of the stimulus (Fig. 4). The optimal frequency for oxytocin release is about 50 Hz, and a 2–4 sec period of stimulation will release a pulse of oxytocin equivalent to that observed during suckling in the rat. The threshold frequency, below which stimulation is ineffective, is about 25–30 Hz. There is no sharp cutoff as the frequency is increased above the optimal but the amount of oxytocin released per stimulus pulse rapidly declines. The lower frequency threshold is thought to relate to a process of calcium-dependent facilitation operating in the release of the neurosecretory granules from the nerve terminals in the neurohypophysis. These stimulation parameters have been explored extensively, both *in vivo* (Harris *et al.*, 1969) and *in vitro* (Ishida, 1970); and in both conscious and anaesthetised animals. Indeed, 10 sec of stimulation (50 Hz) applied to the pituitary stalk of the conscious rabbit on Day 31 of pregnancy promotes labour within 5 min, with an efficiency which is hard to match with injections of exogenous oxytocin (Lincoln, 1971). Such stimulation parameters as these have been selected largely on the basis of trial and error as those which promote the maximum physiological response from the target tissue, in this case the uterus. As has so often been the case with electrical stimulation, experiments such as these have been in advance of our understanding of the normal process of hormone release. For example, stimulation was being employed to release oxytocin in pulses long before it was appreciated that the hormone was naturally released this way. Experiments with stimulation have also forecast developments in another field, for whilst 50 Hz was considered the optimal frequency there was no evidence to indicate that hypothalamic neurones fired at these rates. This dilemma has now been resolved and frequencies of cell discharge in the 40–60 spikes/sec range have recently been recorded.

Intermittent periods of high frequency stimulation ensure the maximum and most efficient response from both the neural and target tissues, at least as far as the hormone oxytocin is concerned. For example, stimulation of the neurohypophysis of the rat for 2–4 sec releases about 1 mU oxytocin (and probably as much vasopressin), which is well within the limits estimated for the readily releasable pool of hormone. A more prolonged period of stimulation will release more oxytocin, but the amount released per stimulus pulse decreases. This fall could be due to ionic effects such as hyper- or hypopolarization or to differences in the "relative" releasibility of the hormone granules. In terms of the target tissue, the rat mammary gland displays its maximum contractile response to about 1 mU oxytocin (Fig. 4). Much larger quantities or infusions of the hormone (2 mU/min) create multiple contractions, and bolus injections at intervals of less than 3 min result in a reduction in the contractile response of the mammary gland to the second and subsequent injection. Thus the neurohypophysis and the mammary gland of the rat display their most efficient response when electrical stimulation is used to release pulses of oxytocin at intervals of about 5 min. This is precisely the pattern of hormone release observed during suckling (Fig. 4).

The stimulation parameters for the release of oxytocin have often been applied to study the release of other hypothalamic hormones, though there has usually been a lack of collateral evidence to indicate that these parameters were in any way optimal. Often these parameters have been "intensified", presumably due to failure, to values which are unlikely to be physiological. Parameters of 100 Hz, 30 sec on and 30 sec off, have commonly been employed, but such parameters have no parallel in recordings of electrical activity, and involve time and frequency considerations that preclude an efficient response from the nervous tissue. There are two problems to be considered when attempting to calculate the parameters of electrical stimulation for the release of a particular hormone: first, where to stimulate and secondly, how to measure, with an accuracy of seconds, the release of the hormone. Some progress has been made and recently observations have been obtained which suggest that 50 Hz is the optimal frequency of stimulation of the preoptic area of the rat for the release of LH (and thus presumably LHRH) (Jamieson and Fink, 1975). This is one step, the next is to determine the pattern of stimulation which mimics the natural ovulatory release of the hormone.

Another way of analysing hypothalamic functions is to electrically stimulate the afferent pathways conveying information to the hypothalamus, but so far this approach has not proved rewarding. The problem is similar to that mentioned earlier with the application of exteroceptive stimuli designed to simulate a component of the natural sensory stimulus. Such inputs are most complex and, even if the nerve tracts controlling the various endocrine events could be isolated in the brain stem and elsewhere, it would undoubtedly prove difficult to elaborate a stimulation programme which would reproduce the endogenous pattern of spike transmission. There may, however, be further problems concerning the central processing of sensory information which are not fully appreciated. For example, electrical stimulation of the endocrine neurones which produce the neurohypophysial hormones almost always promotes hormone release, whereas stimulation of the afferent pathways which impinge on these neurones produces spasmodic results, working one moment and failing the next; this is a disturbing problem when trying to trace afferent pathways. A parallel to this situation does occur in the milk ejection reflex. Rat pups remain attached to the nipples for hours and provide, as far as can be analysed, a sensory stimulus which is both continuous and uniform in intensity, although as previously mentioned the output of oxytocin is pulsatile. These two observations and recordings of electrical activity suggest that a mechanism similar to an electronic gate operates between the sensory input and the activation of the endocrine neurones; and, once activated, this gate becomes refractory to further sensory stimulation until a predetermined time has elapsed. An extension of this approach in the study of neuroendocrine reflexes is the examination of the entire pathway through the stimulation of cutaneous receptors. For example, hypothalamic neurones can be driven in their patterns of spike discharge by electrical stimulation of the reproductive tract (Fig. 5). Such observations immediately prove that information passes from point A to point B, and establishes a multisynaptic connection that would be difficult to trace by the classical neuroanatomical methods of axonal degeneration. Furthermore, the patterns of hypothalamic activity which result reflect the physiology of the entire pathway and its many synapses. In Fig. 5 these properties have been highlighted by connecting the hypothalamic unit into a programme of self-stimulation, such that each time the cell generates an action potential a volley of stimuli are delivered to the reproductive tract. When the synaptology of the pathway is excitatory, the cell, through its spontaneous activity (see Section V) will proceed to fire faster and faster towards an "experimental climax" where through the intervention of a refractory component (= inhibitory synaptic activity or changes in cell polarization) the positive feedback link is broken and a pause in activity follows. Similarly, when inhibitory synapses predominate in the circuit the feedback link will slow the firing of the unit until an equilibrium is reached dependent on the magnitude of the inhibitory component.

Electrical stimulation has been used to analyse many of the non-neuro-

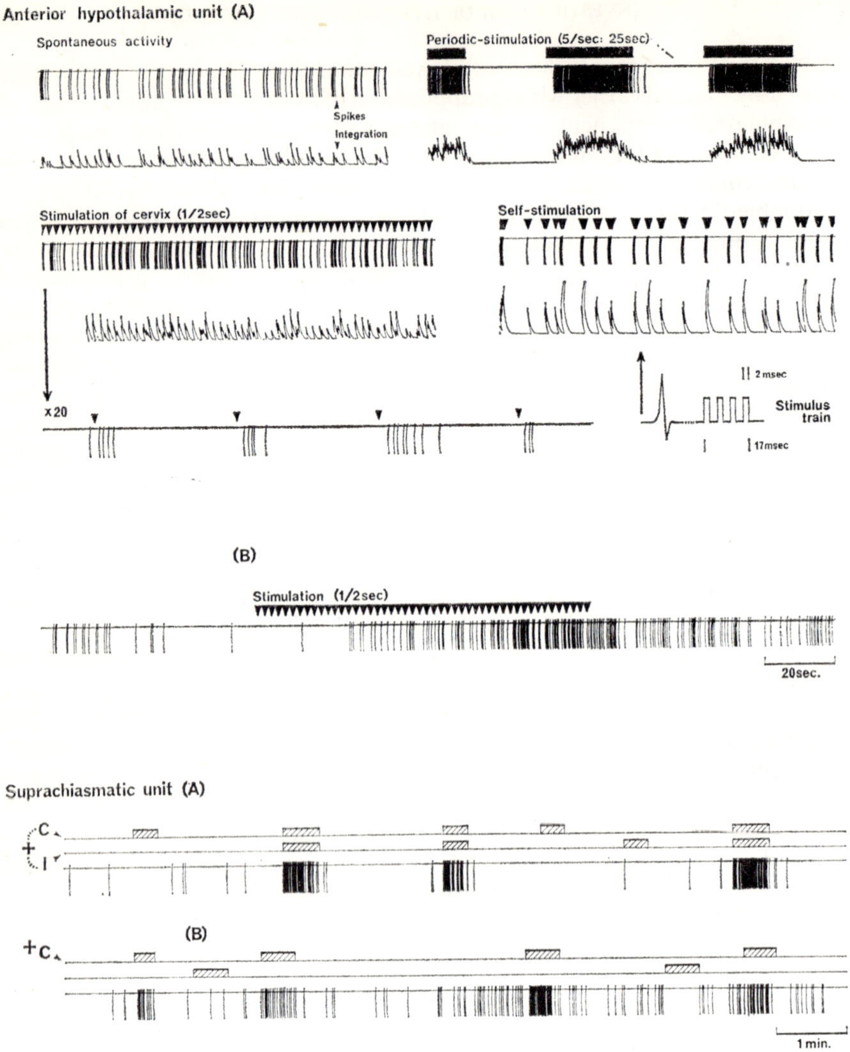

Fig. 5. Spike trains of four hypothalamic units recorded from a female rat anaesthetised with urethane. Each deflection of the trace corresponds to one action potential (spike).

In the upper part of the figure the activities of two anterior hypothalamic units are shown in relation to electrical stimulation of the vaginal parts of the cervix. Unit A, with a typical pattern of background activity, was driven in a very precise way by stimulation. The program of self-stimulation was produced by arranging for the spike activity of the cell to trigger the stimulator. Unit B was also driven by stimulation, but the response was of a different character. With this unit there was no precise stimulus response coupling, though the unit fired progressively faster as the period of stimulation was extended. One explanation for this difference is that Unit B is some steps further removed from the afferent input to the hypothalamus than Unit A.

The lower part of the figure displays the responses of two suprachiasmatic neurones to changes in the illumination of the retina. White light, of natural intensity, was applied to the contralateral (C) or ipsilateral eye (I) of both, as indicated. Unit A responded to illumination of both eyes together, whilst unit B responded to the illumination of the contralateral eye only.

endocrine functions of the hypothalamus. These studies are of interest to the neuroendocrinologist as some have produced observations, and even principles of general importance. Electrical stimulation of the anterior and medial parts of the hypothalamus tends to elicit the activation of components of the parasympathetic division of the autonomic nervous system, whereas stimulation in more caudal and lateral regions incites a generalised sympathetic reaction. The sympathetic system is involved in endocrine functions, both directly and indirectly; first, through the activation of the adrenal medulla and secondly, through the actions of the hormones of the adrenal medulla on other endocrine and target tissues. Reciprocal mechanisms have been observed in the control of other functions by the hypothalamus. Opposite responses in thermoregulation are found between the anterior and posterior hypothalamic areas; the anterior area containing a centre concerned with heat loss and the posterior area a centre for heat gain. With food intake, opposing control mechanisms exist in the lateral and ventromedial hypothalamic regions. Sleep and wakefulness have been induced by hypothalamic stimulation, the response depending on the areas stimulated and the frequencies selected. And not least among this spectrum of observations, studies with electrical stimulation suggest that the hypothalamus contains a centre concerned with "pleasure". A rat, provided with an electrode implanted in the lateral hypothalamus and a means by which it can provide itself with electrical stimulation, will indulge in an act of self-stimulation to the expense of almost every other activity. It is naturally difficult to combine such diverse observations into a fully integrated picture of hypothalamic activity, but these observations serve to emphasise the enormous complexity of the hypothalamus as an area for study when approached by the apparently simple technique of electrical stimulation.

V. Measurements of Electrical Activity

A. Introduction

The endocrine neurones of the mammalian hypothalamus are bioelectrically active, and as far as is known their electrical properties of spike generation and conduction do not differ appreciably from those of hypothalamic interneurones or of small neurones elsewhere in the brain.

Gross recordings of hypothalamic activity, as expressed in electroencephalograms (EEG) and evoked potentials, were among the first electrical measurements to be made, though neither have featured significantly in recent investigations, due to difficulties of interpretation (see Cross and Silver, 1966). Both are recorded from large double electrodes, not unlike those used for stimulation, and the recorded signals represent the potential differences between the exposed tips of the electrode assembly. The precise origin of these potentials remains obscure.

Somewhat more concrete results have been obtained by the measurement of hypothalamic activity with electrodes which monitor the spike activity of single neurones (unit activity) or small groups of neurones (multi-unit activity). Cross and Green (1959) pioneered the first recordings of hypothalamic unit activity. They used the now familiar steel microelectrode (1μ tip) insulated with a lacquer, and though alternative electrodes have since been produced, e.g. electrolyte-filled glass micropipettes, tungsten insulated with epoxylite and platinum insulated with glass, steel electrodes are still preferred by many laboratories for the recording of extracellular spikes. In such recordings, the excellent signal-to-noise performance of steel electrodes, combined with the ability to transmit large spikes, is more important than the faithful reproduction of the spike waveform. Unit recording electrodes generally have impedances of $>10^6$ ohms and must consequently be connected to specialist preamplifiers of input impedance up to 10^{14} ohms. Indeed, it is in the field of preamplification, rather than in electrode design, where the greatest improvements in recording technology have been made in recent years. These developments have involved increases in input impedances, reductions in noise levels, capacitance controls and the application of filters for the removal of all but the biological signals. Steel electrodes, despite their capricious construction and occasional failure for unknown reasons, offer the additional advantage that marker deposits of iron may be placed in the brain at strategic sites in relation to the recorded units. These deposits develop a Prussian blue colour if the brain is treated with potassium ferrocyanide during histological processing. Electrodes for multi-unit recordings are of 20–50 μ in tip diameter, and they are usually constructed in parallel arrays of two or more electrodes. Differential recordings are then taken from whichever pair of electrodes provides the better signal. The unit and multi-unit recordings are entirely different in terms of the observed activity. The unit recording is represented by trains of identical spikes of large amplitude (sometimes of 2 mV or more); while the multi-unit recording is rarely more than 50–100 μV in amplitude, and is composed of the spike trains of several cells. The signal processing of the unit recording is straightforward: the spikes can be photographed by one of several methods, or transformed by a level discriminator into digital pulses for further analysis (Figs 5, 7–9). Sometimes these recordings contain the activity of two units, but differences in spike waveform usually render them easy to identify and separate. The multi-unit recording, where the spikes of individual cells are usually difficult to distinguish, involves a more sophisticated analysis based on the extraction from the background noise of all signals with a frequency characteristic within the action potential range and the expression of this activity in digital or analogue form.

B. SPONTANEOUS SPIKE ACTIVITY

Many neurones in the brain are described as being "spontaneously" active, i.e. they are found to be generating spikes when approached by the microelectrode, and electrophysiologists have relied heavily on this activity for the detection of units. The distribution of the spontaneous firing rates within a population of hypothalamic units is exponentially arranged (Lincoln, 1967). Fifty percent, perhaps 80%, of hypothalamic units fire at <1 spike/sec and few fire at rates >10 spikes/sec. This presents a difficult situation, because the slower a unit fires the more difficult it is to detect and record. Thus, as silent or slow firing units predominate, the sample of units

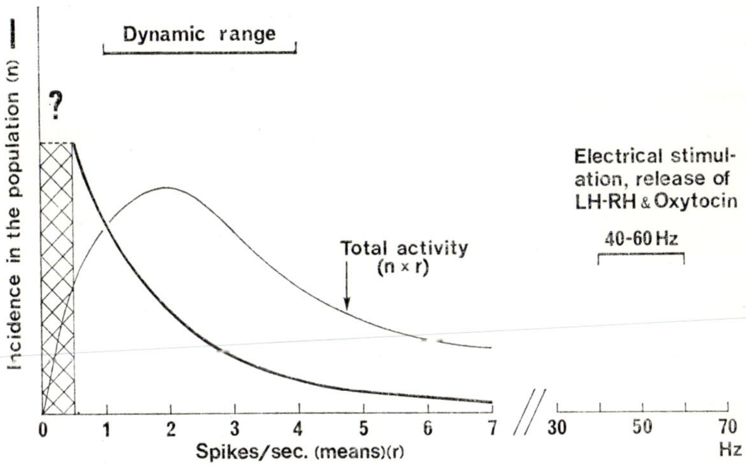

Fig. 6. Spontaneous firing rates of hypothalamic units in the rat (thick line). The "dynamic range" encompasses the various mean and geometric mean firing rates which have been observed under different hormonal (and surgical) situations. The total spike activity curve is the product of the firing rate and the incidence with which that rate is found within the population. The optimal frequencies of electrical stimulation for the release of LHRH and oxytocin are given on the right.

obtained in any study will depend on the skill of the operator to monitor the slow firing category, and in part this will be determined by the quality of the micromanipulator, the microelectrode and the electronic recording apparatus. This dilemma is further emphasised by the possibility that cellular responsiveness is related to spontaneous activity, often slow firing cells are accelerated whilst fast firing cells are inhibited by common stimuli. Mean firing rates and ranges of firing rate have often been quoted in publications and used as measures for scientific comparison. For the most part, too much weight has been given to such figures; they are often mathematically invalid or they simply indicate the developing skill of the operator. Firing rates of cell populations have to be compared by non-parametric statistical tests, or subjected to a mathematical transformation (a geometric mean is a satisfactory expression: Dyer *et al.*, 1972). The distribution of spike activity appears in somewhat better light when translated into an activity curve, i.e. the product of cell number and firing rate, for units firing in the range of 1–4 spikes/sec contribute most to the total spike activity of the hypothalamus (Fig. 6), and these units are easy to record. Silent, or slow firing units, should not in the future present the same problem; they can now be identified either by antidromic activation or by the iontophoretic application of an excitatory neurotransmitter.

For several years a state of silent rivalry has existed between those experimentalists who elected to study hypothalamic units in anaesthetised animals and those who sought to record only from the conscious and unrestrained animal. Ideal as the latter may sound, there remain enormous technical difficulties in precisely orientating microelectrodes, obtaining prolonged unit recordings and preventing stress in the conscious animal, and studies are restricted to the larger and more expensive species. Furthermore, the simple monitoring of the activity of single neurones is no longer sufficient. Stimulating electrodes have to be implanted for the electrophysiological identification of the recorded units, and methods have to be devised for the experimental activation and precise monitoring of the neuroendocrine events under study. The "anaesthetic" approach allows the experimentalist more freedom and provides the opportunity to perform technical manipulations which are far too exacting to apply to a conscious animal. At the same time, there is the additional and often neglected task of analysing the neural and endocrine actions of the anaesthetic employed. Urethane (ethyl carbamate) has been more widely used as an anaesthetic than any other material for experiments involving the acute recording of unit activity in rats; a single i.p. injection of 1·2 g/Kg will immobilise a rat for more than 12 hours. The brain activity of the rat is not noticeably reduced in quantitative terms (Cross and Dyer, 1971b) though there could be subtle changes in the patterns of cell firing. The EEG patterns under urethane anaesthesia are those of sleep (Lincoln, 1969), and reflex nervous functions, whilst reduced, are not eliminated (Barraclough and Cross, 1963). Urethane completely disorganises the neuroendocrine system; the release of hypothalamic hormones appears to be reduced (Blake and Sawyer, 1972a) whilst the corticosteroid levels of the blood are elevated to those associated with stress (Spriggs and Stockham, 1964). To what extent these changes reflect changes in neuronal activity or the flow of portal blood is difficult to determine. Rats will recover from urethane, though a proportion will die later from liver damage and the development of cancers. Therefore, this particular anaesthetic should never be used for recovery experiments. Barbiturates and other drugs are sometimes used for recording of cell activity, but most seem to create variable changes in brain activity and require frequent supplementation.

The spike activity of hypothalamic neurones is closely correlated with the EEG patterns of the cerebral cortex, and 50% (perhaps 80%) of

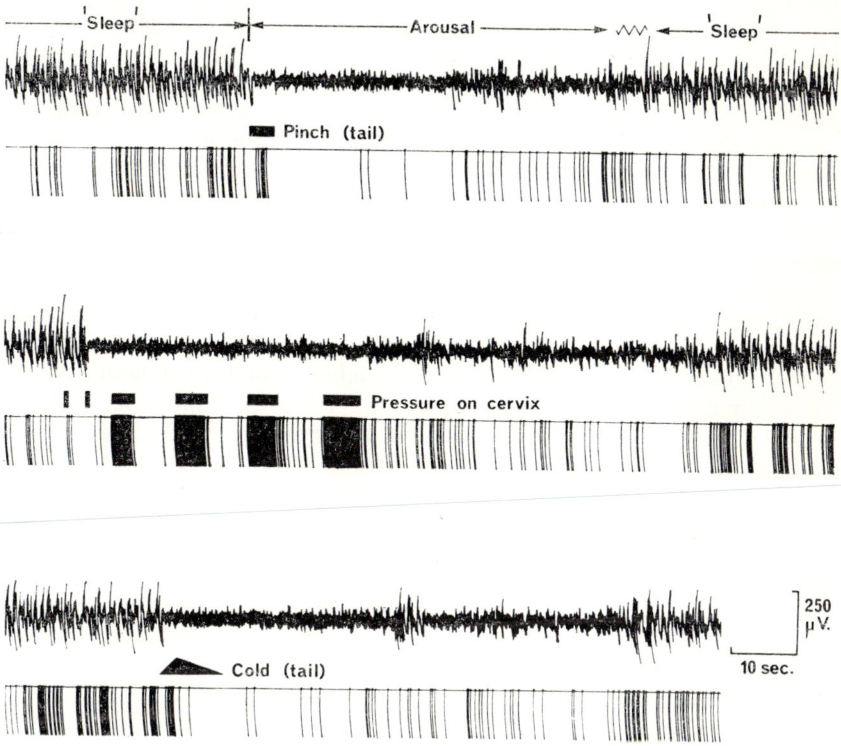

Fig. 7. A simultaneous recording of a cortical EEG and a hypothalamic unit to illustrate the difference between a specific and a non-specific unit response. Pain and temperature produced a mild inhibition of the hypothalamic unit which persisted for the duration of the EEG change. Pressure applied to the cervix with a glass rod produced a marked acceleration of the unit, but only for the duration of the stimulus (a specific response). The two event marks preceding the application of pressure correspond to the lifting of the rat's tail and the insertion of the glass probe into the vagina.

hypothalamic units display a substantial change in firing rate when the cortical EEG changes between slow-wave sleep, paradoxical sleep and a state of arousal (Fig. 7). These changes in firing are not all in the same direction, the slow-firing units tend to accelerate when the EEG switches from a pattern of slow waves to a pattern of higher frequency (arousal) whilst the faster units tend to slow down (Lincoln, 1969). While there are exceptions to this rule, the result of this reciprocal action is to transfer the majority of the hypothalamic activity from one network of neurones to another; the total spike activity does not have to change. This EEG correlation has caused many problems in the interpretation of cell activity. The EEG of the rat lightly anaesthetised with urethane displays cyclical changes, not unlike those of the natural sleep cycle, and the EEG changes over a period of 15 min or longer between slow-wave sleep, paradoxical sleep and arousal, but the application of any strong sensory stimulus (a pinch of a paw) will override the natural cycle and precipitate immediate arousal. This response probably occurs through the actions of sensory collaterals, of pathways running between the spinal cord and the thalamus, impinging on the reticular formation of the brain stem. The resultant change in brain stem activity is then propagated through to the hypothalamus, the non-specific thalamic nuclei and the cerebral cortex. A large proportion of hypothalamic units will therefore appear to be either activated or inhibited by the applied stimuli, but most of the changes will be "non-specific" and will persist for the duration of the EEG change, and this may last from a few seconds to an hour or more. So, as a matter of standard technique, the cortical EEG should be recorded during every study of hypothalamic unit activity.

There are three further questions to be considered concerning the nature and function of background spike activity. (1) Are all classes of hypothalamic units equally active? There is no complete answer to this question, but the neurosecretory cells of the supraoptic and paraventricular nuclei have patterns of background firing similar to those found in the hypothalamus as a whole (Figs 8 and 10). (2) How is this spontaneous activity maintained? Evidence indicates that some neurones in the central nervous system have characteristics of membrane polarization which permit spike generation in the absence of an afferent input. This activity may then, through excitatory synaptic connections, spread through the nervous tissue and even feedback onto the cells of origin in a self-propagating manner. It is interesting that in islands of hypothalamic tissue, neurones continue to generate spike trains, though the patterns are more uniform than those seen in the intact brain (Cross and Kitay, 1967). Presumably inputs from other brain areas normally impinge on the hypothalamus to modulate the intrinsic activity, and so long as activity is present the modulation can involve both excitation and inhibition. (3) Does the spontaneous activity of the hypothalamic units cause a basal release of the hypothalamic hormones? It is impossible at present to provide a conclusive answer to this question, but the rates of spontaneous discharge are 1–2 orders of magnitude slower than the optimal frequencies now being suggested for the electrical stimulation of hormone release (Fig. 6). This point will be further discussed in Section VI.

C. ANTIDROMIC IDENTIFICATION OF HYPOTHALAMIC UNITS

To some extent the axons and axon terminals of endocrine neurones are accessible in the neurohypophysis (= end station) and in the median eminence, and this has led to the development of techniques for the selective identification of neurones with axons projecting to these areas. The important principle, and one fundamental to neuronal theory, is that axons are capable of conducting spikes in both directions (Fig 8). When, for example, the neurohypophysis is stimulated with a single pulse, spikes are generated and each neurosecretory fibre entering the gland conducts a single spike back (antidromically) to its cell of origin. An action potential will then invade the cell body and be recorded by the microelectrode—the antidromic spike. The time taken for the antidromic spike to reach the cell body will depend upon the distance it has to travel and the conduction velocity of the axon. The conduction speeds of endocrine neurones are slow (<1 m/sec) (Sundsten *et al.*, 1970), and it takes 10–14 msec for a spike to pass from the neurohypophysis to the paraventricular nucleus of the rat. Whilst a uniform latency (i.e. $\pm$ 0·1 msec) and a constant response to each and every test stimulus is good evidence of a direct axonal connection, further evidence is required to verify that the spikes are not synaptically mediated. Two additional tests are applied. In the first a train of stimuli is applied at a frequency of 50 Hz or greater, i.e. higher than most of the calculated rates of maximal synaptic conduction, and an antidromic spike is expected to follow each and every stimulus pulse. The second test is that of collision, where an antidromic spike is initiated in the axon terminals by stimulation at the same time as a spike is orthodromically generated in the cell body. This is performed by arranging for the natural (orthodromic) spike to trigger the electrical stimulator. The two spikes then proceed along the axon in opposite directions, meet and mutually extinguish one another. When two antidromic spikes, spaced some 20 msec apart, are triggered by the orthodromic spike, the first will be eliminated by collision but the second of the pair will pass unimpeded to reach the cell body. These are powerful electrophysiological tools and provide the only means of electrophysiologically identifying units, in real time, according to their axonal projections. It is also possible to search for quiescent units by delivering a stimulus pulse, about every 3 sec, as the microelectrode is lowered through the brain. Such cells have, of course, to be activated before the collision test can be applied, but this can be achieved by the iontophoretic application of an excitatory neurotransmitter. Paraventricular and supraoptic neurones are now routinely identified by antidromic stimulation of the neurohypophysis. Similarly, arcuate neurones have been identified by stimulation of the median eminence (Moss *et al.*, 1975b).

Antidromic and trans-synaptic activation has also found favour as an experimental tool for the characterisation of hypothalamic units other than

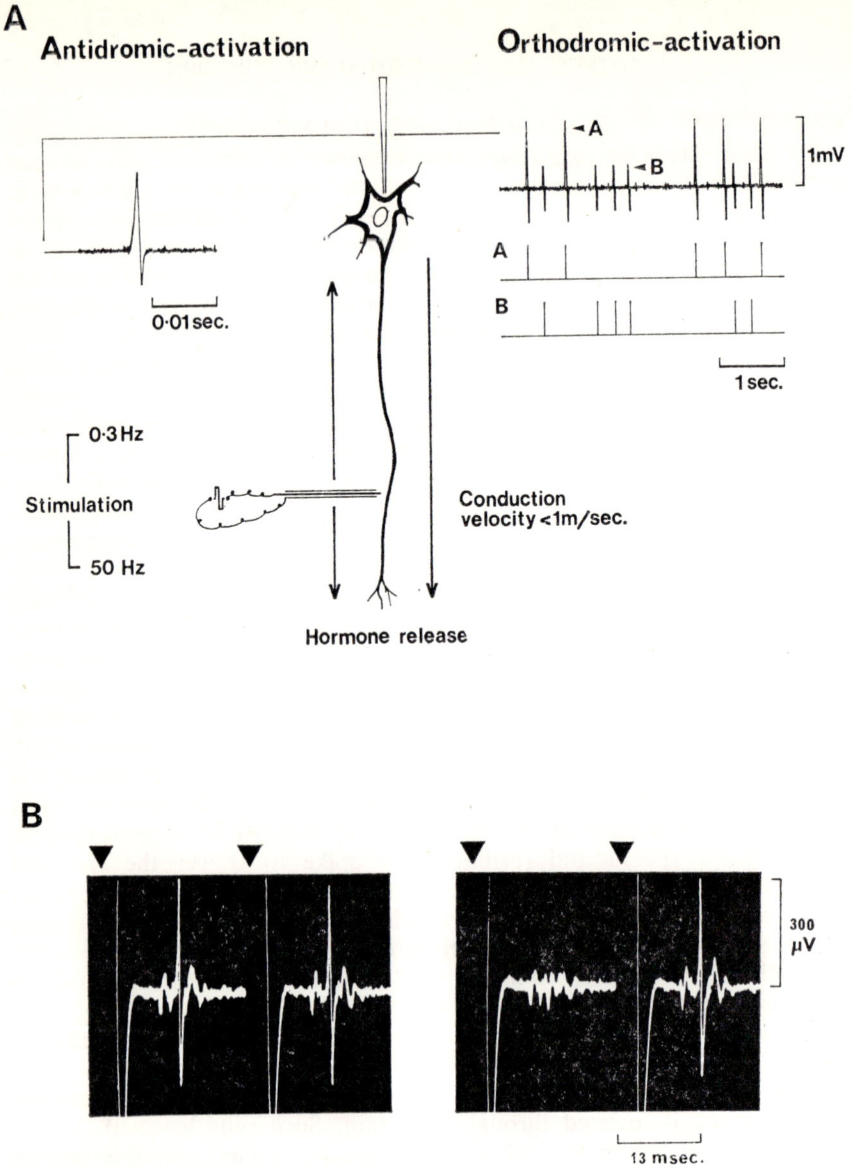

Fig. 8. Techniques for the electrophysiological identification and analysis of endocrine neurones (or hypothalamic interneurones) of known projection.

A. Antidromic activation, by the stimulation of the nerve axon or terminal, is shown on the left. Note, low frequencies (one pulse every 3–5 sec) are used in the search for antidromically activated units and high frequencies (50 Hz) are used for the induction of hormone release. The analysis of the on-going spike activity by the use of level discriminators (or waveform analysers) to separate double and triple unit recordings into individual spike trains, is shown on the right.

B. Traces of an antidromically activated paraventricular neurone in a rat. Each event mark represents the application of a stimulus pulse to the neurohypophysis. On the left is displayed the frequency following two stimuli delivered about 20 msec apart, on the right the first of the evoked spikes has been extinguished by collision (see text). Evoked activity, representing the firing of adjacent neurones, is also visible, especially after the removal of the large spike by collision. Each photograph contains five super-imposed traces.

those already mentioned. Neurones in the preoptic area of the rat have been classified according to their response to stimulation applied to the arcuate nucleus and elsewhere (Dyer, 1973). Whilst this approach is a considerable improvement when compared with the blind sampling technique used hitherto, it does not result in the identification of endocrine neurones *per se*. There exist two problems. Neurones in the preoptic area may be driven by stimulation of the arcuate nucleus, but the axons which are stimulated in the arcuate region may not terminate there (Fig. 9), they could pass on as far as the spinal cord, perhaps to synapse with the cells of the intermediolateral column (the sympathetic outflow). Secondly, both Dyer and others have now observed units in which antidromic activation has been elicited by stimulation in two quite separate sites, e.g. anterior hypothalamic units have been driven from both the preoptic and arcuate regions; this suggests the existence of bipolar neurones. However, this complication has not been fully explored, and until now the stimulating electrodes have only been placed in sites to which the neurones were thought to project.

A phenomenon of post-stimulus inhibition is associated with antidromic activation and this is characterised by a reduction in orthodromic spike generation in the first 50 msec or so after the arrival of an antidromic spike at the cell body (Dreifuss and Kelly, 1972). One widely accepted interpretation is that the axons have collateral branches which feedback directly, or through interneurones, to inhibit the cells under study. Unfortunately, collaterals axons do not appear to be very abundant if one is to judge from the paucity of evidence provided by studies with light and electron microscopy. Furthermore, when the oxytocinergic neurones of the paraventricular and supraoptic nucleus are physiologically activated to high frequencies (see Section VI) no inhibition is observed in the adjacent units either during the response or shortly thereafter: this implies that inhibitory feedback, if it exists, is extraordinarily specific. An alternative hypothesis is that post-stimulus inhibition involves an abnormal change in cell activity. When the neurohypophysis of the rat is stimulated, for example, a synchronous volley of activity is propagated back to the four neurosecretory nuclei. This volley of spikes will arrive over a period of about 5 msec, allowing for differences in conduction velocity. Such a density of spike generation is most abnormal and exceeds by 1–2 orders of magnitude the maximum rate of spike generation in 5 msec under conditions of maximal orthodromic stimulation.

D. IONTOPHORESIS AND THE STUDY OF HYPOTHALAMIC UNITS

Iontophoresis, in conjunction with unit recording, is a technique now widely used to study the sensitivities of individual neurones to chemical substances, with particular reference to neurotransmitters, hormones and antagonists.

Multibarrel micropipettes (tip diameter 3–5 μ) are used for the simultaneous recording of cell activity and drug ejection. The central barrel is usually selected for recording and is filled with electrolyte (NaCl or KCl); the other barrels, up to six in number, are filled with the drugs to be investigated. The concentration and pH value of each drug solution is critical. When solutions are used at concentrations in excess of 1M it becomes difficult to stop the drug leaving the electrode tip by diffusion. At the other extreme, a solution

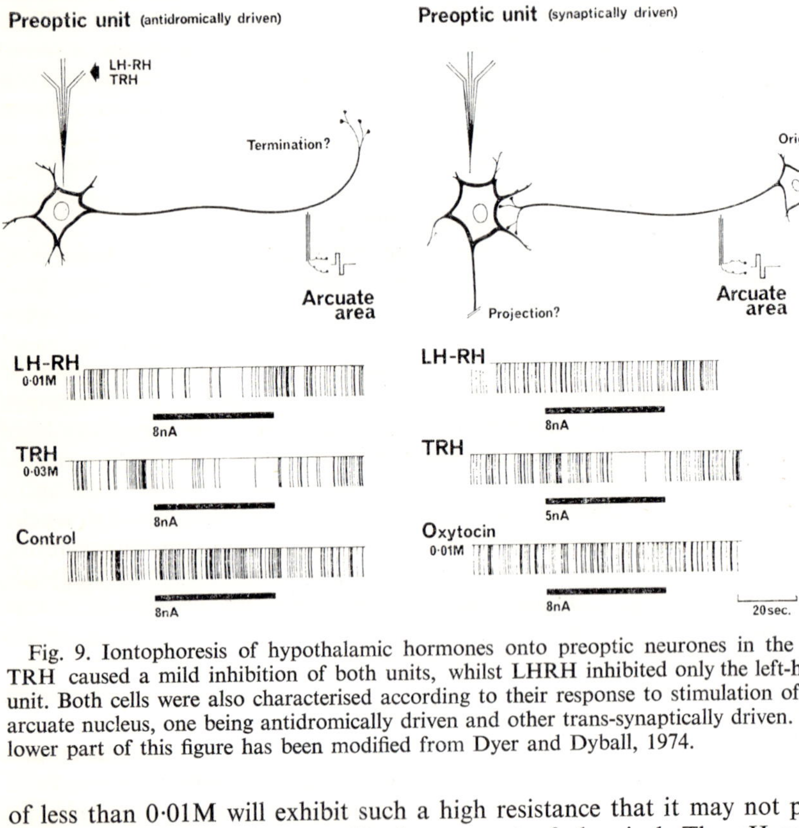

Fig. 9. Iontophoresis of hypothalamic hormones onto preoptic neurones in the rat. TRH caused a mild inhibition of both units, whilst LHRH inhibited only the left-hand unit. Both cells were also characterised according to their response to stimulation of the arcuate nucleus, one being antidromically driven and other trans-synaptically driven. The lower part of this figure has been modified from Dyer and Dyball, 1974.

of less than 0·01M will exhibit such a high resistance that it may not pass sufficient current to eject an effective amount of chemical. The pH value controls the degree of ionisation, and thus cations such as acetylcholine and noradrenaline should be used at pH 3·5, anions such as glutamate at pH 8·0, whilst neutral amino acids such as γ-aminobutyric acid (GABA) can be used at either of these pH values. The immediate problem once the electrode assembly has been filled, tested and inserted into the brain, is the prevention of drug loss by diffusion or through the action of hydrostatic pressure. To prevent these losses retaining currents are applied (up to 20 nA) with the opposite polarity to that required for drug ejection, and the direction of this current will depend upon the predominant ionic species. The currents applied for ejection will range from 1–100 nA and such currents pass sub-picogram amounts of the various drugs onto the surface of the neurone. Electrode impedances can fluctuate widely during ejection, and it is highly desirable, if one is to control the procedure with any accuracy, to use a current source capable of pumping constant current through an impedance which may rise to 200 megohms. Balancing currents are often applied during drug emission to minimise the changes in neuronal activity that result from current flow: this involves an accurate computation of the net current application and the introduction of an equal and opposite current through a barrel filled with NaCl. Alternatively, current controls can be applied subsequent to drug ejection by passing the same net current through a control electrolyte-filled barrel for the same period of time. Current and extreme pH effects are characterised by changes in spike activity which are immediate in onset and coterminous with the period of current flow.

Electro-osmosis is an important variation of iontophoresis and has particular application for the ejection of materials that do not ionise well or which have low solubilities. The materials are made up in solution with an electrolyte such as NaCl and introduced into the micropipette. A current is passed and sodium ions are ejected, but water moves with the sodium ions and with the water moves the drug whose effects are to be examined.

Both iontophoresis and electro-osmosis raise problems of interpretation, assuming one has successfully resolved the technical difficulties already mentioned. Iontophoresis produces a pin-point deposit of a drug, at a high and perhaps unphysiological concentration, on one small part of the cell membrane. Diffusion will occur from this site, but the concentration will decline very rapidly with distance; furthermore, the ejected drug may exert its effects at either a presynaptic or postsynaptic site. The points just considered are founded on the assumption that some of the drug has actually been released; in fact, there is no reliable way of measuring how much—if any—of the drug has been liberated.

Several studies have examined the response of hypothalamic neurones to a range of putative neurotransmitters (Barker *et al.*, 1971; Dyball *et al.*, 1974; Moss *et al.*, 1975b). A broad spectrum of responses has been observed, and in common with results from other parts of the brain, most of the neurotransmitters have produced changes in spike activity of some neurones irrespective of whether or not the materials have been found in the brain areas under study. At best, the technique of iontophoresis provides for a pharmacological categorisation of the neurones—"a pharmacological fingerprint". In no situation within the hypothalamus has it proved possible to associate a morphological or functional cell type with a specific response to one or more of the neurotransmitters. Perhaps this exceeds realistic expectations, considering the multitude of synapses impinging on every individual cell. In a typical situation a cell might be found that was excited by acetylcholine, serotin and glutamate while unresponsive to other agents; whereas the next cell, some 50 μ distant, might be excited by acetylcholine and glutamate, slightly inhibited by noradrenaline and strongly inhibited by GABA. But to record a response as either excitation or inhibition is a gross oversimplification, for every excitation or inhibition is different in its magnitude and uniformity.

TRH, LHRH and vasopressin have all been applied to hypothalamic neurones by iontophoretic methods, and changes in the spike activity of cells have been observed (Moss *et al.*, 1972; Dyer and Dyball, 1974) (Fig. 9). These results could account for ultra-short loop feedback (see Chapter 3), but beyond this there is very little which can be said about the endocrinological significance of these findings. Oxytocin has been applied to the cells of the paraventricular nucleus, that is, to the cells producing oxytocin, and both excitation and inhibition have been reported. Those who have encountered inhibition of cell firing have interpreted the effect as imitating the actions involved in recurrent inhibition, where oxytocin might be released at the terminals of recurrent collateral axons. Others have suggested, where excitation has been the predominant response, that oxytocin has a positive feedback action as might be required in the release of a pulse of hormone. One further postulate to emerge from these observations, and the one which might embody the greatest element of scientific truth, is that neurosecretory cells package their hormone into granules as a means of protecting themselves from the actions of the hormone they produce (Cross *et al.*, 1975).

VI. Relation of Electrical Activity to Hormone Release

A. Early Studies

With the introduction of unit recording, it became possible to search—with some prospect of success—for the neural determinants of hormone release. The recording of spike activity was straightforward, despite the problems mentioned in Section V, but as was found with so many parts of the brain it was the analysis and interpretation of the activity that was to prove so difficult. Fifteen years elapsed between the recording of hypothalamic units by Cross and Green (1959) and the establishment of the first reasonably precise patterns of spike activity for the release of a mammalian hypothalamic hormone.

Throughout the early part of the 1960s several attempts were made to correlate the spontaneous firing rates and responses of hypothalamic units with conditions such as pseudopregnancy, persistent light-induced oestrus, ovariectomy and with the acute and chronic administration of steroid hormones. The validity of much of this work was subsequently questioned on grounds of anaesthesia, insufficient identification of cells, and lack of precise information on the endocrine state of the animal at the actual time of recording. Although an animal may have been in oestrus at the time it was taken from the animal quarters, four hours of anaesthesia and stereotaxic restraint may well have changed "its reproductive feelings". Critical as one may now be in reviewing these early studies, they did establish many of the basic characteristics of hypothalamic activity that we now take for granted and use as a basis for further investigation; they examined the firing patterns of individual neurones, established the dynamics of spontaneous activity, and explored the correlation with the cortical EEG (see Section V).

Two important observations made in these early studies are worthy of consideration. Hypothalamic units, when grouped as a population according to area, were influenced by the endocrine state of the animal at the time of anaesthesia (Barraclough and Cross 1963; Cross and Silver, 1965). Similarly, differences were subsequently found in cell activity recorded from islands of hypothalamic tissue prepared at different stages of the oestrous cycle (Dyer *et al.*, 1972). Both approaches suggested the persistence, despite the trauma of surgery and anaesthesia, of endocrine-induced differences in hypothalamic activity. The differences in firing rate were never large, 2–4 spike/sec at the most, but the full extent of these changes may not have been appreciated, for when units were examined on a population basis an acceleration in the activity of some units may have been counterbalanced by a deceleration in others. The majority of the neurones in these samples must have been interneurones or endocrine neurones which had nothing to do with the specific endocrine problem under investigation, and as such it is difficult to understand the significance of these apparently global changes in background activity. The mechanism responsible for this change in cell activity is not understood; it could relate to an overall change in membrane polarization, or it could reflect a change in the balance of excitatory to inhibitory synaptic activities.

The separation of cell responses into specific and non-specific categories was another important feature to be recognised in the early studies. *Non-specific* responses were simply those that were related to the "arousal" of the brain; they were correlated with the cortical EEG and ranged in duration from a few seconds to thirty minutes or more. *Specific* responses were of an entirely different character. They were rare by comparison with the EEG effects and were characterised by a marked acceleration in spike activity during the application of a stimulus to a particular receptor field of the body. Such activation ceased the moment the stimulus was removed. Specific responses to light are shown in Fig. 5 and to the probing of the vaginal parts of the cervix with a glass rod in Fig. 7. This latter effect is most interesting as it might represent the afferent input for the reflex induction of ovulation and psuedopregnancy. Also, repeated probing of the vagina of the urethane-anaesthetised rat has an effect on the brain which is the opposite to that created by most sensory stimuli, it promotes a synchronisation of the EEG—inducing "sleep" rather than "arousal" (Cross and Silver, 1966).

B. DEVELOPMENT OF A NEW APPROACH

By 1967 or thereabouts, a position of impasse had developed in the electrophysiological investigation of the hypothalamus. The limitations of the techniques previously employed were now apparent, and new approaches were required. Thus, multi-unit recording was introduced with the objective of making long-term measurements of hypothalamic activity in conscious animals; hypothalamic islands were constructed to eliminate the need for anaesthesia and to remove unwanted neural influences; antidromic stimulation was applied for the selective identification of endocrine neurones; and radio-transmitters and hydraulic microdrives were constructed for the recording of unit activity from the conscious animal.

Multi-unit recording from the hypothalamus was pioneered largely in the laboratory of C.H. Sawyer in Los Angeles. This technique solved some of the technical limitations mentioned earlier, but it raised in a new context the problem of how to interpret mass activity. The majority of the activity contained in these recordings must have originated from neurones that were of no immediate interest to the functions under study, and increases in the spike activity of some units may have been counterbalanced (or obscured) by decreases in the activity of others. Although it could be argued, as endocrine neurones represent the motor component of the neuroendocrine system, that only an increase in spike activity could release hormone. Claims have been made that the multi-unit activity of the arcuate region is increased at times of LHRH release (Terasawa and Sawyer, 1969; Kawakami *et al.*, 1970), but it is very questionable whether this activity related specifically to the release of this single hormone. Similar increases in the multi-unit activity of the arcuate region have been observed following stimulation of the hippocampus, without the ovulatory release of LHRH (Gallo *et al.*, 1971).

The introduction of antidromic activation for the selective identification of hypothalamic neurones was a major advance, but it promoted a shift in the research of many laboratories and many electrophysiologists turned their attention for the first time to the analysis of neurohypophysial functions. The reason was simple, electrical stimulation of the neurohypophysis or the neural stalk provided, without question, a means of selectively identifying the neurones sending axons to the neurohypophysis. In addition, much more was known of the hypothalamo-neurohypophysial system in terms of hormone production and release, than was known about adenohypophysis and its control. There were only two hormones to consider, namely oxytocin and vasopressin: these were produced in separate neurones and released independently in response to specific stimuli. Finally, there were *in vivo* bioassay procedures available for the measurement of the hormone released.

C. ELECTRICAL DETERMINANTS FOR THE RELEASE OF OXYTOCIN

The milk-ejection reflex of the rat provided the first situation in which it was possible to establish the precise electrical determinants for hormone release. In unit recordings from anaesthetised lactating rats, about 50% of paraventricular and supraoptic neurones, antidromically identified from the neurohypophysis, were found to accelerate to a rate of 40–60 spikes/sec, for 2–4 sec, some 12–14 sec before the rise in intramammary pressure at each milk ejection (Fig. 10) (Wakerley and Lincoln, 1973; Lincoln and Wakerley, 1975). These brief periods of accelerated activity were stereotyped in intensity and duration, and units displaying such activation remained responsive at each and every milk ejection, i.e. there was no recruitment or loss of responsiveness. The remaining 50% of the neurosecretory cells, whilst also antidromically activated, displayed no detectable change in firing. The synchronous burst of firing in the responsive neurones spanned the optimal frequency band for the electrical stimulation of hormone release (Fig. 4), and was associated with a similar latency to milk ejection (= release time + blood circulation time + myoepithelial cell response time). These parallels, and more recent evidence in which it has been possible to fractionate the periods of neurosecretory activation and obtain proportional reductions in oxytocin release (Lincoln and Wakerley, 1975), provide incontrovertible evidence for the involvement of electrical activity in the release of endogenous hormones.

The responsive neurones, i.e. the oxytocinergic cells, were found in both the magnocellular nuclei, conflicting with the widely held view which links oxytocin solely and specifically with the paraventricular nucleus, and vasopressin with the supraoptic nucleus. However, most recent biochemical and histochemical studies tend towards the view that these nuclei have mixed functions (Mey *et al.*, 1974).

Working from a knowledge of cell numbers, the amount of hormone released, and the number of spikes generated during each period of activation, calculations have been made which equate each spike to the release of about 3 fg oxytocin or the contents of about 40 neurosecretory granules (Cross *et al.*, 1975). Does the background activity release hormone? The responsive neurones are not without background activity (see Fig. 10), but such activity cannot release any appreciable amount of hormone, for at 3 fg/spike this background activity would be sufficient to discharge, in only 2–3 hours, all the 1000 mU of oxytocin and vasopressin stored within the neurohypophysis. Such a release does not occur; in fact, the daily turnover of these two hor-

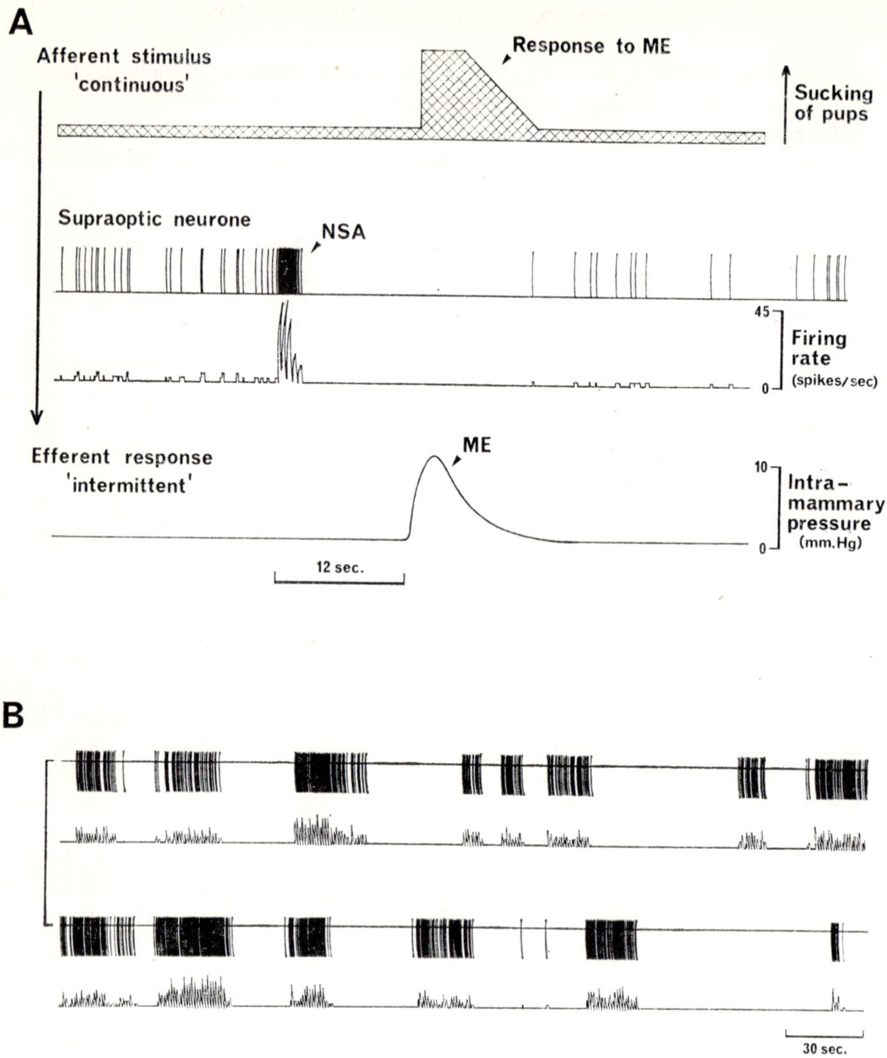

Fig. 10. Polygraph recordings of antidromically identified supraoptic and paraventricular neurones.

A. Illustrates the response of an oxytocinergic neurone in the supraoptic nucleus during suckling. Note, a dramatic burst of electrical activity precedes milk ejection by about 12 sec. A similar latency is observed between electrical stimulation of the neurohypophysis and milk ejection (See Figure 4).

B. A simultaneous recording of two phasically-active paraventricular neurones (vasopressinergic neurones?). Both were recorded from the same microelectrode and their activities were separated by a window discriminator. Note the lack of correlation in the spike trains of these adjacent cells, and the dramatic contrast their activity makes with the background activity of the unit shown in A.

mones in a non-stimulated animal is no more than 50 mU/day (Pickering *et al.*, 1974). This comparison of hormone release, between periods of activation and background activity, provides a physiological parallel to the observations made with electrical stimulation where only high frequencies (> 30 pulses/sec) cause hormone release. It is an efficient system in another respect, for each period of accelerated activity is followed by a marked hiatus in cell discharge (Fig. 10) and thus the total activity of the cells, when averaged over the entire response, is not appreciably greater than that which would have been generated as background activity. In a way, the spike activity has been re-programmed.

There are many more questions to be resolved about the neural determinants for the release of oxytocin. For example, what mechanism determines the simultaneous activation of the oxytocinergic cells in the "four" neurosecretory nuclei? What process determines the regular interval between one period of activation and the next? How is the suckling stimulus analysed within the brain, and why does it not modulate the neurosecretory cells moment by moment? Is the neural mechanism for the pulsatile release of oxytocin common to the release of other hormones in the rat and in other species?

D. ELECTRICAL DETERMINANTS FOR THE RELEASE OF VASOPRESSIN

The neural determinants for the release of vasopressin are still a matter for conjecture, in spite of the numerous studies that have sought an answer. Many investigations have shown that stimuli which release vasopressin also modify the firing patterns of supraoptic neurones (Dyball and Koizumi, 1969), but none have established beyond doubt a causal relationship between the recorded electrical activity and the release of the hormone. Sometimes there were differences in time between the two events which could not be explained, or the neural responses were irregular and complex. Furthermore, the stimuli applied were unphysiological. Injections of hypertonic saline and the substitution of 2% NaCl for drinking water were extensively used, and these must have caused widespread and diverse changes in neural functions.

During the analysis of the electrical determinants for the release of oxytocin a possible contender for the role of the "vasopressinergic neurone" has been discovered. Twenty to thirty percent of antidromically identified neurones in the supraoptic and paraventricular nuclei of the rat are periodically active (Wakerley and Lincoln, 1971), and few of these remarkable cells are associated directly with the burst of activity which precipitates the release of oxytocin. These phasically active cells, as they are termed, are quite unique within the hypothalamus, and they alternate abruptly every 20–60 sec between total or almost total quiescence and a period of moderately fast activity (4–8 spikes/sec). Further, the cells are not synchronised with each other, judged from double recordings in which it has been possible to monitor simultaneously the activity of two phasically active cells.

Two attempts have been made to implicate the phasic cell in the release of vasopressin. Two stimuli have been devised which are thought to promote the selective release of vasopressin, namely bilateral carotid occulsion (Harris, *et al.*, 1975) and controlled haemorrhage in which 4 ml of blood is slowly withdrawn from a rat and then replaced (Wakerley *et al.*, 1975), and both stimuli have been found to modify the activity of the phasic cells. Bilateral occlusion, which is more acute of the two stimuli and which creates a release of vasopressin of sufficient magnitude to contract the myoepithelial cells of the mammary gland, causes the activation within 10 sec of phasic cells which are in their silent mode, and prolongs and increases the firing of already active units. Controlled haemorrhage has a similar effect, though the time course is longer. As haemorrhage proceeds the active periods of the phasic cells increase in length until some cells are fired continuously. Then, the rate of spike discharge also begins to increase, though the very high rates of firing associated with the facilitated discharge of oxytocin are not attained. Both stimuli are somewhat unphysiological; at least, these stimuli are not those associated with the control of the daily fluctuations in the release of vasopressin, and it remains to be determined whether the observed changes in spike activity represent the activation of the normal control process. Critics also contend that there are insufficient phasic cells, at 30% of the population, to account for what is known about the stores of vasopressin and its distribution within the neurosecretory nuclei. Perhaps this group of cells is not as finite as the present studies suggest, maybe the proportion of phasic cells is larger in the conscious animal (Hayward and Jennings, 1973).

VII. Concluding Remarks

Having described the experimental evaluation of the electrical determinants for the release of the neurohypophysial hormones, it would seem logical to attempt a similar analysis of studies designed to establish the determinants for the release of TRH and LHRH, for example. Unfortunately, very little progress has been made in this electrophysiological quest, and this situation is not likely to change until there is definitive information on which cells produce precisely which hormone, and even then we shall require to know much more about the physiological control of these hormones than we do today. To many readers this chapter may have over-emphasised the hypothalamic control of the neurohypophysis, but, placing personal feelings apart, it is within the context of this system that many of our technical approaches to the study of the hypothalamus as a whole have been pioneered. For the moment our knowledge of the neural control of the neurohypophysis is a decade or more in advance of our knowledge of the neural control of the adenohypophysis, perhaps, by 1986, the balance will have been restored.

CHAPTER 3

Control of the Endocrine Hypothalamus

J. S. M. HUTCHINSON

I. Introduction

The hypothalamic hormones are synthesised within the cell bodies either of the neurones of the so-called hypophysiotrophic area (anterior pituitary releasing and inhibiting hormones), or of the supraoptic and paraventricular neurones (oxytocin and vasopressin). These specialised neural structures which possess endocrine functions serve as the major controlling links between the brain and the endocrine system and have been called neurotransducer peptidergic neurones (Wurtman, 1970a; Reichlin, 1974). The functions of these peptidergic neurones are governed directly or indirectly from higher nerve centres by changes in both the internal and external environment (Findlay, 1972; Hayward, 1972). These include higher nerve inputs and time dependent rhythms, reflex responses by the peripheral nerves, hormonal and other chemical and physical changes in cerebrospinal fluid and peripheral and portal blood, and stimuli such as stress, season, light, temperature mediated via the central nervous system (Table 1). This chapter will discuss some of the general aspects of this control of the hypothalamus.

TABLE 1

Inputs to the Endocrine Hypothalamus

Neural Inputs
Special sense organs; smell, hearing, vision.
Exteroceptors; suckling, coitus, pain, temperature.
Interoceptors; uterus, cervix, blood volume, chemoreceptors.
Stress; physical, emotional, overcrowding.
Time dependent rhythms; sleep-wake cycles.
Cerebrospinal Fluid Inputs
Hormones
Ions
Other substances
Vascular Inputs
Hormones; steroids, thyroid hormones, pituitary hormones, angiotensin.
Other substances; glucose, amino acids, blood gases.
Other properties of blood; temperature, osmotic pressure.

It is impossible to define the borderlines between the hypothalamus and the structures that control it, as the hypothalamus is one component of a complex integrated system (Donovan, 1972; Anderson and Haymaker, 1974). Hence any discussion of hypothalamic and other "centres" or studies of the "isolated" hypothalamus should be viewed with caution. However, such concepts and experimental models are useful in trying to understand the relative importance of hypothalamic input mechanisms.

As with other mechanisms there are often marked species differences in the control of hypothalamic function. Much of the data has been derived from studies on the rat and man and extrapolation to other mammals and lower orders is dangerous.

II. Levels of Hypothalamic control

A. Peptidergic neurone autonomy

The hypothalamic peptidergic neurones are capable by themselves of sustaining a certain autonomous function. This has been shown by experiments in which the hypothalamo-pituitary complex has been isolated from the rest of the brain (Halász, 1969, 1972, 1974; Szentágothai *et al.*, 1972; Krey *et al.*, 1975*a*; Sawyer, 1975; see Chapter 2). Many pituitary functions are maintained in both the rat and the rhesus monkey when the hypothalamus is disconnected from the rest of the brain. The basal secretion of GH is unchanged or slightly elevated, and both the basal secretion of prolactin and the release of ACTH in response to insulin hypoglycaemia are unaffected. The pulsatile rhythm of basal FSH and LH release is unaffected, as is the negative feedback action of oestrogen (Blake and Sawyer, 1974; Krey *et al.*, 1975b). However, TSH and vasopressin secretion, the ACTH diurnal rhythm and GH secretion induced by hypoglycaemia, are all reduced or abolished. The ovulatory surge of gonadotrophin is lost in the rat, but may or may not be lost in the monkey (Krey *et al.*, 1975b; Spies and Norman, 1975). These data suggest that though some hypothalamic function can occur in isolation, stimulatory and inhibitory inputs from elsewhere are needed for the full functioning of the hypothalamus; they lend support to the concept that the neural control of the anterior pituitary functions at two levels (Barraclough and Gorski, 1961; Flerkó, 1962; Szentágothai *et al.*, 1972). One, represented by the hypophysiotrophic area of the hypothalamus, regulates basal secretions of the anterior pituitary and neurohypophysial hormone secretion. A second higher level of control includes all other hypothalamic and extrahypothalamic structures which exert stimulatory or inhibitory effects on the secretion of the pituitary. Some of these control areas lie within the hypothalamus. Köves and Halász (1970) have shown that in rats in which a hypothalamic island includes the preoptic area, ovulation can take place. The preoptic area may thus represent a "surge" centre which initiates and integrates signals which impinge on the peptidergic neurone. Such a centre may also be involved in the control of prolactin secretion (Neill, 1974). Special neuronal receptors sensitive to changes in hormones, temperature, glucose and osmolarity, which effect the activity of the peptidergic neurones, are also located in the hypothalamus (see Section IV, A, 4).

B. EXTRAHYPOTHALAMIC AREAS

Many parts of the brain contribute to the input of information to the hypothalamus (Raisman, 1966, 1970a; Nauta and Haymaker, 1969; Raisman and Field, 1971a; Donovan, 1972; Koikegami, 1974; Lammers and Lohman, 1974). Important structures include the septum, hippocampus, anterior thalamus, amygdala, pyriform cortex and midbrain, with additional brain areas influencing the hypothalamus indirectly. Almost every afferent pathway is complemented by a hypothalamic efferent pathway, thus forming a two-way interaction between the hypothalamus and the structures that influence it. Knowledge of how these structures specifically influence neuroendocrine control is complicated by the detailed inter-relationships which exist, and by the fact that anatomical pathways have frequently not been correlated with physiological phenomena. The transmission of stimuli from the special sense organs, exteroceptors and interoceptors to the hypothalamus and sleep-wake related phenomena, probably involve these structures (see Section IV, C). The long-loop feedback control of ACTH, gonadotrophin and prolactin secretion by corticosteroids and sex steroids may occur at the level of the midbrain, hippocampus and amygdala (Section IV, A, 2). Some areas of the amygdala are involved in stimulation of ACTH, GH and gonadotrophin secretion; others may inhibit ACTH and GH output. The hippocampus is probably inhibitory to ACTH and gonadotrophin output and stimulatory to GH. The mesencephalon is stimulatory to ACTH, GH and TSH output and has both stimulatory and inhibitory effects on gonadotrophin release.

III. Neurotransmitters

A. CONTROL OF PEPTIDERGIC NEURONES

The chemical nature of the mediators involved in transferring information from extra-hypothalamic areas to the peptidergic hypothalamic neurones has not been fully elucidated. Since the original discovery showing the importance of central adrenergic mechanisms in the control of ovulation (Sawyer *et al.*, 1949) evidence has accumulated that stimulatory and inhibitory multisynaptic pathways may involve adrenaline, noradrenaline (NA), 5-hydroxytryptamine (5HT), dopamine (DA), acetylcholine (ACh), γ-aminobutyric acid (GABA) and glycine (Fig. 1; Wurtman, 1970a, b; Ganong, 1974; Reichlin, 1975). All these compounds are amines or amino acids and undergo inactivation by enzymes or re-uptake by the neurones of origin. DA, 5HT and NA are referred to as monoamines and DA and NA are catecholamines. ACh is synthesised from choline and acetylcoenzyme A by a reaction catalysed by choline acetylase. Two receptor types have been recognised in cholinergic systems: "muscarinic" and "nicotinic" so-called, because muscarine and nicotine were found to stimulate them. The rapid hydrolytic degradation of ACh is catalysed by acetylcholine esterase. NA and DA are synthesised from tyrosine by decarboxylation and hydroxylation reactions, DA being an intermediate in the formation of NA. 5HT is synthesised from tryptophan by a similar series of reactions. The degradation of the monoamines is partly neuronal, catalysed by monoamine oxidases and partly extraneuronal catalysed by catechol-*o*-methyl transferase. Two receptor types are involved in the adrenergic system, the α and β receptors. These may either be different sites or may represent the same receptor site with properties which vary according to localised conditions.

Under a wide variety of internal and external influences which are known to alter the function of the pituitary, e.g. oestrous cycle, diurnal rhythm,

E.H.

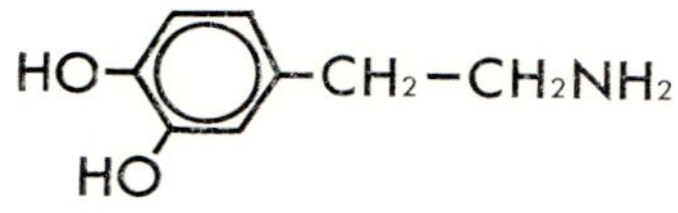

Dopamine

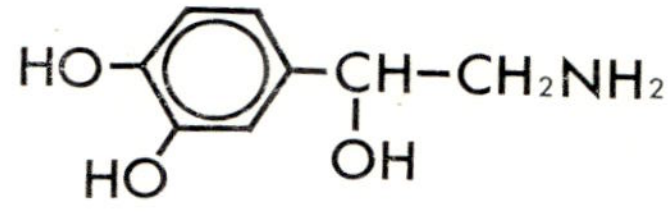

Noradrenaline

5-Hydroxytryptamine

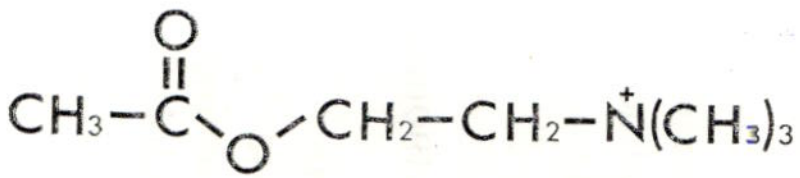

Acetylcholine

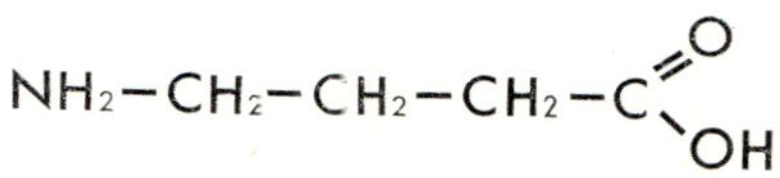

γ-Aminobutyric acid

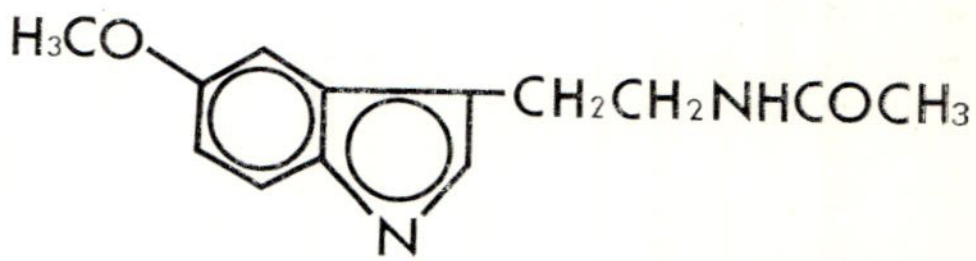

Melatonin

Fig. 1. Structures of neurotransmitters and melatonin which may be involved in regulation of the peptidergic neurones.

Table 2. Some drugs affecting different aspects of neurotransmitter metabolism and action

Site of action	Drug examples	Effect
SYNTHESIS		
Natural precursor	L-dopa	precursor of DA, NA
	5-hydroxytryptophan	precursor of 5HT
Block synthesis	Hemicholinium	blocks choline uptake
	α-methyl-p-tyrosine	block tyrosine hydroxylase; synthesis of DA, NA
	Disulfiram	block dopamine β-hydroxylase: conversion DA→NA
	p-chlorophenylalanine	block tryptophan hydroxylase: synthesis of 5HT
Formation of false neurotransmitter	α-methyl dopa	forms α-methyl DA, α-methyl NA
STORAGE		
Interference with storage	Reserpine	depletes DA, NA, 5HT stores
RELEASE		
Stimulate release	Methylamphetamine, Tyramine	release DA, NA
Block release	Botulinum toxin	blocks ACh release
	Guanethidine	blocks NA release
RECEPTOR		
Mimic neurotransmitter	Methacholine	muscarinic ACh agonist
	Nicotine	nicotinic ACh agonist
	Apomorphine, Bromocriptine	DA agonists
	Methoxamine, Phenylephrine	α-adrenoceptor agonists
	Isoproterenol, Isoprenaline	β-adrenoceptor agonists
Block receptor	Atropine	blocks muscarinic ACh receptor
	Dihydro-β-erythroidine	blocks nicotinic ACh receptor (central)
	Hexamethonium	blocks nicotinic ACh receptor (ganglia)
	Pimozide, Haloperidol	block DA receptor
	Phentolamine	α-adrenoceptor blocker
	Propanolol	β-adrenoceptor blocker

Table 2—*contd.*

Site of action	Drug examples	Effect
RECEPTOR		
Block receptor	Methysergide	blocks 5HT receptor
	Strychnine	blocks glycine receptor
	Bicuculline, Picrotoxin	block GABA receptor
DEGRADATION	Physostigmine	anticholinesterase agent
	Tranylcypromine, Pargyline	monoamine oxidase inhibitors
UPTAKE	Imipramine	blocks DA, NA, 5HT uptake

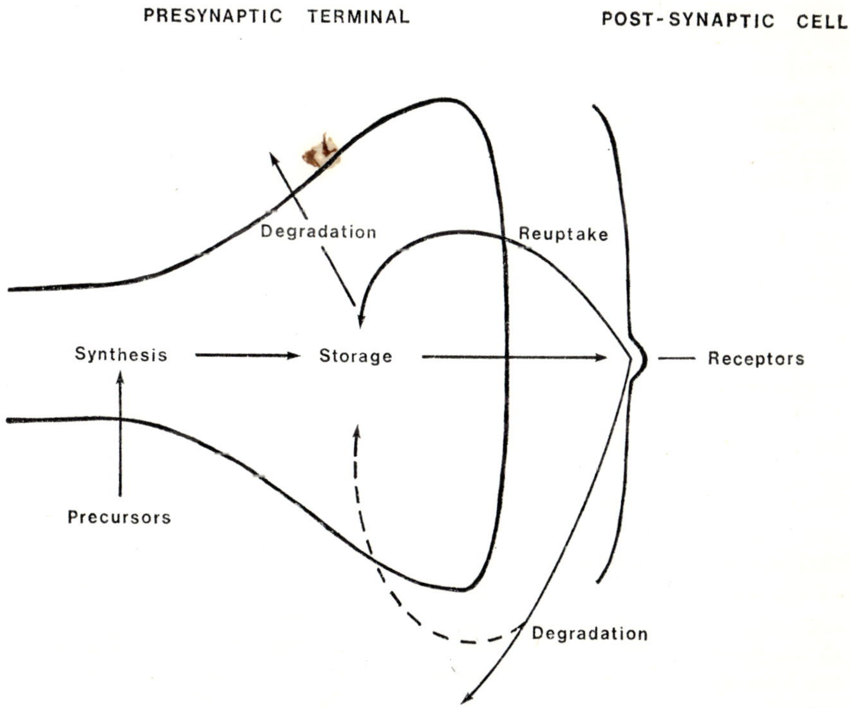

Fig. 2. Schematic representation of neurotransmission indicating the general mechanism of synthesis, storage, release, receptor binding, degradation and re-uptake of a neurotransmitter (after Bachelard, 1974).

ovariectomy, thyroxine and steroid administration, alterations in adrenal activity and stress, marked changes have been shown to take place in neurotransmitter activity in the hypothalamus as determined by alteration in content, turnover, rate of synthesis from labelled precursors, enzyme activity and histochemical fluorescence (Lichtensteiger, 1970; Glowinski, 1970; Fuxe and Hökfelt, 1970; Porter and Ben-Jonathan, 1974; de Wied and de Jong, 1974). A large number of drugs are available which can affect every aspect of neurotransmitter metabolism and action; uptake of precursors, synthesis, formation of a false transmitter, transport, storage, release, receptor binding, degradation and re-uptake (Table 2; Fig. 2; Bachelard, 1974; de Wied and de Jong, 1974; Frohman and Stachura, 1975). Many of these have been used to assess the role of the appropriate amine in hypothalamic control. Methylamphetamine has been shown, for example, to stimulate GH release; the GH response to insulin induced hypoglycaemia is increased by propranolol and reduced by phentolamine; and imipramine inhibits nocurnal GH release. Such studies can clearly help to unravel the neurotransmitter mechanisms involved. However, there can be problems of interpretation (see Chapter 4). The drug may be effective at different points on a multisynaptic pathway. Different forms of hypothalamic stimulation or inhibition, e.g. metabolic, hormonal, stress, sleep may involve a different transmitter. DA is converted to NA by the hypothalamus and one or other of these substances may alter the secretion of 5HT. Finally, the anterior pituitary hormone may be under both stimulatory and inhibitory control from the hypothalamus (e.g. GH); or as in the case of DA, not only act on the peptidergic neurone producing PIF, but also act directly at the pituitary level as a PIF (Rennels *et al.*, 1975).

(1) *Location*

Fluorescent, antibody and enzyme-isotopic techniques have been applied to the localisation of monoaminergic pathways in the brain and their terminal distribution in the hypothalamus (Fuxe and Hökfelt, 1970; Palkovits *et al.*, 1974b; Hökfelt *et al.*, 1975b; Brownstein *et al.*, 1976). There is evidence for a dopaminergic innervation of the external layer of the median eminence, originating from the arcuate nucleus and the anterior paraventricular hypothalamic nucleus. There is a noradrenergic innervation of the internal layer of the median eminence and throughout the hypothalamus and preoptic area in general, from cell bodies farther away in the hypothalamus or outside in the pons or medulla. Noradrenergic fibres from the mesencephalon also reach the hypothalamus, and a serotoninergic (5HT) pathway from the mesencephalon terminates diffusely in the hypothalamus particularly at the suprachiasmatic nucleus, anterior median eminence area and medial retrochiasmatic area. All three types of monoaminergic fibre arise in the mesencephalon and are distributed to the limbic system, the amygdala and hippocampus.

Less is known about possible adrenaline secreting neurones or about the distribution of cholinergic pathways. There is a cholinergic system which arises from the midbrain and projects through the hypothalamus to the basal forebrain area (Shute and Lewis, 1966; Shute, 1970). Some of these fibres appear to enter the hypothalamus anteriorly.

(2) *Sites of action*

There are several sites at which brain amines might participate in the control of pituitary function (Wurtman, 1970a, b; Anton-Tay and Wurtman, 1971; Gay, 1972; Fig. 3). (1) A synapse in a tract that carries inputs from the limbic system or other extrahypothalamic region or sensory input to the peptidergic neurone. (2) A direct synapse of extrahypothalamic fibres with the cell body of the peptidergic neurone. (3) The neurotransmitter may be present in the peptidergic neurone and act within the cell to liberate the releasing factor. (4) The transmitter may be present in fibres running parallel to or in close contact with the peptidergic neurones and act via an axo-axonal mechanism at the nerve terminals. (5) It could act directly on the anterior pituitary to liberate or block the secretion of a hormone or to modify the response of the pituitary to a true releasing or inhibiting factor. In teleost fish there is a specialised system of neurosecretory innervation of adenohypophysial cells. Direct innervation of the adenohypophysis, especially of the pars intermedia, also occurs in other vertebrates (Bern, 1972; Chapter 12).

(3) *Role in hypothalamic hormone release*

The complexity of the neural arrangements and the indirectness of the evidence prevent firm conclusions about the role of specific transmitters in various parts of the hypothalamo-pituitary system. All transmitters probably have multiple effects because of the multisynaptic nature of the input pathways to the system (de Wied and de Jong, 1974; Ganong, 1974, 1975; Imura *et al.*, 1975; Johansson, 1975; Reichlin, 1975; Chapters 7 12).

There is evidence that ACh and DA are involved in the release of vasopressin and oxytocin. In the control of CRF, 5HT is probably involved in circadian activity and in stress-induced ACTH release, whereas, an inhibitory role has been suggested for NA, and ACh has also been implicated. In the control of GH, the pathways mediating stress and metabolic responses may involve NA, DA is concerned in release and 5HT in slow wave sleep-induced release. However, conclusions from the different studies on GH release are limited because of the existence of species differences (see Chapter 10). NA may be involved in TRH release and 5HT in its inhibition. DA is involved in the basal release of PIF, 5HT in its inhibition and ACh in the nocturnal release of prolactin during pseudopregnancy in the rat.

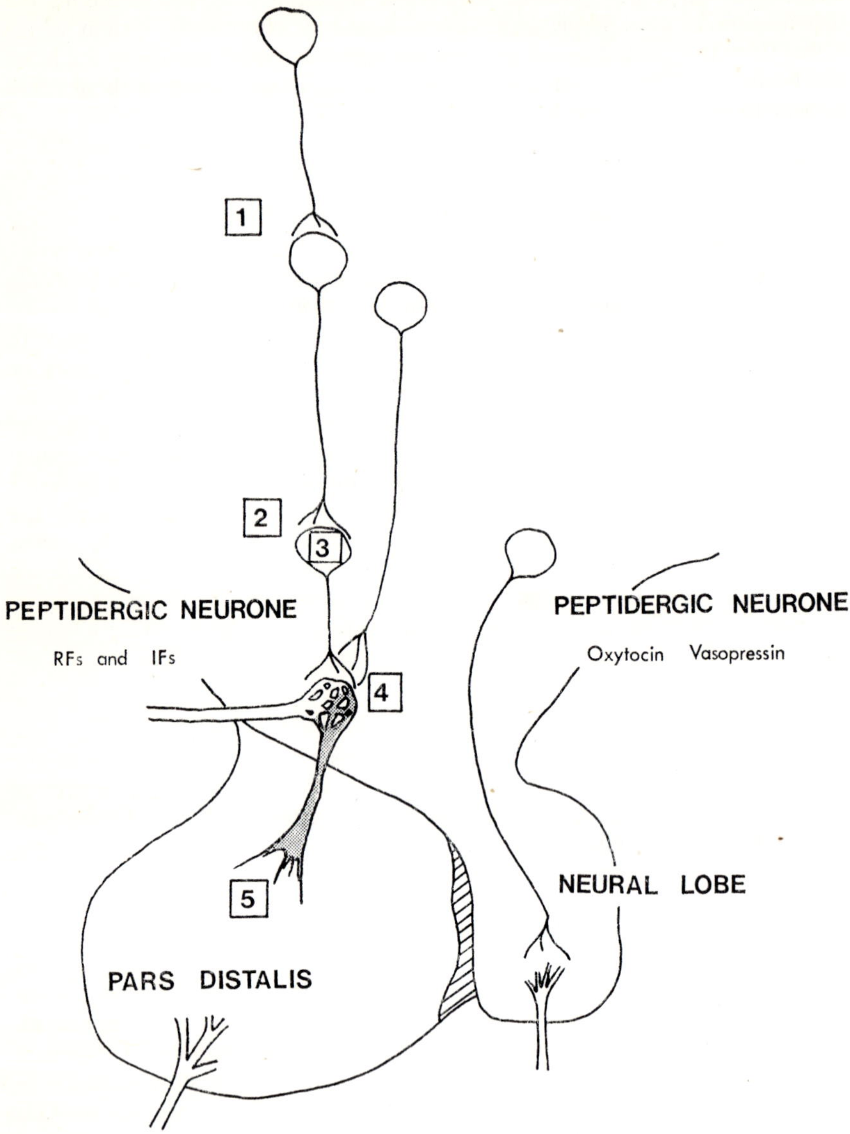

Fig. 3. Possible sites at which brain amines might participate in the control of pituitary function. For explanation of numbers 1–5 see text, Section III, A, 2 (after Wurtman, 1970b; Gay, 1972).

ACh, DA, NA and GABA are concerned with different aspects of gonadotrophin release and 5HT in inhibition.

The feedback action of target gland hormones (see Section IV, A, 2) on the release of anterior pituitary hormones may be mediated by neurotransmitter activity (Kalra and McCann, 1973; de Wied and de Jong, 1974). Oestradiol, for example, is capable of blocking the stimulatory effect of DA on LHRH release and corticosteroid-sensitive neurones in the hypothalamus are inhibited by NA.

B. MECHANISM OF PEPTIDERGIC NEURONE HORMONE RELEASE

(1) *Ions* (Douglas, 1973; Hayward, 1975)

The release of hypothalamic hormones is evoked by a stimulus, believed to be an action potential originating in the cell body of the peptidergic neurone, from input signals reaching it. Evidence relating secretion to action potentials is derived from electrical recording and stimulation experiments (see Chapter 2). When the action potential reaches the nerve ending, depolarisation occurs leading to peptide release. Experimental alteration of the external potassium ion (K^+) concentration, which mimics the depolarising effect of the action potential, has been shown to cause the secretion of vasopressin, oxytocin, CRF and PIF but not LHRH and TRH (Edwardson *et al.*, 1972; Bennett and Edwardson, 1975; Bennett *et al.*, 1975; Johansson, 1975). Calcium ions (Ca^{++}) are important in the secretion process. Membrane depolarisation controls the opening of Ca^{++} channels in the terminal membrane, thus allowing Ca^{++} entry, leading to hormone release.

(2) *Prostaglandins and cyclic AMP*

Recently prostaglandins (PGs) and cyclic AMP (cAMP) have been implicated in anterior pituitary hormone release (Costa *et al.*, 1975; Hafs, 1975; Johansson, 1975; Sandow and Babej, 1975). Both are present in most parts of the brain including the hypothalamus (Usunov and Weiss, 1972), and can be influenced by neurotransmitters (Coceani *et al.*, 1971; Palmer *et al.*, 1973). PGs may act to control the release of anterior pituitary hormones either at the pituitary level or at the hypothalamic level (Eskay *et al.*, 1975b; Hafs, 1975; Sato *et al.*, 1975) where they probably do not function as neurotransmitters but rather, modify or modulate the inhibitory and releasing effect of neurotransmitters on the peptidergic neurones.

Specifically, PGs of both the E and F series can cause ACTH, prolactin and GH release; principally acting on the hypothalamus in the case of ACTH and prolactin, and on the pituitary in the case of GH release. PGs of the F series may not alter gonadotrophin release directly but the E series release both FSH and LH via a hypothalamic action. Steroids may influence this effect. These actions of PGs and neurotransmitters (Section II, A) and the feedback effects of steroids (Section IV, A, 2) may all be mediated by cAMP (Gunaga and Menon, 1973; Palmer *et al.*, 1973; Ojeda *et al.*, 1974b; Reichlin, 1975; Weissman *et al.*, 1975).

IV. Homeostasis, Feedback, Neurally Mediated Stimuli and Cyclical Phenomena

Survival of a species in a variable environment depends upon the ability to adapt. Ideas on how this is achieved have been developed from the concepts of Bernard (1878) and Cannon (1929) on the "constancy of the internal environment" and "homeostasis", through the idea of the "adaptation syndrome" (Selye, 1950) including endocrine regulation, to information theory and cybernetics (Wiener, 1948). There has been the development of the idea of a continuous "seeking" for an equilibrium and a returning to it after a displacement. There are two basic mechanisms of control, rhythms (and the factors that control them) and feedback, either alone or in combination (Sollberger, 1956).

All animals live in an environment with an abundance of external rhythms; diurnal, tidal, seasonal, solar, etc. A capacity to follow these oscillations can ensure survival as shown by poikilotherms, hibernators, twenty-four hour activity patterns, etc. Body functions may also oscillate without the external environment. Such oscillations may be either passive, and be driven by an outside agent or synchroniser, *Zeitgeber* (Aschoff, 1965) or active and represent a spontaneously oscillating system. Frequently such oscillations are approximately equal to an external rhythm and may be genetically selected. The controversy as to whether there are truly endogenous rhythms or whether those that seem to be independent of major external influences such as light, temperature, etc., are in fact being influenced by more subtle regulators remains unanswered (Brown, 1970; Hastings, 1970).

Cannon's term homeostasis included Bernard's idea of a fixed or constant internal environment, but also suggested dynamic self-regulating processes that serve to maintain that constancy, and to return the internal environment to normal should it deviate. This is the concept now called "negative feedback", that is, if there is a deviation in one direction there is a reaction in the opposite direction (Langley, 1965).

Such general concepts can be more directly applied to the study of the various factors which control and modify the functions of the endocrine hypothalamus. Feedback is an important concept in a discussion of the blood-borne hormonal stimuli influencing hypothalamic function. Various other blood-borne and cerebrospinal fluid-borne stimuli will also be discussed. Neurally mediated stimuli, coming from both the external and internal environment will also be examined; and their importance in generating or triggering rhythmical or cyclical function discussed.

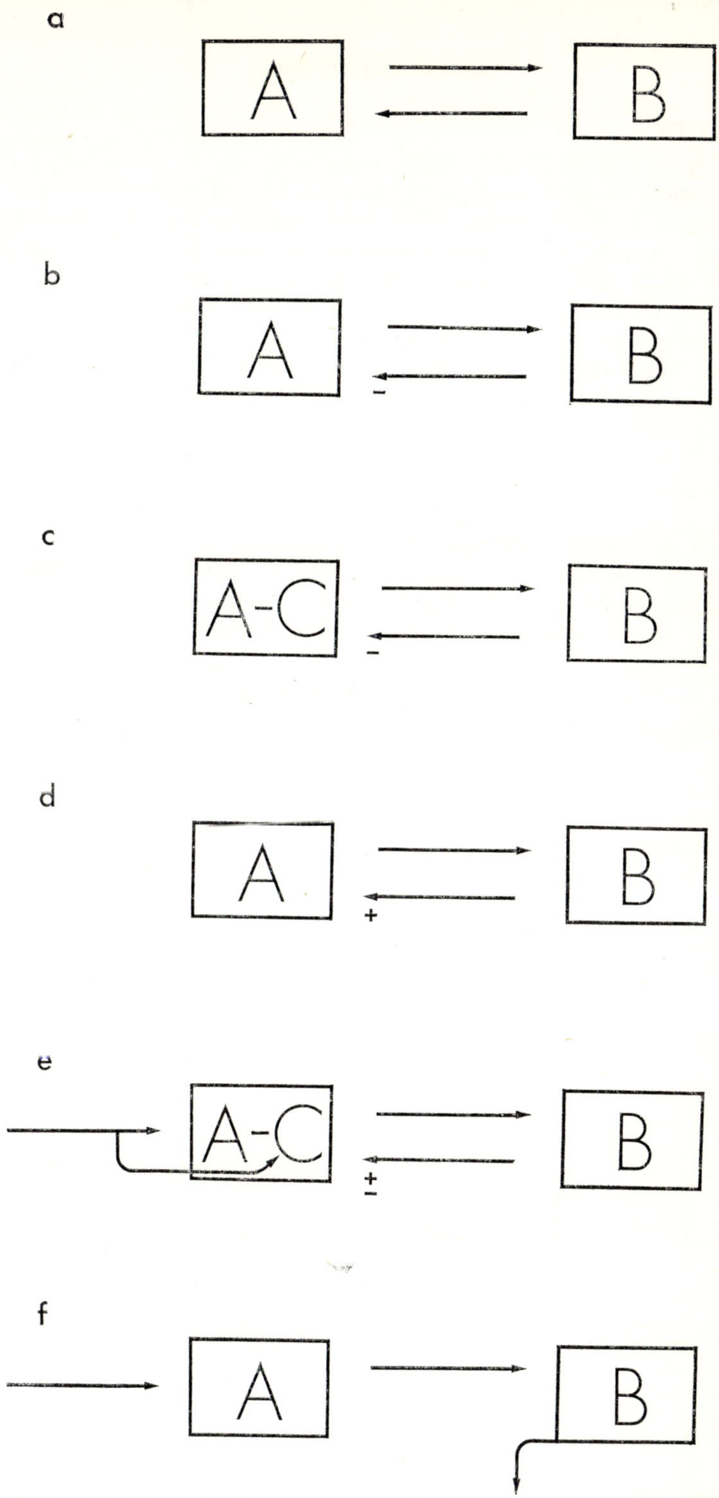

Fig. 4. Types of feedback, for description see text Section IV, A (after Sollberger, 1956).

A. FEEDBACK

The hypothalamo-pituitary-target relationship can be described in terms of interaction between parts analysed as a flow of information, and self regulation by feedback (Riggs, 1970; Szentágothai *et al.*, 1972; Rasmussen, 1974). In a process where information moves from link to link, feedback occurs if a later link acts back on an earlier one to stimulate or depress it, establishing a closed loop as with A and B in Fig. 4a. In negative feedback, if A stimulates B and B in turn *inhibits* A (Fig. 4b) a "standstill" will be achieved. The inhibition will often be only to a reference value or "set-point" C (Fig. 4c). There may also be an "overshoot" oscillation of A before the system returns to normal. In positive feedback (Fig. 4d) there is an increase at each turn of the loop till an upper limit of the system is reached or break-down takes place. Unlike negative feedback which is a dampening process, positive feedback appears inherently unstable and short term but this is not necessarily so (Riggs, 1970). These basic feedback systems may be modified by inputs into A which may inhibit or stimulate the system, alter the setpoint of the feedback (Fig. 4e) or the loop may be completely open as in Fig. 4f, where all control of the system is now independent of B.

(1) *Feedback mechanisms and the control of pituitary function*

Feedback mechanisms play a fundamental role in the control of pituitary function. The original proposal of Moore and Price (1932) for the pituitary-gonadal relationship has been more generally applied to the pituitary-target

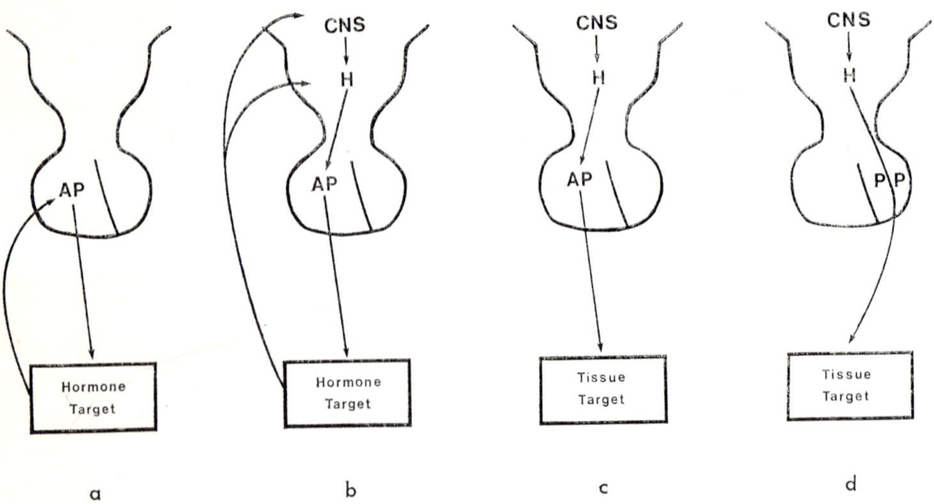

Fig. 5. Types of CNS-hypothalamo-pituitary relationships. For explanation see text Section IV, A, 1.

gland model (Fig. 5a). Increase in the blood level of the hormone se by the target gland (e.g. thyroxine, corticoids, sex steroids) inhibits the secretion of the corresponding pituitary trophic hormone (TSH, ACTH or gonadotrophin). Conversely, a decrease in the blood level of the target gland hormone lowers the inhibition and allows the secretion of the corresponding trophic hormone. There is also positive feedback, first suggested by Engle (1931) where increases in target hormone secretion stimulate trophic hormone release.

At one time feedback was believed to occur at the anterior pituitary (Fig. 5a); however, Hohlweg and Junkmann (1932) suggested that the central nervous system may be involved in the regulation of gonadotrophin secretion by steroids. Such feedback may take place at different levels in the hypothalamus or at certain extra-hypothalamic sites (Fig. 5b).

Attempts to determine whether the pituitary or the hypothalamus is involved in feedback control using hormone implants, has been complicated by the so-called "implantation paradox" (Bogdanove, 1964). The effectiveness of a hormone implant into the hypothalamus may result from transport of the hormone to the pituitary via the portal vessels, while an intrapituitary implant may not diffuse through the whole gland (see Chapter 2).

In the third hypothalamo-pituitary-tissue model (Fig. 5c) which is applicable to GH, prolactin and MSH, there are no direct target glands to take part in the feedback control. It is probably significant that in these cases the hypothalamus appears to exert a general inhibitory influence over the secretion of the pituitary hormones (Chapters 9, 10, and 12). Other factors, such as metabolic products and hormones from other target glands also play an important role in their control (see Sections IV, A, 4 and IV, A, 5).

In the fourth hypothalamic hormone model (Fig. 5d) there is no anterior pituitary target or peripheral target hormone control. This model is applicable to the neurohypophysial hormones, vasopressin and oxytocin.

(2) *Long loop feedback control of TSH, ACTH and gonadotrophins*

The principal negative feedback action of thyroxine and triodothyronine probably takes place at the pituitary, by blocking the pituitary response to TRH (Reichlin, 1975; Chapter 7). Both negative and positive feedback at the hypothalamic level are also possible. The pituitary effect appears to be mediated by the stimulation of a suppressor protein and the positive effect at the hypothalamus by stimulation of TRH synthetase. Thus, feedback effects of thyroid hormones involve the stimulation of protein synthesis as in their actions on other tissues (Malkinson, 1975).

The feedback effects of corticoids are of two types, rate-related fast feedback and slow feedback (Chapter 11). There is evidence that the feedback loop is largely directed at the pituitary, the secretion of ACTH in response to CRF being modulated by steroid levels. Corticoid reduces responsiveness by promoting synthesis of a specific protein and absence of corticoid leads to increased responsiveness by promoting synthesis of a different protein (Brattin and Portanova, 1975). Stimuli, like stress and circadian rhythms release CRF that can "break through" this feedback. Higher doses of corticoid are needed to suppress CRF production. Whether this effect is at the hypothalamus or higher nerve areas is not clear, as corticoids are readily taken up by the septum and hippocampus in addition to the hypothalamus. As with other steroids, the major site of action of corticoids is on the cell nucleus (McEwan, 1973) and their overall action may be to regulate enzymes, which are rate limiting for the synthesis of neurotransmitters.

Under most circumstances sex steroids (oestrogens, progestagens and testosterone) are inhibitory to gonadotrophin secretion, probably both at the level of the hypophysiotrophic area and at the pituitary level (Davidson, 1969; Sawyer and Hilliard, 1971). Under certain circumstances oestrogens may exert a stimulatory effect on LH and FSH release, also acting at either pituitary, hypothalamic or higher brain levels. This positive feedback is important in the hastening of puberty and in the induction of ovulation (see Davidson, 1969; Smith and Davidson, 1974; Chapter 8). The stimulatory action of oestrogen on LH release appears to depend on the presence of progesterone (Krey *et al.*, 1973). Actions of sex steroids at the pituitary level alter both the magnitude of the response and the ratio of FSH and LH secreted (Fink *et al.*, 1975; Greeley *et al.*, 1975). In so doing they make a non-specific stimulus by gonadotrophin releasing hormone specific for FSH and LH at the pituitary level. There is also evidence that the effect of oestrogen on pituitary gonadotrophin secretion is biphasic (Vilchez-Martinez *et al.*, 1974a; Libertun *et al.*, 1974) being initially inhibitory and later stimulatory. This may explain how both positive and negative feedback effects can follow oestrogen administration. There may also be at least two populations of oestrogen sensitive neurones. One group may be sensitive to brief intense oestrogen stimulation and the other to long term low levels (Reichlin, 1975).

There is increasing evidence that conversion of the secreted steroid to other metabolites takes place before active feedback occurs. Androgen is aromatised to oestrogen (Naftolin *et al.*, 1971, 1972; Weisz and Gibbs, 1974) or 5α-reduced to 5α-androstan-17β ol-3-one or 5α-androstan-3α, 17β-diol (Denef *et al.*, 1973) for feedback effects at both hypothalamic and pituitary levels (Motta *et al.*, 1973). The conversion of progesterone by the hypothalamus or pituitary to 5α-dihydroprogesterone and 3α-hydroxy-5α pregnane-20-one may play a role in progesterone feedback (Cheng and Karavolas, 1973; Karavolas *et al.*, 1976). Specific steroid binding (Vreeberg *et al.*, 1975) is probably a prerequisite of action, with localisation at the cell nuclei, leading to alterations in RNA synthesis and protein synthesis. The fe of steroids probably involves an interaction with the neurotransmitter system (Section III, A, 3).

The role of "inhibin", probably a polypeptide, which has been postulated as a specific tubular feedback mechanism for FSH in the male (Setchell and Main, 1974), is probably at the pituitary level (see Chapter 8).

(3) *Short loop feedback*

In addition to the long loop feedback of target hormones, there is evidence that certain pituitary hormones are capable of regulating their own secretion (Collin, 1928; Halász and Szentágothai, 1958; Motta *et al.*, 1969; Szentágothai *et al.*, 1972). Such feedback may take place at the pituitary level, at the hypothalamus or at higher nerve levels. It may have a special role in the control of GH, prolactin and MSH where there is no specific target hormone feedback. The system apparently has a "high setting" as high blood levels of anterior pituitary hormones produced by target gland removal or nephrectomy (Gay, 1974) remain high. There is no evidence to suggest that such a feedback acts at physiological levels to rapidly adjust the secretion of releasing factors or pituitary hormones.

Feedback at either pituitary or hypothalamic level may be achieved via the peripheral circulation. Additionally anterior pituitary hormones are localised in the median eminence region (Harris, 1970; Davies *et al.*, 1975a) and observations by Török (1962) that the blood flow in certain circumferential portal vessels is toward the median eminence may be related to the short loop feedback phenomenon.

FSH, LH, GH, ACTH, prolactin and perhaps TSH can inhibit their own release at the hypothalamic level, and prolactin, also at the pituitary level. There is also evidence of a positive feedback of FSH at the hypothalamic level in immature rats, which may play a role in the onset of puberty.

It is not known how the short loop feedback operates at either level. It may interfere with the production of releasing factors or pituitary hormones or modify the response of the hypothalamus or pituitary to other inhibitions or stimuli. Like other feedback systems, short loop feedback may act through the cyclic AMP mechanism (Section III, B, 2).

(4) *Non-target feedback* (Findlay, 1972; Hayward, 1972)

Vasopressin and GH are both controlled by special mechanisms, osmoreceptors and glucoreceptors, respectively. The growth hormone responses to hypoglycaemia and hyperglycaemia are mediated by glucose sensitive neural structures. It seems likely that the lateral hypothalamus is sensitive to falling levels of glucose and the ventral medial hypothalamus to rising levels of glucose. In addition, aminoacids and fatty acids play a role in GH secretion (Glick, 1969). Vasopressin is controlled, in part, by osmoreceptors in the hypothalamus which are sensitive to osmotic stimuli such as hypertonicity of blood perfusing the head. It also seems probable that central thermoreceptors are a feature in the control of TSH secretion (Reichlin, 1975) and vasopressin secretion (Hayward, 1972). The possibility that similar receptors may play a role in the control of other hormones has been suggested by Cramer and Porter (1973).

(5) *Interaction of different feedback and hormone control systems*

The relative importance of the long, short and ultra short (see Section IV, B, 2) feedback loops in different physiological states is not clear, although, as considered in Section IV, A, 3, the short loop system probably has a limited function. There are, however, important interactions between the feedback loops of the different hypothalamo-pituitary-target hormone systems.

Short loop feedback is usually specific, but prolactin may stimulate gonadotrophin release at puberty and GH and prolactin may influence TSH release; the latter effect, perhaps, being related to the effect of TRH on prolactin release. The amplitude of the TSH and prolactin responses to TRH can be modulated by stimulatory and inhibitory influences of oestrogens and thyroid hormones (Dronin *et al.*, 1975).

Thyroid hormones may influence pituitary ACTH release, and the concurrent inhibition of TSH secretion and stimulation of ACTH secretion, as a result of non-specific stress, may be related to altered hippocampal activity (Fortier, 1973). Thyroid hormones may also be involved in the GH release system; whether at the level of the anterior pituitary, releasing hormone or neurotransmitter is not clear.

Oestrogens have an effect on several systems. They have a complex effect at the pituitary level in the control of TSH release; low levels being stimulatory and high levels inhibitory. They stimulate the pituitary directly to synthesise and release prolactin, inhibit PIF release from the hypothalamus, stimulate the "surge" centre release of prolactin and sensitise the prolactin release system to exteroceptive stimuli. Both oestrogens and androgens stimulate GH release probably by increasing the pituitary response to GHRF. Progesterone inhibits some GH secretory responses.

Glucocorticoids appear to inhibit GH secretion but may facilitate the TRH stimulated synthesis of TSH. It has been suggested that glucocorticoids and progesterone may serve to link the adrenal and ovarian systems in ovulation control (Rodgers *et al.*, 1974). Adrenalectomy results in the loss of a factor, probably progesterone, involved in the initiation of the preovulatory LH surge (Mann *et al.*, 1975).

Apart from anterior pituitary hormones other polypeptides may be important in "feedback" control. Angiotensin II stimulates the release of vasopressin. However, angiotensin may not be a blood-borne hypothalamic

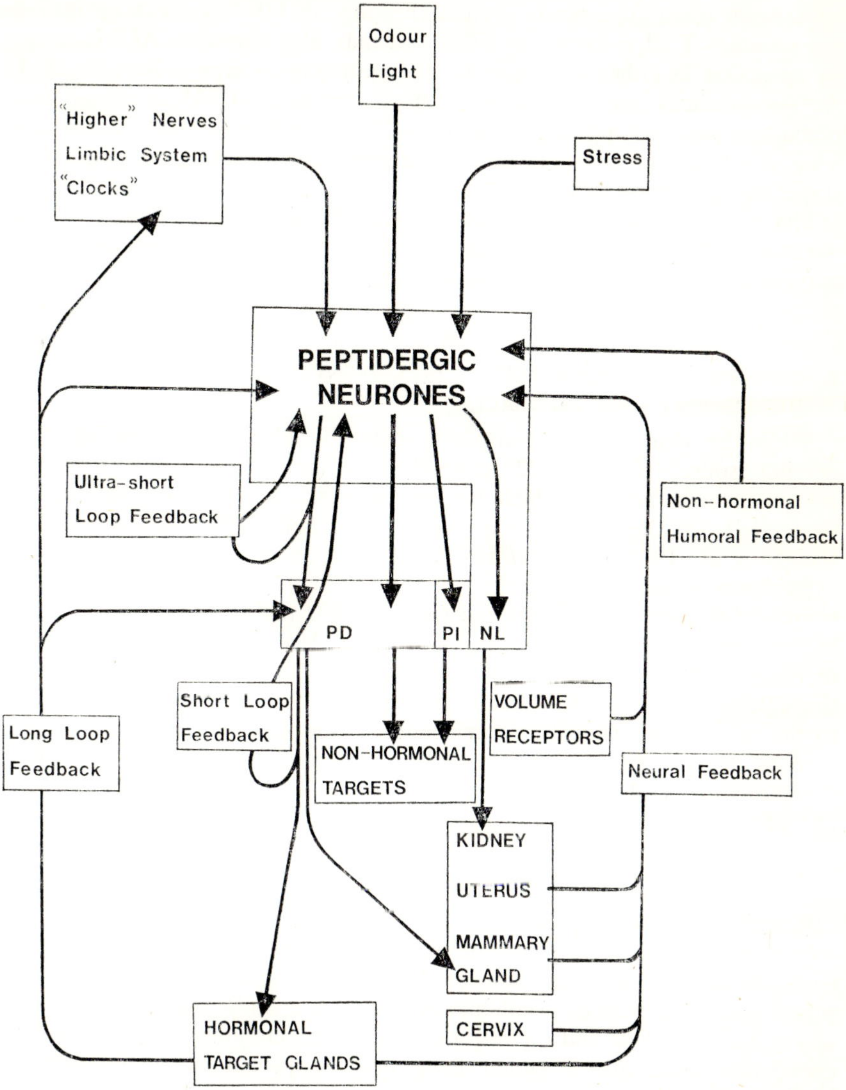

Fig. 6. Summary of the control of the hypothalamic peptidergic neurone system. PD, pars distalis; PI, pars intermedia ; NL, neural lobe.

input since renin also stimulates vasopressin release suggesting that there may be a brain renin-angiotensin system (Findlay, 1972). Somatostatin inhibits the pituitary TSH response to TRH and may also stimulate ACTH release by an action at either the pituitary or hypothalamic level (Moldow, 1975). At the pituitary level vasopressin may stimulate ACTH synthesis from precursors and potentiate the action of CRF, while at the hypothalamic level, it may act to stimulate CRF secretion. Large doses of vasopressin can stimulate the release of GH.

The interaction of feedback mechanisms illustrated here may arise from the complexity of the neural inputs into the hypothalamus. Many systems may be effected as a result of messages received from one specific receptor site.

B. CEREBROSPINAL FLUID (CSF), CIRCUMVENTRICULAR STRUCTURES AND ULTRASHORT LOOP FEEDBACK

While the idea of the hypothalamo-hypophysial portal system has been gaining ground (Harris, 1972), it has become increasingly shown that there are other ways through which the central nervous system might influence the activity of the pituitary complex (Zuckerman, 1970; Knowles, 1972; Knigge *et al.*, 1973; Porter *et al.*, 1975; Scott, 1975).

Evidence is accumulating that CSF may be involved as a pathway for transmitting chemical information from various regions of the brain to the hypothalamus (Findlay, 1972; Hayward, 1972). Hormones, neurotransmitters or other substances may be released into the CSF and carried by the CSF through the ventricles of the brain to be extracted by the brain at a site distal to the release, or concentrated by neural tissue and later released into the vascular system (Porter *et al.*, 1975). Biogenic amines, steriods (Anand Kumar, 1973) and hypothalamic hormones (Chapter 4) have been found in CSF.

(1) *Circumventricular structures*

Certain areas of the ventricles of the brain are lined by specialised ependymal cells which have morphological characteristics generally associated with considerable secretory and/or absorptive function. These include the pineal gland, derived embryologically from the ependymal cells of the roof of the third ventricle and a group of more or less circumscribed areas of the ependymal wall, all of which have been suspected of involvement in pituitary function (Knowles, 1972).

(a) *The pineal.* There are two groups of potential pineal hormones: melatonin (Fig. 1) and other indoles and a number of polypeptides (Reiter, 1973; Wurtman and Cardinali, 1974). They have been shown to have a variety of functions, including modifying adrenal and thyroid function, and inhibiting gonadal maturation, gonadal function and cyclicity. Pineal hormones may be released either into the general circulation or into the CSF. The two groups of hormones may act via different mechanisms. Melatonin has been shown to be concentrated in the hypothalamus and midbrain and may act by modifying neurotransmitter inputs or the secretion of hypothalamic hormones. The secretion of melatonin is inversely dependent on light and may be particularly involved in the control of diurnal phenomena (Fraschini and Martini, 1970). The polypeptides may act directly on the pituitary like the hypophysiotrophic hormones (Reiter, 1973).

It has been claimed that there are high concentrations of hypothalamic hormones in the pineal (Chapter 4), which may provide a source of these hormones to the portal vessels via the CSF and ependymal cells (Section IV, B, 1, b).

(b) *Other ependymal structures.* Lining the roof and walls of the preoptic recess of amphibia are certain *liquor-contacting neurones*, which lie in the ependymal wall, project into the CSF, make synaptic connection with other neurones and may contain catecholamines (Vigh-Teichman, 1968). Whether these cells are involved in LHRH release is not known, but they are close to the area known to be involved in the ovulatory gonadotrophin surge in the rat. The ependymal cells in the floor of the preoptic recess, the *prechiasmatic gland*, are secretory in appearance and extend from the third ventricle to the capillary network of the *organum vasculosum* of the terminal lamina (OVLT) suggesting a possible secretory route. These cells have been implicated in LH release (Leveque, 1972) and LHRH has been found in the OVLT (Weiner *et al.*, 1975). However, this mechanism may not be primarily involved in LH release because, in the main, venous drainage of the OVLT does not reach the pars distalis. The specialised ependymal cells which lie in the lateral walls of the third ventricle along the upper margin of the infundibular recess have been termed the *paraventricular organ* (Knowles, 1972). These cells contain catecholamines and may be involved in the control of the pars intermedia, particularly in fish and amphibia.

Specialised ependymal cells lying in the floor of the ventrolateral walls of the infundibular recess (the *X-area*) are of particular interest. Cells have been described in amphibia, reptiles and mammals which terminate on the pituitary portal vessels or, in some instances, on pituitary cells and are in intimate contact with the CSF (Knowles, 1972; Mitchell, 1975). Other cells apparently link the CSF and the arcuate nucleus (Bleier, 1972). There is evidence that such cells take up releasing hormones (Chapter 4) and neurotransmitters and that they respond to changes in amounts of circulatory steroids (Anand Kumar, 1973). They appear to act to "recover" hypothalamic hormones or neurotransmitters from the CSF and to deliver them to the portal vessels, peptidergic neurones or pituitary cells (Knigge *et al.*, 1973; Porter *et al.*, 1975; Scott, 1975).

(2) *Ultrashort loop feedback*

The hypothalamic hormones have been shown to influence their own release by an ultrashort negative feedback loop (Motta *et al.*, 1969; Dyer and Dyball, 1974). This may be via the peripheral circulation, ascending portal vessels and ependymal cells (Section IV, B, 1, b), or via secretion from branches of the peptidergic neurones which do not terminate near the portal vessels but on other peptidergic neurones. The presence of releasing hormones in various extrahypothalamic sites additionally suggests that these may have a role as neurotransmitters (Chapter 4). Such a neurotransmitter role for releasing hormones could provide a possible mechanism for co-ordination of different aspects of hypothalamic function. However, there is evidence that there are several extra-endocrine actions of hypothalamic hormones on the central nervous system which could explain the wide distribution of these peptides in the brain (Kastin *et al.*, 1975; Moss *et al.*, 1975a).

C. NEURALLY MEDIATED HYPOTHALAMIC INPUTS

Marshall (1942) was the first to document the wide association between a multiplicity of environmental and exteroceptive factors and sexual function in animals. Subsequently, it was recognised that a variety of internal stimuli mediated by the nervous system can influence the secretion of hormones from the hypothalamo-pituitary system (Green, 1969; Findlay, 1972; Hayward, 1972). These stimuli may be received through: the special sense organs, e.g. visual, olfactory, auditory; various exteroceptors, e.g. suckling, coitus; various interoceptors, e.g. reproductive tract stimulation, target glands, blood volume. In addition various ill-defined stimuli such as stress, sleep-wake rhythms and overcrowding play a role (Table 1).

Light and vision have proved to be major external stimuli particularly in the control of reproduction (Wurtman, 1967). The duration of the photoperiod is important in the timing of the breeding season in many species. There is also a well-documented dark/light stimulus involved in the timing of ovulation in the rat (Everett, 1970). The dark/light cycle is also involved in the entrainment of the circadian hypothalamo-pituitary-adrenal rhythm (Krieger, 1974) and the hypothalamo-pituitary-thyroid rhythm. In addition the secretion of several anterior pituitary hormones, including ACTH, is related to sleep as a whole or to phases of the sleep cycle (Dunn, 1974; Weitzman, 1974; Rubin, 1975). Whether the hypothalamus itself is involved in changes in the sleep/wake state or whether the primary control is from elsewhere in the brain is not clear (Findlay, 1972). Also, the relative importance and possible interaction between light and sleep rhythms in their control over the endocrine hypothalamus is not known (Findlay, 1972).

An important variable in the manifestation of neurally mediated stimuli controlling the endocrine hypothalamus is the presence of other individuals of the same species, particularly young and members of the opposite sex. Olfactory stimuli have marked effects on reproduction. These effects are mediated by pheromones, substances secreted into the environment by an individual and received by other members of the same species (Bruce, 1970). Examples are the pregnancy block seen in mice placed in the presence or odour of a strange male; and probably the increasing synchronisation of menstrual cycles over a six-month period amongst room-mates and close friends at a college (McClintock, 1971). Suckling and the presence of the young causes the release of prolactin and other galactopoietic hormones, ACTH and GH and may concomitantly inhibit gonadotrophin release. Oxytocin is also released in response to suckling by the milk ejection reflex (Tindal and Knaggs, 1970), and during labour and copulation (Cross, 1966). In certain species the rabbit, the cat, etc., ovulation is caused by copulation or vaginal stimulation (Jöchle, 1975). In addition, copulation may hasten or facilitate ovulation in spontaneously ovulating mammals: the rat, the sheep and even women. Also in the rat and some other spontaneous ovulators, copulation leads to an adequate release of prolactin required for the development of a fully functional corpus luteum.

Changes in blood volume can change the rate at which vasopressin leaves the neurohypophysis (Hayward, 1972) and unilateral adrenalectomy is signalled to the hypothalamus via a neural mechanism (Halász and Szentágothai, 1959; Engeland and Dallman, 1975).

The response to stress is associated with the release of ACTH and vasopressin, leading to the production of glucocorticoids and the retention of water and salt (Smelik, 1970). It has also been shown that both physical and emotional stress affect the secretion of TSH, GH, MSH, gonadotrophins, prolactin and oxytocin. The possible role of temperature in the seasonal control of gonadotrophin release, the effect of sound on reproductive function (Zondek and Tamari, 1967) and the effect of increased population density on the increased activity of the hypothalamo-pituitary-adrenal system associated with suppressed reproductive activity and increased mortality (Christian, 1971) may all be part of the stress response.

(1) *Pathways of neurally mediated stimuli*

The elucidation of the pathways and mechanisms by which the various neurally mediated stimuli bring about changes in endocrine function is still incomplete, and the extent to which the various pathways to the hypothalamus (Section II, B) may represent the routes over which specific information gains access to the peptidergic neurones is not known (Raisman, 1970a). It seems likely, however, that the various stimuli discussed above do act via the hypothalamo-pituitary axis.

The first step is the conversion of the stimulus into a nervous impulse, and receptors have been described in the special sense organs, the mammary gland teats (Findlay, 1966) and the vagina (Bower, 1959).

Olfactory structures of the brain have a large access route to the hypothalamus (Raisman and Field, 1971a). The olfactory bulb receives the fibres of the olfactory nerves and projects to the olfactory tubercle and pyriform cortex and also to the amygdala. From these areas come two extensive projections to the hypothalamus, the ventral amygdalofugal pathway and the stria terminalis.

The brain of some species, rabbit, sheep, duck, may be directly sensitive to light and in the duck the hypothalamus itself may be sensitive. Despite controversy over the years (Raisman, 1966) there is evidence for a direct retinal projection to the suprachiasmatic region of the hypothalamus (Campbell and Ramaley, 1974). There is also evidence that a pathway via the pineal is important (Section IV, B, 1, a) and that the final link to the hypothalamus is a humoral one. However, where the visual input is associated with social and recognition inputs the route to the hypothalamus may be more complex (Raisman, 1970a), involving the neocortex. Direct neocortical-hypothalamic projections are uncertain, but hippocampal projections to the hypothalamus could provide an indirect pathway, via the entorhinal area. It seems that a primary function of the hippocampus may be to channel information from the neocortex to the hypothalamus. The olfactory system also has access to this route by means of fibre connections from the pyriform cortex to the entorhinal area. The common use of the route through the entorhinal area, hippocampus and septum to the hypothalamus, suggests this system may be an area in which information from different areas is converted into a form necessary for hypothalamic control (Raisman, 1970a).

The major afferent fibres from the female genital tract run via the pelvic and pudendal nerves and suckling activates the segmental mammary nerve (Cross, 1966; Swaab, 1972). Some details of the pathways in the spinal cord and brain stem have been described (Tindal and Knaggs, 1970; Swaab, 1972).

D. ENDOCRINE RHYTHMS

Pincus (1943) demonstrated that there was a daily 24-hour pattern of 17-hydroxycorticoids in the urine of healthy adults. Since then several endocrine rhythms have been shown. All pituitary hormones are secreted in an episodic fashion with intervals of about 1–3 hours and several hypothalamo-pituitary systems, ACTH, prolactin, GH and TSH and under some circumstances gonadotrophins, have temporal patterns close to 24 hours. A cycle of menstrual or oestrous cycle length has been shown in some species for FSH, LH, prolactin and GH and longer annual or seasonal changes in ACTH, the gonads and thyroids have been demonstrated (Ferin *et al.*, 1974a; Halberg, 1974; Dunn, 1974).

The importance of such single rhythms in endocrine control is far reaching. Two or more may, however, influence or control a variety of functions. Thus, the time of greatest release of prolactin in birds is important in the regulation of several seasonal conditions (Dunn, 1974; Chapter 9), and depending on its temporal relationship with corticosterone prolactin may induce fattening, loss of fat, crop sac proliferation or some combination of these.

It is clear from previous sections that the various endocrine rhythms are under the influence or control of a variety of humoral, CSF and neural inputs emanating from both the internal and external environment. The extent to which the rhythms are dependent on these inputs, or whether there might be an internal rhythmic controller which is simply triggered or entrained by the inputs, is not clear.

The possibility that the hypophysiotrophic area may show evidence of pulsatile gonadotrophin release without humoral or neural inputs (Blake and Sawyer, 1974; Krey *et al.*, 1975a, b; Section II, A) suggests that there may be one or several neuroendocrine generators; these may be located either in an assembly of neurones each of which has a definite intrinsic rhythmic activity, or in the cell population of a neurotransducer or endocrine gland. Such a generator may be an oscillatory biochemical process in the cell in which the oscillations are the length of the required rhythm, or perhaps much shorter, triggering the rhythm after many oscillations, like the minute and hour hands of a clock. Ehret and Trucco (1967) have proposed the so-called "chronon" hypothesis in which all or part of the DNA is organised into chronons and the synthesis of much, if not all new cellular material, is programmed on a circadian basis. Changes in the quantity, location or activity of an enzyme, changes in the synthesis of inhibitors or substrates, a protein subunit association—dissociation cycle, or changes in protein tertiary structure could be involved (Hastings, 1970). Whether any of these ideas are applicable to the study of hypothalamic rhythms is unknown.

V. Influence of Age

A. The Foetus

There is much evidence that the foetal endocrine glands are active at least towards the end of foetal life (Jost *et al.*, 1970; Grumbach and Kaplan, 1973; Jost *et al.*, 1974; Monroe and Paull, 1974; Villee, 1975). The time of development of a definitive adult-like hypothalamo-hypophysial relationship varies according to species. It occurs before birth in the rabbit, man and the guinea-pig, but after birth in the rat and the mouse (Jost *et al.*, 1970; Monroe and Paull, 1974; Chapter 1), although some neurovascular link is already established in these latter species during the last days of foetal life (Halász, 1974).

Gonadotrophins, ACTH, TSH, prolactin, oxytocin and vasopressin have been shown to be present in both foetal pituitary and blood of a variety of species (Grumbach and Kaplan, 1973; Jost *et al.*, 1974; Monroe and Paull, 1974; Villee, 1975; Pecile and Bossa, 1976). The precise role of these hormones and their method of control is not clear. Evidence particularly from anencephaly studies in man and the rat, suggest that GH secretion shows some, and ACTH secretion complete dependence on the foetal hypothalamus, whereas TSH secretion is largely independent of foetal hypothalamic control. TSH may be responding to feedback control at the pituitary level, but it has also been shown that TRH is able to cross the placenta from the mother to the foetus.

Although observations on congenital abnormalities and experiments, particularly on sheep (Liggins *et al.*, 1973), have suggested an essential role for the foetal hypothalamo-hypophysial-adrenal system in the initiation of labour, in the rat and man initiation can occur independently of the foetal hypothalamus (Swaab and Honnebier, 1974). The foetal hypothalamus may play a role during the course of labour rather than in its initiation. The demonstration of high concentrations of oxytocin in foetal blood apparently produced by the foetus, suggests that the foetal hypothalamo-neurohypophysial system is involved.

B. DIFFERENTIATION OF THE HYPOTHALAMUS

Dörner (1974) has stressed the probable critical organisational phase of the hypothalamus around the time of parturition, when a placenta-foetal target gland-foetal pituitary-foetal hypothalamus relationship is replaced by a hypothalamo-pituitary-target gland relationship.

Since the pioneering studies of Pfeiffer (1936) there have been numerous studies on the sexual differentiation of the neuroendocrine axis in mammals (Gorski, 1971, 1973). The pattern of gonadotrophin release in the female is cyclic—the ovulatory surge—whereas the release of gonadotrophins in the normal male is non-cyclic. Pfeiffer (1936) showed that ovarian cycles could be induced in adult male rats, castrated immediately after birth and implanted with ovaries. In contrast, ovulation did not occur in either female rats neonatally implanted with testes, or male rats castrated after a critical neonatal period, then implanted with ovaries. It seems, then, that the undifferentiated system is female; the development of a male pattern of gonadotrophin regulation being initiated by the presence of a testis or androgens (or oestrogens) during a critical neonatal period in the rat. A similar situation probably occurs in the mouse, the hamster, the guinea-pig and the sheep (Short, 1974). In the latter two species, the critical period is probably during the foetal life. The influence of androgen is thought to be on the hypothalamus (Barraclough and Gorski, 1961), probably on the preoptic/

anterior hypothalamic area known to be involved in ovulation control, at least in the rat. The androgen which may be first converted *in situ* to oestrogen for action (see above; Section IV, A, 2) probably acts by altering hypothalamic RNA metabolism and protein synthesis (Salaman, 1974), thus affecting the formation and maintenance of steroid receptors and rendering them insensitive to the positive oestrogen feedback which causes an ovulatory surge of gonadotrophin after puberty (Chapter 8).

There is evidence that the female pattern of gonadotrophin release is not eliminated by exposure to androgen during early life in primates. Females with congenital adrenocortical hyperplasia, with virilisation as a result of exposure to large amounts of androgens *in utero*, subsequently develop normal ovulatory cycles if treated adequately with cortisol (Villee, 1975); also, both men and male monkeys respond to an acute rise in serum oestradiol concentration with a surge of LH. Experiments on surgical isolation of the hypophysiotrophic area from the rest of the brain in the rhesus monkey, indicate that the preoptic/anterior hypothalamic area may not be an obligatory component of the positive feedback system (Krey *et al.*, 1975a, b; Spies and Norman, 1975).

There is evidence that in addition to sexual differentiation, similar effects of target hormones may be involved in the hypothalamic organisation of other systems (Dörner, 1974). The mechanics for the regulation of prolactin secretion differ between the male and the female (Neill, 1974); this may be differentiated by androgens in a similar way to the organisation of gonadotrophin secretion. Thyroid overdoses during the critical neonatal period in the rat result in altered thyroid function (Bakke *et al.*, 1972). Overdosage with corticoids neonatally can abolish the circadian plasma corticoid rhythm in rats at 30 days (Krieger, 1974). This effect is not permanent and other aspects of the hypothalamo-pituitary-adrenal axis are unaffected. However, such corticoid administration neonatally does lead to an inhibition of development associated with hypothalamic disturbance (Sawano *et al.*, 1969).

C. POSTNATAL AGEING

After birth and the development of the definitive hypothalamo-hypophysial system, age related changes occur in the endocrine hypothalamus throughout life. There are also changes in the pituitary sensitivity to releasing factors and the hypothalamic sensitivity to feedback (Dilman, 1971; Gregerman and Bierman, 1974; Everitt, 1976b; Everitt and Burgess, 1976).

In the neonate, a relative inability to control urine output in response to restricted fluid intake, compared to an adult, may in part be due to a failure of vasopressin release in response to an effect on osmotic receptors.

In neonatal children there is, apparently, a different release mechanism for GH from that in older children. Thus in the newborn, GH concentrations are found to be in the acromegalic range. Infants do not respond to insulin hypoglycaemia as do older children and the sleep related release of GH is not developed until 4–5 years of age (Brasel and Blizzard, 1974).

Changes in the hypothalamus are probably of major importance in the timing of puberty (Donovan and van der Werff ten Bosch, 1965; Donovan, 1974). The characteristic of the prepubertal state is that, though the gonads and secondary sexual structures are capable of being stimulated by exogenous hormones, and the pituitary is capable of stimulation by releasing hormones and the negative feedback system of pituitary-gland secretion occurs, the adult pattern of hormone secretion does not occur. A study of hypothalamic disease and experimental brain lesions has shown that the hypothalamo-pituitary axis is under inhibition before puberty. The principal change at puberty is a change in hypothalamic sensitivity to feedback by gonadal steroids. The sensitivity to both negative and positive feedback changes (Taleisnik *et al.*, 1971), thus allowing both an increased basal gonadotrophin secretion for follicle growth and oestrogen secretion and the ability to respond to increased oestrogen with an ovulatory gonadotrophin surge. The organisation of these mechanisms has been discussed in Section V, B.

During senescence some hypothalamic functions and actions alter, although the application of a series of dynamic tests has not revealed marked age-related defects in feedback or neural control of hypothalamo-pituitary function in man (Lazarus and Eastman, 1976). In aged rats vasopressin secretion is impaired and the secretion of vasopressin in response to dehydration is reduced and delayed (Turkington and Everitt, 1976). However, in man the blood vasopressin concentration may increase with age (Everitt, 1976b). GH production rates are decreased in elderly human subjects, probably due to a decreased GH secretion during sleep (Carlson *et al.*, 1972). There is a reduction of hypothalamic GHRF and a decreased pituitary response to GHRF in aged rats (Pecile *et al.*, 1965; Pecile and Bossa, 1976). TSH levels found during ageing are conflicting but there is a decrease in the TSH response to TRH in elderly men (Gregerman and Bierman, 1974; Everitt, 1976b). In the aged rat, there is a hypersecretion of prolactin (Aschheim, 1976), perhaps indicating PIF failure and there may be a decreased sensitivity of the hypothalamus to corticoids (Riegle, 1976). In both elderly men and women the gonadotrophin response to LHRH may be accentuated (Vermeulen, 1976; Lazarus and Eastman, 1976). The possible involvement of the hypothalamus and altered sensitivity to oestrogen in the reduction of ovarian function in aged rats and menopausal women has been suggested (Aschheim, 1976; Everitt, 1976a, b).

VI. Disease and Hypothalamic Function

A. Abnormal Hypothalamic Inputs

Several endocrine diseases probably involve abnormal hypothalamic inputs. The hypothalamus may be affected by altered feedback produced by removal or malfunction of a target gland (Labhart, 1974; Reichlin, 1974). In such conditions of hypoadrenal function, hypothyroidism and hypogonadism, high blood concentrations of ACTH, TSH and gonadotrophins respectively, are present due to lack of the negative feedback loop (Section IV, A, 2). Such a release of the hypothalamo-pituitary axis from normal feedback inhibition is frequently followed, in animals, by the development of pituitary tumours (Furth, 1955). Similar examples can be found in man (Peck, 1975), although patients with primary adrenal failure or gonadal failure (e.g. Addison's disease; Turner's syndrome), though developing excessive secretion of stimulating hormones, rarely develop tumours. Pituitary hypersecretion of ACTH, in certain patients with Cushing's syndrome (Chapter 11) may be due to an excessively high threshold for feedback by cortisol. Some patients treated for Cushing's syndrome by total adrenalectomy develop pituitary tumours. These may have been present previously, blocked by high cortisol levels and may be stimulated by cortisol withdrawal. Pituitary tumours producing GH and prolactin may be influenced by hypothalamic inputs. GH secretion may be modified in some such patients by hypoglycaemia and alterations in adrenergic and dopaminergic states (Peck, 1975). Lack of GH secretion in deprivation dwarfism, polycystic ovarian hyperplasia, Cushing's syndrome and hypothalamic hypothyroidism, may also be associated with altered neurotransmitter control (Frohman and Stachura, 1975).

As indicated in Section IV, C, various external and neural factors may influence hypothalamic function. These include certain physical and emotional stress situations. In anorexia nervosa there is a cessation of normal menstrual cycles in young women due to psychogenic factors resulting in lowered food intake (Russell, 1969). Alterations in sleep/wake and light/dark rhythms and travel across time zones also have a profound effect on hormone output (Ferin *et al.*, 1974a; Section IV, C). Diffuse brain disease may also lead to a loss of normal diurnal rhythms of adrenal and GH function (Reichlin, 1974). In addition, a variety of neural factors may be responsible for the inappropriate ADH syndrome, and the development of an excessive output of vasopressin.

B. Tests of Endocrine Hypothalamic Function

The integrity of the endocrine hypothalamus can be tested by three main approaches: (a) Directly, by the measurement of hypothalamic hormones in body fluids, both basally and after appropriate stimulation and inhibition; (b) Indirectly, by the measurement of adenohypophysial (or target organ) hormones in body fluids, in conjunction with appropriate tests for the integrity of the adenohypophysis and target gland by stimulation and inhibition; (c) By testing the integrity of the various humoral and neural inputs into the hypothalamus.

The direct measurement of neurohypophysial and hypophysiotrophic hormones in body fluids is possible (see appropriate Chapters 4, 7, 8). However, since the hypophysiotrophic hormones are essentially local hormones operating within the hypothalamo-pituitary portal system, only small amounts appear in the general circulation. In addition there are problems of rapid clearance, identity and pulsatile secretion (Chapter 4).

Sensitive radioimmunoassays are available for the adenohypophysial hormones. On demonstrating a "low" concentration, a differential diagnosis between hypopituitarism and a hypothalamic endocrinopathy can be made, in theory, by testing the ability or otherwise of the adenohypophysis to respond to TRH, LHRH or vasopressin, the latter thought to release ACTH at the pituitary level. A normal adenohypophysial response indicates a lesion in or "above" the hypothalamus. In practice, patients with hypothalamic disease often exhibit a delayed and sustained TSH response to TRH and the interpretation of the TSH responses to TRH, in patients with secondary hypothyroidism, is complicated further by the coexistence of both pituitary and hypothalamic disease in many cases (Lazarus and Eastman, 1976).

Various tests have been suggested to test the integrity of the humoral inputs to the hypothalamus and its ability to respond. Dexamethasone, a potent synthetic glucocorticoid and metyrapone, an 11β hydroxylase inhibitor, which block and indirectly stimulate the hypothalamo-pituitary-adrenal axis respectively, can be used to assess the integrity of the long negative feedback loop (Chapter 11). Oestrogens may be used to test the integrity of the positive feedback loop which initiates the ovulatory surge of gonadotrophins. Certain synthetic anti-oestrogen/weak oestrogens (e.g. clomiphene) also stimulate the hypothalamo-pituitary axis to secrete gonadotrophins. They appear to act either by blocking an oestrogen negative feedback at the hypothalamus or pituitary, or by direct stimulation. A failure to suppress elevated basal plasma GH concentrations with glucose can help to establish the diagnosis of a pathologic hypersecretion of GH. Amino acid infusion and insulin induced hypoglycaemia have been used to stimulate GH secretion; and insulin hypoglycaemia has been used as a non-specific stimulant for ACTH and prolactin release. Water deprivation and infusion of hyperosmotic saline have been used to raise plasma osmolarity and stimulate vasopressin release.

As discussed in Section III, many drugs are available which can affect almost every aspect of neurotransmitter metabolism and action. Many of these have been used clinically (Frohman, 1972; Reichlin, 1974). L-dopa, for example, which enters the brain and is rapidly converted to dopamine stimulates GH release and inhibits prolactin release. Chlorpromazine, which depletes hypothalamic catecholamines causes a rise in prolactin secretion. Recently, there has been particular interest in bromocriptine, which probably activates receptors of dopamine, in the pituitary and hypothalamus. Prolactin levels are lowered and fertility restored in patients with hyperprolactinaemia and infertility (Thorner *et al.*, 1975a). GH is also reduced in some patients with acromegaly but not in normal men and women (Sachdev *et al.*, 1975).

Studies of normal and altered diurnal and sleep patterns of pituitary hormone release are now being extensively studied since hormone determination methods now allow the continued monitoring of plasma levels (Dunn, 1974; Ferin *et al.*, 1974; Weitzman, 1974).

VII. Systems Analysis

A number of neuroendocrine control systems have now been mathematically analysed and corresponding model systems devised (Donovan, 1970; Inoué and Sekaguchi, 1973; Yates, 1974). These are often very complex involving many differential equations and non-linear functions. Such studies represent a method of gathering all information together in an integrated form, with appropriate assumptions. They may not bring direct answers as to what occurs in the real system but they can often reveal what is lacking in understanding and point to new experimental approaches.

VIII. Summary and Conclusions

Rather than attempt a mathematical approach the information discussed in this chapter will be summarised by a simple link diagram (Figure 6). This chapter has attempted to show how the multiple endocrine functions of the hypothalamus are integrated and affected by information mediated through other parts of the nervous system, the bloodstream and cerebrospinal fluid.

The endocrine hypothalamus is apparently capable of sustaining a certain autonomy of function without these inputs (Section II, A), but it is impossible to think of the hypothalamus in isolation because it is clearly only one component of a complex integrated system. The method of transfer of extrahypothalamic neural information probably involves a wide array of neurotransmitters (Section III, A) which act at several loci to influence hypothalamic and pituitary function (Section III, A, 2). The neurotransmitter effects are probably mediated by the cyclic AMP system (Section III, B, 2). Neurotransmitters are probably involved in the feedback action of target and adenohypophysial hormones on the hypothalamus. Such hormone feedback at the hypothalamus, as well as at extrahypothalamic and pituitary sites plays a large role in modifying the release and actions of hypothalamic hormones. The hypothalamic hormones themselves, the pituitary hormones, the target hormones and apparently unrelated hormones may be involved in a complicated, though ill-understood integrated mechanism (Sections IV, A, 1; IV, A, 2; IV, A, 3; IV, B, 2). In addition, various non-hormonal blood-borne stimuli (Section IV, A, 4) and perhaps cerebrospinal fluid-borne stimuli (Section IV, B) have important influences on the hypothalamus. Many interoceptive and exteroceptive neurally-borne stimuli serve to integrate the hypothalamo-pituitary system with the whole internal and external environment (Section IV, C). Whether in addition to such triggering mechanisms there is also an internal rhythm within the system or whether an observed rhythmical phenomenon (Section IV, D) is simply a result of a complex integration of humoral and neural inputs is unknown (Section IV, D). The function of the endocrine hypothalamus is influenced by age, sex and disease (Sections V; VI) and a variety of clinical tests is available to test the integrity of its function (Section VI, B).

The control systems for the individual hypothalamic endocrine functions are becoming better understood and mathematical models are being devised (Section VII). The relative importance of the various control factors for an individual system and the interrelationships between different systems awaits further study.

CHAPTER 4

The Biochemical Investigation of the Endocrine Hypothalamus

S. L. JEFFCOATE

I. Introduction

The investigation of the function of the endocrine hypothalamus involves the widely differing backgrounds and conceptual approaches of, among others, anatomists, neurophysiologists, chemists, biochemists, zoologists and clinical endocrinologists. This diversity compounds the already enormous difficulties associated with the study of the hypothalamus. It is an inaccessible and heterogeneous organ with complex neural and vascular connections. It contains and is influenced by a variety of biochemical substances including hormonal and non-hormonal peptides, neurotransmitters, prostaglandins, thyroid and steroid hormones. Its secretions, their functions, metabolism and control can in many instances only be studied indirectly. Finally, it is becoming apparent that the peptides until recently thought of as *hypothalami* hormones have a wider distribution and, presumably, function than mere releasers of adenohypophysial hormones (see Section V, D).

In the study of the endocrine hypothalamus, as in other scientific fields, the validity of the knowledge gained from the different experimental approaches depends upon a critical understanding of the value and limitations of the methods used. In Chapter 2 the anatomical and physiological techniques are discussed and in this chapter, the many biochemical techniques are described and evaluated and examples of their application given. This coverage is not expected to be exhaustive (or exhausting!) but it is hoped that it will be broad enough to deal, however thinly, with all the methodological approaches that have been adopted in the attempt to unravel the mysteries of the hypothalamus.

II. Chemistry and Immunology of the Hypothalamic Hormones

A. Chemistry

The general procedure in the study of the chemistry of the hypothalamic hormones is as follows: extraction of hypothalamic tissue; progressive purification culminating in isolation of a pure peptide; then analysis of its amino acid sequence; decision on its nomenclature; preparation of synthetic peptide; comparison of the chemical and biological properties of natural and synthetic hormones; synthesis of analogues and fragments for investigation of structure=activity relationships. The subject is considered in detail in Chapter 5 and then in subsequent chapters on specific hormones. In this section a few special points concerning the chemistry of these hormones will be highlighted.

(1) *Extraction and purification*

The methods used for extraction of hormones from the hypothalamus do not differ essentially from those used in other situations. A major consideration is the lability of the peptides and the need to protect them from the peptidases present in many tissues (see Section VI, B). Dilute acetic or hydrochloric acid, aqueous methanol or alcohol and more recently alcohol-chloroform (Burgus *et al.*, 1975) have been successfully used.

The precise anatomical region taken for extraction is important and often not clearly defined in the literature. The contamination of neurohypophysial extracts with pituitary stalk has, for instance, caused confusion in the study of CRF (see Chapter 11). Extraction of specific hypothalamic nuclei is now becoming possible with the availability of more sensitive assay methods and this is covered further in Section III, E.

Apart from a failure to define anatomical areas accurately a number of other factors may lead to difficulties and discrepancies between different investigators attempting to extract and isolate hypothalamic hormones.

(a) There may be *species differences.* This is clear for the neurohypophysial hormones (Chapter 6) but the hypophysiotrophic hormones appear to be the same in many classes of animals (see especially, TRH discussed in Chapter 7).

(b) The *state of the animal* before removal of the tissue may lead to changes in the amount and/or nature of the hormones.

(c) Differences in *extraction technique* may lead to a failure to denature peptidases or to the extraction of substances interfering with assay procedures.

(d) *Chromatographic* artefacts can be induced by, for instance, inadequate desalting. All the common chromatographic systems for peptide separation are used: countercurrent distribution; gel filtration; ion exchange chromatography; electrophoresis; organic phase gel filtration (sephadex LH20); adsorption chromatography on thin-layers of alumina or silica gel or columns of cellulose; affinity chromatography using specific solid-phase-coupled binding reagents. Each of these has its own advantages and disadvantages and examples are given in specific chapters. The use of a combination of techniques is important as this allows different physico-chemical characteristics of the molecule to be utilised for separation.

(e) *Assays* that are used to monitor purification procedures may be non-specific. Thus, bioassays for releasing or inhibiting hormones, whether conducted *in vivo* or *in vitro*, will be affected by the ions, neurotransmitters, steroid hormones and prostaglandins and other regulatory peptides which may be present in crude extracts and have biological effects which are unpredictable and either synergistic with, or antagonistic to, those expected. Immunological assays are also prone to interference by non-specific factors and, more specifically, there may be cross-reaction with larger or smaller peptides sharing an antigenic determinant with the peptide under investigation. Assay methods for the hypothalamic hormones are discussed further in Section III.

(2) *Sequence analysis and chemical synthesis*

The hypothalamic hormones so far isolated have either been cyclic with an –S–S– bridge (oxytocin, vasopressin and their related peptides, and somatostatin) or have had chemically blocked terminal residues (LHRH, TRH). They, and related analogues have then been synthesised either by the classical peptide synthesis or by the more recent solid phase methods (see Chapter 5). It might be possible with suitable preparations of hypothalamic tissue, e.g. synaptosomes, to make small amounts of pure peptides biosynthetically.

Large amounts of the synthetic peptides are required for clinical purposes (diagnosis and treatment), for the preparation of antibodies (see Section II, B) and for use as reference material in chemical, biochemical and physiological studies.

Synthesis, even that using a solid-phase technique and even of small peptides (and the problem increases exponentially with increasing peptide chain-length), involves the removal of side-products formed at each stage. A further difficulty with somatostatin has been poor yields and polymer formation at the final stage, that of cyclisation (see Chapter 5).

(3) *Structure-activity relationships*

Knowledge of the primary structure of a peptide allows study of structure-activity relationships after the synthesis of fragments and analogues with one or more deletions, additions or replacements. The neurohypophysial nonapeptides have been extensively studied (Walter, 1971 and see Chapters 5 and 6) and more recently, the search for potent or longer-acting or inhibitory analogues has stimulated the study of the releasing hormones. This has resulted in the development of inhibitory and superactive analogues of LHRH and somatostatin (see Chapter 5). Apart from these applications, analogues are useful in the investigation of the important sites in the molecule for biological function, binding to receptors (see Section V, A), specificity of antisera (see Section II, B) and metabolism *in vivo* (Section VI, A) and *in vitro* (Section VI, B).

(4) *Nomenclature*

The name of the hypothalamic peptides is a difficult and controversial problem and is discussed elsewhere (see Preface and Chapter 5). One guiding principle (followed for instance in the chapters on the control of corticotrophin and of the gonadotrophins) has been that a releasing or release-inhibiting *factor* becomes promoted to *hormone* when it has been fully characterised and synthesised, the term factor being limited to the description of more or less purified biological extracts. There is an even earlier phase, that of releasing "activity" which describes the biological property of an extract before it is dignified with the title of factor. It is debatable whether these peptides should be called hormones in view of the increasing evidence for non-hypophysiotrophic roles (e.g. TRH and somatostatin).
whether peptides should be called hormones in view of the increasing evidence for non-hypophysiotrophic roles (e.g. TRH and somatostatin).

It is customary to specify in its name the adenohypophysial hormone that is controlled by the peptide. The original concept of specificity has been eroded however so that there is for instance one releasing hormone for two gonadotrophins and TRH releases both TSH and prolactin.

No less than eight variant names and/or abbreviations have been used for the gonadotrophin releasing decapeptide (Jeffcoate, 1975a). To these must be added the WHO proposal, *gonadorelin* when it is used as a pharmaceutical product and the IUPAC–IUB commission (1975) proposal, *gonadoliberin* (see Chapter 5).

B. IMMUNOLOGY

Following the isolation and structural analysis of a hypothalamic peptide, its availability in a synthetic form makes possible its use as an immunogen for the preparation of antisera. These antibodies are used as binding reagents of high avidity and specificity and represent the single most valuable tool in the elucidation of hypothalamic function that has been applied in recent years.

(1) *Preparation of antisera to hypothalamic peptides*

Because of their small size and lack of species-specificity, these peptides are of low immunogenicity. Experience with the neurohypophysial peptides is about ten years older than that of the releasing hormones and has been reviewed by Chard (1973).

Antibodies can be prepared against unconjugated oxytocin, vasopressin and LHRH (Arimura *et al.*, 1973; Chard, 1973; Kerdelhué *et al.*, 1973) particularly if the peptide is physically adsorbed onto macromolecular particles such as carbon, alumina or polyvinylpyrollidone (PVP). There is no doubt, however, that antibody responses are more consistent following immunisation with conjugates of the peptide covalently linked as a hapten to a protein carrier.

For each conjugate it would be useful to know: (i) the site of conjugation in the peptide hapten; (ii) the number of moles of peptide attached per mole of carrier protein; (iii) whether any alteration has occurred to the configuration of the hapten during conjugation; (iv) relative exposure of the antigenic determinant on the surface of the carrier.

The nature of the three components: the hapten, the linkage and the carrier will determine the characteristics of the resulting antiserum and thus need to be carefully considered.

(a) *The hapten.* The site at which a small hapten is linked to a carrier will determine the specificity of the resulting antibodies. These tend to be "far-sighted" being directed towards the parts of the hapten furthest from the point of conjugation (Crumpton, 1974). TRH and LHRH both have their C- and N-terminals blocked and unavailable for conjugation. TRH can be linked through its central histidyl residue, but for LHRH the site of conjugation is not always known with certainty (Jeffcoate *et al.*, 1976a). Where it is known (e.g. Koch *et al.*, 1973b) the character of the antiserum is more predictable. For somatostatin too there may be antiserum heterogeneity because linkage via an amino group could involve the

α-amino group or either of its two ε-amino groups (Arimura and Schally, 1975).

(b) *The linkage.* The techniques for conjugation of peptides (Likhite and Sehon, 1967) include carbodiimide, mixed anhydride, glutaraldehyde and diazonium compounds. The bis-diazotised benzidene method has been particularly successful for TRH (Bassiri and Utiger, 1972a). A two-stage coupling procedure, such as that described for LHRH by Koch *et al.* (1973b) should avoid the possibility of oligomerisation of the hapten.

(c) *The carrier.* Although a synthetic carrier such as poly-L-lysine may give a more uniform linkage, there is evidence that natural proteins are better as immunological carriers.

(2) *Testing of antisera*

In order to determine the possible usefulness of an antiserum and to interpret the results obtained using it the following characteristics need to be tested (Jeffcoate *et al.*, 1976a).

(a) *Titre* which is a measure of the concentration of antibody binding sites but gives no indication of their quality.

(b) *Affinity* which is a measure of the energy of the binding between the antibody and its antigen.

(c) *Specificity* which is a measure of the degree to which related peptides will react with the same binding site. It is important to state that specificities determined in one system (e.g. radioimmunoassay) may not be exactly applicable to another system (e.g. immunofluorescence) (see for instance Swaab and Pool, 1975).

(d) *Endogenous hormone* may be present in large amount bound to the antibodies in the serum. This has been demonstrated for vasopressin (Edwards *et al.*, 1972) and for LHRH (Jeffcoate *et al.*, 1976a) as well as for non-hypothalamic hormones. The presence of this bound hormone may reduce the usefulness of the antiserum in binding the peptide in experiments either *in vivo* or *in vitro* especially as the binding sites of the highest affinity are the most likely to be occupied.

(3) *Uses of antisera* in vitro

(a) *Immunohistochemistry.* This is undoubtedly a technique that yields much precise anatomical information and its introduction to neuroanatomy in 1969 has led to many applications (Hökfelt *et al.*, 1975b, f). These have utilised antisera raised not only against the hypothalamic peptide hormones but also against neurophysins and the enzymes involved in the biosynthesis of neurotransmitter amines; correlations between these have been valuable. The most widely used techniques are the indirect (Coon's) technique with a second antibody labelled with a fluorescent marker (Hökfelt *et al.*, 1973) or an immunoenzymatic method (e.g. Zimmerman *et al.*, 1974).

The exclusion of non-specificity has been a fundamental consideration. It is common to incubate the antiserum with a large excess of the antigen but this is not always adequate; presumably some of the antigen-antibody complexes dissociate on the tissue section. A method using solid phase-coupled antigens as immunabsorbents to actually remove anti-oxytocin and anti-vasopressin antibodies from a serum has been used to solve this problem (Swaab and Pool, 1975).

Antibodies raised against the catecholamine (CA) synthesising enzymes (prepared from adrenal tissue) have been used to demonstrate their distribution in the central nervous system (Fuxe *et al.*, 1975; Hökfelt *et al.*, 1975f). The implication is that the neurotransmitters are present and active in the same sites. Wherever LHRH containing terminals are found, large numbers of DA nerve endings are also (Hökfelt *et al.*, 1975f). This supports the concept that DA mediates LHRH release—though whether the effect is stimulatory or inhibitory is disputed (see Section IV, B).

Apart from LHRH (Barry *et al.*, 1973a; Arimura and Schally, 1975; Ramirez and Kordon, 1975) immunolocalisation in the hypothalamus has been described for somatostatin (Hökfelt *et al.*, 1974; Pelletier *et al.*, 1974a) and most recently, TRH (Hökfelt *et al.*, 1975c). The distribution is illustrated in Fig. 3 of Chapter 2.

One controversy in this area is whether the peptides are found in nerve-terminals (and perhaps axons) or in the cell-bodies in addition. The general view is that unless treatment with, e.g. colchicine, is used to block axonal transport artificially, peptides cannot readily be located in cell bodies.

The median eminence localisation of these peptides is in keeping with their postulated role as hypophysiotropic hormones; they have also been located in (i) *hypothalamic nuclei* outside the median eminence (Hökfelt *et al.*, 1974; Pelletier *et al.*, 1975), (ii) *extrahypothalamic brain tissue* (Hökfelt *et al.*, 1975c, e,; Pelletier *et al.*, 1975), (iii) *brain stem and spinal cord* (Pelletier *et al.*, 1975; Hökfelt *et al.*, 1975d, e), (iv) *gastrointestinal tissues* (somatostatin) (Hökfelt *et al.*, 1975a). These findings form part of the evidence indicating a wider physiological role for these peptides (Section V, D).

The localisation of neurophysins by immunofluorescence in the cell bodies of the supraoptic and paraventricular nuclei, axons in the internal layer of the median eminence and in nerve terminals of the neurohypophysis (Pelletier *et al.*, 1974b) provides elegant confirmation of the association between these proteins and vasopressin and oxytocin and of the role played in their transport and secretion.

(b) *Affinity chromatography* utilises antibodies coupled to a solid phase for separating complex mixtures of antigens. It has been applied to the assay of TRH (Montoya *et al.*, 1975) and might be a valid approach to the preparation of peptide-specific synaptosomes for instance (Section IV, A).

(c) *Radioimmunoassays* using antibodies as specific high affinity binding reagents *in vitro* are discussed in Section III, C.

(d) *Inhibition of hormone action.* Antibodies can be used to inhibit the hypophysiotropic action (although they could also be applied to the other functions) of the hypothalamic hormones. The design and interpretation of these experiments requires care as evidenced by the fact that using anti-LHRH sera and rat median eminence extracts, similar experimental results led to the conclusion that there is (Shin and Kraicer, 1974) and that there is not (Shahmanesh and Jeffcoate, 1976) a separate FSH-RF!

(4) *Uses of antisera* in vivo

The classical ablation approach in endocrinology has been widely applied to the study of the hypothalamus (Chapter 2). There is no doubt, however, that it is a blunt instrument in the investigation of an organ as complex, heterogenous and inaccessible as the hypothalamus. Antibodies can act as ablation instruments *in vivo* of such specificity that a single hormonal peptide is removed and the effects of this studied in isolation. This can be done by passive transfer of antibodies or during active immunisation. Apart from the development of diabetes insipidus in a rabbit immunised against vasopressin (Edwards *et al.*, 1972) all reported studies to date have been with LHRH.

(a) *Passive transfer* of anti-LHRH serum during the critical period of the oestrous cycle of the female rat blocks ovulation (Fraser and Gunn, 1973) and the LH and FSH surges are blocked (Koch *et al.*, 1973a; Arimura *et al.*, 1974a). These experiments are used to delineate the necessity of LHRH in the control of pituitary gonadotrophin secretion. In an extension of this technique Fraser (1976) used the injection of large amounts of a biologically inactive but immunologically active fragment of LHRH (the 3–10 octapeptide) to specifically relieve the blockade at known times by swamping the antibody binding sites; by this means LHRH inhibition was studied over specified time intervals. It was shown that relief of the blockade was dependent on newly secreted LHRH from the hypothalamus and not on LHRH freed from the antibody by the octapeptide, because it could be prevented by pentobarbital.

(b) *Active immunisation* against LHRH produces the morphological and biochemical changes associated with decreased LH and FSH secretion (Fraser, 1976). Of particular value is the sequential study of these changes (see Fig. 1 of Chapter 8). Animals in whom an adequate endogenous titre of antibodies has been induced can also be used in experimental situations where the secretion of that peptide might be increased, e.g. the effect of exposure to cold or of thyroidectomy in animals chronically immunised against TRH.

(5) *Conclusions*

Provided care is taken in the control of experiments using antisera, they are potentially instruments of high precision and specificity in the biochemical study of hypothalamic hormones.

III. Assay of the Hypothalamic Hormones

A. GENERAL PRINCIPLES

The quantitative study of the activity of the endocrine hypothalamus depends on the ability to detect and measure the minute amounts of the hypothalamic peptides in tissues and biological fluids. In general, two types of assay are used: *biological*, in which an activity of the hormone is measured *in vivo* or *in vitro*; and methods depending upon *saturation analysis*, radioimmunoassays and receptor binding assays. Some of the principles governing their use are common to both sorts of assay and will be discussed before more specific consideration is given to the techniques and their application to biological fluids.

The design, validation and statistical analysis of bioassays has been extensively covered in the past. Many of these are applicable also to saturation analysis which in addition has its own theoretical and mathematical bases (Ekins, 1974).

Some points need special mention in the context of the assay of hypothalamic peptides: the first is the necessity for standardisation. As with other polypeptide hormones, there is a continuing need for international standards or reference preparations, whether these are derived from natural or synthetic peptides, in order to facilitate meaningful communication between laboratories. There are at present international standards available for oxytocin and vasopressin derived from neurohypophysial extracts and defined on the basis of worldwide collaborative bioassays. Synthetic peptides are not necessarily more satisfactory than natural ones; there is a possibility of synthetic peptide byproducts and a variable non-peptide (e.g. water and acetic acid) content.

Comparison of like with like is an important assay concept; so is parallelism of dose-response curve between unknown and standard. It is a *sine qua non* of a valid assay but is not in itself proof of identity, a point which is frequently misunderstood. Ekins (1974) explains how parallelism in radioimmunoassays can be misleading; it can certainly escape detection if precision is poor. Furthermore, in many assays for hypothalamic peptides the assay sensitivity is being pushed to the limit so that parallelism cannot even be investigated, let alone established.

Other assay characteristics such as specificity, sensitivity, accuracy and precision need to be studied in the same way as for other substances. Ekins (1974) makes one more point that is worth stressing: statements of the sensitivity of an assay are usually optimistically based on the appearance of a standard curve without statistical consideration of the precision.

B. BIOLOGICAL ASSAYS

(1) *Neurohypophysial peptides*

The bioassays for oxytocin and vasopressin are highly sensitive and though this is not sufficient to measure circulating levels in man, for instance, the sensitivity of radioimmunoassay is no better.

(2) *Hypophysiotrophic peptides*

The biological actions of hypothalamic releasing hormones *in vivo* and *in vitro* are described in detail in Section V and Chapters 7–12. The usefulness of assays based on these actions is limited by several considerations. Firstly, bioassays tend to be inconvenient in terms of cost, time and the expertise required. They are also, in general, less precise and sensitive than immunological assays. Furthermore, the end-point of the assay is often the assay of another (adenohypophysial) hormone either by bioassay or radioimmunoassay and the overall characteristics of the assay will include those of the additional assay. Where this is particularly sensitive, as in the cytochemical assay for TSH, a highly sensitive (0·1 pg/ml) bioassay for TRH can be developed (Gilbert *et al.*, 1975). In some situations (e.g. growth hormone, discussed further in Chapters 5 and 10) bioassay and immunoassay of the adenohypophysial hormone give conflicting results.

(3) *Are bioassays "better" than immunoassay?*

It is generally considered that bioassays are more meaningful than immunoassays because they are measuring a response dependent upon a biological portion of the molecule which may not be the same as the antigenic part. The latter may be present as metabolites in biological fluids and be measured in radioimmunoassays. Further consideration of this is given in sub-Sections C, D and E but it should be remembered that bioassays may also respond non-specifically to other components in biological fluids.

C. RADIOIMMUNOASSAYS AND RECEPTOR ASSAYS

Reviews of these techniques are covered in books (Kirkham and Hunter, 1971; Odell and Daughaday, 1971) and chapters on basic principles, standardisation, labels, preparation of binding reagents and separation techniques are collected in one volume (*Brit. Med. Bull.*, 1974). The radioimmunoassay of neurohypophysial peptides has been reviewed by Chard (1973) and Chard and Martin (1974) and of the hypophysiotrophic peptides by Arimura and Schally (1975) and Jeffcoate *et al.* (1975).

(1) *Preparation of binding reagents*

The preparation and properties of antisera to the hypothalamic peptides has been discussed in Section II, B. Tissue hormone receptors, like antibodies, bind with high avidity and specificity. DePierre and Karnovsky (1973) have reviewed the isolation and properties of mammalian plasma membranes and pointed out some of their potentialities and pitfalls. Their use in receptor assay for the hypothalamic peptides is described by Grant and Vale (1975) and Marshall and Odell (1975).

Compared with antibodies, the preparation and purification of plasma membranes present problems. They are unstable; but future developments with solubilised receptors may avoid this. A further disadvantage is that *in vitro* receptor binding assays tend to reveal high capacity but low affinity receptors (see Section V, A) with a consequent assay of low sensitivity.

(2) *Preparation of labelled ligands*

Hypothalamic peptides are available commercially with (^{3}H) labels and these have been used successfully in the study of receptor binding (Section V, A) and of metabolism of the peptides (Section VI). Their specific activity, of the order of 20 mCi/μM, is, however, much too low for use in assays needing only femtomolar amounts of labelled ligand. A radioimmunoassay for LHRH, for instance, if it is to be sensitive enough to measure the levels in biological fluids, uses less than 10 pg (or 10 fm) of labelled LHRH. This amount of (^{3}H)-LHRH yields 20 dpm approximately. A further point, frequently overlooked, is the need to check radiochemical purity by chromatography, etc. McKelvy (1975) has described a valuable method for the purification of small amounts of (^{3}H)-TRH.

Specific activities up to 100 times greater than (^{3}H) can be obtained using radioiodination by any of the common techniques (Hunter and Bolton, 1974). Lactoperoxidase has been particularly advocated as a mild technique giving minimal damage to the peptide (Arimura *et al.*, 1973; Miyachi *et al.*, 1973; Arimura and Schally, 1975). The results with chloramine T are similar though, and it is not so much the iodination technique that determines the quality of the label as the subsequent purification which at its optimal can be designed to remove not only unreacted iodide but also damaged peptide, di-iodinated peptide and even unlabelled peptide yielding carrier-free, mono-iodinated peptide of high specific activity. The separation of unlabelled from labelled hormone has been described for LHRH, TRH, oxytocin and vasopressin.

Iodine atoms are most readily incorporated into tyrosine residues in peptides, but the histidine of TRH can be iodinated without changing the usual conditions and ^{125}I-TRH is highly immunoreactive despite the fact that the ^{125}I atom increases the molecular weight by 35%. Somatostatin presents a problem because it has no tyrosine or histidine residue; a tryosine$_1$ analogue has been synthesised which after iodination binds to anti-somatostatin antibodies (Arimura *et al.*, 1975b) and a tyrosine$_{11}$ analogue is also available.

A limited shelf-life has been found to be a problem with LHRH and somatostatin. It is possible to re-purify these (^{125}I)-peptides by cellulose adsorption chromatography (LHRH) or adsorption onto glass particles (e.g. Vycor).

D. PROBLEMS ASSOCIATED WITH ASSAYS IN BIOLOGICAL FLUIDS

Although sensitive (sub-picogram) assays for the hypothalamic hormones can be readily established in buffer solution using the approaches mentioned above, their application to body fluids and the study of hypothalamic function has been limited by a number of factors.

(1) *Low concentrations*

In the median eminence, ng amounts of the releasing hormones are found and these act, in pg amounts within the hypophysial portal system. Sampling *in vivo* from peripheral sites involves a further dilution so that only fg amounts would be expected. (These calculations are approximate only). Where reliable measurements are available, and these are few, it is found that even the neurohypophysial peptides which are secreted into the general circulation are in the low pg or fg/ml range. Only with the neurophysins (ng/ml) is sensitivity a minor problem.

Highly sensitive assays are clearly required and all current assays are pushing the sensitivity potential of the assays to their limit with all that that implies for poor precision and difficulty in establishing specificity. There are two main approaches to increasing assay sensitivity.

(a) *Refinement of the reagents.* For bioassays, one approach is pretreatment of the animals, e.g. the oestrogen-progesterone treated ovariectomised rat for LHRH. For assays based on saturation analysis, sensitivity can be increased by increasing the specific activity of the labelled ligand, by selecting out high-affinity binding sites, by reducing errors which effectively limit sensitivity and by "optimisation" of the assay conditions (Ekins, 1974).

(b) *Extraction and concentration* of samples has been extensively used for the bioassay (Fitzpatrick and Bentley, 1968) and the radioimmunoassay (Chard, 1973) of oxytocin and vasopressin. More recently it has been applied to LHRH and TRH (Jeffcoate *et al.*, 1975; Arimura and Schally, 1975) using organic solvents (ethanol, methanol, acetone) or adsorption onto glass (e.g. Florisil, Vycor) followed by elution. The final redissolved extract can be as much as 50–100 times more concentrated than the original sample. Important considerations are: that the recovery of the peptide should be at least 50% and reproducible; and that non-specific interfering substances will be concentrated to the same degree and may be actually added by the extraction procedure (Chard, 1973). As always, a rigorous check on parallelism is essential.

(2) *Episodic secretion*

It is becoming evident that for many hormones, there must be added to the known seasonal, cyclical and daily changes in secretion, variations that occur hour-to-hour and minute-to-minute. This means that sampling has to be sufficiently frequent to ensure that the periodic fluctuations do not give a misleading view of the secretion pattern. Evidence for the hypothalamic peptides is scanty but the LHRH-like immunoreactivity in sheep jugular vein (Foster *et al.*, 1976), LHRH bioactivity in human peripheral blood (Seyler and Reichlin, 1974a) and TRH *in vitro* (McKelvy and Grimm-Jørgensen, 1975) appear to be secreted in a pulsatile manner. For this reason studies using single daily samples (e.g. Arimura *et al.*, 1974c), are open to doubt. In our own work, in the sheep, we have progressively reduced sampling intervals from 120 to 15 to 4 minutes. It may be that continuous sampling will be required. Assays in urine offer potential advantage here, in that pulses of secretion are integrated so that overall rates of secretion are more accurately reflected.

(3) *Enzymic degradation*

As described in Section VI, A, enzyme degradation systems for the hypothalamic peptides are found in many tissues; these may complicate assays unless precautions are taken. Thus all the hypothalamic hormones are destroyed by hypothalamic extracts though at different rates. TRH (very rapidly), oxytocin (especially in late pregnancy), somatostatin and LHRH (slowly) are destroyed by serum. Urine, on the other hand, seems not to inactivate any of these peptides (Boyd *et al.*, 1972; Bassiri and Utiger, 1973; Jeffcoate *et al.*, 1975).

The degradation can have three effects on assays: (a) the hormone present will disappear; (b) degradation may yield immunoreactive, biologically-inactive fragments causing discrepancies between immuno- and bio-potency estimates; (c) if the enzyme is present in a radioimmunoassay tube it will degrade the (^{125}I)-peptide. This will result in an artefact whereby an apparent high hormone concentration is deduced on the basis of low antibody-bound (^{125}I). There are three methods that have been used to avoid or minimise degradation: (a) addition of enzyme inhibitors, e.g. BAL (2, 3 dimercaptopropanol). In the case of TRH in whole blood this must be present in the syringe (Jeffcoate *et al.*, 1975); (b) extraction after addition of a solvent, e.g. methanol or acetone which denature the enzymes; (c) heating to 60° for 40 min to denature the enzymes. This is claimed to give a better recovery for TRH than methanol (Eskay *et al.*, 1975a).

(4) *Doubts about identity*

All assays, especially those based on immunological properties of a molecule must be carefully validated before identity of an unknown substance with a known standard is established with any degree of certainty. Four main approaches have been used in assays of the hypothalamic hormones.

(a) *Biological: immunological correlations.* These fall into two categories. (i) Parallel estimates of potency against a standard in a bioassay and an immunoassay should give identical results. Several instances of a dissociation have been described. Oxytocin degraded by late pregnancy plasma loses biological activity faster than immunoreactivity (Chard, 1973). Extracts of chicken hypothalamus contain much more (50 times) LH-releasing activity than can be accounted for its content of immunoreactive LHRH decapeptide (Jeffcoate *et al.*, 1975). LHRH-like immunoreactivity in sheep jugular vein and urine has no measureable biological activity. TRH levels in the mammalian hypothalamus are about 20 times higher by immunoassay than by bioassay (see Jeffcoate and White, 1975b for refs). In many instances it is not possible to validate an immunoassay against a bioassay because of its greater sensitivity. In any event, failure to obtain equal results does not necessarily mean that the immunoassay is not measuring something of relevance. A final point, mentioned earlier (Section B), is that bioassays are just as likely to be subject to interference as radioimmunoassays. (ii) Do the immunoassay data correlate with expected changes in physiological and pathological conditions? This concept is widely used in the validation of the immunoassays and is of great value where the physiology of the gland and its hormone is clearly understood. For the hypothalamus, where the anatomy is awkward, physiology in an early stage of understanding and knowledge of biochemical pathology almost non-existent, attempts to validate assays by reference to accepted physiological dogma are dangerous.

Thus, a failure to demonstrate an increase in LHRH secretion at the time of the pre-ovulatory LH surge does not necessarily invalidate the LHRH assay. It may be that it is wrong to expect such an increase: indeed there is some evidence that this is so (Chapter 8). Again, for TRH, assays might be expected to reveal changes in TRH levels in diseases of the thyroid gland, yet it is increasingly evident that control of TSH secretion is only one function of TRH (see Section V, D and Chapter 7) and any change that might occur in thyroid disease might be obscured by lack of change in "non-hypothysiotrophic" TRH levels.

Validation of one assay by another is also possible. Thus changes in neurophysin levels by radioimmunoassay paralleled those of either vasopressin or oxytocin in response to appropriate stimuli (Forsling *et al.*, 1973).

(b) *Dose-response curves parallel* to the standards are an absolute requirement (see Section A). Parallel dose-response curves can be given by closely related peptides in both biological and immunological assays. Thus the interpretation must be based upon a detailed knowledge of the assay specificity: this can be defined with some exactness for the hypothalamic peptides because of the availability of synthetic analogues and fragments. This knowledge may be particularly important in situations where there is metabolism of the peptide *in vivo* or *in vitro* yielding metabolites that may or may not cross-react with antisera depending on the specificity of the antiserum (Section E).

(c) *Chromatographic identity* of unknown and standard peptide can only be established if sufficient material is available and is strong evidence of identity, particularly if several different chromatographic systems are used relying on different physico-chemical properties of the molecule. It is important to avoid chromatographic artefacts induced by contaminants in biological extracts. Addition of standard material to the extract will test for this and for the possibility of binding or metabolism that might alter the peptide's mobility.

A combination of different separation systems and immunological and biological assays is more reliable and informative than dependence on a single technique. Thus the LHRH-like immunoreactivity in urine is indistinguishable from the decapeptide on gel filtration but separable by ion exchange chromatography (Jeffcoate and Holland, 1975). Chromatography has also revealed another LH-releasing activity in avian (Jackson, 1971) and mammalian (Fawcett *et al.*, 1975) hypothalamus.

In steroid analysis, the concept of derivative formation, in which a specific chemical reaction induces a specific change in chromatographic mobility, is widely used for identification. That this important tool can also be applied to small peptides has been shown by the dinitrophenylation of TRH followed by chromatography (McKelvy and Grimm-Jørgensen, 1975).

(d) *Enzymic analysis of substrate identity*. The hypothalamic peptides are subject to degradation by enzymes present in biological fluids and tissues (see Section VI, B); this is accompanied by loss of immunochemical and biological activities. Since the substrate-specificity of enzymes is akin to that of antibodies and dependent on the conformation of the peptide substrate, the kinetics (e.g. with respect to time or enzyme concentration) of degradation of an unknown and standard peptide can be compared.

Chemical (non-enzymic) reduction of the disulphide bridge of oxytocin and vasopressin has been used to validate bioassays. The oxytocinase present in late pregnancy plasma inactivates bio- and immuno-activity at different rates however and could not be used to validate an immunoassay.

Serum inactivation of the biological and immunochemical activities of TRH, on the other hand, are identical (Bassiri and Utiger, 1972b) and can be accurately quantitated (Jeffcoate and White 1975b, c) and used to validate the immunoreactive TRH isolated from various tissues.

When combined with the biological, immunochemical and chromatographic approach, enzymic analysis furnishes strong evidence of identity.

E. SIGNIFICANCE OF ASSAY RESULTS

Despite the problems outlined in the preceding sections, assays for the hypothalamic peptides have been widely applied to biological situations, with varying degrees of success. At the present time, many of the doubts are still about the qualitative nature of the material being assayed; confidence about quantitation is even lower.

(1) *Hypothalamus*

Peptide hormones have been measured in hypothalamic tissue for many years by bioassay, but in most instances the rigorous validation required has not been possible. More recently radioimmunoassay has been applied and attempts made to validate the measurements (Bassiri and Utiger, 1974; Jeffcoate *et al.*, 1975) and hypothalamic LHRH, TRH and somatostatin are indistinguishable from the synthetic peptides using a combination of different criteria. As regards quantitation in either type of assay, there are a number of factors to be considered when tissue levels in different situations are being compared (cf. Section II, A).

(a) Have factors such as area of tissue being excised, time of day, pretreatment of animals, etc., been adequately standardised?

(b) Is the same substance being measured in each instance? One set of conditions may alter the amounts of interfering substances, such as prohormones, peptidases, neurotransmitters.

(c) Quantitation can be expressed as a concentration (e.g. pg/mg tissue weight) or as a total amount in the gland. Each may be revealing when used for comparison with hypothalami from other situations or with other tissues. The concentration will clearly depend on how wide a definition of "hypothalamus" is taken.

(d) The *level* of a hormone in the hypothalamus (or other tissue) is a resultant of the opposing forces of synthesis on the one hand and of degradation and secretion on the other. Nothing can be firmly concluded about these parameters merely by measuring the concentration in the tissue. Thus it is meaningless to write of a "paradoxical" effect of gonadectomy on hypothalamic LHRH levels (Shin and Kraicer, 1974).

(2) *Organ distribution of hypothalamic hormones*

(a) *Within the hypothalamus.* Except for the neurohypophysial peptides, discrete centres with well-defined functions cannot readily be demonstrated for the hypothalamic hormones: attempts to do this have been discussed in Chapter 2. The geographical distribution as assessed by bioassay was introduced by McCann; more recently immunoassay has been used, which complements the data obtained by immunohistochemistry, and this has revealed a selective distribution of LHRH (Palkovits *et al.*, 1974a), TRH (Brownstein *et al.*, 1974; Jackson and Reichlin, 1974a, b; Winokur and Utiger, 1974) and somatostatin (Arimura and Schally, 1975).

(b) *Other CNS regions.* Although there are some early data indicating extrahypothalamic distribution of "hypothalamic" hormones by bioassay (e.g. Endöczi and Hilliard, 1965) the advent of radioimmunoassays has brought evidence of the presence of these peptides in neural tissue from the cerebral cortex to the lumbar region of the spinal cord. This has been particularly true for TRH which has been located in mammalian cerebral cortex, cerebellum, pineal, brain stem and spinal cord (Jackson and Reichlin, 1974a, b; Oliver *et al.*, 1974c; Winokur and Utiger, 1974; Winters *et al.*, 1974; Jeffcoate *et al.*, 1976b). In the case of the cerebrum and cerebellum, only low concentrations (1–2 pg/mg) are found and it is only by multiplying this concentration by the large mass of tissue that it becomes apparent that there are large amounts of TRH in these tissues (e.g. Winokur and Utiger, 1974). In view of the doubtfulness of the assays at these low levels these calculations are not entirely convincing.

Very large amounts of immunoreactive TRH and LHRH were found in the pineal gland by White *et al.*, (1974). This has not been confirmed by others, e.g. Ramirez and Kordon (1975) who found very little and Jeffcoate and White (1975a) who found none.

There are three situations in which extrahypothalamic TRH is well substantiated. These are: in sub-mammalian chordates (e.g. reptiles, fish) in which large amounts are found throughout the brain (Jackson and Reichlin, 1974b); in gastropod circumoesophageal ganglia (Grimm Jørgensen *et al.*, 1975); and in mammalian brain stem and spinal cord (Jeffcoate *et al.*, 1976b). LHRH has not been convincingly found in tissues outside the hypothalamus but somatostatin, like TRH, has been identified by radioimmunoassay throughout the mammalian central nervous system.

(c) *Other tissues.* Somatostatin has been also identified by immunohistochemistry and radioimmunoassay in the pancreas and upper intestine (see Section V, D).

(d) *Subcellular localisation.* Cell fractionation studies confirm those using immunohistochemistry and show that most of the hormone in the hypothalamus is found in the nerve terminals (Martin *et al.*, 1975; Ramirez and Kordon, 1975). Nerve terminals ("synaptosomes") isolated by the method of Whittaker (1959) contain CRF, PIF, LHRH, somatostatin and TRH and this preparation can be used to study their synthesis and secretion (Section IV, A). Synaptosomes prepared from both brain stem and spinal cord also contain and secrete both TRH and somatostatin *in vitro*.

(e) *Cerebrospinal fluid* (*CSF*). The assay of releasing hormones in CSF should be considered as part of the evidence implicating this fluid in their transport to the pituitary (Chapters 1 and 3). TRH, in particular has been measured by radioimmunoassay in both ventricular and spinal CSF (Knigge and Joseph, 1974; Oliver *et al.*, 1974b; Eskay *et al.*, 1975b). The tanycytes lining the third ventricle take up (^{3}H)-labelled TRH (Joseph *et al.*, 1973) and LHRH (Vaala and Knigge, 1974) and LHRH immunofluorescence is found in these cells (Zimmermann *et al.*, 1974). Others (e.g. Cramer and Barraclough, 1975; Jeffcoate *et al.*, 1975) failed to find LHRH in CSF. In any case the possible hypophysiotrophic significance of the TRH in spinal CSF is called in question by the fact of its presence in the spinal cord itself (Jeffcoate *et al.*, 1976b).

Finally, GH-releasing activity has been measured by bioassay in CSF (Barbato *et al.*, 1974) but whether this represents a peptide GH-RF is uncertain.

(f) *Hypophysial portal blood.* The detection and measurement of releasing hormones in portal blood is obviously a direct approach that avoids problems of dilution, degradation and non-hypothalamic sources of secretion. There are, however, problems associated with its collection. Fink and Harris (1970) emphasised that it is a stressful procedure that may alter hormone secretion; evidence for this was later presented by Eskay *et al.* (1975a). Contamination with blood from elsewhere cannot be excluded.

The data so far available whether based on bioassays or radioimmunoassays are inconclusive. Changes in LHRH during the rat oestrous cycle have been examined by Eskay *et al.* (1975a) and Fink *et al.* (1975). Eskay *et al.*, (1975b) found no apparent change in LHRH secretion in response to electrical stimulation but that prostaglandin (PG) E_2 caused an increase (see Section IV, B). Portal blood TRH was also measured by Eskay *et al.* (1975a) but as no precautions were taken to prevent degradation until after plasma separation nothing can be concluded from their results.

(g) *Systemic blood.* The assay of the hypothalamic peptides in the general circulation has been severely limited by the problems mentioned above and has caused much controversy and confusion.

(i) Neurohypophysial peptides. Chard and Martin (1974) have recently reviewed this subject. Basal levels (a few pg per ml) are not convincingly measurable but rises are seen in response to appropriate physiological stimuli. Neurophysins are easier to measure and seem to be secreted in parallel to the neurohypophysial peptides mole for mole, though the calculations (Forsling *et al.*, 1973) are somewhat indirect. There is a requirement for assays specific for the individual neurophysins and for species-specific assays.

(ii) TRH (see also Chapter 7). Several reports of immunoreactive TRH in peripheral blood have been published but without adequate inhibition of peptidases interpretation is difficult. Basal levels are not well authenticated in any situation; there are some reports of changes in thyroid disease and in response to cold stress but in view of the technical difficulties and the non-hypophysiotrophic role of most of the TRH in the body, their significance is doubtful.

(iii) LHRH (see also Chapter 8). The assay of LHRH in the circulation by bioassay or radioimmunoassay has been reviewed recently by Arimura and Schally (1975); Jeffcoate (1975a); Jeffcoate *et al.* (1975) and is a confused field which is difficult to summarise. A few points stand out. Data based on infrequent sampling (e.g. Arimura *et al.*, 1974c) are unreliable; bioassayable LHRH in human peripheral circulation reaches levels (Seyler and Reichlin, 1973; 1974a, b) much greater than those measured by radioimmunoassay; very large pulses of LHRH-like immunoreactivity are found in sheep jugular venous blood; the specificity of the antiserum used is a critical factor determining whether immunoreactive metabolites are detected or not.

(iv) Somatostatin. No firm reports are available at the time of writing.

(h) *Urine*. Vasopressin, oxytocin, TRH and LHRH-like peptides have been measured by radioimmunoassay in urine into which they are rapidly excreted. Urine assays have some advantage over those in blood: there are no peptidases; there is integration of pulses of secretion; there is less non-specific interference in the assay.

(i) Vasopressin (Oyama *et al.*, 1971) and oxytocin (Boyd *et al.*, 1972) can be readily measured. The identity of the immunoreactive oxytocin in urine was checked in two chromatographic systems. Their assay is of interest because the *concentration* in the urine appears to be a better guide to secretion rates than total excreted *amounts* per unit time because the latter increases with increasing water load.

(ii) TRH (see also Chapter 7) has been identified in rat (Jackson and Reichlin, 1974a) and human (Oliver *et al.*, 1974a) urine. Its identity can be checked by chromatography and enzyme inactivation and on the basis of this, Jeffcoate and White (1975a) were doubtful about the nature of this immunoreactivity. Rather high (ng) levels have been reported in some studies. In any case the widespread distribution of TRH militates against a urine TRH assay being of value as an index of its role in the control of TSH secretion.

(iii) LHRH-like immunoreactivity is excreted in urine after exogenous administration of the decapeptide. It is not, however, LHRH itself but a C-terminal metabolite(s) (Jeffcoate and Holland, 1975) which will only be measured by non-specific antisera which cross-react with these metabolites. The significance of the small amounts excreted in basal urine samples has yet to be established.

IV. Synthesis and Secretion of the Hypothalamic Hormones

The biochemical study of the synthesis and secretion of the hypothalamic hormones is not only made difficult by the inaccessibility and heterogeneity of the organ and the problems associated with their assay, both of which have been mentioned earlier, but also by the wide range of substances that affect hypothalamic function: ions, neurogenic amines, prostaglandins, steroids, thyroid hormones and peptide hormones (Chapter 3). The discussion is subdivided into two sections: the method of assessing the synthesis and secretion of the hormones *in vitro* and *in vivo*; methods of studying the role of controlling factors.

A. Methods of Study *In Vitro* and *In Vivo*

(1) *Assay in hypothalamic tissue*

The hypothalamic hormones can be measured in extracts of the median eminence (and non-hypothalamic tissue) by biological or immunological methods (Section III, E). Changes in the assayable levels have been reported after such procedures as removal of target organs, or the administration of drugs or stressful stimuli, or in association with circadian or sexual cycles. Such measurements, even assuming they are valid in other respects, are *static* and cannot in themselves be used as indices of synthesis or secretion.

(2) *Biosynthesis of hypothalamic hormones*

The study of the biosynthesis of these peptides is in its infancy. The general procedure is incubation of suitable tissue preparations with (^{3}H)-labelled amino acid precursors followed by exhaustive radiochemical purification of the resulting labelled peptide. McKelvy (1974, 1975) in studying the biosynthesis of TRH, has particularly stressed the need for rigorous criteria to be applied before accepting that radioactivity has actually been biosynthetically incorporated into TRH.

For a small peptide like TRH, four possible biosynthetic routes are possible (McKelvy and Grimm-Jørgensen, 1975): (i) *de novo* ribosomal synthesis of the tripeptide; (ii) ribosomal synthesis of a precursor "prohormone" with cleavage of the tripeptide; (iii) aminoacyl t-RNA synthesis; (iv) RNA-independent synthesis. There is evidence that the MSH-regulating peptides may be formed by method (ii) by cleavage of the octapeptide, oxytocin (see Chapters 5 and 12) and for the non-ribosomal synthesis of TRH and LHRH (Reichlin and Mitnick, 1973a; McKelvy and Grimm-

Jørgensen, 1975). In the newt the presence of RNA is necessary for TRH biosynthesis since this can be inhibited by an RNase; method (iii) may be in operation here. Ribosomal inhibitors, e.g. cycloheximide have no effect.

Simultaneous degradation of the synthesised peptides by peptidases needs to be considered and was found to be a particular problem for TRH by Bauer (1974). His approach was to add pGlu.His.OCH_3 which prevents proteolysis but not deamidation; deamido TRH was therefore obtained as the biosynthetic product. Labelled proline, but not histidine, was incorporated into this; Bauer considers that this is by reversal of the peptide hydrolysis reaction.

Long-term organ cultures appear to be the best preparations for demonstrating biosynthesis though it can also be demonstrated in synaptosomes and short-term superfusion systems. These approaches will allow study of the subcellular localisation of the biosynthetic processes as well as the factors controlling them.

(3) *Hypothalamic secretion* in vitro

Whole (undissected and unfractionated) preparations of median eminence or hypothalamus can be incubated *in vitro* in either static or perfusion systems and shown to secrete the releasing hormones. The effect of stimuli, such as K^+, electric currents and chemical factors can then be studied. This approach seems to have been particularly valuable in the study of CRF (see Chapter 11). A number of factors that may affect interpretation of the results need consideration:

(a) The tissue is removed from higher neural influences and from its physiological connection with the portal vessels.

(b) The tissue is heterogeneous.

(c) The polysynaptic nature of the pathway complicates the study of the actions of neurotransmitters.

(d) Maintenance of tissue viability during incubation may be difficult.

(e) Substances added to the medium (e.g. neurotransmitters) may not readily diffuse into the tissue.

(f) Diffusion of released hormone into the medium is equally uncertain.

(g) Tissue enzymes may degrade either neurotransmitters or the secreted peptides.

All these considerations have limited the usefulness of hypothalamic tissue incubations.

(4) *Incubation of isolated nerve terminals (synaptosomes)*

The use of hypothalamic nerve terminals (synaptosomes) isolated by differential centrifugation on sucrose gradients (reviewed by de Belleroche and Bradford, 1973) allows the study of the factors controlling neurosecretion without many of the disadvantages listed above. They have been shown to secrete CRF, vasopressin and PIF (Edwardson *et al.*, 1972; Bennett and Edwardson, 1975) and LHRH and TRH (Bennett *et al.*, 1975) and somatostatin (unpublished results). Calcium-dependence is an important feature and they respond to electrical stimulation, K^+ depolarisation and neurotransmitters. A direct modulating effect of corticosteroids on CRF secretion (Edwardson and Bennett, 1974) and of ovarian steroids on LHRH secretion (Bennett *et al.*, 1975) has been shown.

Synaptosomes appear to be a preparation of great potential. One problem is that the synaptosome population in the hypothalamus is a mixed one and the ability to separate out, say TRH-synaptosomes from a median eminence extract using affinity chromatography, for example, would be of great benefit. They can of course be prepared from regions of the CNS other than the hypothalamus and for instance brain stem and spinal cord synaptosomes have been shown to secrete TRH and somatostatin.

(5) *Methods of study* in vivo

Assessment of the secretion of the hypothalamic hormones *in vivo* involves either direct assay of that hormone in a body fluid or indirectly by the effect on (a) the secretion of its adenohypophysial hormone; (b) the function, histology or hormonal secretion of the final target gland, e.g. ovulation, cortiscosteroid secretion; (c) a direct biological response, e.g. oxytocin and milk secretion.

The problems associated with the direct assay of the hypothalamic hormones are great and fully discussed in Section III. Any indirect measure will involve consideration of the responsiveness ("sensitivity") of the target organ to a given stimulus. It is important, for instance, to distinguish between the effect of a given treatment on the pituitary from that on hypothalamic function. To take an example: Blake (1974) showed that urethane and phenobarbitone given to ovariectomised rats lowered LH secretion. Exogenous LHRH resulted in an unimpaired LH response; thus he was able to conclude that these drugs reduce LHRH synthesis or release.

B. CONTROLLING FACTORS

(1) *General considerations*

Before considering individually the methods of studying the factors controlling the secretion of hypothalamic hormones, some general comments should be made:

(a) The possible role of a given factor can be studied by *administration* of (i) the factor itself; (ii) a substance which increases the endogenous levels of the factor by increasing its synthesis or decreasing its breakdown; (iii) a substance which decreases the endogenous levels by decreasing its synthesis or increasing its breakdown; (iv) a substance which increases or reduces the *effectiveness* of a given level of the factor.

(b) The *dose* of the substance given should be in the physiological range if possible; pharmacological doses may have non-specific effects. Depending on the route of administration, doses higher than physiological may have to be used to ensure adequate levels in the hypothalamus.

(c) The *route* of administration will also affect the response to a given factor. (i) Peripheral administration of a substance involves consideration of its adsorption, clearance, degradation, binding to carrier proteins and tissue receptors, non-hypothalamic actions, degree of penetration into the CNS. (ii) Local administration within the CNS clearly ensures that the substance gets closer to the hypothalamus; microiontophoretic application is the most precise method used. It is important to remember that there is a possibility of the substance diffusing elsewhere, to the pituitary for instance; and that the hypothalamus is a bilateral organ. Precise application can only be one-sided.

(d) The *end-points* used include: (i) Uptake of the substance in the hypothalamus. (ii) Effect on the secretion of the hypothalamic peptide measured either directly or indirectly (see Section III). (iii) Effect on other supposed factors in the control system. (iv) Measurement of changes in the levels of the factor itself in situations in which there are changes (real or presumed) in the secretion of the peptide.

(e) A final general consideration affecting the interpretation of studies on the role of factors in the control of the hypothalamus is that of *timing*; the timing of the collection of the samples (in conjunction with the dose and route of administration) may determine the effect observed. The biphasic effect of oestrogen on LH secretion (Chapter 8) is a well-established example of this. Anti-cholinergic drugs have, also, been shown to induce an initial decline in LH and prolactin followed by an increase (Libertun and McCann, 1974).

(2) *Neurotransmitters*

A close relationship between neurotransmitters (mainly the biogenic amines, adrenaline, noradrenaline, serotonin and dopamine) and the hypothalamic peptides has been clearly established (Chapter 3). The subject and the methods used to study it have been reviewed by Kamberi (1973), van Loon (1973), Snyder *et al.* (1973), de Weid and de Jong (1974), Burn (1975), Ganong (1975), Lawson and Gala (1975), Weiner (1975).

(a) *Histochemical methods* have been extensively used to localise the amines and for mapping the hypothalamic neurone systems. This is essentially a static and semi-quantitative approach but, particularly when correlated with localisation of the hypothalamic peptides, it is valuable. Until recently it was limited to the formaldehyde fluorescence technique (references in

Hökfelt *et al.*, 1975b, f). A very sensitive enzymic-isotopic method can be used for noradrenaline and dopamine (Palkovits *et al.*, 1974b) and as discussed in Section II, B, immunohistochemistry using antibodies against the biosynthetic enzymes has led to a refined analysis of the distribution of the amines (Fuxe *et al.*, 1975; Hökfelt *et al.*, 1975b, f) and their association with the hypothalamic peptides. The distributions of dopamine- and LHRH-containing terminals are close, for instance, supporting the evidence for a role of dopamine in LHRH secretion. A lack of correlation between these two has also been reported however (Kizer *et al.*, 1975) and the fact that the formaldehyde fluorescence and enzymic-isotopic localisation give rather different pictures of dopamine distribution (Palkovits *et al.*, 1974b) indicates the need for cautious interpretation.

(b) *Degeneration of pathways.* Degeneration of specific monoamine pathways follows the intracerebral administration of 6-hydroxy analogues, e.g. 6-hydroxy dopamine and 5, 6, dihydroxytryptophan which cause the degeneration of NA/DA—neurones and serotoninergic pathways respectively. Thus the effects of lack of these pathways can be observed in isolation, though rather widespread cerebral effects will also be produced.

(c) *Pharmacological methods.* Drugs are available that can effect every aspect of amine metabolism, from synthesis through release, re-uptake and binding to receptors, to enzymic degradation (see Table 1 in Chapter 3). Despite (or perhaps because of) the number of different approaches used, opposite conclusions are frequently drawn, e.g. on the role of dopamine in LHRH secretion (see Chapter 8). It is worth recalling the general considerations listed above in A3 and B1, which affect the interpretation of such studies.

(d) *Stimulation with neurotransmitters* can be studied *in vivo* (where dose, route of administration, metabolism and timing are critical) or *in vitro*. In the latter the effectiveness of the amines on the system will be affected, for instance, by the levels of degrading enzymes (e.g. monoamine oxidase) and the penetration of the tissue by the added amine. Synaptosomes are a useful model for studying amine action, but as stated earlier, they represent nerve terminals isolated from a polysynaptic control system.

(3) *Ions*

(a) Potassium ions stimulate synaptic transmission because of a depolarising action (Goldman equation) at concentrations of around 55–60 mmol/l. Hypothalamic synaptosomes incubated at such a concentration release increased amounts of CRF, PIF, vasopressin (Bennett and Edwardson, 1975) but not LHRH or TRH (Bennett *et al.*, 1975). The latter was surprising since electrical depolarisation induced 5–7 fold increases and the difference may lie in the fact that electrical depolarisation is transitory, whereas that produced by K^+ is tonic. K^+ reduces MSH secretion in the superfused pars intermedia probably via depolarisation of CA fibres (Tilders *et al.*, 1975).

(b) *Calcium.* The K^+ effect is dependent on calcium ions (Ca^{++}) and is accompanied by an influx of Ca^{++}. Calcium is involved in the *action* of many hormones including the hypophysiotrophic hormones in association with the cAMP system but also appears essential for the *release* of these hormones at least *in vitro* (de Belleroche and Bradford, 1973; Bennett and Edwardson, 1975).

(4) *Prostaglandins*

There is evidence that prostaglandins are involved in both the synthesis/release of the hypothalamic peptides in the median eminence and in their mode of action on the anterior pituitary (Section V). Since the evidence (see (d) below) is firmly in favour of a major hypothalamic site of action, the general methods of investigating the role of prostaglandins will be considered here and only the interrelationships with cAMP covered in Section V.

(a) *Measurement of prostaglandins.* Prostaglandins have been located in the hypothalamus (Holmes and Horton, 1968) but there is much more that needs to be known, perhaps using immunological methods, of the precise anatomical distribution of specific prostaglandins.

(b) *Pharmacological studies.* Prostaglandin (PG) precursors, e.g. dihomo-γ-linolenic acid or arachidonic acid administered *in vivo* increase endogenous PG synthesis but the use of inhibitors of PG synthesis has been the most widely applied method. Salicylates and aspirin-like drugs, notably indomethacin (Vane, 1971) appear from kinetic studies to be non-competitive with arachidonic acid, whereas another anti-inflammatory agent, fenoprofen, inhibits competitively. The PG synthetase inhibitors may have non-specific effects particularly in high doses. It would be valuable for measurements of PG levels in the median eminence to be made in association with these studies to confirm that effective inhibition of synthesis has occurred. 7-oxy-13-prostanoic acid acts in a different way, as a competitive antagonist of prostaglandins.

(c) *Prostaglandin administration.* The effect of PG on the hypothalamus can be studied *in vitro* or *in vivo* after general or local intracerebral injection. The half-life of peripherally administered PG is very short, and since there also appears to be a blood-brain barrier the peripheral route is of little value (Labhsetwar, 1974). They can be given intracerebrally into the third ventricle (e.g. Ojeda *et al.*, 1974b) or by iontophoresis (Poulain and Carette, 1974). Labhsetwar (1974) emphasises the importance of *bilateral* injections (see Section B1). Even with the most precise techniques there is the possibility of diffusion into, and effects on, other tissues. Very little gets from the third ventricle to the pituitary.

(d) *Site of action of PG.* The well-established effects of PGs on adenohypophysial hormone secretion could in theory be mediated via a direct effect on the pituitary cells as well as via an effect on the hypothalamus. The evidence against this comes from different methodological approaches of which the following are examples: (i) Intra-pituitary administration of PGs is less effective than intra-cerebral on, e.g. LH secretion (Ojeda *et al.*, 1974b). (ii) Indomethacin reduces the PGE_2 level in the adenohypophysis but LH release induced by LHRH *in vitro* is not reduced (Naor *et al.*, 1975). (iii) Indomethacin does not reduce LHRH-induced LH release in man (Craig *et al.*, 1975). (iv) PG *in vitro* increases cAMP levels but not TSH releases, whereas TRH increases TSH release but not cAMP levels (Tal *et al.*, 1974). Similar results are obtained with LHRH (Chobsieng *et al.*, 1975).

(5) *Hormones*

The hormonal control of hypothalamic function (Chapter 3) is a large and poorly understood subject and the methods used can only briefly be discussed.

(a) *Binding and metabolism of hormones in the hypothalamus:* (i) *In vivo* binding of radioactive hormones after intravenous injection can be assessed by auto-radiography of appropriate tissues. (^{3}H)-labelled testosterone and oestradiol are bound in this way by hypothalamic receptors. More recently (^{125}I)-FSH has been shown to be concentrated in certain parts of the median eminence after intravenous injection (Davies *et al.*, 1975a). The highest amounts were in areas of high LHRH content.

(ii) Binding of radioactive hormones *in vitro* by receptors partially purified from hypothalamic tissue can be studied in the standard way (see Section V, A) and their concentration, affinity and specificity assessed. This approach has been limited so far to the binding of (^{3}H)-labelled steroids, e.g. oestradiol. Variations in the abundance of oestradiol receptors during the oestrous cycle of the rat have been reported (Ginsburg *et al.*, 1975).

(iii) Metabolism of steroids by hypothalamic tissue *in vitro* has also been studied in attempts to delineate their role in feedback processes. Thus testosterone is converted into 5α-reduced metabolites (Zanisi *et al.*, 1973); androstenedione can be aromatised to oestrogens (Naftolin *et al.*, 1971); oestradiol converted into 2-hydroxyoestradiol (Fishman and Norton, 1975). In each case it is proposed that it is the metabolic product that exerts the feedback. The oestrogen receptors in the hypothalamus also bind 2-hydroxyoestrogens (Davies *et al.*, 1975b) and it has been proposed that the inhibition of the catechol O-methyl transferase enzyme by 2-hydroxyoestrogens may be a means whereby oestrogens affect catecholamine function in the hypothalamus.

(b) *Administration of hormones* (i) The route of administration of hormones as with other factors, is of importance. Systemic administration, e.g. of steroids is associated with a number of uncertainties, particularly when different steroids are being compared. Clearance, binding to carrier proteins, metabolic conversions are all processes intervening between the site of injection and the hypothalamus.

For this reason, in experimental animals, hypothalamic implantation, particularly by iontophoresis (see also Chapter 2), has many advantages and the latter has been used for steroids, adenohypophysial hormones and hypothalamic peptides themselves though this approach can only be used for acute studies.

(ii) The dosage used should be "physiological" in so far as this can be defined. Many earlier experiments, e.g. with steroid implantation, used doses far higher than those encountered naturally. The physiological significance of results may depend on what appears to be the relatively minor differences between physiological and non-physiological levels. Thus Stewart-Bentley *et al.* (1974) in a careful study of the feedback control of LH and FSH in normal men, showed that LH was suppressed by both testosterone and and oestradiol when infused at physiological rates. Only at rates five times the physiological could FSH be suppressed by these steroids; thus it is doubtful whether FSH is controlled by them physiologically as suggested by previous studies.

It is important to remember that what is physiological for one species may not be for another. Thus the circulating levels of oestradiol in the sheep and rat appear to be about ten times lower than those in the human.

(c) *Removal of hormones* can be done in four ways. (i) Surgical ablation. This has the disadvantage that other hormones are removed in addition, e.g. hypophysectomy. (ii) Chemical ablation. This can be done where a specific reagent is available, e.g. thiouracil. (iii) Antibodies to hormones can be used as a highly specific version of the classical endocrinological ablation approach. The endocrine effects of active and passive immunisation with adenohypophysial and steroid hormones are being increasingly studied particularly in the field of reproduction (Edwards and Johnson, 1975). (iv) Administration of anti-hormones can also produce specific reduction in one hormonal activity. For example, clomiphene has been widely used as an anti-oestrogen both experimentally and therapeutically.

V. Actions of the Hypothalamic Hormones

The functions of the hypothalamic peptides fall into three categories: those of the neurohypophysial peptides which are discussed in Chapters 5 and 6; the non-hypophysiotrophic actions of the releasing hormones which are considered in section D below and the actions on the adenohypophysis considered in sections A to C.

A. BINDING TO PITUITARY RECEPTORS

(1) *Introduction*

The concept has grown in recent years of specific receptor sites for peptide hormones located on the plasma membrane of the cells in the target organ. This membrane receptor is associated (but not identical) with the enzyme adenyl cyclase and the altered state of the membrane is followed by changes in intracellular cAMP.

The formation of the complex between the peptides and its receptor is reversible and the law of mass action can be applied. The receptors can be isolated, shown to bind radioactive peptides *in vitro* and their specifications and affinities measured.

(2) *Preparation of pituitary receptors*

This has been mentioned earlier (Section III, C). Two considerations affect interpretation of *in vitro* binding experiments using isolated receptors. First, the heterogenous nature of the adenohypophysis means that specific receptors cannot be prepared. The use of tissue culture yields a more pure population of cells for receptor binding studies (Hinkle *et al.*, 1974; Tixier-Vidal *et al.*, 1975). The second factor is that the behaviour of membrane fragments *in vitro* may be rather different from those *in situ*. The ionic milieu, for instance, is of critical importance (Vale *et al.*, 1973).

(3) *Binding of labelled peptides*

Peptides labelled with (^{3}H), whilst they might be expected to retain their binding properties when compared with the unlabelled peptide, can only be used to study receptors of relatively low affinity and are of too low a specific activity to be of use in the study of high affinity receptors. Labelling with (^{125}I) is superior, provided it can be established that the iodinated peptide retains full biological (Miyachi *et al.*, 1973) or binding (Marshall and Odell, 1975) activity.

(4) *Specificity of receptors*

Competitive inhibition of binding by peptide analogues has been widely used to study structure-activity relationships and receptor specificities. For TRH (Vale *et al.*, 1973; Hinkle *et al.*, 1974) there appears to be a stringent conformational preference for both binding and biological activity and there is a high correlation between these over a 10,000 fold range of activity. The 3-methyl analogue is about 8 times more active than TRH itself. It was also interesting to note that the specificity of the "prolactin receptors" was the same as the "TSH receptors".

Vale *et al.* (1973) also studied the specificity of the LHRH receptors, finding that single amino acid replacements resulted in a reduction in the binding affinity and in some instances also a competitive antagonism to LHRH binding. Thus (gly_2)-LHRH is a partial (50%) agonist and a competitive antagonist.

(5) *Affinity of receptors*

For both LHRH (Vale *et al.*, 1973; Spona, 1973b; Marshall and Odell, 1975) and TRH (Hinkle *et al.*, 1974; Tixier-Vidal *et al.*, 1975) binding of labelled peptides *in vitro* reveals two binding sites, one of high capacity but low affinity (K of the order of 10^{-8} mol/l) and a second of lower capacity and higher affinity (K of the order of 10^{-9} to 10^{-10} mol/l). These affinities seem low when compared with the physiological concentrations of these peptides which are of the order of 10^{-12} mol/l. It seems likely (Vale *et al.*, 1973) that *in vitro* binding experiments only reveal low affinity receptors; the receptors of highest affinity being saturated with tightly bound endogenous hormone.

(6) *Abundance of receptors*

The modulation of releasing hormone action on the adenohypophysis by target organ hormones may involve changes in the abundance of receptors. Thus the number of TRH receptor sites is related to thyroid status being increased in hypothyroidism (Labrie *et al.*, 1975) and gonadal steroids affect the numbers of LHRH receptors. These results must be interpreted with caution however, in view of the unphysiologically low affinity of the bindings studied.

B. CYCLIC NUCLEOTIDES AND PROTEIN SYNTHESIS

(1) *Cyclic nucleotides*

The activation of adenyl cyclase and the second messenger role of cAMP (or other cyclic nucleotides) has become established as a general mechanism of hormone action and the hypothalamic peptides are no exception. The following four criteria (Peake, 1973) need to be satisfied before a cyclic nucleotide can be accepted as a second messenger and these represent the four main methodological approaches.

(a) The releasing hormone when added to pituitary cell preparations *in vitro* increases (or in some cases, reduces) the formation of the nucleotide. This was first shown after stimulation with hypothalamic extracts (Zor *et al.*, 1969) and subsequently with LHRH and its analogues (Labrie *et al.*, 1975; Lippmann, 1975). No increase is found with TRH (Tal *et al.*, 1974); this negative finding may be a result of the mixed cell population, only a small fraction of which are thyrotrophes so that changes are hard to demonstrate. The same point was made by Borgeat *et al.* (1974) who found a reduction of cAMP levels by somatostatin but only to about half the control levels. Since about half the adenohypophysis is composed of gonadotrophes, lactotrophes and corticotrophes which are probably unaffected by somatostatin, they concluded that somatostatin reduced cAMP to very low levels in the other trophic cells. Somatostatin also produces an increase in cyclic GMP (Kaneko *et al.*, 1974). The levels of cGMP in the pituitary are about one tenth of those of cAMP.

(b) In intact pituitary tissue, the levels of the nucleotide change in proportion to the amount of releasing hormone added and should correlate with the amount of adenohypophysial hormone released.

(c) Phosphodiesterase inhibitors, e.g. theophylline, should increase the release of adenohypophysial hormones and, particularly, to potentiate the effect of the releasing hormones.

(d) Exogenous nucleotides should reproduce the effects of the releasing hormones. Thus cAMP or cGMP (or the dibutryl derivatives which are less easily destroyed by phosphodiesterase) can be administered *in vitro* and *in vivo* though the amounts required are several orders of magnitude (10^5) higher than the physiological levels which may be due to poor cell penetration. In the case of GH and cGMP conflicting results have been obtained. Thus theophylline increases cGMP levels and GH release and added cGMP increases GH release, which are consistent. On the other hand somatostatin has divergent effects on the levels of cGMP (increased) and GH (reduced).

(2) *Prostaglandins*

Prostaglandins have been implicated in the cAMP second messenger mechanism for hormones (Chapter 5) and since the demonstration by Zor *et al.* (1969) that PGE could stimulate adenyl cyclase activity in the pituitary, it has been suggested that PGs are involved also in the action of the hypophysiotrophic hormones. Recent evidence (see Section IV, B) indicates however that PG do not mediate these actions, since the effects of PGs and drugs such as indomethacin can be disassociated from the cAMP-mediated stimulation of pituitary hormone release by hypothalamic hormones. Thus, to take an example (Naor *et al.*, 1975), indomethacin reduces the intra-pituitary levels of PGE_2 but does not reduce the stimulation of adenyl cyclase nor the LH release produced by addition of LHRH.

(3) *Protein synthesis and hormone release*

The final step in the cAMP mediated mechanism of hormone action is the binding of cAMP by a protein kinase and the initiation of protein synthesis. The investigation of this has generally involved the use of substances that disrupt nucleic acid and protein synthesis. Non-specific effects may be associated with the use of these inhibitors; Newton (1970) has reviewed their use in general.

It is clear that pituitary hormone release can be stimulated independently of protein synthesis. Thus puromycin reduces pituitary GH synthesis but not the release of GH induced by PG, dibutyryl cAMP or theophylline (MacLeod and Lehmeyer, 1970). This question can be dissected more precisely in superfusion systems. In an investigation of the "priming effect" (see Section C below) of LHRH in a rat pituitary superfusion system and using cycloheximide to inhibit synthesis (Edwardson and Gilbert, 1975), it was found that the initial release of LH is not dependent on new protein synthesis but the response to subsequent doses is. The conclusion is that the initial action of the releasing hormone is to release pre-formed pituitary hormone and to stimulate the synthesis of hormone (or of some other protein involved in its release) to be released in response to subsequent stimulation.

C. SECRETION OF ADENOHYPOPHYSIAL HORMONES

(1) *Studied* in vitro

(a) *Pituitary cell preparations.* A number of different *in vitro* pituitary preparations have been used to study the action of hypothalamic peptides on pituitary hormone release. Some of their advantages and disadvantages have been discussed by Guillemin and Vale (1970) and Vale *et al.* (1973). The major problems have been a lack or reproducibility in the magnitude of the response and the cellular heterogeneity of pituitary systems. A method of differential separation of trophic cells would be advantageous, e.g. by sedimentation or centrifugation, by affinity chromatography, by tissue culture or by *in vivo* treatment to increase one cell type relative to others. The different systems used are now briefly described and as an illustration the results obtained with one particular action: the stimulation of rat GH release by TRH. It will be seen that different conclusions may be drawn from the use of different techniques. (i) *Pituitary pieces*, e.g. hemipituitaries, have the disadvantage of being a static system and of giving poor reproducibility (Vale *et al.*, 1973). TRH has *no effect* on GH secretion in this system. (ii) *Dispersed cell preparations* using collagenase or trypsin may give a low response unless the enzymic dispersion is carefully controlled. When used in a static incubation, again TRH is *without effect* on GH secretion. (iii) *Cells in tissue culture* give a more reproducible response and are also more uniform in type, e.g. the GH_3 cell type. TRH causes a *reduction* in GH release in this system (Hinkle *et al.*, 1974). (iv) *Perfusion systems* whether using pituitary segments (Carlson *et al.*, 1974) or dispersed cells (Lowry, 1974) have advantages over static systems. They more resemble the physiological situation and allow sequential treatments and priming effects to be studied (Edwardson and Gilbert, 1975; Osland *et al.*, 1975). TRH causes an *increase* in GH release from perfused adenohypophyses (Carlson *et al.*, 1974). The perfusion of specific areas of the pituitary, e.g. the pars intermedia (Tilders *et al.*, 1975) allows some specificity of cell type to be introduced. (v) *Portal vien* administration perhaps approaches closest to the physiological state. TRH infused in this way *increases* rat GH release (Takahara *et al.*, 1974a).

(b) *What is measured?* The choice of hormone to be measured and the nature of the assay may be critical. The disparity between the release of bioassayable GH and immunoassayable GH for instance, has caused great confusion (see Chapters 5 and 10). The root of the problem is the lack of knowledge regarding the chemical nature of the hormones as they are synthesised and released from the adenohypophysis, in particular the possible existence of prohormones, subunits and fragments which may cross-react in some but not all assays.

(c) *Factors modulating hormone secretion.* (i) The effect of hypothalamic extracts, natural peptides and synthetic hormones and analogues can be compared. (ii) Substances such as neurotransmitters, prostaglandins, cyclic nucleotides, steroid hormones, thyroid hormones can be added *in vitro* to test their interaction with the releasing hormone at the level of the pituitary. (iii) The same substances or drugs affecting their synthesis, breakdown or efficacy (see also Section IV, B above and paragraph 2 below) can be administered *in vivo* at a suitable time before the pituitary is removed. Removal of possible modulators may be done by surgical or chemical means. (iv) Immunological inhibition using antibodies to the hypothalamic peptides has been mentioned earlier (Section 1, B).

(2) *Studied* in vivo

The hypothalamic control of adenohypophysial hormone secretion has been extensively studied after *in vivo* administration both in man (clinical aspects are covered in paragraph f) and in other animals. This avoids some of the doubts and problems encountered in *in vitro* methods but adds others. Many of these, concerning the route of administration, dosage etc., are similar to those discussed in Section IV, B.

(a) *Route of administration.* Intravenous injection has been most widely used. Since these peptides are rapidly cleared and inactivated (Section VI) and the blood flow to the adenohypophysis is so small, only a tiny fraction of the injected dose is able to reach the appropriate cells. Intramuscular and subcutaneous injection delay the clearance and prolong the action of the peptides. All these have been compared for LHRH by Mortimer *et al.*, (1974a). LHRH is also active when given intranasally (London *et al.*, 1973) though much higher doses are needed. Injection into the carotid arteries (Krause, 1975) can produce higher intracerebral concentrations. TRH is even active when given orally. Presumably, being a small peptide, sufficient is absorbed unmetabolised.

(b) *Timing of administration.* Because single injections produce only a transitory response, use is frequently made of repeated or constant infusions. Constant infusions of LHRH and TRH lead to a rapid (within 15 min) attainment of a plateau concentration in the blood. The secretion of the pituitary hormones is not, however, maintained for more than about one hour but this is not necessarily due to the obvious explanation, i.e. depletion of pituitary stores. Pituitary stores of LH and TSH are not depleted though the remainder may be in a "non-readily-releasable" form. Furthermore, for TRH, the falling TSH response to repeated oral doses (Rabello *et al.*, 1974) is associated with a rise in serum T_3 and T_4 levels which are known to the reduce pituitary responsiveness to TRH.

The pituitary gonadotrophes respond best to pulsed injections of LHRH *in vivo*. This so-called "priming effect" has been very clearly demonstrated in the rat (Fink *et al.*, 1975) and sheep (Crighton *et al.*, 1975). The optimal time interval between injections is similar in the two species being about 90 mins.

(c) *Dosage*. Where the responses of two different groups are being compared the dosage of the hormone given may be important. This must not be so large as to obscure differences between the groups. Varying the dosage of TRH has been used to support the contention that TRH is the physiological PRH in man, since the sensitivity of the pituitary to TRH is the same for prolactin secretion as for TSH.

The phrase "pituitary sensitivity" is frequently misapplied to the situation where there is an *altered* response to a given dose, e.g. the exaggerated LH response to LHRH administered in the preovulatory period. Ideally it should be restricted to situations in which it is demonstrated that a *different* dose of releasing hormone is required to give the *same* response. Whilst these definitions will coincide in most situations it may not invariably be the case.

(d) *End point*. The end-point will vary from semi-quantitative distant target organ phenomena like ovulation, through other tissue and histological responses to the measurement of hormones from the target organs and from the adenohypophysis itself. Clearly the measurement of the pituitary hormones themselves removed many intermediate doubts.

The timing of sample collection should be sufficient to delineate accurately the secretion pattern. Measuring the relative potencies of analogues of LHRH is complicated by the fact that the time-course as well as the magnitude (peak-height) of the responses differ.

(e) *Modulation by other factors*. The study of the factors modulating the responsiveness of the adenohypophysis to the hypophysiotrophic hormones *in vivo* has accumulated a large recent literature which can only be briefly covered here. The general approach is to administer the peptide (bearing in mind the considerations (a)–(d) above) in conditions in which the milieu of the pituitary is changed either naturally or experimentally.

(i) Developmental, cyclical, seasonal factors may all affect pituitary hormone secretion (Chapter 3). This has been particularly studied with respect to LH and FSH secreted in response to LHRH stimulation which is dependent on the stage of sexual development and temporal (within and between days and seasonal) factors (see Chapter 8).

(ii) Hormonal milieu. The concentration of the final target hormone in body fluids is often part of the mechanism controlling the hypothalamic-adenohypophysial axis. The long-term negative feedback control of TSH secretion, for instance, is exerted predominantly at the level of thyrotrophes whose response to TRH is controlled by the level of thyroid hormones (see Chapter 7). The control of FSH also is mediated at the pituitary where LHRH action is selectively controlled by a secretion of testicular seminiferous tubules, inhibin. This has been demonstrated in man in cases in which tubular function alone is reduced, there is an exaggerated FSH response to LHRH (Zarate *et al.*, 1974; Bramble *et al.*, 1975).

Sex steroids also affect responses to LHRH. There is a biphasic effect of oestrogen in female animals with an initial reduction followed by an increase (Libertun *et al.*, 1974; Shaw *et al.*, 1975). The mechanism of the second, positive feedback, effect of oestrogen has been thoroughly studied and shown to be due physiologically to ovarian secretion of oestradiol. Thus in the rat, it is abolished by ovariectomy and restored by oestradiol (Aiyer and Fink, 1974) and in the menstrual cycle in women, the anti-oestrogen, clomiphene has been used experimentally to negate the cyclical variations in LHRH responsiveness (Wang and Yen, 1975).

The negative feedback effects of sex steroids appear to be both on the hypothalamus and the pituitary, the latter being inferred from a reduced response to administered LHRH. It could also be explained by a hypothalamic site of action however, if it is remembered that the response to LHRH is dependent on prior exposure to LHRH. Thus the response to exogenous LHRH might be reduced in situations where endogenous LHRH secretion has been lowered by steroids with no need to invoke a direct pituitary site of action.

In addition to the effect of hormones within a given hormonal axis, there are several instances of effects of other hormones and it is important not to think of each axis in isolation without cross-links. Thus TSH secretion is modulated not only by thyroid hormones but also by adrenal corticoids, oestrogens and growth hormone (see Chapter 7).

(iii) Neurotransmitters and prostaglandins. The methods of studying the effects of these have been reviewed above (Sections IV, B and V, B). They are equally applicable to the study of pituitary hormone secretion *in vivo*.

(f) *Clinical aspects*. Following the wide availability of synthetic LHRH, TRH and somatostatin an enormous amount of work has been done on their use in the diagnosis and treatment of hypothalamic-hypophysial disorders. Much of this is summarised in the relevant chapters and only a brief survey is given here.

(i) Diagnosis. Despite earlier claims made, the administration of the hypothalamic hormones has little or no place in routine clinical diagnosis. This apparently sweeping statement should not obscure the fact that a good deal of information concerning the pathophysiology of these disorders has been gained.

(ii) Treatment. The major drawbacks to releasing hormone therapy are the expense and the short duration of the therapeutic response. Some experience with LHRH therapy has been summarised by Mortimer *et al.* (1975). Therapeutic effects can be prolonged by subcutaneous, intranasal or oral administration. There is evidence that repeated doses give better responses. Another approach is to prepare longer-acting preparations with delayed absorption, e.g. protamine-zinc somatostatin or analogues, e.g. the D-ala$_6$ethylamide$_9$ nonapeptide analogue of LHRH which has a much longer time-course of action than the decapeptide (see Chapter 5) and is even active orally (Gonzalez-Barcena *et al.*, 1975).

There appears to be no therapeutic indication for TRH therapy (in thyroid disorders at least) because the thyroid hormones themselves are readily and safely available. The use of TRH is psychiatry is discussed below.

Although somatostatin reduced GH levels in acromegaly, again the effect is temporary (Besser *et al.*, 1974) and medical treatment with a dopamine agonist like ergocryptine is likely to be of more value in long-term therapy.

(iii) Hormone interactions in disease. The use of somatostatin has revealed some unexpected hormonal interactions in diseases such as acromegaly (Gomez-Pan *et al.*, 1975b). Thus both TRH and LHRH will increase GH levels in some acromegalics; in the case of TRH this can be reduced by somatostatin but not by thyroid hormones. Somatostatin also reduces the TSH response to TRH and the secretion of ACTH in Nelson's syndrome (Tyrrell *et al.*, 1975). These effects are thought to be due to a loss of receptor specificity in the pituitary cells.

D. NON-HYPOPHYSIOTROPHIC ACTIONS

Voltaire neatly epitomised the Holy Roman Empire by writing that it was "in no way Holy, nor Roman, nor an Empire". The study of the hypothalamic releasing hormones has recently given rise to comparable statements, these peptides being not necessarily hypothalamic, nor releasing, nor hormones. Although this section is relatively short it may be that in the future the non-hypothalamic, non-hypophysiotrophic roles of these peptides will assume increasing significance.

(1) *LHRH*

Of the known hypophysiotrophic peptides, there is least evidence for LHRH having non-hypophysiotropic actions. In rats, LHRH in large doses increases the lordosis response in hypophysectomised-ovariectomised female rats (Pfaff, 1973; Moss *et al.*, 1975a). This suggested a direct effect on the CNS, and there is some evidence to suggest the presence of LHRH outside the median eminence area for instance above the optic chiasma and around the third ventricle and in extra-hypothalamic areas such as the organum vasculosum and the subfornical organ (Hökfelt *et al.*, 1975f).

(2) *TRH*

(a) *Localisation in the CNS.* TRH has been localised by both radio-immunoassay (see Section III, E) and immunofluorescence (Hökfelt *et al.*, 1975c, d) throughout the mammalian CNS including the brain stem and spinal cord and the total amounts outside are greater than those within the hypothalamus.

(b) *Psychopharmacological effects.* Many CNS effects of TRH in man and other animals have been described in the past two years. They (and those of LHRH) have been summarised recently by Martin *et al.* (1975). They include an enhancement of cerebral noradrenaline turnover (Constantinidis *et al.*, 1974; Keller *et al.*, 1974) and of the hyperactivity induced by 5-hydroxytryptamine accumulation (Green and Grahame-Smith, 1974) and DOPA (Plotnikoff *et al.*, 1974) and evidence for a depressant action on certain central neurones (Renaud *et al.*, 1975). There was also evidence for an alleviation of the symptoms of depression in man but later controlled trials have failed to substantiate this (Mountjoy *et al.*, 1974).

One noteworthy feature of these experiments is the large doses of TRH used, e.g. 2–10 mg/kg in rats. Despite this, TRH has actions on, and is found throughout, the mammalian CNS and there is a strong implication that it has roles there other than control of TSH secretion (see also Chapters 5 and 7).

(c) *Is TRH a neurotransmitter?* It has been recently suggested on the basis of its wide neuronal distribution that TRH is a neurotransmitter in the CNS. The criteria for acceptance of a substance as a neurotransmitter (Eccles, 1964; Salmoiraghi *et al.*, 1965) include the following: (i) The substance is present in the nerve terminals; (ii) appropriate systems for its biosynthesis must be present; (iii) there must be mechanisms for its enzymic degradation and/or synaptic re-uptake; (iv) the substance must be released when the pre-synaptic axon is stimulated and shown to be identical with standard material; (v) the isolated substance and the standard material must have identical actions on the post-synaptic cell.

Whilst some of these criteria might seem to be unduly severe, it is clear that for TRH and the other hypothalamic peptides, evidence for their neurotransmitter role is still in its infancy.

(3) *MIF*

The tripeptide, pro-leu-glyNH_2, like TRH, enhances the behavioural effects of DOPA and 5-hydroxytryptamine in animals including man (Barbeau, 1975). It has no effect on brain amines and its effect is probably post-synaptic.

(4) *Somatostatin*

(a) *Localisation.* Like TRH, somatostatin has been located by immunofluorescence and by radioimmunoassay throughout the CNS, but unlike TRH it is also found in non-neural tissue (Arimura and Schally, 1975; Hökfelt *et al.*, 1975a, b; Polak *et al.*, 1975). Thus apart from the median eminence it is found in the ventro-medial hypothalamic nucleus, posterior pituitary, spinal ganglia and certain gastro-intestinal cells including the α_1-cells of the pancreas, and secretory cells of the upper gastrointestinal tract including the stomach. The finding of hypothalamic peptides in gastrointestinal cells supports the idea of the neuro-ectodermal origin of the APUD cells (Polak *et al.*, 1975). Substance P, a peptide found in the hypothalamus but with no known hormonal function (Leeman, 1974); has a similar distribution.

(b) *Non-hypophysiotrophic actions.* As well as reducing GH (and prolactin and TSH) secretion somatostatin has suppressive effects on non pituitary hormones (insulin, glucagon and gastrin) and on non-hormonal functions such as the secretion of gastric acid and pepsin. The physiological significance of these remains to be established. *In vitro* pancreatic systems, e.g. the isolated perfused pancreas (Efendic and Luft, 1975a) and monolayer cell culture (Fujimonto *et al.*, 1974) are now being applied to the study of the effects on insulin and glucagon release.

VI. Metabolism of the Hypothalamic Hormones

The final facet of the biochemical investigation of the hypothalamic hormones is their metabolism; the distribution and clearance from body fluids of either endogenous or exogenous peptide and the enzymic degradation mechanisms studied *in vitro*.

A. METABOLISM *IN VIVO*

(1) *Distribution and clearance of labelled hormone*

Hypothalamic peptides can be labelled with (^{14}C), (^{3}H) or (^{125}I) with retention of their biological characteristics. Such tracers can be administered to study tissue distribution and uptake and routes of clearance. It is important to establish that the radioactivity detected, whether by autoradiography or after tissue extraction and isotope quantitation, is still associated with the peptide molecule. Immuno-precipitation is a widely used technique for testing this; chromatography can be used for confirmation.

Using (^{125}I)-peptides, uptake of radioactivity into the thyroid gland is an indication that de-iodination has taken place, though Steiner (1974) has reported a large uptake of (^{3}H)-TRH into the thyroid!

After a single intravenous injection the concentration of labelled hormone in the circulation can be used to calculate the following clearance parameters: the half-time of disappearance ($T^{\frac{1}{2}}$); the volume of distribution; metabolic clearance rate (see Jeffcoate *et al.*, 1974a). The initial half-time for these peptides is of the order 3–4 minutes (Redding *et al.*, 1973a; Isidori *et al.*, 1974). Kastin *et al.* (1974a) using (^{14}C)-MIF have reported a delayed disappearance of MIF after hypophysectomy. The half-life of neurophysin has also been measured (Forsling *et al.*, 1973); the use of (^{125}I) and (^{35}S)-peptides were compared and immunoprecipitation shown to be superior to TCA precipitation which includes breakdown products.

(2) *Clearance of unlabelled hormones and analogues*

Measurement of the unlabelled hormone by either radioimmunoassay or bioassay after it has been injected intravenously can also be used to calculate the clearance parameters (Jeffcoate *et al.*, 1974a). For TRH (Bassiri and Utiger, 1973) and LHRH (Virkunnen *et al.*, 1974) the results of biological and immunological assays were identical.

This technique can be used to compare: (a) The levels reached after different routes of administration, e.g. of LHRH (Fink *et al.*, 1974); this may have therapeutic implications. (b) the clearance of analogues compared with the native hormones. The latter would not necessarily be expected to reveal differences in metabolism at the pituitary for instance since a large part of the initial $T^{\frac{1}{2}}$ reflects distribution of the injected peptide within and between the extracellular pools.

Clearance can also be measured during a constant infusion of the peptide. TRH and LHRH rapidly attain a constant plasma concentration and the metabolic clearance rate (MCR) can be calculated from the formula:

$$\text{MCR (1/day)} = \frac{\text{infusion rate } (\mu\text{g/day})}{\text{equilibrium concentration } (\mu\text{g/1})}$$

For TRH and LHRH it is about 2000 1/day in human subjects.

(3) *Urine excretion of metabolites*

After injection of either labelled or unlabelled peptides, the hormone or related metabolites may be excreted in urine. For the neurohypophysial peptides (see Boyd *et al.*, 1972) the reported amounts excreted have varied widely in different studies. This may in part be due to the assay methods used, some metabolites being immunoreactive but not biologically active and in part due to the fact, mentioned earlier, that the amount of these peptides excreted is proportional to the urine flow.

For TRH about 5–10% of an injected dose is excreted unchanged as assessed by bioassay, radioimmunoassay, chromatography and enzymic inactivation (Bassiri and Utiger, 1973; Jeffcoate and White, 1975d). About 1% of an injected dose of LHRH is excreted as metabolites that are not identical with the decapeptide chromatographically (Jeffcoate and Holland, 1975) and are only immunoreactive in some assay systems (see Jeffcoate *et al.*, 1973). The excretion of biologically-active LHRH shown by Virkunnen *et al.* (1974) is an anomalous finding; no other evidence for intact LHRH in urine has been reported. Using (^{3}H)-LHRH, Redding *et al.* (1973a) showed that whilst 70% of the (^{3}H) is excreted, none is in the form of the intact decapeptide. That LHRH is metabolised before excretion is also indicated by its slower rate of excretion—over 4 to 8 hours (Jeffcoate *et al.*, 1974a) compared with ½ hr for TRH (Bassiri and Utiger, 1973; Jeffcoate and White 1975d).

The isolation and identification of metabolites after injection of peptides with radioactive labels in different portions of the molecule should be a useful approach to the study of the molecular sites of metabolism.

(4) *Circulating metabolites*

The possible occurrence of circulating metabolites of endogenous hypothalamic hormones has been little studied, mainly because of a lack of available techniques. As mentioned above, exogenous LHRH retains the immunological, biological and chromatographic properties of the decapeptide after injection in man.

In the jugular vein of the sheep large amounts of endogenous LHRH-like immunoreactivity have been found (Section III, E) but because of the inaccessibility of the hypothalamo-pituitary system it is difficult to study this metabolic process *in vivo*.

(5) *Factors affecting metabolism* in vivo

Whilst an increasing amount is known about the factors affecting *in vitro* metabolism of the hypothalamic hormones (see Section B below) nothing is known about its possible physiological significance. With the exception of MIF (see above) there are no reports of their *in vivo* clearance being related to pituitary (or gonadal, thyroid, etc.) function.

The placental enzyme oxytocinase increases in late pregnancy but is this associated with increased oxytocin breakdown *in vivo?* There are marked changes in the *in vitro* serum TRH-degrading activity in thyroid disease in the rat, but the significance of this is unknown.

B. METABOLISM *IN VITRO*

(1) *General*

(a) *Tissue preparation.* Tissue preparations for enzyme studies *in vitro* are frequently very simple. They vary from dilution of plasma, homogenisation of tissue with or without subsequent dialysis, to fractionation by centrifugation to yield particulate (microsomes and mitochondria) and supernatant (soluble cytoplasmic) fractions. The general procedures for handling enzymes need to be observed.

(b) *End-point.* (i) The formation of constituent amino acids can be measured on an amino acid analyser. This is the crudest technique using nM amounts of peptides, e.g. somatostatin (Marks and Stern, 1975); (ii) Loss of biological activity, e.g. LHRH and oxytocin (Griffiths and Hooper, 1974), TRH (Redding and Schally 1969a; Bassiri and Utiger, 1972b); (iii) Loss of immunoreactivity, e.g. LHRH (Griffiths *et al.*, 1974), TRH (Bassiri and Utiger, 1972b; Griffiths *et al.*, 1975a; Jeffcoate and White, 1975c). In this instance it is important to confirm that the known or presumed products of the enzyme action do not cross-react in the radioimmunoassay. Because of the sensitivity of these assays, low (e.g. pM) quantities of the peptides can be used. Correlation of the loss of biological and immunological activity is important, e.g. for plasma TRH degradation (Bassiri and Utiger, 1972b); (iv) Degradation of other substrates has been used in the hope or expectation that this correlates with the breakdown of the peptide under study. Thus oxytocin and LHRH appear to share the same peptidase system (Griffiths and Hooper, 1974) and so for a time oxytocin was used as a substrate in the study of LHRH inactivating enzymes (Hooper, 1968). Kuhl and Taubert (1975a, b) have used a completely different enzyme, L-cystine arylamidase as an index of LHRH degradation but the validity of this is not established; (v) Identification of peptide metabolites can be done by chromatography. Thus, thin-layer chromatography of LHRH metabolites with standard peptide fragments followed by colorimetric localisation has suggested a metabolic split between residues 6 and 7 (Kochman *et al.*, 1975); (iv) (^{3}H)-lebelled peptides used as substrates allows easier localisation of peptide fragments on chromatography, e.g. TRH (Bauer, 1974) or electrophoresis, e.g. LHRH (Koch *et al.*, 1974).

(c) *Enzyme Kinetics* can be studied in the usual way: dependence on temperature and pH, time and enzyme concentration (e.g. Bassiri and Utiger, 1972b).

(d) *The molecular specificity of the inactivation* can be studied in a number of ways: (i) Isolation of unlabelled and labelled products as described above; (ii) Competitive inhibition by peptide analogues, e.g. for TRH (Vale *et al.*, 1971; Bauer, 1974); (iii) Effect of specific pathophysiological variations on the enzyme activity (see below).

(e) Use of enzymes to identify peptides in biological fluids has been described in Section III, D.

(2) *Hypothalamic enzymes*

These have been described for oxytocin, vasopressin, TRH, LHRH and somatostatin.

(a) *LHRH*. Rapid degradation of LHRH has been shown using all the techniques listed above. The enzymes are mainly in the supernatant fraction, show sex differences, diurnal and cyclical variations and are affected by gonadal status (e.g. Griffiths *et al.*, 1975c). They may represent one mechanism by which the negative feedback control of LHRH secretion is mediated. Loss of biological and immunological activity proceeds in parallel with production of the 1–6 and 7–10 peptides.

(b) *TRH* inactivation by the hypothalamus is not so rapid as that of LHRH. Analogue inhibition and chromatographic studies indicate that the major site of cleavage (in contrast to the serum enzyme) is between the histidyl and prolyl residues (Bauer, 1974). No indication of alterations with thyroid status have been reported.

(c) *Oxytocin* degradation by the hypothalamus has two interesting features: (i) the products, a tripeptide, MIF and a pentapeptide which acts as an MRF are themselves peptides with hormonal activity (see Chapters 5 and 12); (ii) oxytocin competes with the LHRH-degrading enzyme and the hydrolysis of the p-nitroanilides of amino acids by the aryl amidase studied by Kuhl and Taubert (1975a). TRH and vasopressin also compete with the latter.

(3) *Other brain areas*

Coincident with the localisation of hypothalamic peptides in extra-hypothalamic tissue and evidence for a wider physiological role, there has been a demonstration of enzyme degradation mechanisms. Thus in cortex, thalamus and cerebellum, there are peptidases that degrade LHRH (Griffiths *et al.*, 1975b) and TRH (Griffiths *et al.*, 1976). Several questions are raised by the presence of these enzymes: (a) Does the existence of a degradation system in a tissue imply a role for the peptide there? (b) Is the mechanism of inactivation the same as for the hypothalamic peptide? (c) Are they subject to the same alterations in pathophysiological conditions? It is known, for instance, that the sex difference for LHRH peptidase in the cerebral cortex is the same as that found in the hypothalamus (Griffiths *et al.*, 1975b).

(4) *Adenohypophysis*

Pituitary degradation of the hypothalamic peptides has been indicated by a number of studies. (a) The occurrence of metabolites of LHRH in sheep jugular vien and produced by pituitary preparations *in vitro* (Kochman *et al.*, 1975). (b) Degradation of TRH by pituitary cells in tissue culture (Hinkle and Tashjian, 1975). (c) L-cystine arylamidase activity is also present in pituitary tissue and its activity is related to the oestrous cycle and steroid therapy (Kuhl *et al.*, 1975).

(5) *Blood*

The degradation of TRH by blood was first reported by Redding and Schally (1969a, b) and has been studied intermittently since. Its enzymic character was firmly established by Bassiri and Utiger (1972b). There are important species differences; the TRH-degrading activity of rat plasma is some 5 times greater than that of human plasma for instance (White *et al.*, 1976). The use of analogues and (^{3}H)-TRH indicates that de-amidation is an important mechanism.

Redding and Schally (1969b) first suggested that this activity might be related to thyroid status. This was not confirmed in the rat by Vale *et al.* (1971) nor in the human by Jeffcoate and White (1975a). More recently it has been confirmed using a radioimmunoassay that in the rat there is an increase of TRH-degradation in hyperthyroidism and a decrease in hypothyroidism (White *et al.*, 1976). The significance of this enzyme and the changes in thyroid disease is unclear. It could certainly contribute to difference in the response to administered TRH.

(6) *Other tissues and body fluids*

Homogenates of many tissues, e.g. liver and kidney can degrade peptides like TRH and LHRH (Jeffcoate *et al.*, 1975) (see also Chapter 5). All the peptides studied are stable in urine and cerebrospinal fluid which lack the required peptidases.

VI. Conclusions

This chapter, together with Chapter 2, illustrates the enormous variety of the methods used to study the endocrine hypothalamus. Each of these has its own strength and weaknesses and gains considerably from correlations with other methods. Whilst there have been advances in methodology which have produced exciting increases in knowledge, the study of the hypothalamus is complicated by the many factors detailed in this chapter, not least of which, is the accumulating evidence for the wider physiological role of these so-called hormones.

CHAPTER 5

Chemistry of the Hypothalamic Hormones

J. SANDOW AND W. KÖNIG

I. Introduction

A. Historical Notes

Early biological evidence for the existence of hypothalamic neurohumours (or hormones) was based on the observation that following pituitary stalk section, the secretion of several pituitary hormones was altered, but the final chemical identification of these substances has been hampered by both anatomical and chemical difficulties.

This evidence for the existence of hypothalamic regulatory hormones has been reviewed by Schally *et al.* (1968a), McCann and Porter (1969), Burgus and Guillemin (1970a), Schally *et al.* (1973a) and Besser and Mortimer (1974). In fact, corticotrophin releasing factor (CRF), the first neurohumour to be partially purified, was isolated from posterior pituitary tissue, and the neurohypophysial and hypothalamic regulatory hormones share many characteristics. However, it has been considered evidence for the hypothalamic origin of a releasing factor that it is not extractable from the posterior pituitary (McCann and Porter, 1969).

The amount of hypothalamic hormones per unit of tissue is low (e.g. 12 ng of LHRH and 30 ng of TRH per sheep hypothalamus) and the starting material in terms of hypothalamic fragments (stalk-median eminence, SME) for a yield of sufficient amounts of purified natural hormone for structural analysis was of the order of 200,000 for TRH and 250,000 for LHRH.

The chemical structure of three hypothalamic regulatory hormones has been established and confirmed by synthesis (LHRH, TRH and somatostatin) and tentative structures have been published for other hypothalamic hormones (GHRH and CRF). The derivation of regulatory hormones for the control of MSH secretion from oxytocin and vasopressin suggest the existence of prohormones or hormonogens (Rudinger *et al.*, 1972) for hypothalamic hormones. There is strong evidence in favour of the peptide nature of the remaining hypothalamic principles of unknown structure and the demonstration of neurotransmitter action at the pituitary level does not exclude the existence in addition of peptides regulating e.g. prolactin secretion.

B. NOMENCLATURE

The nomenclature of the hypothalamic neurohormones has been discussed by Schally *et al.* (1968a). When extracts at different stages of purification containing a specific type of releasing or release-inhibiting activity are referred to, the term "hypothalamic factor" is used. For hypothalamic peptides of known structure the term hypothalamic regulatory hormone is generally accepted. This term covers both the effects on release and synthesis of pituitary hormones (for stimulatory hypothalamic hormones) and on release-inhibition and synthesis-inhibition (by release-inhibiting hypothalamic hormones). The nomenclature for hypothalamic hormones is still debated, synonyms for TRH being protirelin (WHO) or thyroliberin (IUPAC/IUB). For LHRH, gonadorelin (WHO), gonadoliberin (IUPAC/IUB) or GnRH have been proposed. Growth hormone release-inhibiting hormone is either abbreviated GHRIH or designated somatostatin (IUPAC/IUB). The nomenclature used here is summarised in Table 1.

C. ISOLATION PROCEDURES

The isolation procedures currently used for hypothalamic neurohormones were adapted and modified from the procedures previously developed for neurohypophysial hormones. The material to be extracted is subjected to preparatory steps, such as acid extraction in hydrochloric or acetic acid or acidified methanol. After defatting, various chromatographic procedures

Table 1. Nomenclature of hypothalamic regulatory hormones

Trivial name (IUPAC/IUB)	Other names	Current abbreviations
Vasopressin	Adiuretin	VP, ADH
Ocytocin (Oxytocin)	———	OT, OXT
Thyroliberin	Thyrotrophin releasing hormone	TRH, TRF
Gonadoliberin	Luteinising hormone releasing hormone, Gonadotrophin releasing hormone	LHRH, LH/FSHRH GnRH, LRF
Somatostatin	Somatotrophin release-inhibiting hormone	GHRIH, SRIF
Somatoliberin	Somatotrophin releasing factor	GHRH, GRF
Prolactostatin	Prolactin release-inhibiting factor	PIF, PRIH
Prolactoliberin	Prolactin releasing factor	PRF
Corticoliberin	Corticotrophin releasing factor	CRF
Melanostatin	Melanotrophin release-inhibiting factor	MIF, MSHRIF
Melanoliberin	Melanotrophin releasing factor	MRF, MSHRF

J. Biol. Chem. **250,** 3215 (1975)

Note: The term "factor" of the IUPAC/IUB recommendations has been replaced by "hormone" in case of unequivocal structural elucidation. The hypothetical folliberin or FSH-RF has not been included.

are used (gel filtration, partition chromatography or thin layer chromatography, free flow electrophoresis, high voltage electrophoresis). Countercurrent distribution is an efficient way to concentrate small amounts of starting material. The methods used for isolation may lead to concentration of substances other than peptides, as recently demonstrated by the identification of catecholamines in highly purified fractions of hypothalamic peptides.

D. ASSAY SYSTEMS FOR PURIFICATION

The purification of hypothalamic hormones is critically dependent upon sensitive and reliable assay systems. These should be specific and sufficiently sensitive to require only small amounts of purified hormones. In general, *in vitro* systems seem to be more sensitive because of their amplifier characteristics. Isolated anterior pituitaries are incubated and the content of hormone after stimulation assayed either by bioassay or by radioimmunoassay (RIA).

However, the relevance of the RIA systems used has to be ascertained by comparison with bioassays (see Chapter 4). *In vitro* systems can also be designed to distinguish between release and *de novo* synthesis of pituitary hormones. Due to the accumulation of hormones in the medium, results obtained *in vitro* are quite different from the kinetics of release and/or synthesis in the living animal.

In vivo assays are often designed to have some amplifier characteristics, but more often the initial level of circulating hormone is adjusted so as to allow maximal release or release-inhibition. Methods relying on changes in tissue content of pituitary hormones (pituitary depletion methods) are insensitive and may be misleading.

Several contaminants of partially purified extracts can induce spurious changes in responsiveness of the test preparation and thus obscure the result (e.g. metal ions, catecholamines or vasopressin) (see also Chapter 4).

E. ANALOGUES

The activity of synthetic analogues of peptide hormones has usually been studied in a limited number of assay systems, which may not cover the full activity spectrum and the changes in activity and specificity related to chemical structure should be studied in as wide a set of assay systems as possible. The receptor affinity of an analogue (receptor binding) and its ability to stimulate the receptor (intrinsic activity or efficacy) do not necessarily correlate. In the case of inhibitory analogues, the formation of a drug-receptor complex does not induce a hormonal stimulus (efficacy) and may diminish or block the response to the parent hormone by receptor occupancy. Both the analogue alone and the combination of parent hormone plus analogue have to be studied, to detect possible competitive inhibition or antagonism. In such experiments the inhibitory activity is usually expressed as the molar ratio of peptide analogue to parent hormone required for 50% inhibition, or as pA_2 ratio (Schild, 1969).

II. VASOPRESSIN AND OXYTOCIN

A. ISOLATION AND PURIFICATION

(1) *Extraction*

Vasopressin and oxytocin are extracted for chemical studies from their storage site, the neurohypophysis, rather than the hypothalamus because of their low concentration in the latter. Methods for isolation and purification have been reviewed by Acher (1966) and Ressler and Popenoe (1973).

Both hormones can be extracted from acetone-dried, defatted pituitaries by dilute acetic acid under heat (Kamm *et al.*, 1928). The protein-bound hormones are salted out by ammonium sulphate and extracted with glacial acetic acid. The hormones are then precipitated by ether, followed by petroleum ether. Extraction of the hormone-protein complex can also be carried out in the cold by sulphuric acid. Direct extraction of the hormone free of neurophysin can be achieved by homogenising the pituitary tissue with cold trichloroacetic acid. Another extraction method for vasopressin is percolation of pituitary powder with ethanol and increasing concentrations of acetic acid.

(2) *Purification of vasopressin*

Purification can be achieved by a variety of methods, e.g. counter-current distribution, partition chromatography on Sephadex, ion exchange chromatography on Amberlite IRC-50 or carboxylmethyl-cellulose, serial paper chromatography, thin layer chromatography (TLC) and high voltage electrophoresis. Portanova and Sachs (1967) used affinity chromatography, adsorbing vasopressin on bovine neurophysin coupled covalently to cellulose.

The purification of vasopressin can be followed by the rat pressor assay (Dekanski, 1952) and the rat antidiuretic assay (Jeffers *et al.*, 1942) using USP posterior pituitary powder as reference standard or the Third International Standard for oxytocin and vasopressin. Synthetic vasopressin preparations stored under suitable conditions may also be used. Vasopressin can oxidise and polymerise on storage.

(3) *Purification of oxytocin*

Purification of oxytocin by countercurrent distribution (CCD) or partition chromatography is effective because of its solubility in organic solvents. In contrast, vasopressin has a high positive charge and low solubility in organic solvents and is purified more successfully by ion exchange chromatography. In addition to CCD, oxytocin has been purified by partition chromatography on Hyflo Supercel and by gel filtration on Sephadex G-25 in 0,2 N-acetic acid or 0,1 M-pyridine acetate pH 4,9. These methods are also suitable for the purification of synthetic analogues of oxytocin.

Affinity chromatography with neurophysin as described for vasopressin is also suitable for the purification of small quantities of oxytocin-like principles from pituitaries of rare species.

B. STRUCTURE, CHEMICAL SYNTHESIS AND ANALOGUES

(1) *Structure*

The structure of oxytocin was independently elucidated by du Vigneaud *et al.* (1953a) and Tuppy (1953) by structural analysis of fragments obtained by cleavage with bromine water, partial acid hydrolysis and by chemical studies on native, performic acid oxidised and desulphurised oxytocin. The structure of arginine vasopressin (AVP) was elucidated at the same time by du Vigneaud *et al.* (1953b). The structures were subsequently confirmed by synthesis (du Vigneaud, 1956). The conformations of oxytocin (Urry and

Walter, 1971) and lysine vasopressin (LVP) (Walter *et al.*, 1972a) have been studied and a number of analogues have been developed from their findings.

(2) *Chemical synthesis*

(a) *Oxytocin and vasopressin.* In most methods, cysteine is introduced as S-benzyl-cysteine (Bodanszky and du Vigneaud, 1959). After assembly of the peptide chain, both benzyl groups are cleaved off with sodium in liquid ammonia. Oxidation is carried out by ferricyanide or oxygen in dilute solutions. In addition to the cyclic disulphide, dimers and polymers are formed at this stage and have to be removed by chromatography and CCD. Another synthesis of this type was recently described by Walti and Hope (1973).

Different types of synthesis have been developed using a preassembled disulphide ring. Muhlemann *et al.* (1972) synthesised oxytocin and deamino-oxytocin by coupling X-Tyr-Ile-Gln-Asn-Cys-OH and H-Pro-Leu-GlyNH_2 (X = Boc-Cys or β-mercaptopropionic acid) by the DCCI-HOBt method (DCCI = dicyclohexylcarbodiimide, HOBt = 1-hydroxybenzotriazole). Jones *et al.* (1973) similarly used a preassembled disulphide ring for synthesis of AVP by reacting BOC-Cys-Tyr-Phe-Gln-Asn-Cys-Pro-ONSu with H-Arg-Gly-NH_2 (ONSu = N-hydroxysuccinimide ester). After removal of the BOC group and purification by ion exchange chromatography a product with 454 IU/mg of anti-diuretic activity in the ethanol-water loaded rat was obtained.

A new method (catalytic hydrogenation over palladium in liquid ammonia) for the selective cleavage of the benzyloxy-carbonyl-group (Z-group) in the presence of the S-benzyl-group has been successfully used in an oxytocin synthesis of Kuromizu and Meienhofer (1974).

(b) *Labelled peptides.* Synthesis of oxytocin deuterated in the half-cysteine positions 1 or 6 has been reported by Spatola *et al.* (1974). Solid phase synthesis was found to be superior to the solution method. [DL-($\alpha.\beta.\beta^2H_3$)Cys^1] oxytocin and [DL-($\alpha.\beta.\beta^2H_3$)Cys^6]-oxytocin were resolved into their diastereomers by partition chromatography. After purification of the peptides by gel filtration the L-isomers had the same activity as oxytocin, while the D-isomers had lower activity.

Synthesis of labelled vasopressin has been described by Neer (1973). Tryptic digestion of LVP resulted in cleavage of the Lys^8-Gly^9-bond. The resulting 9-des-Gly-NH_2-LVP was coupled with ^{14}C-Gly-NH_2 to give [9-^{14}C-GlyNH_2] LVP. The labelled hormone had full biological activity and specific activity was 100 mCi/mg. However, in this method racemisation of Lys^8 cannot be excluded. Synthesis of ^{3}H-labelled AVP was described by Holton *et al.* (1973) by a solid phase method. The end product was purified by gel filtration on Sephadex G-25. The labelled hormone had full biological activity and specific activity was about 1 mCi/mg.

(3) *Analogues*

The pharmacology of synthetic analogues of oxytocin and vasopressin has been reviewed recently by Rudinger *et al.* (1972) and Sawyer and Manning (1973). The characteristic activity spectrum of the natural neurohypophysial hormones can be modified by chemical substitution (Table 2). Vasotocin, the hybrid of vasopressin and oxytocin can be considered as the common chemical denominator of all analogues. While substitution of Ile^3 by Phe^3 leads to increase of pressor activity and decrease of uterotonic activity, the reverse is seen with substitution of Arg^8 by Leu^8. This is illustrated by natural evolution in oxytocin and vasopressin (see Chapter 6).

For the hormonal activity of oxytocin and vasopressin analogues the moieties Ile^3, the amide in Asp^5, the C-terminal glycinamide and the ring structure are of special importance. The disulphide bridge is not essential and can be readily substituted by a carbon or selenium atom. In fact, 1-carba-oxytocin is more active than oxytocin but further substitution (as in dicarba-oxytocin) leads to a decrease in activity because of a conformational change due to the smaller ring structure. Carba-analogues are to some extent protected against inactivation by reduction of the S-S bond. Oxytocic activity can be further enhanced by removal of the N-terminal amino group, while pressor activity is reduced. Desamino-1-carba-oxytocin and desamino-1-seleno-oxytocin are about 3 times more potent than natural oxytocin (uterotonic activity).

Substitution in position 2 or 4 can also increase oxytocic activity; [Thr^4] oxytocin and [p-fluor-Phe^2] oxytocin are 2–3 times more active but the corresponding desamino compounds are less active. Selective substitution of position 4 can change the ratio of milk ejection: uterotonic activity in favour of milk ejection. Thus [Val^4] oxytocin and [Asn^4] oxytocin show a relative increase in milk ejecting activity and both analogues have only minor pressor and antidiuretic activity. Further elongation of the side chain in this position, e.g. [Leu^4] oxytocin, produces a vasopressin antagonist with diuretic activity. Substitution of sulphur for oxygen in the C-terminal amide of desamino-oxytocin reduces the oxytocic and avian vasodepressor activity.

The ring structures in oxytocin and vasopressin (tocinamide and pressinamide) retain some uterotonic and antidiuretic activity, but no pressor activity. Tocinamide, [Ser^4] tocinamide and [Ser^4] tocinoic acid have residual milk ejection activity, but are otherwise inactive.

A structure-activity relationship between the conformation of oxytocin and biological efficacy has been proposed by Walter *et al.* (1971). Three types of structural modification were considered: (i) changes in the stabilisation of the peptide-backbone (affecting both affinity and efficacy), (ii) alteration in the steric environment and charge distribution of limited surface areas (affecting affinity and efficacy differentially), and (iii) changes in the

Table 2. Changes in activity spectrum of synthetic analogues of oxytocin and vasopressin

Substitution		Enhancement of
Arg[8]-Vasotocin H-Cys-Tyr-Ile-Gln-Asn-Cys-Pro-Arg-Gly-NH_2 → Arg[8]-Vasopressin H-Cys-Tyr-Phe-Gln-Asn-Cys-Pro-Arg-Gly-NH_2	[D-Arg[8]]	antidiuresis: pressor activity
	[Orn[8]]	pressor activity: antidiuresis
	[Val[4]]	antidiuresis activity
	[Ile[3]]	pressor activity: antidiuresis
	[Phe[2]] combined with [Orn[8]]	pressor activity: antidiuresis
	[Tyr(Me)[2]]	antidiuresis: pressor activity
	Des-amino	antidiuresis activity
Arg[8]-Vasotocin → Oxytocin H-Cys-Tyr-Ile-Gln-Asn-Cys-Pro-Leu-Gly-NH_2	[Val[4]], [Asn[4]]	milk ejection
	[Thr[4]]	oxytocic activity
	[Leu[4]]	(with antivasopressin activity)
	[p-Fluor-Phe[2]]	oxytocic activity
	[Tyr(Me)[2]]	inhibitor of OT and VP
	1-desamino	oxytocic activity
	1-carbamoyl combined with [Tyr(Me)[2]]	inhibitor of OT and VP

steric and electronic requirements of moieties comprising the active surface (affecting efficacy without altering affinity).

Competitive inhibitors of oxytocin have been obtained in various ways. Substitution of residues 1, 2 or 4 have resulted in partial agonists or inhibitors. The [Leu^4]-analogues of oxytocin and mesotocin act as antagonists of the natural hormones in kidney and amphibian bladder. [Ile^2, Leu^4] oxytocin and [Phe^2, Leu^4] oxytocin are weak inhibitors. Other inhibitors of oxytocin action are [3-Iodo-Tyr^2] oxytocin, which inhibits the uterotonic effect of oxytocin and a group of analogues, substituted in residue 1 (Chan *et al.*, 1974). An irreversible inhibitor, [N-bromoacetyl]oxytocin (Walter *et al.*, 1972b) completely blocks the stimulation of adenylcyclase.

Much synthetic effort has been directed towards obtaining a split between the pressor and antidiuretic activities of vasopressin. Antidiuretic activity can be enhanced by modifications in position 1, 2, 4 and 8. 1-des-amino-compounds have increased antidiuretic activity and this is further enhanced by increased lipophilicity in position 4, pressor activity being reduced. Substitution by Val^4 is most efficient (Manning *et al.*, 1973). Substitution of Arg^8 by D-Arg^8 leads to further potentiation of antidiuretic activity and 1-desamino [Val^4, D-Arg^8] vasopressin (DDAVP) is the most active antidiuretic peptide so far obtained and is almost devoid of pressor activity. Antidiuretic activity is also increased by alkylation of the phenolic hydroxyl group, e.g. [Tyr(Me)^2] LVP. Conversely, substitution of Arg^8 by Orn^8 as well as Phe^3 by Ile^3 results in an increased pressor: antidiuretic ratio which is further elevated by the combination of Orn^8 with Phe^2. [Phe^2, Ile^3, Orn^8] vasopressin, incorporating all three modifications has a pressor: antidiuretic ratio of 220.

C. BIOSYNTHESIS AND SECRETION

(1) *Vasopressin*

The biosynthesis of vasopressin by the neurosecretory neurones of the nucleus supraopticus and paraventricularis has been reviewed by Sachs *et al.* (1969) and Lederis and Jarhysena (1970). Vasopressin-synthesising neurones are different from oxytocin-synthesising neurones. Vasopressin is initially synthesised as part of a larger molecule (prohormone or hormonogen) and the final formation of the peptide occurs in the neurosecretory granules. Identification of vasopressin in its precursor form by immunohistochemistry has been reported (Swaab and Pool, 1975). The rate of hormone biosynthesis *in vivo* has been estimated to be 10^{-4}–10^{-6} μ moles/h and requires a ribosomal mechanism, not being seen in cell-free systems. The release of vasopressin from the precursor hormone within the neurosecretory granule is observed during axonal transport to the neurohypophysis which proceeds at a rate of several mm/h.

(2) *Neurophysins*

The biosynthesis of the transport proteins for vasopressin and oxytocin, the neurophysins, also involves the formation of precursor molecules (Sachs *et al.*, 1969), and this process may be coordinated with the biosynthesis of

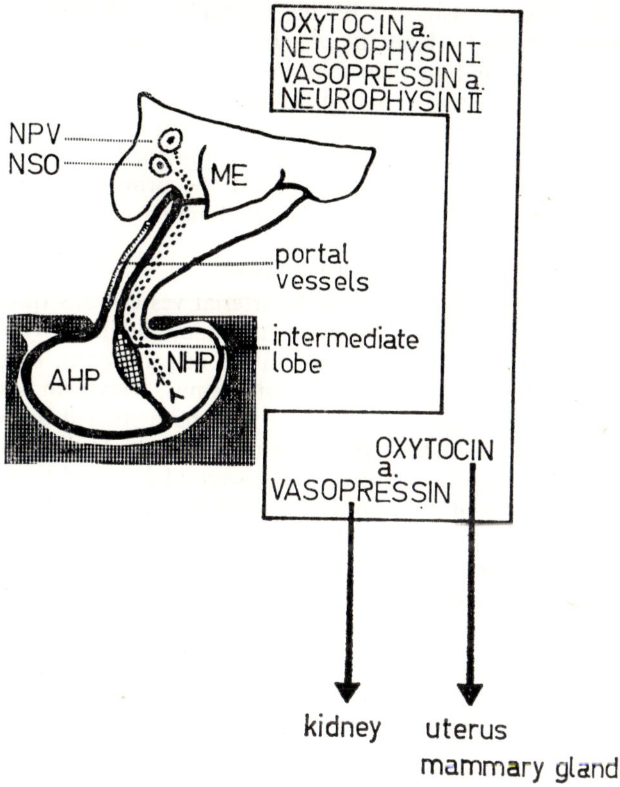

Fig. 1. Hypothalamic neurosecretion. Oxytocin and vasopressin are synthesised in hypothalamic nuclei together with their transport proteins neurophysin I and II. They migrate to the posterior pituitary by axonal transport and are released into the bloodstream upon appropriate stimulation.

NPV = nucleus paraventricularis. NSO = nucleus supraopticus. ME = median eminence. AHP = adenohypophysis. NHP = neurohypophysis.

vasopressin. The neurophysins exist in two major components (I and II) and a minor component (III) (Wuu *et al.*, 1971), They have a molecular weight of about 10,000, consisting of 90–100 amino acids, Neurophysins I and II will bind both vasopressin and oxytocin equally well *in vitro*, but in the hypothalamus oxytocin is associated with neurophysin I and vasopressin with neurophysin II. It is not known whether the neurophysins serve only as transport and storage proteins or exert independent hormonal actions. The pool of readily releasable vasopressin is only about 15% of the total content, (Sachs and Haller, 1968) indicating different types of neurophysin complexes.

The binding characteristics of bovine neurophysin II have been studied with analogues of LVP (Kluh *et al.*, 1973). Analogues extended at the N-terminus do not bind to neurophysins, while analogues deleted at the C-terminus were consistently bound, including des-Lys^8 Gly^9-vasopressin. The deamino^1-analogue of oxytocin does not form a complex with neurophysin. Similarly, hydroxy analogues with a modified N-terminus do not bind to neurophysin II, demonstrating that the primary amino group in oxytocin is necessary for the affinity for the binding protein.

D. RECEPTOR BINDING AND MODE OF ACTION

Vasopressin receptors are localised in arterial vessel walls (probably also in the capillaries), in the distal tubules of the kidney and in the hypothalamus. Oxytocin receptors have been localised in the target organs, kidney, uterus and mammary gland and in the hypothalamus. Oxytocin receptors in particulate fractions of the human uterus have been studied by Soloff *et al.* (1974). The receptor bound ^{3}H-oxytocin which could be displaced by unlabelled peptide. Analogues of oxytocin were also bound to a significant extent, whereas LVP had a low affinity. Oxytocin receptors in the mammary gland have been identified by Egan and Livingstone (1973). The requirements for receptor binding in the rat uterus have been studied in detail by Krejci *et al.* (1973), who found that the ring structure of oxytocin, tocinoic acid, is not sufficient for the formation of a ternary complex between the hormone, the binding site of the receptor and Mg^{++} ions. The group involved in magnesium potentiation appears to be located in the terminal tripeptide portion of the oxytocin molecule.

The mechanism of action of oxytocin and vasopressin is related to the adenylcyclase system and requires the presence of Ca^{++} and Mg^{++} ions. Isolated receptors from homogenates of renal medulla were stimulated by AVP and LVP and both hormones produced maximal but not additive stimulation of adenylcyclase activity. A single population of receptors was suggested (Steward, 1974). Ferguson and Twite (1974) have demonstrated an effect of vasopressin on toad bladder membrane proteins. The phosphorylation of protein D, identified by gel electrophoresis, was increased.

There appears to be an interaction between prostaglandins and the neurohypophysial hormones. For instance, the physiological response of the uterus to oxytocin seems to depend on the presence of prostaglandins. After oxytocin stimulation, isolated uteri release prostaglandin-like activity into the bath fluid. During the oestrous cycle of the rat, contractility can be stimulated by oxytocin, angiotensin II or $PGF_{2\alpha}$. The prostaglandin antagonist polyphloretin phosphate reduced uterine sensitivity to all three hormones. The uterotonic effect of oxytocin is increased, and the activity of inhibitory analogues diminished, by Mg^{++} (Chan *et al.*, 1974). The oxytocin analogue (1-L-penicillamine-oxytocin) inhibits the uterotonic effect of oxytocin, but not the stimulation by prostaglandin.

There is also evidence for the participation of a prostaglandin receptor in the effect of vasopressin on hydro-osmotic regulation in the mammalian kidney (Berl *et al.*, 1974).

E. INACTIVATION, METABOLISM AND EXCRETION

Neurohypophysial hormones and hypothalamic releasing hormones share certain features of kinetics and metabolism due to chemical similarity and evolutionary analogies. The subject has been reviewed by Lauson (1974).

(1) *Plasma inactivation*

Plasma inactivation of vasopressin and oxytocin is slow. Binding to plasma proteins does not significantly contribute to the half-life; only a small fraction of vasopressin is bound to plasma proteins (Cohn fractions II and III). Vasopressin and oxytocin are cleared from plasma by an early rapid phase ($T\frac{1}{2} = 1$–2 min) and a later prolonged phase. The distribution volume is larger than the plasma volume. The large initial distribution volume can be accounted for if muscle tissue were a major early repository. In normal human subjects, the plasma clearance of AVP is 11 ml kg^{-1} min^{-1}; that of oxytocin is similar.

(2) *Organ inactivation*

For the localisation of the inactivating organ, the method of organ exclusion has been extensively used. Mainly, the kidneys (43% of total clearance) and the liver (50%) contribute to the irreversible removal of vasopressin and oxytocin in rats. When physiological quantities of the hormone are infused, hepatic extraction appears to be lower than after pharmacological doses. The lungs and muscle do not contribute to removal of vasopressin and oxytocin. About 6 to 30% of exogenous vasopressin is excreted in the urine in a biologically active form. In patients with nephrogenic diabetes insipidus this fraction is increased (55%), indicating impaired binding to renal receptors. Urinary excretion of oxytocin is much lower than of vasopressin (0·2–0·3% of the dose).

Oxytocin is not only inactivated by the liver and kidneys, but also by the mammary gland. Half-life in lactating rats is shorter and the steady state total clearance higher than in control animals.

(3) *Isolated organs and tissues*

In studies with isolated, perfused organs, both the liver and kidneys extract vasopressin. Hepatic inactivation is not dependent on oxidative metabolism and non-viable livers extract the hormone at the same rate as intact livers. Studies with tissue slices, homogenates or extracts, confirm the importance of the kidneys and liver for inactivation of oxytocin and vasopressin. In addition, significant inactivation is produced by extracts of spleen, pancreas and the small intestine (mainly by aminopeptidases). Oxytocin is also inactivated by homogenates of the mammary gland and uterus. The inactivating capability of the pregnant human uterus is located in the soluble fractions of homogenates. In the placenta, two separate enzyme systems are found for the degradation of neurohypophysial hormones, in the soluble fraction and in the mitochondrial and microsomal fraction. Extracts of the ovary and of striated muscle can also inactivate oxytocin. Enzymes in the nervous system have been found capable of destroying both vasopressin and oxytocin. Enzyme activity is highest in the hypothalamus, but for vasopressin is also demonstrable in brain cortex.

(4) *Metabolism*

Some of the enzyme systems participating in the inactivation of oxytocin (Fig. 2) and vasopressin are identical. Studies with synthetic analogues have been of particular value in the elucidation of the structural requirements of enzyme inactivation.

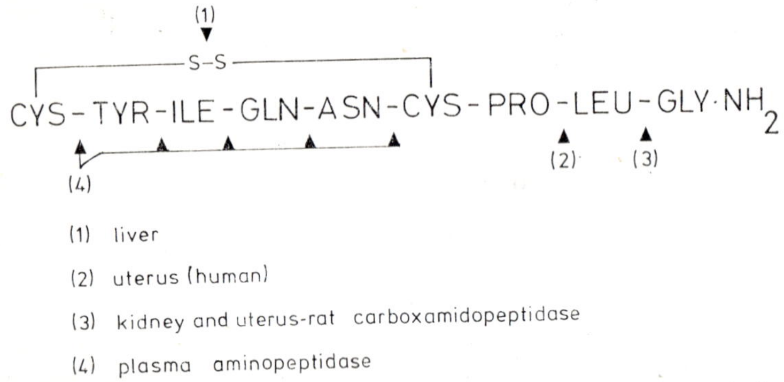

Fig. 2. Metabolic inactivation of oxytocin: the sites of enzymatic cleavage are indicated by arrows. (4) The initial split between Cys[1] and Tyr[2] is followed by successive breakdown of the peptide chain.

The enzyme system in pregnancy plasma (oxytocinase) cleaves between Cys[1] and Tyr[2] both in oxytocin and vasopressin. This enzyme cannot degrade deamino-oxytocin. A further action of oxytocinase is to cleave between

Tyr^2 and Ile^3 in oxytocin and between Tyr^2 and Phe^3 in vasopressin. Inactivation stops at the tetrapeptide stage. Oxytocinase is also capable of degrading angiotensin II and angiotensin II-amide. Other enzyme systems have been characterised; one cleaving the Leu^8-$GlyNH_2^9$ bond. Inactivation of oxytocin by this enzyme from rat uterus proceeds only under nitrogen, and not under oxygen or air.

The main mechanism for liver inactivation appears to be a two-stage system: firstly, reductive cleavage of the disulphide bond and secondly, aminopeptidase degradation. Identity of the liver enzyme inactivating oxytocin with leucine aminopeptidase from the kidney has been suggested. *In vivo*, reduction of the disulphide ring does not prevent further degradation. The carba-analogues of oxytocin (where CH_2-CH_2 is substituted for the S-S ring) are inactivated at the same rate as cyclic oxytocin. Similarly, 1-desamino-oxytocin is inactivated at the same rate as oxytocin, although the lack of the N-terminal amino group affords protection against attack by aminopeptidases.

There are trypsin-like and chymotrypsin-like peptides in many tissues, the trypsin-type cleaving the peptide bond between Arg^8 (or Lys^8) and the C-terminal $GlyNH_2$ in vasopressin and the chymotrypsin-type cleaving the corresponding bond between Leu^8 and $GlyNH_2$ in oxytocin. A purified enzyme from kidney homogenates also inactivates oxytocin by cleaving the Leu^8- $GlyNH_2$ bond. This type of enzyme is widely distributed (spleen, lung, heart, stomach, small intestine, testes and muscle). Highest activity is found in liver and spleen. Such enzyme systems may contribute both to the inactivation of neurohypophysial hormones, peptides involved in MSH regulation (MIF, MRF) and some other releasing hormones (LHRH, somatostatin) having a limited substrate specificity. Inactivation of neurohypophysial hormones appears not to involve deamidation of the C-terminal $GlyNH_2$.

(5) *Organ distribution*

The distribution of ^{125}I-labelled oxytocin in male rats was studied by Ågmo (1974). Oxytocin was preferentially accumulated in the adeno- and neurohypophysis. No accumulation in other organs of male rats was seen. The uptake of ^3H-oxytocin into different areas of the brain was studied by Stibal *et al.* (1974); they also found higher concentrations in the anterior and posterior pituitary, with lesser activity in the olfactory region, motor cortex and tegmentum.

F. COMPARATIVE ASPECTS OF STRUCTURE AND FUNCTION

The comparative endocrinology of neurohypophysial hormones has been reviewed by Acher (1974a, b) and is covered in Chapter 6. There is an evolutionary pedigree of hormones with considerable structural uniformity. Very efficient purification techniques had to be developed for the minute amounts of hormone of rare species. The ancestral molecules appear to be vasotocin for the vasopressin line of hormones and isotocin for the oxytocin line. All these hormones contain a pentapeptide ring, joined by a disulphide bond between two half cysteine residues at position 1 and 6. The physiological and pharmacological activity spectrum is qualitatively similar, with quantitative differences. Glumitocin was identified in hypotremata (rays) and valitocin as well as aspargtocin have been isolated in pleurotremata (sharks). Arginine vasotocin is also found in the pineal gland of several species (Cheeseman and Farris, 1970). After treatment of mice with arginine vasotocin (AVT), the development of reproductive organs is inhibited (Vaughan *et al.*, 1974). AVT was also reported to reverse the effects of pinealectomy by inhibiting synthesis and/or release of prolactin inhibiting factor (Pavel *et al.*, 1975).

III. TRH (Thyroliberin)

A. Isolation and Purification

A purified preparation of hypothalamic TSH-releasing factor was first obtained by Guillemin *et al.* (1962). After extraction with 2 N acetic acid and preliminary purification by gel filtration on Sephadex G-25, further purification was by ion exchange chromatography on carboxy-methylcellulose (CMC). Guillemin *et al.* (1965) obtained 400 μg of TRH from 70,000 sheep hypothalami. Schally *et al.* (1969) obtained 900 μg TRH from 20,000 porcine hypothalami again using acetic acid for extraction followed by phenol extraction, CMC chromatography and CCD and free-flow electrophoresis. Final purification included partition chromatography, adsorption chromatography on charcoal, analytical gel filtration and finally paper chromatography. Higher yields were subsequently obtained, e.g. 7·2 mg of pure TRH was isolated from 265,000 porcine hypothalami. The extraction yielded a biologically active material which was not firmly bound to CM-Sephadex, in contrast to acid extracted material. It was speculated, that the changes in adsorptive behaviour during the early stages of purification of TRH may have resulted from binding of the low molecular weight material to a larger molecule whose size and/or charge governed the behaviour of the complex. This observation is in accordance with the findings of Tsuji *et al.* (1968), who observed retention of TRH on anion exchange columns. The yield of Burgus and Guillemin (1970b) was 1 mg TRH from 270,000 sheep hypothalami.

B. Structure, Chemical Synthesis and Analogues

(1) *Structural analysis*

The structure of TRH of ovine and porcine origins were reported in 1969. In highly purified preparations, the amino acids Glu, His, Pro were found in equimolar ratios and accounted for 81% of the preparation (Schally *et al.*, 1969; Burgus and Guillemin, 1970b). With this information the tripeptide structure of TRH was synthesised by Bøler *et al.* (1969), and confirmed by structural analysis of natural TRH (Nair *et al.*, 1970) after enzymatic degradation, as well as by infrared spectroscopy, nuclear

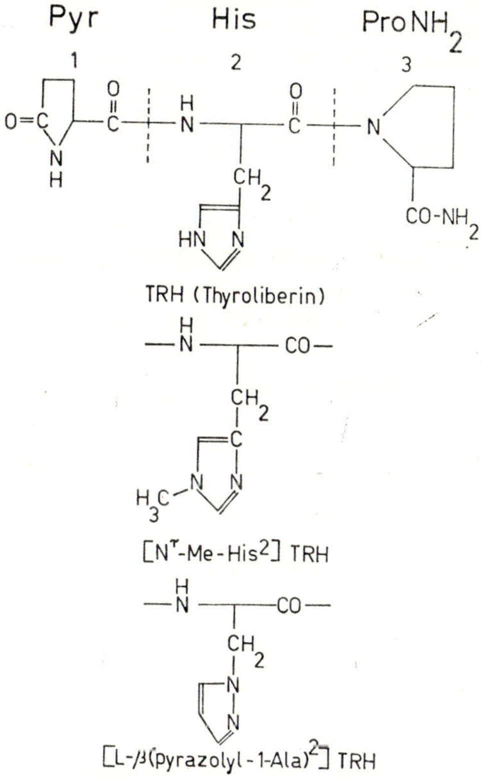

Fig. 3. Structure of TRH (thyroliberin) and two analogues with high biological activity: Pyr = Pyroglutamic acid. Substitution of His² by N^{τ}Me-HIS² or L-β-(pyrazolyl-l-Ala)² results in increased biological activity.

magnetic resonance and mass spectroscopy. Pyroglutamyl peptidase (Doolittle, 1969) is able to destroy the biological activity of TRH. This enzyme appears to be specific for the cleavage of pyroglutamic acid from Pyr-terminal peptides and has an important function in the inactivation of TRH and LHRH.

TRH is a tripeptide with protected N-and C-terminal amino acids (Fig. 3) which explains its resistance to most proteolytic enzymes. The biological activity of highly purified ovine TRH and synthetic TRH was found to be identical *in vivo* and *in vitro* (Burgus *et al.*, 1970). The chromatographic identity of natural and synthetic TRH was shown by Bøler *et al.* (1969) by comparison in 17 different TLC systems.

(2) *Synthesis*

Chemical synthesis of TRH can be achieved by a large variety of methods. For a classical synthesis, some possible approaches are:

(a) Incremental chain elongation or fragment condensation (e.g. Pyr-His-OH + H-Pro-NH_2). Only those coupling methods which do not lead to considerable racemisation are useful, e.g. the azide method (Gillessen *et al.*, 1971).

(b) Of considerable advantage are procedures in which the tripeptide can be crystallised in its protected form without the necessity for extensive purification. 4.4′-dimethoxybenzhydryl (Mbh) has been used as a protective group with lipophilic and crystallisation promoting characteristics in different methods of TRH synthesis (König and Geiger, 1970), e.g.

(i) Z-Gln (Mbh)-His-Pro-NH_2
(ii) Z-Gln (Mbh)-His-Pro-NH-Mbh
(iii) Z-Gln-His-Pro-NH-Mbh

$$\xrightarrow[\text{anisole}]{CF_3COOH} \text{Pyr-His-Pro-NH}_2$$

The glutaminyl compound is cyclised to the pyroglutamyl compound during cleavage of the protective groups in boiling trifluoroacetic. TRH cannot be crystallised as the free base and is stable and nonhygroscopic after freeze drying. The various salts of TRH, however, are consistently hygroscopic. Hatanaka *et al.*, (1974) have found that TRH can be crystallised as its tartrate from aqueous ethanol.

(c) Direct synthesis of the pyroglutamyl peptide or synthesis of glutaminyl peptide with consecutive cyclisation. Cyclisation of the glutaminyl peptide is easily achieved, e.g. by glacial acetic acid, trifluoroacetic acid as above or mixture of water and organic solvents.

Numerous methods have been developed for *solid phase synthesis*. Coupling can be by carbodiimide, mixed anhydride and activated esters. Racemisation during coupling of BOC-Cys (Bzl)-OH with prolineamide can be prevented by using the DCCI/HOBt method (König and Geiger, 1970). Pietta *et al.* (1974) have used the polybenzhydryl resin, where TRH can be cleaved off by acid treatment. Gross *et al.* (1973) used dehydro-alanine solid phase synthesis to form TRH amide which is released from the Boc-Pyr-His-Pro-dehydroalanine polymer after treatment with 1 N HCl/acetic acid. Glass *et al.* (1972) coupled Boc-His to the polymer by a N^{Im}-dinitrophenyl-Gly bridge to the imidazolyl ring of histidine. After assembly of the peptide by elongation at the C- and N-terminus, TRH was cleaved off the polymer by 2-mercaptoethanol.

Table 3. Analogues of thyroliberin (TRH): biological activity (TSH-release).

Structure	Biological activity (TRH = 100)
Pyr-His-Pro-NH_2 (= TRH)	100
D-Pyr-D-His-D-Pro-NH_2	inactive
———His-Pro-NH_2	0·1
Pyr-D-His-Pro-NH_2	2–5
Pyr-His-D-Pro-NH_2	0·1
———Pro-OH	0·02
———OMe	20
———NH-NH_2	14
———NH-Me	20
———NH-Et	14
———$N(Me)_2$	0·5
———$N(Et)_2$	0·05
———NH-CH_2-CH^2-OH	16
———piperidine	0·2
———Gly-NH_2	35
———Ala-NH_2	0·5
Pyr-His pyrrolidine	0·2
———prolinol	1·2
———NH_2	0·003
———OMe	inactive
———Gly-NH_2	<0·02
——— Ala-NH_2	0·1
———Leu-NH_2	0·04
———Val-NH_2	0·1
———Abu-NH_2	0·1
———Ile-OMe	inactive
———Ile-NH_2	inactive
———Thr-NH_2	inactive
———Met-NH_2	inactive
———Phe-NH_2	inactive
———Trp-NH_2	<0·02
Pyr-Gly-Pro-OMe	inactive
————NH_2	inactive
Pyr-Ala-Pro-NH_2	inactive
Pyr-Leu-Pro-OMe	inactive
————NH_2	0·2
Pyr-Met-Pro-NH_2	1·0
Pyr-N^{π} MeHis-Pro-NH_2	0·04
Pyr-N^{τ} MeHis-Pro-NH_2	800
Pyr-N^{π} MeHis-OMe	<0·0005

Table 3. (*continued*)

Structure	Biological activity (TRH = 100)
Pyr-N^{τ} MeHis-OMe	0·02
Pyr-Phe-Pro-OMe	inactive
——————NH_2	10
Pyr-Phe-3-Hyp-NH_2	0·14
Pyr-Tyr-Pro-NH_2	0·084
Pyr-Trp-Pro-NH_2	inactive
Pyr-Thi-Pro-NH_2	~0·2
Pyr-D-Thi-Pro-NH_2	~0·2
Pyr-Arg-Pro-NH_2	0·05
Pyr-Lys-Pro-NH_2	0·02
Pyr-Orn-Pro-NH_2	0·025
Pyr-His (Bzl)-Pro-NH_2	0·2
Pyr-β-3-pyrazolyl-Ala-Pro-NH_2	5
Acetyl-Glu-His-Pro-NH_2	inactive
H-Glu-His-Pro-OH	inactive
H-Glu-His-Pro-NH_2	5
N-Me-Pyr-His-Pro-NH_2	1·2
H-Pro-His-Pro-OH	inactive
———————NH_2	0·01
Thiophenyl-CO-His-Pro-NH_2	0·2
Furanyl-CO-His-Pro-NH_2	0·01
Cyclopentyl-CO-His-Pro-NH_2	<0·01
Cyclobutyl-CO-His-Pro-NH_2	0·016
H-Gly-His-Pro-NH_2	inactive
H-Glu-Pro-His-OH	inactive

(3) *Analogues*

Many analogues of TRH have been synthesised (Table 3) but few of them have significant biological activity (in this context it should be kept in mind that TSH release was mostly used for evaluation, prolactin was not always determined).

Substitution of the N-terminal usually leads to a marked decrease in activity. The replacement of $ProNH_2$ may retain some activity (e.g. Pyr-His-ProOMe with 20% TRH activity).

Only two analogues of higher activity than the parent molecule have been found (Fig. 3): [$N^{\tau}MeHis^2$]-TRH with 8x the activity of synthetic TRH and [L-β-(pyrazolyl-1)-Ala^2] TRH with 1·5x (Coy *et al.*, 1975). It is interesting, that the homologue of the former, [$N^{\pi}MeHis^2$] TRH has only 0·04% TRH activity; this confirms the structural model proposed for the tertiary structure of TRH (Blagdon *et al.*, 1973), where the π-hydrogen of the imidazole ring is linked to the α-NH of histidine by hydrogen bonding. A methyl group in this moiety will change the tertiary structure completely and abolish biological activity. In contrast, a methyl group replacing the τ-hydrogen does not interfere with the tertiary structure.

Some peptides have been described with competitive antagonism to TRH (Virkuunen *et al.*, 1973). Cyclopentylcarbonyl-histidyl-pyrrolidine inhibits TRH-induced TSH release at a molar ratio of 1 : 30 (TRH : antagonist). Apparently, this peptide binds to the pituitary TRH receptors and makes them refractory to TRH.

C. BIOSYNTHESIS AND SECRETION

(1) *Biosynthesis*

Biosynthesis of TRH *in vitro* has been studied by Mitnick and Reichlin (1972), Reichlin and Mitnick (1973b) and McKelvy and Grimm-Jørgensen (1975). TRH appears to be synthesised *de novo* by a non-ribosomal enzymatic mechanism rather than split off from a prohormone: it has been shown by an increase of bioassayable TRH as well as by incorporation of labelled precursors. There is evidence in brain tissue for an enzyme catalysing cyclisation of amino-terminal glutamic acid moieties of synthetic substrates (Ortowski *et al.*, 1969) which may explain why there is incorporation of labelled glutamic acid into TRH, while pyroglutamic acid is not incorporated. TRH synthetase activity is localised in specific brain areas and requires ATP and Mg^{++} ions; it is stimulated by thyroxine and it appears that thyroxine exerts a positive feedback on hypothalamic TRH synthesis, in contrast to its negative feedback effect on TSH-secretion (see Chapter 7).

(2) *Secretion*

The possible factors controlling TRH secretion are covered in Chapter 7. The detection limit of TRH assays does not allow measurement of levels of TRH in plasma or assessment of fluctuations in portal plasma of animals. TRH is secreted in an oscillatory pattern by isolated hypothalami of the newt (McKelvy and Grimm-Jørgensen, 1975), so that TSH secretion by TRH may be frequency modulated. TSH secretion is also regulated by variable pituitary sensitivity.

D. RECEPTOR BINDING AND MODE OF ACTION

TRH is active by various routes, parenterally as well as nasally and orally (Redding and Schally, 1970; Gordon *et al.*, 1972). Nasal absorption is 5% and oral absorption 1% of the parenteral dose. TSH release is seen in many species (see Section F). TRH is relatively non-toxic, the LD_{50} being about 1·5 g/kg in the dog as compared to a therapeutic dose of 10 μg/kg in the human.

Specific pituitary receptors for TRH have been widely studied *in vitro* (e.g. Grant *et al.*, 1972; Barden and Labrie, 1973a; Wilber and Seibel, 1973a; Eddy *et al.*, 1973). The pituitary TRH receptor is modulated by circulating thyroid hormone levels (Wilber and Seibel, 1973a). Receptor binding is increased in hypothyroid rats and diminished in hyperthyroid rats. Synthetic analogues of TRH fail to bind to the receptor, and the lack of biological activity may be due to a reduction in binding affinity. Prostaglandin (PGE_1) does not bind to this receptor, although it is capable of stimulating TSH secretion in interaction with TRH. The separate receptors for TRH and prostaglandins seem to function in a complementary or cooperative manner.

Table 4. Toxicity enhancement by TRH of drugs acting on the central nervous system

Drug (i.p.)	LD_{50} (95% confidence limits) drug + saline	drug + TRH 25 mg/kg s.c.	Increase in Toxicity (× —fold)
Neuroleptic drugs:			
Chlorpromazine	227 (205–251)	6 (4 – 9)	38·5
Thioridazine	455 (420–493)	28 (14 – 54)	18·2
Haloperidol	42 (33– 52)	15 (11 – 20)	2·8
Antidepressive drugs:			
Imipramine	164 (145–186)	45 (32 – 63)	3·7
Amitriptyline	87 (80– 94)	12 (5 – 29)	7·4
Nomifensine	341 (301–386)	8 (6 – 12)	42·1
Other drugs:			
Amphetamine	18 (14– 22)	6 (4 – 8)	3·1
Caffeine	264 (245–285)	119 (96–148)	2·2
Morphine	393 (353–439)	136 (93–198)	2·9

TRH action at the pituitary cell level may be mediated by the adenyl cyclase system, but this hypothesis has been challenged by Tal *et al.* (1974), PGE_2 being much more effective on cAMP formation than TRH. Thyrotrophe cells make up a small percentage of the cell population of the adenohypophysis, and it may thus be difficult to measure changes in one cell population only (see Chapter 4). Measurements of such non-specific adenohypophysial parameters may be misleading and a satisfactory technique for concentration of homogeneous pituitary cell populations may be needed to extend these studies. It is not fully understood whether the TSH release by TRH is energy dependent, or whether it is TSH synthesis that is energy dependent. Yamamoto *et al.* (1972) have shown stimulation of glucose oxidation by TRH; this is not inhibited by cyanide, 2-desoxy-glucose, ouabain, arsenate, fluoride or dinitrophenol in a glucose-free medium. The TSH-release is blocked by N-ethylmaleimide or iodoacetamide (Guillemin *et al.*, 1968). The *in vivo* secretion of TSH stimulated by TRH is impaired by inhibitors of protein synthesis like cycloheximide, puromycin or actinomycin; it would appear that the acute release is not energy dependent in the presence of sufficient oxygen concentrations.

Table 5. Neurotropic effect of analogues of TRH in the DOPA-potentiation test (TRH activity = 100)

Structure	Relative potency vs. TRH neurotropic activity	TSH-release
Pyr-His-ProNH$_2$ (TRH)	100	100
Pyr-His-ProOH	16	0·02
Pyr-His-Pro-n-hexylamide	33	0·05
Pyr-His-Pro-cyclohexylamide	42	0·07
Pyr-His	19	inactive
Pyr-His-pyrrolidid	14	0·2–0·8
Pyr-Phe-ProNH$_2$	7	10
Pyr-Met-ProNH$_2$	22	1·0
Pyr-Lys-ProNH$_2$	3	0·02
Pyr-His-TrpNH$_2$	16	0·02
Pyr-4-amino-butyric acid	18	inactive
AcPro-Leu-ProOH	9	inactive
Pyr-Ala-ProOH	21	inactive

Neurotropic activity

Values are means of 4 or more assays, 4 mice per group.
TRH test dose 5 and 25 mg/kg

TSH-release

TRH tested on anterior pituitaries *in vitro*, 2 or 3 dose levels.

The effects of TRH on the central nervous system are much more disseminated than was initially suspected, e.g. Tables 4 and 5 (see Chapter 7). These are more pronounced under chlorpromazine treatment, probably due to hypothalamic blockade. TRH has a definite analeptic effect on the dog, although in doses far in excess of any dose required for TSH release.

E. INACTIVATION, METABOLISM AND EXCRETION

(1) *Plasma inactivation*

TRH is inactivated by incubation in plasma or serum from many species (Redding and Schally, 1969a; Nair *et al.*, 1971b; Bassiri and Utiger, 1972b; Utiger and Bassiri, 1973). In the rat the rate of plasma inactivation is increased by pretreatment with triiodothyronine. Inactivation is temperature and pH dependent and the inactivating factor, which is associated with the globulin fraction, is not dialysable, indicating an enzyme system (Vale *et al.*, 1971). TRH analogues can antagonise plasma inactivation. Collection of intact TRH in blood requires the addition of enzyme inhibitors or extraction with ethanol (Vale *et al.*, 1971). (See also Chapters 4 and 7.)

(2) *Plasma elimination and organ distribution*

After injection, ^{3}H-labelled TRH is rapidly cleared from plasma and accumulated in liver, kidney and pituitary gland. ^{3}H-TRH has a plasma $T\frac{1}{2}$ of 5 min (Leppaluoto *et al.*, 1972), and is rapidly excreted in the urine.

Studies with unlabelled TRH in physiological concentrations became possible following development of a radioimmunoassay (Bassiri and Utiger, 1972a). Half-life estimates by this method correlate well with the result of isotope studies. With the initial plasma $T\frac{1}{2}$ of 5·3 min, the distribution volume was found to be 15·7 litres and immunoreactive TRH is excreted in the urine for up to 90 min after injection.

Impaired renal or hepatic function will result in reduced TRH clearance. There is for instance a substantially higher TSH response to TRH in patients with impaired renal clearance (Gonzalez-Barcena *et al.*, 1973). The kinetics and distribution have also been studied using ^{125}I=TRH in mice (Rothenbucher *et al.*, 1973): it has full biological activity and is stable *in vivo*. Plasma elimination was rapid ($T\frac{1}{2} = 2$ min). Maximal uptake by the kidney was at 30 min and by the liver at 15 min; 38 % of the injected dose were excreted in the urine within 1 hour. Studies of the organ distribution of the ^{14}C and ^{3}H-labelled TRH by Redding and Schally (1971) indicated accumulation in the anterior pituitary, liver and kidney after 5–30 min. No accumulation was seen in cerebral, cortical and hypothalamic tissue. A similar pattern was seen in rats and mice.

(3) *Metabolism*

The metabolism of TRH *in vitro* (Fig. 4) has been studied with labelled preparations. Extraction of ^{14}C-TRH from anterior pituitaries of rats yielded Pyr-His-ProOH (TRH-free acid) as the main metabolite (Nair *et al.*, 1971b; Redding and Schally, 1971). It was concluded that *in vitro*, TRH is inactivated by deamidation, not by enzymic cleavage into amino acids. The products of TRH degradation by serum and hypothalamic tissue were isolated by Bauer and Kleinkauf (1974). The main metabolite was deamidated TRH, further steps observed were proteolytic cleavage yielding free Pro, and

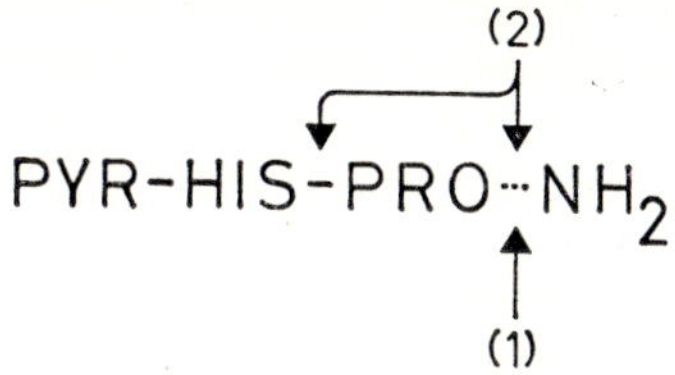

Fig. 4. Metabolic inactivation of TRH (thyroliberin): Sites of enzymatic cleavage indicated by arrows. Main metabolite is Pyr-His-ProOH.

then hydrolysis of the dipeptide. The TRH analogue Pyr-His-OCH_3 could inhibit degradation by serum, but not by brain tissue.

F. COMPARATIVE ASPECTS OF STRUCTURE AND FUNCTION

The biological effect of synthetic TRH is identical in all species tested and suggests that endogenous TRH may be identical with the tripeptide Pyr-His-ProNH_2. Since substitution of single amino acids in synthetic analogues leads to drastic decreases of activity, there appears to be not much scope for evolutionary modification (see Chapter 7).

The large amounts of hypothalamic tissue required for isolation of TRH have limited the structural determination to that of ovine, porcine or bovine origin. Using radioimmunoassays (which require much less tissue) extensive characterisation of the cross reactivity of TRH antisera with synthetic TRH analogues is necessary.

IV. LHRH (Gonadoliberin)

A. ISOLATION AND PURIFICATION

Regulation of gonadotrophin secretion by the hypothalamus was an early recognised phenomenon (see Chapters 1 and 8). The first demonstration of an extractable neurohormone from the hypothalamus was by McCann *et al.*, (1960). In different assay systems and by means of differential localisation in the hypothalamus, two separate regulatory hormones for LH and FSH secretion were postulated. However, final purification of LH- and FSH-releasing activity from hypothalamic tissue has led to the identification of only one hormone controlling the secretion of both gonadotrophins.

Partial purification of LH- and FSH-releasing activity from hypothalamic tissue has been carried out by numerous authors (Nikitovitch-Winer *et al.*, 1965; Fawcett *et al.*, 1968; Schally *et al.*, 1968a; Amoss *et al.*, 1971). Various extraction methods were used: either acetic acid or acidified organic solvents will result in a good yield of LHRH. Further purification was generally carried out by gel filtration on Sephadex G 25, phenol extraction, ion exchange chromatography on CM-cellulose, thin layer chromatography, followed by further steps including free-flow electrophoresis and counter-current distribution. The finding of separate fractions with LH- and FSH-releasing activity (White *et al.*, 1968), was later shown to be due to the presence of polyamines (putrescine, cadaverine etc.), which release FSH *in vitro*, but not *in vivo*. Evidence for the identity of LH- and FSH-releasing hormone in the human was presented by Kastin *et al.* (1969a), when highly purified LHRH was shown to release both LH and FSH in clinical trials.

B. STRUCTURE, CHEMICAL SYNTHESIS AND ANALOGUES

(1) *Structure*

The structure of porcine LHRH was elucidated by Schally *et al.* (1971a), Matsuo *et al.* (1971) and Baba *et al.* (1971). The procedure for structural analysis was as follows. Highly purified LHRH was treated with proteolytic enzymes (chymotrypsin and thermolysin). The free C-terminal of fragments were identified by selective tritium labelling after exposure to acetic anhydride and pyridine in the presence of ^3H_2O: The N-terminal residues of the proteolytic peptides were identified by microdansylation. The amino acid sequence was established by successive Edman degradations and the presence of tryptophan was ascertained by alkaline hydrolysis. No free N-terminus or C-terminus was detected in the native peptide by enzymatic degradation, but there was the characteristic peak of pyroglutamic acid in mass spectrographic analysis. The structure obtained initially for the porcine hormone has been confirmed for the bovine (Currie *et al.*, 1971) and ovine (Amoss *et al.*, 1971) hormones. Studies using specific antisera raised against LHRH indicate that a sequence closely related or identical to the decapeptide is found in a great number of species, with the possible exceptions of the frog and quail.

(2) *Synthesis*

Numerous methods of synthesis for LHRH have been described. Five of ten amino acids in the decapeptide are trifunctional and the side-reactions produced by these amino acids make synthesis and especially purification of synthetic LHRH difficult. Two extremes of the synthetic approach have to be distinguished:

(a) Maximal protection of the functional groups and subsequent cleavage of the protective groups (mainly in methods of solid phase synthesis).

(b) Minimal protection. The ultimate step of synthesis is carried through with unprotected side-chains, the advantage being that cleavage of protective groups is not necessary.

Most types of synthesis are intermediates between these two extremes. In the methods of classical synthesis, intermediate fragments with protective groups on the side-chains are used, being de-protected before the last coupling step. De-protection at the end of synthesis frequently results in numerous by-products, which have to be removed by extensive purification. The advantage of easy synthesis is thus balanced by the difficulty of final purification. The conditions of cleavage for the protective groups used in solid phase synthesis for histidine, serine, and tyrosine and the guanido-protective groups (tosyl or nitro groups) are unfavourable for LHRH, e.g. Na in liquid NH_3 (Manning *et al.*, 1972) and liquid HF. Studies on the chemical stability of LHRH under these conditions of de-protection (König and Geiger, 1974) have revealed that LHRH is partly decomposed by both methods. With Na in NH_3 there was almost complete destruction by reductive peptide cleavage between arginine and proline. Under conditions of de-protection in liquid HF 25% of LHRH is broken down to yield about seven new products by decomposition of tryptophan plus cleavage of the peptide bonds between Pyr^1 and His^2 and between His^2 and Ser^3. In contrast to Arg (NO_2), the tosyl-protected guanido group is well suited for de-protection in liquid HF. Arnold *et al.* (1973) obtained a yield of 50% pure LHRH from tosyl-protected LHRH, but only 25–30% from the corresponding nitro-derivative. Benzyl-oxycarbonyl and nitro groups on the guanido moiety of arginine may be removed by catalytic hydrogenation, but there is the possible formation of by-products during the reduction of the nitro-group and Bajusz *et al.* (1973) have reported, that during the hydrogenation of [Arg $(NO_2)^8$] LHRH there was saturation of Trp^3.

Another difficult step in the synthesis of LHRH is the C-terminal sequence Pro-Gly, which very readily forms the corresponding diketopiperazin. This is mainly observed in the Merrifield method, where the resin in the form of the benzylester facilitates cyclisation. In addition, H-Pro-Gly-NH_2 may decompose to the corresponding diketopiperazin by elimination of ammonia. H-Pro-Gly-NH-Mbh where the amide function is protected by the 4.4′.-di-methoxybenzylhydryl-group (Mbh), is not susceptible to this reaction. The corresponding benzhydryl resins are also quite stable.

(3) *Analogues*

Numerous analogues of LHRH have been prepared (e.g. Fig. 5) and studied since 1971 and certain general structure-activity relations can be inferred (Table 6).

(a) Pyr^1 seems to be crucial for biological activity. The most effective substitution was by formyl-sarcosine (Okada *et al.*, 1974) (60% activity as compared to LHRH).

Table 6. Analogues of LHRH; examples of modifications retaining considerable biological activity

Reference	Structure	Biological activity (LHRH = 100)
LHRH	Pyr-His-Trp-Ser-Tyr-Gly-Leu-Arg-Pro-$GlyNH_2$ 1 2 3 4 5 6 7 8 9 10	100
Okada *et al.*, 1974	formyl Sar^1	$53^{(2)}$
Okada *et al.*, 1974	$NMePyr^1$	$48^{(2)}$
Prasard *et al.*, 1975	3-(2-napthyl)-Ala^3	$52^{(2)}$
Coy *et al.*, 1974	pentamethyl Phe^3	$69^{(2)}$
Künzi *et al.*, 1974	Phe^5	$64^{(2)}$
Fujino *et al.*, 1974	D-Ala^6	$2830^{(1)}$
Fujino *et al.*, 1974	D-Leu^6	$2900^{(1)}$
Ling and Vale 1975	$N^{\alpha}MeLeu^7$	$102^{(3)}$
Geiger *et al.*, 1974	$homoArg^8$	$22^{(2)}$
	Pro^9 ethylamide	$670^{(1)}$
Fujino *et al.*, 1973	Pro^9 propylamide	$380^{(1)}$

$^{(1)}$Ovulation test. $^{(2)}$LH release *in vivo* $^{(3)}$LH release *in vitro*

Table 7. Competitive peptide inhibitors of LHRH

Structure	Sequence						Inhibitory activity in vivo[2]	Inhibitory activity in vitro[1]
LHRH	Pyr-	2 His	-Trp-Ser-Tyr-	6 Gly	-Leu-Arg-Pro-	10 $GlyNH_2$	—	—
des His^2	—	des His	—	—	—	—	Ø	Ø
des His^2 EA^{10}	—	des His	—	—	—	Pro-NH-C_2H_2	40·6	—
des His^2 D-Ala^6	—	des His	—	D-Ala	—	—	44·7	1000
des His^2 D-Lys^6	—	des His	—	D-Lys	—	—	32·5	—
des His^2 D-Leu^6	—	des His	—	D-Leu	—	—	79·0	—
D-Phe^2	—	D-Phe	—	—	—	—	30·4	2000
D-Phe^2 D-Leu^6	—	D-Phe	—	D-Leu	—	—	60·9	1300
D-Phe^2 D-Phe^6	—	D-Phe	—	D-Phe	—	—	77·4	—

EA = LHRH$^{(1-9)}$-nonapeptide-ethylamide

[1]Ferland *et al.*, 1975a: molar ratio of inhibitor vs. LHRH, recalculated.

[2]Vilchez-Martinez *et al.*, 1975: % inhibition at a ratio of 2500: 1 (inhibitor vs. LHRH)

(b) His[2] seems to be indispensable for biological activity, depletion or substitution of this amino acid leading to complete loss of intrinsic activity. Receptor affinity of inhibitors however, is preserved, especially in combination with substitution in position 6 by D-amino acids (Table 7). These analogues are thus competitive antagonists of LHRH: des-His[2]-LHRH, des His[2] [D-Ala[6]] LHRH and its ethylamide-derivatives (Coy *et al.*, 1974a; Ferland *et al.*, 1975), [D-Phe[2], D-Leu[6]] LHRH are typical examples. One of the most potent inhibitors is [D-Phe[2], D-Ala[6], N-Me-Leu[7]] LHRH (Rivier *et al.*, 1975b).

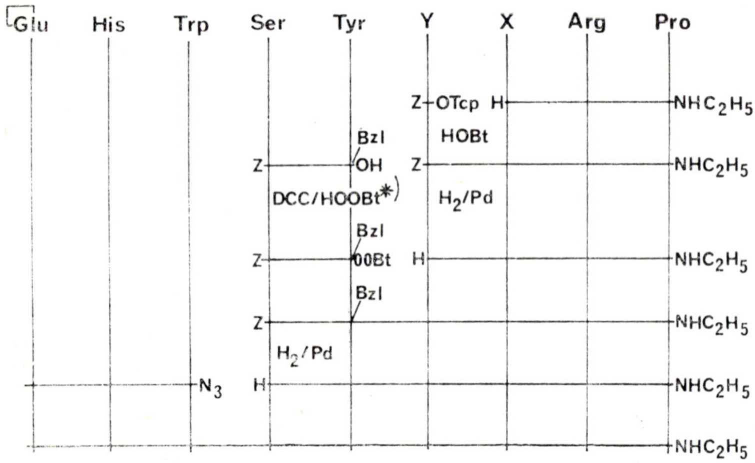

Fig. 5. Synthesis of LHRH analogues by fragment condensation : Substitution of position 6 and 7 in LHRH (1–9)-nonapeptide-ethyl-amides. HOOBt = 3-hydroxy-4-oxo-3, 4-dihydro-1,2,3-benzotriazine.

(c) Trp[3] may be substituted by pentamethyl-Phe with retention of almost all of its biological activity, but other substitutions of Trp[3] lead to marked reductions in potency and have been used as additive substitutions for LHRH inhibitors (Coy *et al.*, 1974b).

(d) Ser[4] and Tyr[5] seem to be of minor importance for the biological activity.

(e) The Gly[6]-Leu[7]-bond appears to be crucial for stability *in vivo* (Koch *et al.*, 1974). Substitution of Gly[6] either by D-amino acids or by protected D-amino acids leads to significant increase of biological activity (Fujino *et al.*, 1973a,b; Vilchez-Martinez *et al.*, 1974b; Coy *et al.*, 1974a; König *et al.*, 1975) (Fig. 6). This is further increased if in addition there is replacement of the C-terminal glycinamide by ethylamine, propylamine or related fluorinated compounds. Analogues with substitutions in positions 6 and 7 can be synthesised by fragment condensation as shown in Fig. 5. Substitution with

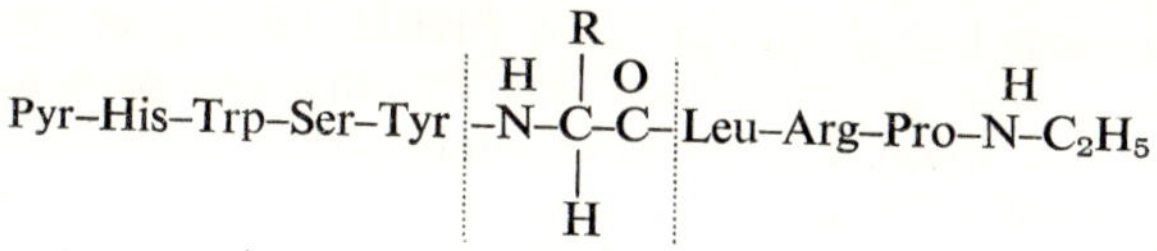

Residue	Name	Biological Activity (LHRH=1)
R = –H	LHRH (1–9)-nonapeptide-ethylamide	2·8
R = $-Ch_3$	D-Ala[6] ,, ,,	70·9
R = $-CH_2-OH$	D-Ser[6] ,, ,,	65
R = $-CH_2-O-C(CH_3)_2-CH_3$	D-Ser(Bu[t])[6] ,, ,,	172
R = $-CH_2-C(CH_3)_2-H$	D-Leu[6] ,, ,,	102
R = $-CH_2-CH_2-C(=O)-OH$	D-Glu[6] ,, ,,	28
R = $-CH_2-CH_2-C(=O)-O-C(CH_3)_2-CH_3$	D-Glu(OBu[t])[6] ,, ,,	44
R = $-CH_2-CH_2-CH_2-CH_2-NH_2$	D-Lys[6] ,, ,,	17·3
R = $-CH_2-CH_2-CH_2-CH_2-NH-C(=O)-O-C(CH_3)_2-CH_3$	D-Lys(BOC)[6] ,, ,,	38

Fig. 6. Contribution of the side chain in position 6 (Gly) of LHRH analogues to biological activity. D-amino acids strongly enhance biological activity (tested by superovulation in immature rats). Bu[t] = tert. butyl, OBu[t] = tert. butyl-ether, BOC = tert. butyloxy-carbonyl.

D-amino acids methylated in the β-position of the side chain appears to be important, since branching substituents as in tert Leu[6], D-Penicillamine[6] or D-Thr(Bu[t])[6] retain only a minimum of activity.

(f) Leu[7] is not essential for biological activity and may be replaced, e.g. by [Ser (Bu[t])[7]] or [Glu (OBu[t])[7]]. The [Cys (Bu[t])[7]] LHRH(1–9)ethylamide analogue has almost 10 times higher activity than LHRH (König *et al.*, 1975). Of special interest is substitution of Leu^7 by N^{α}methyl-Leu without loss of activity (Ling and Vale, 1975). This modification has been successfully incorporated into inhibitors of LHRH.

(g) Arg^8 can be replaced by homoArg^8 with 20% of the original activity (Geiger *et al.*, 1974).

(h) If Pro^9 is replaced there is considerable loss of biological activity and this part of the molecule seems to be crucial for hormonal expression.

(i) The C-terminal glycinamide can be replaced by a number of aliphatic alkylamines. As mentioned above, replacement by ethylamine and propylamine (Fujino *et al.*, 1972), leads to increased stability and prolonged biological activity. The mechanism of enhancement of biological activity is not fully understood. Plasma elimination proceeds at the same rate as for LHRH but inactivation by brain enzymes *in vitro* is reduced (Marks and Stern, 1974).

A consistent feature of the biological evaluation of LHRH analogues is the stability of the activity spectrum compared to the decapeptide. All analogues tested so far have a similar ratio of LH:FSH release. This would mean that the chemical structure of a still hypothetical FSH releasing factor cannot be derived from studies of analogues of LHRH decapeptide.

Inhibitors of LHRH (competitive peptide antagonists) have been developed by Coy *et al.* (1974b) and others. Such inhibitors are capable of suppressing the release of LH induced by LHRH, as well as spontaneous LH-release (Vilchez-Martinez *et al.*, 1975; Ferland *et al.*, 1975). Further development of such analogues may eventually result in a peptide contraceptive. No peptides unrelated to the decapeptide have been found to inhibit LHRH at the hypothalamic or pituitary level.

C. BIOSYNTHESIS AND SECRETION

The synthesis of LHRH in particulate-free extracts of hypothalamic tissue analagous to the biosynthesis of TRH has been reported by Reichlin and Mitnick (1973a, b), Moguilevski *et al.* (1973), Johansson *et al.* (1973), but the findings are still controversial.

The specific areas for synthesis and/or storage are the arcuate nucleus and the median eminence (Palkovits *et al.*, 1974a). LHRH in median eminence tissue has been identified by immunohistochemistry (Leonardelli *et al.* 1973; Baker *et al.*, 1974; Zimmermann *et al.*, 1974; Sétáló *et al.*, 1975). Subcellular localisation of LHRH in the hypothalamus has been reported by Taber and Karavolas (1975). The activity was determined *in vitro* by its LH-releasing activity. The elements containing LHRH were synaptosomes or mitochondria. No LH-releasing activity was found in cerebellum and cerebral cortex.

The presence of a higher molecular weight precursor of LHRH in hypothalamic extracts has been recently claimed by Millar *et al.* (1975). Immunoreactive LHRH was also reported to exist in the pineal gland (White *et al.*, 1973), but this finding has not been confirmed by bioassay.

The secretion of LHRH has been studied by an indirect approach, measuring fluctuations in plasma gonadotrophins. However, the kinetics of gonadotrophin release give little information about the actual hypothalamic secretion of LHRH. At the present time, direct measurement of LHRH either by biological or immunological assays does not give reliable estimates of the secretion of LHRH *in vivo* (see Chapter 4).

D. RECEPTOR BINDING AND MODE OF ACTION

LHRH secreted into the portal vessel system is bound to pituitary receptors located on plasma membranes. Two types of receptors have been identified (Grant *et al.*, 1973a; Spona, 1973a), a high affinity, low capacity population and a low affinity, high capacity population. Only the high affinity receptors bind LHRH without competition from LHRH analogues, whereas the low affinity receptors have no preference for biologically active or inactive LHRH analogues. Another explanation for these different receptor populations was given by Spona (1973b), who claimed to find preferential FSH release by the low affinity population. However, this receptor population binds LHRH analogues independently of their biological FSH-releasing activity.

LHRH is active in all species studied so far including non-mammalian species. It is non-toxic, the dose required for gonadotrophin release being of the order of 2 μg/kg, while the toxic dose is 1·5 g/kg. The identical biological effects of natural and synthetic LHRH were demonstrated *in vivo* by Schally *et al.* (1971b). Although all fractions of highly purified natural LHRH so far examined and all preparations of synthetic LHRH stimulate the release of both LH and FSH, the time course of the response *in vivo* is different. The differential effect of one hypothalamic hormone on both gonadotrophins is explained by the modulatory effect of steroids, prostaglandins and inhibin (see Chapter 8). Regulation of gonadotrophin secretion by LHRH is further complicated by a direct feedback effect of LHRH on its own effect. From animal experiments, it would appear that for chronic treatment with LHRH, there is an optimal low dose treatment leading to stimulation of pituitary hormone synthesis, while excessive doses may limit the effect of LHRH (Sandow and Heptner, 1974).

The mode of action of LHRH includes release and *de novo* synthesis of gonadotrophins. The release of LH and FSH is mediated by the cAMP-mechanism (Makino, 1973). Both rat hypothalamic extract and synthetic LHRH increase intrapituitary levels of cAMP. The extrusion mechanism is stimulated by dibutyryl cAMP and/or theophylline and both can stimulate the release of LH and FSH into the medium without LHRH. The next step is stimulation of an endogenous protein kinase in ribosomes. Histological correlates by electron microscopy were demonstrated: newly synthesised proteins passed from the rough endoplasmic reticulum into the Golgi complex within 10 min and into the secretory granules within 30–40 min. Ishikawa *et al.* (1973) have confirmed the mechanism of action of LHRH on synthesis of new protein and have shown a lag period of up to 4 days after blocking mRNA by actinomycin D. New DNA synthesis was not essential for the stimulatory activity of LHRH. Sundberg *et al.* (1973), have studied the effects of colchicine on the release of LH, FSH, TSH and GH from isolated rat pituitaries. They found potentiation of the effect of hypothalamic extract by colchicine and conclude that microtubules may be involved in the regulation of hormone distribution within the pituitary cell. Colchicine also enhanced the effect of K^+ stimulation. Prostaglandins can modulate the gonadotrophin release induced by LHRH (Sandow and Babej, 1975). PGE_2 and LHRH have an additive effect *in vitro*.

E. INACTIVATION, METABOLISM AND EXCRETION

(1) *Plasma inactivation*

LHRH is only slowly inactivated by plasma *in vitro*. In plasma of several species inactivation of up to 40% during 1 hour incubation was found (Sandow *et al.*, 1973b) but immunoreactive LHRH appears to be stable in human plasma (Jeffcoate *et al.*, 1974b).

(3) *Plasma elimination and organ distribution*

After intravenous injection of LHRH, the initial half-hife in plasma is short (3–6 min) (Fig. 7), followed by a second slower component (20–30 min). In experiments with labelled preparations (^{125}I- and ^{3}H-LHRH), a very long third component of more than 100 min is also observed. It is not known whether this finding represents an artefact due to circulation of metabolites, or whether LHRH can persist in the circulation for longer times. Plasma half-life is identical in all species tested (rat, rabbit, dog, sheep, pig, man) (Redding and Schally, 1973; Arimura *et al.*, 1974b). The correlation between plasma LHRH levels and LH-release after intranasal administration was studied by Bourguignon *et al.* (1974). The peak in serum LHRH preceded the release of LH, 1·5% of the dose of LHRH being absorbed.

When labelled LHRH is injected, there is rapid accumulation of the label in liver, kidney and pituitary gland. Essentially, the same pattern is seen with labelled analogues of LHRH (Sandow, unpublished observa-

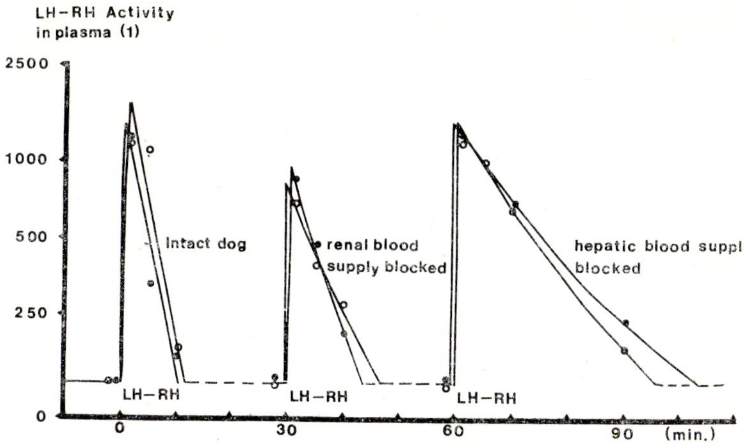

Fig. 7. Inactivation of LHRH *in vivo*. Plasma disappearance in the dog after renal and hepatic occlusion (open circles = dog 1, closed circles = dog 2). LHRH activity tested by bioassay. After hepatic occlusion, plasma elimination is significantly slower.

tion). Autoradiographic studies in the rat and mouse (Dupont *et al.*, 1974) have confirmed the sites of accumulation.

(3) *Homogenates or tissue slices*

Homogenates and tissue slices of liver, kidney and ovary or testis inactivate LHRH very rapidly while other tissues, e.g. muscle or lung, have no effect. LHRH is also inactivated by homogenates or supernatant or particulate fractions of neural tissues, including hypothalamus, (Griffiths *et al.*, 1974, 1975a), anterior and posterior pituitary (see Fig 8). Inactivation was demonstrated both by bioassay and by radioimmunoassay.

One group of enzymes (arylamidases) involved in the *in vitro* degradation of LHRH by the hypothalamus has been characterised by Kuhl *et al.* (1974). L-Cystine arylamidase activity in the rat hypothalamus is stimulated by steroids, there is a sex-dependent response, and luteinising hormone exerts a positive feedback (Kuhl and Taubert, 1975a, b). The enzyme is associated with the 105,000 g supernatant fraction of hypothalamic homogenates and requires the presence of metal ions and free thiol groups. This enzyme is competitively inhibited by TRH, neurohypophysial hormones and LHRH.

Enzyme activity in the supernatant fraction of hypothalamic and cerebrocortical extracts has also been studied by Koch *et al.* (1974). Inactivation of LHRH was measured by bioassay; more than 80% of the initial activity were lost 5 min after incubation. Incubation was inhibited by the kallikrein inhibitor, trasylol. Degradation products were identified on thin layer chromatograms as the fragments $LHRH^{1-6}$ and $LHRH^{7-10}$ (Fig. 9).

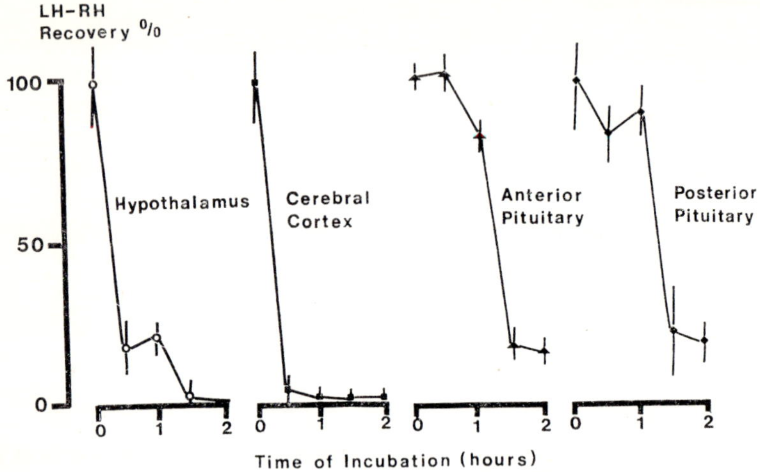

Fig. 8. Inactivation of LHRH *in vitro*. Incubation with homogenates of neural tissue. Recovery of LHRH tested by bioassay. Values are means (± S.E.M.)

The hypothalamic enzyme system in the supernatant and particulate fractions inactivating oxytocin was similarly shown to be under the control of oestradiol (Frith and Hooper, 1974). Addition of oestradiol-17β *in vitro* had a biphasic stimulatory effect on enzyme activation, cycloheximide and actinomycin D partially blocked the activation in the particulate fraction. Oxytocin and LHRH compete for the same inactivating enzyme, whereas TRH does not.

Pyr | His | Trp – Ser – Tyr – Gly | Leu – Arg – Pro | $GlyNH_2$

Fig. 9. Metabolic inactivation of LHRH: vertical lines indicated clearance sites. Main metabolites are LHRH (1–6) hexapeptide and LHRH (7–10) tetrapeptide.

Several metabolic pathways have been identified for the enzymatic breakdown of LHRH (Fig. 9). After injection of ^{3}H-LHRH in the human the main urine metabolites are pyroglutamic acid and Pyr-His (Redding *et al.*, 1973a). A pyroglutamyl peptidase in animal tissue was isolated by Szewczuc and Kwiatkowska (1970). A similar brain enzyme was blocked by substitution of Trp^3. The enzyme releasing Pyr-His has not been identified in rat brain *in vitro*. However, an endopeptidase was found, cleaving the Gly^6-Leu^7 bond with consecutive breakdown and release of internal amino acids. This type of degradation is blocked by substitution with D-amino acids in Gly^6. The concept is supported by the finding that enzymatic inactivation of LHRH leads to preferential release of internal amino acids (Ser, Tyr, Gly, Leu) by a brain endopeptidase (Marks and Stern, 1974). The inactivation of analogues protected in Gly[6] by substitution with D-amino acids and in the C-terminal part by alkylamine substitution of Gly NH_2 is significantly reduced.

F. COMPARATIVE ASPECTS OF STRUCTURE AND FUNCTION

The presence of extractable LH-releasing activity in hypothalamic tissue of many species has been established. Indirect evidence for the identity of the LH-releasing hormone in these species was provided by the effect of synthetic LHRH in rats, hamsters and rabbits (Humphrey *et al.*, 1973), the vole (Charlton *et al.*, 1973), pigs (Chakraborty *et al.*, 1973), ponies (Ginther and Wentworth, 1974), sheep (Jones *et al.*, 1973), domestic fowl (Furr *et al.*, 1973), goats (Pelletier, 1974), guinea-pigs (Donovan and ter Haar, 1974), monkeys (Mori and Hafez, 1973), frogs (Licht, 1974), quail (Smith and Follet, 1972), carp (see Fig. 3 in Chapter 8) (Breton and Weil, 1973). Taking into account the known specificity of the LHRH decapeptide molecule, it is likely that an identical or a closely related peptide is present in these species.

Radioimmunoassays indicate the presence of immunoreactive LHRH in several species. Arimura *et al.* (1973), have shown, however, that the immunoreactive material may be different in frogs and quails.

Immunohistochemical localisation of LHRH decapeptide has been obtained in the median eminence of the rat (see Section C), mink (Sorrentino and Sundberg, 1975), toad (Doerr-Schott and Dubois, 1975). The induction of ovulation by pituitary perifusion has been shown in a salamander, the crested newt (Mazzi *et al.*, 1974). There is immunoreactive LHRH decapeptide in the hypothalamus of *Xenopus laevis* (Deery, 1974), but not in the dogfish and the goldfish (Deery, 1974). However, synthetic LHRH is active in these species and it is possible that the active principle is related to the decapeptide.

V. GHRIH (SOMATOSTATIN)

A. ISOLATION AND PURIFICATION

The presence of two antagonistic principles with GH-releasing and release-inhibiting activity was discovered in 1968 by Krulich *et al.* Stimulatory and inhibitory activity from hypothalamic extracts from sheep and rats were separated by chromatography and demonstrated by their effects on isolated pituitaries *in vitro*. Inhibitory activity is localised in the posterior median eminence and in an area between the anterior hypothalamus and preoptic region (Krulich *et al.*, 1972). In 1973, Guillemin and his colleagues isolated a peptide inhibitor of GH-release, after chloroform-methanol-glacial acetic acid extraction of ovine hypothalamic tissue followed by ion exchange chromatography, gel filtration and liquid partition chromatography with a final yield of 8,5 mg of hormone from 500,000 ovine hypothalami. The structure was then elucidated (Brazeau *et al.*, 1973). The peptide was designated somatostatin, a term now widely accepted and included in the recommendations of IUPAC (1975) for hypothalamic hormone nomenclature. The abbreviation GHRIH (GH-release-inhibiting hormone) is also frequently used. Somatostatin is a cyclic tetradecapeptide containing a 38-membered ring and a disulphide bridge; the structure was confirmed by synthesis (Rivier *et al.*, 1973). An identical structure was found for somatostatin from porcine hypothalamic extracts.

B. STRUCTURE, CHEMICAL SYNTHESIS AND ANALOGUES

(1) *Structural analysis*

The primary structure of ovine somatostatin was derived by stepwise degradation of the intact carboxy-methylated peptide and fragments obtained by trypsin-treatment (Burgus *et al.*, 1973), and mass spectrometry of N-acetylated and permethylated derivatives (Ling *et al.*, 1973). The free C-terminal amino acids were determined by hydrazinolysis and dansylation.

(2) *Synthesis*

Synthesis of somatostatin (Fig. 10) is possible by conventional methods, but the cyclisation of linear somatostatin is difficult, and a considerable amount of polymerised peptide is formed. For protection of the SH-groups of cysteine the p-methoxybenzyl residue was used both in solid phase synthesis (Coy *et al.*, 1973; Yamashiro and Li, 1973; Rivier, 1974) and in classical synthesis (Sarantakis and McKinley, 1973; Fuji and Yajima, 1975). In all methods this residue is removed together with other protective groups by HF/anisol after complete assembly of the chain. The linear peptide is oxidised either by air at a neutral or slightly basic PH, or by potassium ferricyanide.

Another approach was selected by Immer *et al.* (1974), who used the trityl residue to protect the SH-function of cysteine. The remaining functional groups of the peptide were blocked with protective groups of the tertiary butyl type. This fully protected peptide assembled by fragment condensation was reacted with iodine (Kamber and Rittel, 1968) the trityl groups being removed by subsequent cyclisation. The remaining protective groups were removed by cold concentrated HCl. In all these methods polymerised peptide and other by-products are also formed due to de-protection under acid conditions. Extensive purification by chromatography on Sephadex G 25 and carboxymethylcellulose or by partition chromatography is then required.

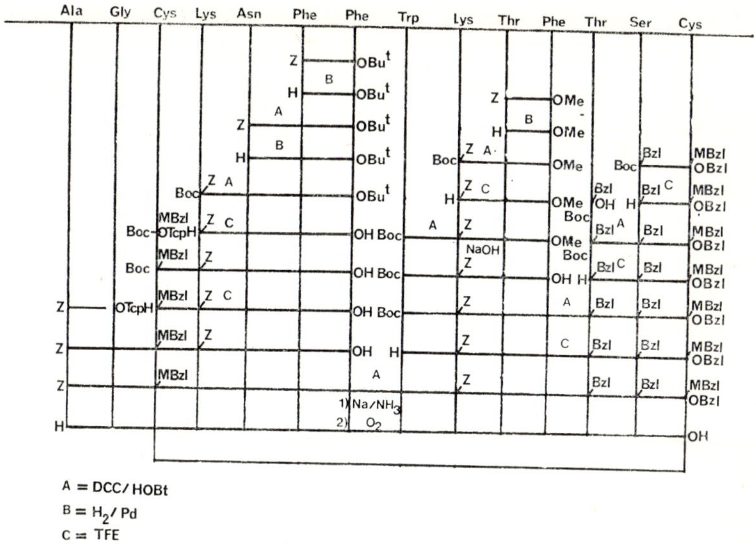

Fig. 10. Synthesis of somatostatin (GHRIH) by fragment condensation. (From Geiger *et al.*, 1975).

(3) *Analogues*

There is some controversy about the relative biological activity of linear somatostatin (also designated dihydrosomatostatin) and cyclic somatostatin. It has been suggested that oxidation and cyclisation of linear somatostatin can occur both *in vivo* and *in vitro*, Linear somatostatin is active in dogs, suppressing GH-release after L-dopa (Lovinger *et al.*, 1974) and linear somatostatin is only slightly less active than the cyclic hormone on inhibition of insulin and glucagon secretion in the human.

Structure-activity studies with synthetic analogues of somatostatin (ST) (Table 8) indicate, that the disulphide bridge as well as the N-terminal dipeptide Ala-Gly are not required for biological activity. Whereas [$Ala^{3,14}$] ST, [SMe-$Cys^{3,14}$]ST have low activity, the reduced forms of des-(Ala^1, Gly^2) ST and des-(Ala^1, Gly^2) [AcCys^3]ST are almost fully active (Rivier *et al.*, 1975b). The reduced linear derivatives of these two analogues are possibly activated *in vivo* by oxidation and subsequent cyclisation. A series of dodecapeptides acylated in Cys^3 were reported to have a prolonged action in the rat (Brown *et al.*, 1975) but not in man (Evered *et al.*, 1975). The analogues des(Asn^5)ST, des(Ser^{13})ST, des(Ala^1,Gly^2,Lys^4)ST, des(Ala^1,Gly^2,Asn^5)ST, [$Tyr^{3,14}$]ST and [(Gly-Gly-Tyr-Ala)^1]ST retain significant biological activity, as well as [NLe^9]ST and [D-Lys^9]ST. It appears that neither the size of the disulphide ring or the length of the peptide chain nor Lys^4, Asn^5 or Ser^{13} are

Table 8. Analogues of Somatostatin (GHRIH); summary of biological activities.

Structure	Biological activity	
	in vivo	*in vitro*
H Ala-Gly-Cys-Lys-Asn-Phe-Phe-Trp-Lys-Thr-Phe-Thr-Ser-Cys-OH 1 2 3 4 5 6 7 8 9 10 11 12 13 14	100	100
AcGly^{2}———	—	65 (1)
Cys^{3}———	—	39 (1)
desamino Cys^{3}———	—	60 (1)
———( )———	20 (2)	20 (2)
———( )———	15 (2)	15 (2)
——( )———	12 (2)	12 (2)
Thr^{1}———	—	37 (1)
——— Ala^{2} ———	135	190 (1)
——D-Ala^{2} ———	227	55 (1)
———Ala^{5}———	112	130 (1)
———Ala^{6}———	<10	1 (1)
———Ala^{7}———	<10	3 (1)
———Ala^{8}———	<10	<0·5 (1)
———Nle^{9} ———	0·5 (2)	—
———DLys^{9} ———	0·5 (2)	—
———Ala^{10}———	<10	25 (1)
———Ala^{11} ———	29	2 (1)
———Ala^{12}———	22	4 (1)
———Ala^{13} ———	<10	6 (1)
———( ) ———	55 (2)	—
desamino carba^{3}———carba^{14}	50 (3)	—
desamino carba^{3}———decarboxy-carba^{14}	50 (3)	—

(1) *In vivo* inhibition of arginine-induced insulin release
In vivo inhibition of GH release in pituitary cell cultures (Rivier *et al.*, 1975b)
(2) *In vivo* inhibition of GH release in the rat (Sarantakis *et al.*, 1975)
(3) *In vivo* suppression of pentobarbital-induced GH release (Veber *et al.*, 1971)

critical for biological activity. [Ala^{2}], [D-Ala^{2}], [Ala^{5}] and [Tyr^{11}]ST were reported to have an activity similar to the parent peptide (Rivier *et al.*, 1975b). [D-Trp^{8}]ST is about 8 times more active than somatostatin (Rivier *et al.*, 1975a) due to reduced enzymatic inactivation (see section F below).

C. BIOSYNTHESIS AND SECRETION

GHRIH has been identified in discrete brain areas by immunohistochemistry (Brownstein *et al.*, 1975). Immunoreactive GHRIH of different molecular sizes has also been identified in the pancreas, stomach and gut (Arimura *et al.*, 1975a). It is not known whether GHRIH is synthesised as a prohormone in the gastrointestinal tract and released locally. If so, local diffusion might be sufficient for adequate tissue distribution. It has not been possible to detect circulating GHRIH by radioimmunoassay (Arimura *et al.*, 1975a) and nothing is known about its secretory dynamics. GHRIH has been measured by radioimmunoassay in cerebrospinal fluid (Kronheim *et al.*, 1977; Patel *et al.*, 1977).

D. RECEPTOR BINDING AND MODE OF ACTION

The pituitary receptors have not yet been identified. The mechanism of action appears to involve inhibition of the adenyl cyclase system. Thus GHRIH added to rat anterior pituitary inhibits accumulation of cAMP during the first minutes of incubation (Borgeat *et al.*, 1974) and the stimulatory effects of theophylline and dibutyryl-cAMP on GH synthesis are antagonised by GHRIH *in vitro* (Vale *et al.*, 1972); this is associated with inhibition of the release of GH and TSH. These findings were confirmed by Kaneko *et al.*, (1974), who found a dose-related increase in cGMP together with a decrease of cAMP. GHRIH inhibits the release of GH in the presence of $BaCl_2$. By electron microscopy, $BaCl_2$was shown to stimulate the secretory pathway controlling emiocytosis of GH granules. Ba^{++} and Ca^{++} appear to interact with the same stimulatory mechanism (Schofield and Moser, 1974). The inhibitory action on GH secretion includes antagonism to pharmacological stimuli (morphine, pentobarbital) and prostaglandin E_2 (Szabo and Frohman, 1975). GHRIH can also antagonise the stimulatory effect of Ca^{++} on insulin secretion (Curry and Bennett, 1974). The effect on insulin secretion in man does not involve α-adrenergic receptors (Eféndic and Luft, 1975b).

The effects on pituitary hormones are suppression of GH release *in vivo* and *in vitro*, including supression of basal secretion and pulsatile GH release in the rat (Martin *et al.*, 1974), an inhibitory effect on TSH and FSH release induced by TRH (Besser *et al.*, 1974) and suppression of ACTH release in patients with corticotrophin-secreting tumours (Tyrell *et al.*, 1975). The effect on prolactin secretion is variable; *in vitro* there is a reduction of prolactin secretion.

The effects on gastrointestinal hormones include inhibition *in vivo* and *in vitro* of insulin secretion (Alberti, 1973; Curry *et al.*, 1974; Hansen *et al.*, 1974); glucagon secretion (Hansen *et al.*, 1974); gastrin and gastric acid secretion (Bloom *et al.*, 1974; Boden *et al.*, 1975; Hayes *et al.*, 1975), and secretin release (Boden *et al.*, 1975). The question again arises, as with neurohypophysical hormones, whether receptor evolution or adaptation can lead to spreading of hormonal control extending a previously limited activity spectrum.

The effects of GHRIH are short-lived, requiring infusions or repeated injections of high doses to suppress secretion of GH or pancreatic hormones. When infusions of somatostatin are stopped, there is a "rebound" with excess secretion of GH, insulin and gastrin. Side effects have been noted, when high concentrations are infused. A sedative effect, somnolence and a pronounced vasodepressor effect is observed in rats. Prolonged treatment with somatostatin was toxic in baboons (Koerker *et al.*, 1975) due to thrombocytopenia and reduced platelet aggregation. Inhibition of platelet aggregation *in vivo* is also seen in rabbits (Chiang *et al.*, 1975) and man (Besser *et al.*, 1975; Mielke *et al.*, 1975).

E. INACTIVATION, METABOLISM AND EXCRETION

The short plasma half-life of injected somatostatin ($T\frac{1}{2} = 4$ min) has been used as an explanation for the short-lived effect (Pimstone *et al.*, 1975), but in fact the $T\frac{1}{2}$ is no shorter than that of other hypothalamic hormones studied so far and is due to equilibration with the large distribution volume. There is no difference between the suppression of GH during somatotatin infusion in acromegalics or normal volunteers. In patients with liver disease, $T\frac{1}{2}$ of GH after suppression by somatostatin infusion is 27 min as compared with 18 min in chronic renal failure (Pimstone *et al.*, 1975). This would indicate a major role of the kidney in somatostatin elimination. It is not known whether somatostatin is excreted into the urine. Estimates of the receptor stimulation based on the decay of the inhibitory effect after cessation of an infusion lead to an estimate of 10 min for half-reversal of inhibition.

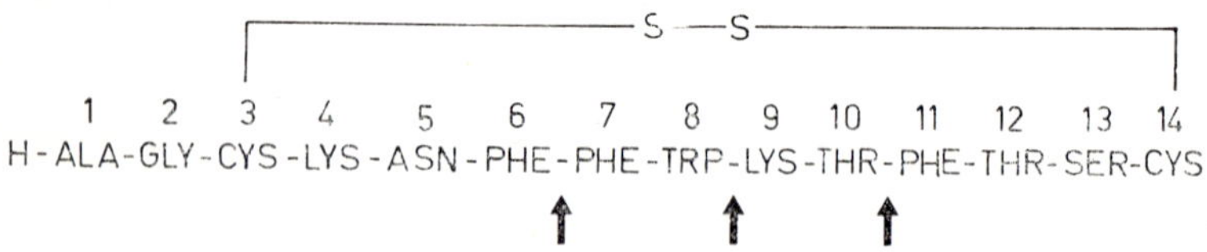

Fig. 11. Metabolic inactivation of somatostatin (GHRIH). Sites of enzymatic breakdown are indicated by arrows. The regulatory enzyme is not limited to the median eminence area. Release of free "internal amino acids" was measured by amino acid analysis.

The inactivation of cyclic somatostatin by brain extracts starts with cleavage near Trp^8 and leads to preferential release of Phe (Phe^6, Phe^7) and Lys^9 (Marks and Stern, 1975) (Fig. 11). Linear somatostatin may have an additional cleavage site at Thr^{10}-Phe^{11}. The ring structure and N-terminal do not seem to be of major importance, because [N^{α}-$AcCys^3$]-somatostatin gave rise to similar release of internal amino acids as in the parent hormone. These results agree with the prolonged activity of [D-Trp^8]-somatostatin (Rivier *et al.*, 1975a).

F. COMPARATIVE ASPECTS OF STRUCTURE AND FUNCTION

No comparative studies on the chemical identity of somatostatin in species other than sheep and pig are available. It is active in rats, dogs, chickens (e.g. Fig. 12), rabbits and man. Support for the occurrence and

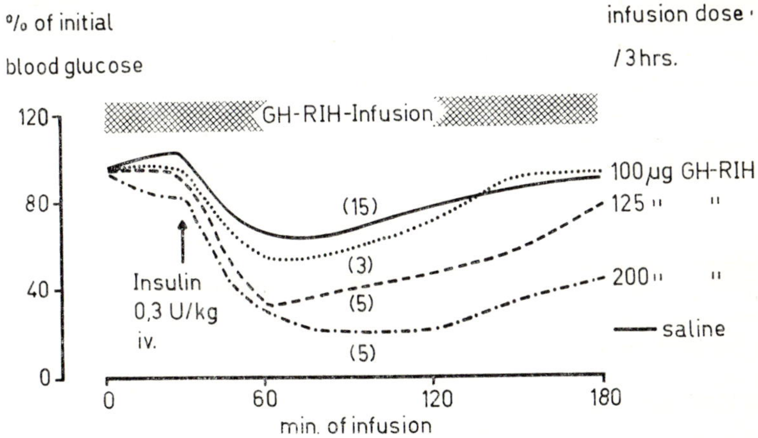

Fig. 12. Effect of somatostatin on blood glucose regulation in chickens: During infusion of somatostatin, the effect of insulin 0·3 U/kg is enhanced by somatostatin in a dose-dependent manner. The counter-regulation of hypoglycaemia by glucagon is blocked by the hypothalamic hormone (Geisen, unpublished observation).

distribution of this hormone in several species may come from histochemical identification by specific antisera and somatostatin radioimmunoassay.

VI. GHRH (SOMATOLIBERIN)

A. ISOLATION AND PURIFICATION

Research on hypothalamic GH-releasing activity has been predominantly a problem of relevant assay systems. The depletion of pituitary GH following intra-carotid injection of hypothalamic extracts has been questioned for its specificity after re-examination of changes in pituitary GH content by disc-electrophoresis and measurement of corresponding plasma levels by GH radioimmunoassay (Rodger *et al.*, 1969). Studies with other releasing hormones indicate that changes in pituitary hormone content do not adequately reflect the functional state of a pituitary cell population. The releasable pool of GH appears to be considerably lower than estimated by the pituitary depletion method (Nicoll, 1972). Measurement of circulating plasma levels of GH by epiphyseal cartilage width in the rat (Greenspan *et al.*, 1949) suffer from the interference of TSH and ACTH. Neurohypophysial hormones and MSH do not seem to interfere in this assay. When specific radioimmunoassays for GH of several species became available, the active fractions with GH-releasing activity identified by bioassay did not coincide with those found by radioimmunoassay.

The presence of GH-releasing activity by bioassay has been demonstrated in hypothalami of cattle, pigs, sheep, rabbits, dogs, mice, frogs, cats and humans (see review by Schally *et al.*, 1968c).

A growth hormone releasing factor (GRF) was first isolated by Deuben and Meites (1964). Release and synthesis of GH in tissue cultures were stimulated by rat hypothalamic extracts. Semi-purified preparations of GRF were obtained from porcine, bovine and human hypothalamic tissue by Franz *et al.* (1962), Dhariwal *et al.* (1965a, b), Schally *et al.* (1968a, b) and Wilber *et al.* (1971). The technical aspects of purifications have been reviewed by Schally *et al.* (1968a, b, c, 1970). The purification of two separate fractions (A-GHRH and B-GHRH) from porcine hypothalami has been reported (Currie *et al.*, 1974). After extraction with methanol acetic acid, the extracts were purified on Bio-Gel P2 and Sephadex G25 in two partition chromatography systems. GH-releasing activity was assayed on isolated rat pituitaries using GH radioimmunoassay. The two fractions were separated during the Bio-Gel run. The activity of factor A-GHRH was inhibited by somatostatin. No confirmation of the *in vivo* activity of these fractions has been reported. Several groups of investigators have shown that peptide material stimulating the release of immunoreactive GH can be isolated from hypothalamic tissue (Frohman *et al.*, 1971; Wilber *et al.*, 1971; Sandow *et al.*, 1972, 1973a; Worthington *et al.*, 1972). Material with GH-releasing activity was isolated from rat and sheep hypothalami and from human tumours (Beck *et al.*, 1973).

In newer attempts to isolate purified GHRH from hypothalami, fractions different from the original starting material leading to the tentative structure of GHRH (a decapeptide, see below) have been obtained with GH-releasing activity in RIA systems (Sandow *et al.*, 1973a; Takahara *et al.*, 1975). Sawano *et al.* (1974) have shown that purified GHRH from bovine hypothalami stimulated release of immunoreactive GH *in vitro* and *in vivo* in the conscious dog, and also synthesis of GH in isolated rat pituitaries. However, their material did not require stimulation of protein biosynthesis for GH release as demonstrated by the lack of inhibition after actinomycin D or cycloheximide.

B. STRUCTURE, CHEMICAL SYNTHESIS AND ANALOGUES

Purification of porcine GHRH by bioassay culminated in the isolation of a decapeptide by Schally and Arimura (1972) (Fig. 13). This peptide was able to stimulate biological events associated with GH release *in vivo* and *in vitro* as measured by the tibia test and sulphation factor assay, but was inactive in test systems using GHRIA. The peptide was later shown to be almost identical with the N-terminal decapeptide sequence of the β-chain of porcine haemoglobin, Gln^9 being replaced by Glu (Veber *et al.*, 1971). The peptide sequence from the β-chain of porcine haemoglobin (H-Val-His-Leu-*Ser-Ala*-Glu-Glu-Lys-*Gln*-Ala-OH and also the decapeptide of the β-chain of human haemoglobin (H-Val-His-Leu-Thr-Pro-Glu-Glu-Lys-Ser-Ala-OH) did not stimulate growth in rats and had no effect on immunoreactive GH in either rats or man.

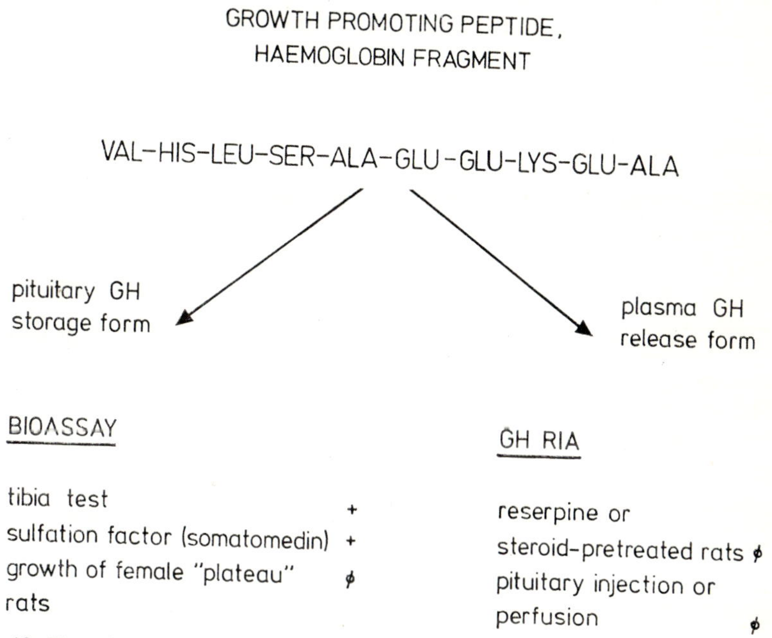

Fig. 13. Tentative structure of a growth hormone releasing factor: This peptide was active in biological assay based on cartilage growth (+), but inactive on release of immunoreactive GH (ɸ) and body weight of female rats (From Schally *et al.*, 1971a).

The situation is complicated by the fact that the action of GHRH is probably mediated by the somatomedins, a group of peptides isolated from liver, plasma and muscle (Uthne *et al.*, 1974). The possibility that the decapeptide isolated by Schally has a somatomedin-like activity has not been investigated in detail.

Reports of GH-releasing activity of a synthetic peptide, Pyr-Ser-Gly-NH_2, isolated from bovine hypothalami (Yudaev *et al.*, 1973) have not been confirmed (Schally *et al.*, 1973b). The presence of GH-releasing activity in plasma or acromegalic patients was shown by Hagen *et al.* (1972b). This factor is heat stable and releases immunoreactive GH from rat pituitaries *in vitro*. The release of GH in humans by α-MSH was reported by Strauch *et al.* (1973), but the specificity of this effect is debatable.

C. BIOSYNTHESIS AND SECRETION

The biosynthesis of GHRH has been studied by Reichlin and Mitnick (1973c) who have claimed a nonribosomal mechanism as for TRH.

The episodic secretion of GH has been shown in many species, including men, monkeys, rabbits and the rat (see Chapter 10). This secretory pattern in the absence of a short-term feedback effect of GH can be explained either by short bursts of GHRH secretion from the hypothalamus or by tonic GH secretion periodically inhibited by somatostatin. This dual control mechanism appears to be very unstable and can be upset by stress, hypoglycaemia, exercise, sleep and many other stimuli (see Chapter 10). Secretory dynamics are further complicated by the influence of neurotransmitters, namely adrenergic impulses from suprahypothalamic centres. GH-releasing activity has been demonstrated in the circulation of long-term hypophysectomised rats (Müller *et al.*, 1967), but these studies need confirmation by chemical identification of the circulating substance. Evaluation of GHRH effects is made difficult by the interposition of somatomedins and peptides with non-suppressible insulin-like activity (NSILA).

The evidence for a hypothalamic peptide stimulating the release of immunoreactive GH rests on electrical stimulation of hypothalamic areas (ventromedial nucleus and hippocampus) as well as the elevation of RIA-GH in rats pretreated with steroids or after pituitary injection or pituitary perfusion by hypothalamic extracts (Bernadis and Frohman, 1971; Sandow *et al.*, 1972; Martin *et al.*, 1973; Takahara *et al.*, 1975). The presence of a GH "synthesising factor" in hypophysial portal blood has been demonstrated by Worthington *et al.* (1972). The addition of hypophysial portal plasma to incubated pituitaries increased to incorporation of ^{14}C-leucine into GH.

D. RECEPTOR BINDING AND MODE OF ACTION

Purified porcine GHRH binds to pituitary plasma membranes and activates the adenyl cyclase system (Labrie *et al.*, 1973a; Peake *et al.*, 1973b). There is some controversy whether the energy system involved is cyclic AMP or cyclic GMP. A preparation of purified ovine GHRH was shown to increase the uptake of ^{45}Ca from the incubation medium into the rat pituitary concurrent with a release of GH (Milligan and Kraicer, 1972). This finding was explained by the hypothesis that an increase in intracellular Ca^{++} leads to the release of preformed GH (Douglas, 1968). Prostaglandin E_2 is also able to stimulate GH release *in vitro* (Dormer *et al.*, 1974).

E. INACTIVATION, METABOLISM AND EXCRETION

The inactivation and metabolism of GHRH have not been studied, although GH-releasing activity in hypothalamic extracts appears to be unstable during prolonged storage at –20°C (Sandow *et al.*, 1972; Beck *et al.*, 1973). The enzymatic inactivation of highly purified GHRH has been studied by Schally *et al.* (1968a), but these studies will have to be extended to fractions with RIA-GH releasing activity.

F. COMPARATIVE ASPECTS OF STRUCTURE AND FUNCTION

Hypothalamic extracts of several species contain factors depleting bioassayable GH from pituitaries *in vivo* and *in vitro* and also factors elevating circulating immunoassayable GH. In the absence of relevant information about the chemistry of these factors no conclusions can be drawn about the species specificity.

VII. PIF (Prolactostatin) and PRF (Prolactoliberin)

A. ISOLATION AND PURIFICATION

Prolactin secretion is under the predominant inhibitory control of the hypothalamus in mammals and under stimulatory control in birds. The prevalence of stimulatory or inhibitory control may be related to supra-hypothalamic regulation by neurotransmitter substances or to hypothalamic peptides (see Chapter 9). Isolation and purification of PIF and PRF activity has been hampered by a lack of appropriate assay systems. The isolated agents, despite changes in prolactin secretion measured by specific radio-immunoassay, may not reflect physiological phenomena related to hypo-thalamic influences. A large number of pharmacological agents can increase prolactin secretion from the isolated pituitary, possibly by changes in membrane permeability. A dual site of action is known for catecholamines and ergot alkaloids, at the pituitary and hypothalamic level.

(1) *PIF (Prolactostatin)*

Prolactin release inhibiting factor activity (PIF) has been purified from crude hypothalamic extracts of rats, sheep, cattle and pigs and the activity of this material demonstrated *in vivo* (Pasteels, 1961; Meites *et al.*, 1963; Kuroshima *et al.*, 1966; Meites and Nicoll, 1966; Grosvenor *et al.*, 1967) and *in vitro* (Talwalker *et al.*, 1963). Schally *et al.* (1974) have purified PIF-activity from 470,000 porcine hypothalami. The material was defatted with acetone and petroleum ether and extracted with 2N acetic acid. Purification was by gel filtration on Sephadex G 25, followed by phenol extraction and chromatography on CM-cellulose; further purification was by counter-current distribution (CCD). This procedure led to the concentration of PIF activity in fractions with low amino acid content and high catecholamine content. Noradrenaline and dopamine were both found to have PIF-activity in various assay systems *in vitro*, a finding earlier described by Birge *et al.* (1970). Furthermore, the infusion of noradrenaline and dopamine in 5% glucose into hypophysial portal vessels of rats strongly inhibits the release of prolactin (Takahara *et al.*, 1974a). This appears to be good evidence for a direct effect of catecholamines on the pituitary gland.

The existence of a specific PIF of peptide nature was indicated by the activity of certain hypothalamic fractions of low catecholamine content in the perphenazine-blocked rat. The partial purification of a porcine PIF of peptide nature has been described by Greibrock *et al.* (1974). This material also inhibited release of immunoreactive GH from isolated rat pituitaries.

(2) *PRF* (*prolactoliberin*)

Prolactin releasing activity has been demonstrated both in birds (Kragt and Meites, 1965) and in rats (Meites *et al.*, 1961). Although synthetic TRH is able to release both TSH and prolactin at equivalent concentrations, a PRF separate from TRH seems likely due to the different secretion rates of prolactin and TSH and to the phase shifts observed in the circadian secretion of the two hormones. Evidence for the non-identity of TRH and PRF in the human was provided by Vanhaelst *et al.* (1973), who found that the nocturnal peaks of TSH and prolactin secretion were dissociated in normal subjects. Evidence in favour of PRF is separation of TRH and PRF in hypothalamic extracts by chromatography. The isolation of PRF activity from the hypothalamic-posterior pituitary complex was reported by Dular *et al.* (1974). After extraction with acetic acid followed by gel filtration and ion exchange chromatography a preparation with PRF activity distinct from TRH was obtained.

B. STRUCTURE, CHEMICAL SYNTHESIS AND ANALOGUES

The chemical nature of the prolactin regulating hypothalamic factors is not known. Among the synthetic analogues of TRH, no specific prolactin releasing peptides have been found.

C. BIOSYNTHESIS AND SECRETION

The biosynthesis of PRF has been studied with the same methodology as for TRH by Mitnick *et al.* (1973). They found synthesis of a PRF by cell-free extracts of hypothalamic tissue, synthesis being dependent on the presence of ATP and amino acids. PIF activity has been described in rat portal plasma by Kamberi *et al.* (1971a, b, c).

In the absence of clearcut evidence about the physiological importance of either catecholamines or peptides in the regulation of prolactin secretion, any conclusions about the secretion of PIF and PRF can only be based on experiments relating to suppression or stimulation of prolactin secretion.

These studies are complicated by the fact that many stimuli act both at the hypothalamic and pituitary level. The most important pharmacological suppressive agent is the ergotamine alkaloid, 2-α-bromoergocryptine (CB-154 or bromocriptine).

D. RECEPTOR BINDING AND MODE OF ACTION

The pituitary receptors for PIF and PRF have not been characterised. The mechanism of activation of prolactin secretion induced by hypothalamic hormones involves receptors localised on pituitary cell membranes, participation of calcium, potassium and magnesium ions, activation of the adenyl cyclase system and a specific protein kinase. Isolated rat pituitaries can be stimulated by theophylline to secrete prolactin and this effect is neutralised by hypothalamic extracts with PIF activity. Since theophylline stimulates the adenyl cyclase system, it could be inferred that PIF acts by lowering intracellular levels of cAMP. The presence of calcium is required for potassium induced release of prolactin and GH (Gautvik and Tashjian, 1973). Participation of the adenyl cyclase system in prolactin release from rat pituitaries was postulated by Pelletier *et al.* (1972), prolactin being released after addition of monobutyryl-cAMP. On the other hand, dibutyryl-cAMP decreases plasma prolactin after injection into the third ventricle (Ojeda *et al.*, 1974b). Such differences in activity may be related to a biphasic action of cyclic nucleotides. The role of PRF on pituitary release is difficult to assess, due to the existence of several prolactin pools. In lactating rats, three such pools were identified by Nicoll (1971). The first pool was readily discharged, the second one was slowly releasable and the third one essentially non-depletable. The pool of non-depletable prolactin was estimated at 50% of the pituitary content. Involvement of microtubules in the process of prolactin and GH release was shown by Labrie *et al.* (1973b). Hormone release is inhibited by destruction of microtubules by deuterium oxide and vincristine.

E. INACTIVATION, METABOLISM AND EXCRETION

No studies on the metabolic fate of either of the proposed prolactin regulating factors have been carried out.

F. COMPARATIVE ASPECTS OF STRUCTURE AND FUNCTION

Evidence for the predominant stimulatory or inhibitory control of prolactin secretion has been reviewed by Nicoll (1971) (Chapter 9). In mammals, a high prolactin turnover rate is found, whereas in the birds, a low turnover rate is associated with the predominant stimulatory control of prolactin secretion (Tixier-Vidal and Gourdji, 1972). The low *in vitro* turnover of prolactin by the isolated pituitary of birds indicates a very limited capacity for independent prolactin secretion. Hypothalamic extracts with inhibitory effect on prolactin secretion *in vitro* have been prepared from the rat (Pasteels, 1961), sheep (Schally *et al.*, 1965), cow (Grosvenor *et al.*, 1965), pig (Kuroshima *et al.*, 1966) and human (Pasteels *et al.*, 1963). The predominantly stimulatory effect of hypothalamic extract of lactating pigeons (Kragt and Meites, 1965), the pekin duck (Gourdji and Tixier-Vidal, 1966a, b) the tricoloured blackbird (Nicoll, 1965), quails, chickens and turkeys (Chen *et al.*, 1968) is ascribed to an avian PRF. In fishes, hypothalamic extracts of goldfish have been shown to contain PIF-activity (Leatherland and Ensor, 1974). Localisation of the neurosecretory areas for PIF in the nucleus lateralis tuberalis and PRF or its afferent pathways in the medial thalamic and anterior hypothalamic region in the goldfish was reported by Peter and McKeown, (1974).

VIII. CRF (Corticoliberin)

A. Isolation and Purification

Hypothalamic corticotrophin releasing factor (CRF) was the first neurohormone to be isolated from hypothalamic and neurohypophysial tissue, CRF activity being found in extracts from posterior pituitaries (Saffran and Schally, 1955). However, hypothalamic lesions did not reduce adrenal activity to the same extent as hypophysectomy (Fortier and de Groot, 1964), and it was concluded that an additional factor of

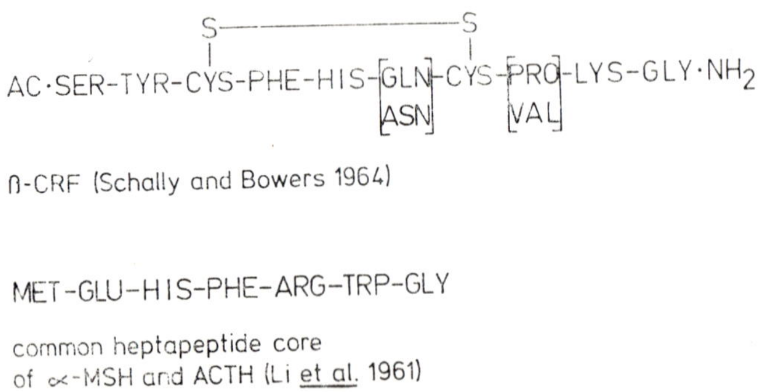

Fig. 14. Corticotrophin-releasing activity of synthetic peptides. Both structures are related to MSH (melanotropin). Brackets indicate probable residues with unknown position, e.g. Gln-Asn or Asn-Gln.

pituitary origin might participate in the regulation of adrenal function: this factor was identified as vasopressin. The corticotrophin releasing activity of vasopressin in hypothalamic extracts interferes with the assay systems of CRF and the ACTH-like activity of hypothalamic extracts, and may be due to the presence of peptides related to the common heptapeptide core of MSH and ACTH, (Met-Glu-His-Phe-Arg-Trp-Gly) (Fig. 14) (Chapters 11 and 12). Animals with hypothalamic lesions or treated with hypothalamic blocking agents can be used for specific CRF assay. The non-specific stress response can be blocked by pretreatment with adrenal steroids or chlorpromazine-morphine-nembutal (CPZ-M-N) as described by Arimura *et al.* (1967).

The evidence for a hypothalamic CRF different from vasopressin has been reviewed by Martini (1966). Vasopressin secretion and ACTH release are dissociated under a variety of experimental conditions and the finding of adrenal hypertrophy in diabetes insipidus is particularly relevant. Hypothalamic CRF is resistant to inactivation by n-bromo-succinimide (Leeman and Patton, 1962) or thioglycollate (Ramirez and McCann, 1964), reagents which inactivate vasopressin. In partially purified hypothalamic extracts, CRF activity is obtained in fraction devoid of vasopressin and MSH or ACTH.

A major problem in the isolation of CRF appears to be the conditions for collection of median eminence tissue. This should be done under minimal stress, e.g. nembutal anaesthesia. The circadian rhythm of CRF in the brain should also be considered. Circadian fluctuations of CRF-content in the pigeon (Sato and George, 1973) and in the rat (Hiroshige *et al.*, 1973) have been reported. Another source of error in the purification of CRF is the presence of ammonium ions from ammonium acetate buffers used in column chromatography (Wei *et al.*, 1973) which stimulate ACTH release.

Schally *et al.* (1960) and Schally and Guillemin (1963) obtained two types of CRF from porcine posterior pituitary powder by ion exchange chromatography and countercurrent distribution. Semi-purified CRF as isolated by Schally *et al.* (1968a) is not affected by thioglycollate, pepsin or chymotrypsin, but inactivated by trypsin. Hypothalamic CRF of ovine origin was purified by gel filtration on Sephadex G-25 and counter-current distribution, and also contained two zones of CRF-activity. The preparations corresponded to the previously obtained neurohypophysial α- and β-CRF (Guillemin *et al.*, 1960). CRF from human hypothalamic and neurohypophysial extracts was purified by Schally *et al.* (1967) and from human hypothalamic tissue by Schilling and Blobel (1973). CRF activity was found distinct from vasopressin and ACTH fractions. Tumours removed from patients with ectopic ACTH syndrome are rich in peptides with ACTH-like activity, but peptides with CRF activity can also be found (Amatruda and Upton, 1974).

Chan *et al.* (1970) extracted CRF from rat median eminence tissue with HCl, 0·1 N NaOH or NaCl. The activity in these crude extracts was stable for long periods, while semi-purified extracts show a marked decrease in activity. Further purification was carried out by thin layer chromatography (TLC) in butanol: acetic acid: water. The active fractions had very low pressor activity. Further purifications by paper electrophoresis revealed one slightly anionic zone, which was ninhydrin-positive. CRF is easily extracted from rat median eminence tissue by extractants of high ionic strength.

Salem *et al.* (1970) have purified CRF of chickens and have shown a possible identity with an ACTH-like substance; its inactivation characteristics resemble those of α-MSH, the ACTH activity of which is resistant to treatment with acid, alkali or boiling at acidic pH and on this basis Salem *et al.* have suggested that α-MSH may be present in chicken hypothalami. The amount of vasotocin present in the chicken hypothalami accounts for only 5% of the total CRF activity.

B. STRUCTURE, CHEMICAL SYNTHESIS AND ANALOGUES

The structure of CRF is not known. Several synthetic peptides have been suggested as potential candidates, but their activity is inconsistent. The partial structure of a β-CRF related to α-MSH and vasopressin was elucidated by Schally and Bowers (1964) (Fig. 14). Saffran *et al.* (1972)

Table 9. CRF-activity of vasopressin and oxytocin-Analogues

Part 1

Structure	Relative CRF-activity %		
	CPZ-M-N	PB-M	Dexa
Derivatives of Arginin-Vasopressin (AVP):			
AVP	—	100·0	100·0
[Phe²]AVP	—	23·0	—
[(His-Ser-Cys)¹]AVP	—	3·5	—
[Cys(Bzl)¹, Cys(Bzl)⁶]AVP	—	—	Ø
[Ala¹, Ala⁶]AVP	5·5	—	—
[β-Ala¹, Ala⁶]AVP	—	—	31·5
[Gly,¹ Gly⁶]AVP	—	—	28·4
[β-Ala¹, Gly⁶]AVP	—	—	7·2

CPZ-M-N: rats blocked with chlorpromazine-morphine-nembutal
PB-M: rats blocked with pentobarbital-morphine
Dexa: rats blocked with dexametasone
—: not determined

(see also parts 2 and 3)

Table 9

Part 2

Structure	Relative CRF-activity % CPZ-M-N	PB-M	Dexa
Derivatives of Lysin-Vasopressin (LVP):			
LVP	—	78·00	88·0
[Phe^2]LVP	—	10·00	14·0
1-desamino-LVP	—	—	39·0
[Ile^3]LVP	—	—	35·0
[Val^7]LVP	—	0·01	—
[(Ac.Ser-Tyr-Ser-Cys)^1]LVP	—	Ø	—
[(His-Ser-Cys)^1]LVP	—	7·50	—
[(Ser-His-Cys)^1]LVP	—	1·30	—
1-desamino-4-decarboxamido-LVP	100·0	—	—
4-decarboxamido-LVP	16·0	—	—
[Ala^1, Ala^6]LVP	1·8	—	16·0
[β-Ala^1, Ala^6]LVP	—	—	—
[D-Ala^1, Ala^6]LVP	—	—	1·0
[Ser^1, Ala^6]LVP	Ø	—	—
[D-Ser^1, Ala^6]LVP	Ø	—	—
[Ser^1, Ser^6]LVP	Ø	—	—
[Ala^1, Ser^6]LVP	1·2	—	2·0
[β-Ala^1, Ser^6]LVP	Ø	—	—
[D-Ser^1, Ser^6]LVP	Ø	—	—
cyclo-[Pro^1, Glu^6]LVP	—	—	1·8
[Gly^1, Glu^6]LVP	Ø	—	5·9
cyclo [Gly^1, Glu^6]LVP	20·1	—	22·0
cyclo [β-Ala^1, γ-Glu^6]LVP	Ø	—	4·0
cyclo [Sar^1, γ-Glu^6]LVP	Ø	—	3·8
cyclo [β-Ala^1, β-Asp^6]LVP	Ø	—	6·0

( ) = extended at N-terminal Cys^1
Ac = acetyl

(see also part 3)

have claimed that pressinoic acid (the ring structure of vasopressin and oxytocin) is a corticotrophin releasing peptide. The findings were based on *in vitro* studies, in which pressinoic acid had CRF activity in doses of 3 to 30 ng/ml. The structural similarity of corticotrophin and α-MSH was taken as an indication that fragments of vasopressin might control both corticotrophin and α-MSH secretion, but confirmation of this hypothesis is lacking.

Table 9

Part 3

Structure	In vitro[1]	Relative CPZ-M-N	CRF-activity PB-M	% Dexa
Derivatives of desCys¹-Lysine-Vasopressin (LVP):				
des Cys¹-[Ala⁶] LVP	—	Ø	—	—
des Cys¹-[propionyl-[Ala⁶] LVP	—	Ø	—	—
des Cys¹-[metoxicarbonyl-[Ala⁶] LVP	—	Ø	—	—
Other vasopressin analogues:				
DDAVP 1-desamino-[D-Arg⁸] vasopressin	Ø	—	—	—
MSH-and ACTH-Fragments:				
ACTH$^{(5-10)}$hexapeptide	1	10·0	—	—
ACTH$^{(4-10)}$heptapeptide	40	18·6	—	—
ACTH$^{(3-10)}$octapeptide	—	46·0	—	—
ACTH$^{(1-10)}$decapeptide	40	—	—	—
des-acetyl-α-MSH	—	45·6	—	—
α-MSH	—	0·2	—	—
Oxytocin and derivatives:				
Oxytocin	Ø	—	1·9	Ø
[(His-Ser-Cys)¹]OT	—	—	Ø	—

[1] Kappeler and Schwyzer (1960) *in vitro* test procedure, CRF standard = 1

The CRF-activity of median eminence extract and various synthetic neurohypophysial polypeptides was compared by Portanova and Sayer (1973). ACTH secretion by isolated rat pituitary cells was maximal after addition of median eminence extracts, whereas synthetic peptides (arginine-vasopressin, homolysine vasopressin, lysine-vasopressin and oxytocin) were of lower activity.

Many vasopressin analogues have been studies for their CRF activity, but only a few of them reach the same level of activity as vasopressin (Table 9). The evaluation has been complicated by calculations in terms of pressor activity of these peptides. It is now known that the CRF activity of vasopressin analogues is neither related to the pressor nor antidiuretic activity. The only analogue with a CRF activity in the same order of magnitude as arginine-vasopressin or lysine-vasopressin is 1-desamino-4-decarboxamido-LVP (Arimura *et al.*, 1969).

C. BIOSYNTHESIS AND SECRETION

CRF is presumably synthesised in the hypothalamic hypophysiotrophic area. Lesion of the medical basal hypothalamus eliminate CRF secreting neurones (Brodish, 1973). Electrical stimulation of different pathways from the amygdala and tuberomamillary hypothalamus evokes ACTH release (Redgate and Fahringer, 1973). CRF biosynthesis appears to be unrelated to vasopressin biosynthesis, because CRF is present in the hypothalami of rats with hereditary diabetes insipidus (McCann *et al.*, 1966), whereas vasopressin is barely detectable. A morphological correlate of CRF has been identified by Bock (1972) in the outer layer of the median eminence (zona externa infundibuli) in the rat. Gomori-positive granules of the CRF type respond to suppression with adrenal steroids or osmotic stimuli in a manner different from that of vasopressin.

CRF activity has been identified in portal plasma of dogs (Porter and Rumsfeld, 1956) and rats (Brodish and Long, 1962). The rate of CRF release and/or biosynthesis is modulated by neurotransmitters, adrenal steroids and prostaglandins (see Chapter 11). Neurotransmitter control involves cholinergic and dopaminergic mechanisms. *In vitro*, acetylcholine and 5-hydroxytryptamine stimulate CRF release from isolated rat hypothalami, whereas noradrenaline, γ-aminobutyric acid (GABA) and melatonin inhibit these effects. Acetylcholine is also active on isolated rat and sheep synaptosomes (Edwardson and Bennett, 1974). A central inhibiting effect on ACTH-secretion was observed by arginine vasotocin (AVT) (Pavel *et al.*, 1973). The effect was specific for AVT and could not be obtained by AVP or oxytocin.

D. RECEPTOR BINDING AND MODE OF ACTION

Pituitary receptors have not been so far identified. CRF regulates the release of ACTH on a minute-to-minute basis (Porter *et al.*, 1967) and stimulates ACTH synthesis. ACTH is synthesised *de novo* in a cell-free system (Ishikawa and Goto, 1973) and the rate of biosynthesis can be stimulated 15-to 26-fold by addition of CRF. Biosynthesis can be inhibited by puromycin and by actinomycin D. The stimulatory effect of rat CRF on ACTH synthesis *in vitro* was confirmed by incorporation of ^{14}C-phenylalanine (Porter, 1974). Studies with purified CRF have shown that ACTH release can occur in the presence or absence of pituitary glycogenolysis: this was interpreted as ACTH release with or without concomitant *de novo* synthesis. One possible mode of action of CRF might be depolarisation of the cell membrane resulting in the release of ACTH independently of its re-synthesis (Hiroshige *et al.*, 1969). ACTH release by CRF is calcium and potassium dependent (Samli and Geschwind, 1968) and ouabain stimulates ACTH release from rat and guinea pig pituitaries by increased Ca^{++} influx

(Fleischer *et al.*, 1972). CRF in rat hypothalamic extract can interact with PGE_2 *in vitro* to enhance ACTH release (Sadow and Penn, 1973).

E. INACTIVATION, METABOLISM AND EXCRETION

The chemical inactivation of CRF is different from that of vasopressin, CRF being resistant to thioglycollate treatment. There are no reports on its metabolic fate or excretion.

F. COMPARATIVE ASPECTS OF STRUCTURE AND FUNCTION

The presence of corticotrophin-releasing activity attributed to a hypothalamic CRF has been demonstrated in hypothalamic extracts by effects in the dog (Ganong, 1959), rat (Hiroshige *et al.*, 1969), chicken (Salem *et al.*, 1970) and pigeon (Sato and George, 1973). Semi-purified extracts of bovine, porcine, ovine and human origin were active in several test systems. Neurohypophysial hormones can activate adrenal function in amphibia (Jørgensen and Nielsen, 1958; Jørgensen and Larsen, 1960). Electrical stimulation of the hypothalamic induces ACTH release in cats (Redgate, 1970) and fish (Redgate, 1974). A circadian periodicity of cortisol secretion was observed in carp and goldfish. Goldfish hypothalamic extract stimulates ACTH release from isolated goldfish pituitaries (Sage and Purrott, 1969). In *Poecilia latipinna*, the transplanted heterotopic pituitary responds to rat hypothalamic extract (Hawkins and Ball, 1970). Indirect evidence for hypothalamic control of adrenal function is given by the atrophy of autotransplanted pituitaries in the toad, *Bufo bufo* (Jørgensen, 1970) and of atrophic changes in the interenal gland of the crested newt after pituitary autotransplantation (Mazzi, 1970). The findings confirm the predominant stimulatory control of adrenal function by the hypothalamus in mammalian and submammalian species.

IX. MIF (Melanostatin) and MRF (Melanoliberin)

A. ISOLATION AND PURIFICATION

The secretion of MSH is controlled both by neural elements and neurohormones. Peptides have been extracted from hypothalamic tissue and obtained by enzymatic cleavage of neurohypophysial hormones with both stimulatory and inhibitory effects *in vitro* and *in vivo* on MSH release and synthesis.

The physiological identification of peptide hormones regulating pituitary MSH secretion is complicated by modulatory neurotransmitters, and by pituitary MSH interfering with the assay systems. For purification of MIF or MRF activity, catecholamines, α-MSH, β-MSH, ACTH and lipotrophic hormones have to be removed. Hypothalamic extracts of frogs, rabbits, kittens, sheep, pigs and man were found to increase pituitary MSH content in the rat by inhibition of MSH release. Following earlier reports, Taleisnik and Tomatis (1967a) have confirmed the presence of both MSH-releasing and MSH release-inhibiting activity in hypothalamic tissue.

(1) *MIF*

Purification of melanotrophin-inhibiting factor (MIF) was reported by Schally and Kastin (1966). Five hundred bovine hypothalami were lyophilised and defatted with acetone and petroleum ether. Two extractions with 2 N acetic acid and subsequently with glacial acetic acid were performed, followed by purification on Sephadex G 25.

MIF activity of the purified material was assayed by pituitary MSH depletion in rats followed by the frog skin assay (Shizume *et al.*, 1954). The extracts were free of a α-MSH, β-MSH and vasopressin.

(2) *MRF*

Dhariwal *et al.* (1966b) purified MRF activity from bovine hypothalami extracted with glacial acetic acid followed by gel filtration on Sephadex G 25. Retrospectively, these experiments have to be viewed with caution because it has been shown that such isolation procedures can lead to the concentration of catecholamines in certain fractions (Takahara *et al.*, 1974a).

B. STRUCTURE, CHEMICAL SYNTHESIS AND ANALOGUES

(1) *MIF*

In 1971, Celis and Taleisnik reported the enzymatic formation of a MIF by hypothalamic extracts of rats. Both oxytocin and vasopressin were suitable

1 2 3 4 5 6 7 8 9
Cys-Tyr-Ile-Gln-Asn-Cys-Pro-Leu-GlyNH$_2$

Pro-Lys-GlyNH$_2$

Pro-Arg-GlyNH$_2$

1 2 3 4 5
Cys-Tyr-Ile-Gln-Asn

Cys-Tyr-Phe-Gln-Asn

Melanostimulating | Melanostatic

Fig. 15. Hypothalamic peptides regulating MSH (melanotropin) secretion: Right = tripeptides derived from oxytocin and vasopressin. Pro-Leu-GlyNH$_2$, Pro-Arg-GlyNH$_2$ and Pro-Lys-GlyNH$_2$ have significant melanostatic activity. Left = Both oxytocin (1–5) pentapeptide and vasopressin (1–5) pentapeptide have MSH-releasing activity.

substrates for formation of this MIF-activity (Fig. 15). One incubation product was identified as Pro-Leu-GlyNH_2 (Celis and Taleisnik, 1971a, b). The isolation of natural MIF from bovine hypothalamic tissue by Kastin and Schally (1966), Schally *et al.* (1968a) led to the structural identification of the same tripeptide (Nair *et al.*, 1971a). The amino acid sequence was

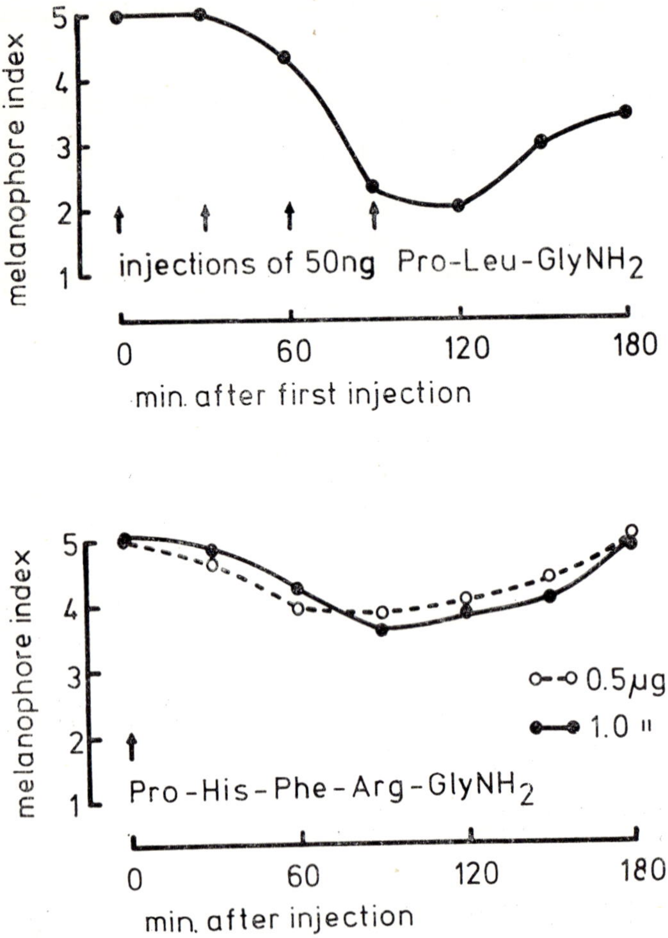

Fig. 16. Melanostatic (MIF)-activity of two hypothalamic peptides. MIF-activity was tested by injection into dorsal lymph sac of frogs (*Rana esculenta*) with hypothalamic lesions. Melanophore index (5) is complete darkening of frog skin, (1) complete absence of pigmentation. Pro-Leu-GlyNH_2 = MIF I; Pro-His-Phe-Arg-GlyNH_2 = MIF II.

determined by Edman degradation, combined with the microdansyl method; the findings were confirmed by mass spectroscopy.

The chemistry of peptides with MIF-activity is still controversial. The initial observation by Celis and Taleisnik (1971b) that the MIF tripeptide is released from oxytocin by incubation with the microsomal fraction of hypothalamic extracts, has been complicated by further reports. The ring structure of oxytocin was also reported to have MIF activity in frogs, while Pro-Leu-GlyNH_2 was inactive (Hruby *et al.*, 1972a). The synthesis of tocinoic acid and deaminotocinoic acid was reported by Hruby *et al.* (1972b). Both compounds have low oxytocic activity and are devoid of vasopressor activity. Bower *et al.* (1971) have claimed that the ring structure of oxytocin (tocinoic acid) or oxytocin (1–5)-pentapeptide inhibits MSH release from rat pituitaries, but not from frog pituitaries, and they find Pro-Leu-GlyNH_2 to be inactive *in vivo* both in the rat and frog. Grant *et al.* (1973b) were unable to show an effect of either tocinioc acid or Pro-Leu-GlyNH_2 on inhibition of MSH release in rat pituitary cell cultures; they were also unable to confirm an effect of Cys-Tyr-Ile-Gln-Asn-OH (MRF) on the release of MSH.

Another peptide with MIF activity has been isolated from bovine hypothalamus (Nair *et al.*, 1972) and the structure Pro-His-Arg-Phe-GlyNH_2 derived from structural analysis and mass spectral identification. The identification of different peptides with MIF activity (Fig. 16) is interpreted as an indication that more than one peptide may control MSH secretion. Celis *et al.* (1972) have studied structure:activity relationships of MIF related peptides: they found that tocinoic acid was without effect on MSH secretion. However, the analogues Pro-Lys-GlyNH_2 and Pro-Arg-GlyNH_2 were found to be inhibitors of MSH release, as well as MIF. All three peptides were able to antagonise the effect of MRF (Cys-Tyr-Ile-Gln-Asn-OH).

(2) *MRF*

The *in vitro* formation of an MSH-releasing factor (MRF) was reported in 1971 by Celis and Taleisnik (1971b) and was identified as a pentapeptide derived from the ring structure of oxytocin (Cys-Tyr-Ile-Gln-Asn-OH). (Celis *et al.*, 1971a) (Fig. 15). The peptide was able in nanogram amounts to increase plasma MSH concentrations and to decrease the MSH content of the rat pituitary. In the frog (*Rana esculenta*) N-terminal pentapeptides derived from vasopressin but not from oxytocin appear to release MSH (Fig. 17) (Sandow *et al.*, 1975). It is not known whether equivalents of such synthetic peptides exist naturally.

C. BIOSYNTHESIS AND SECRETION

The regulation of MSH secretion appears to be related to the metabolism of neurohypophysial hormones. Formation of MIF and MRF occur in the

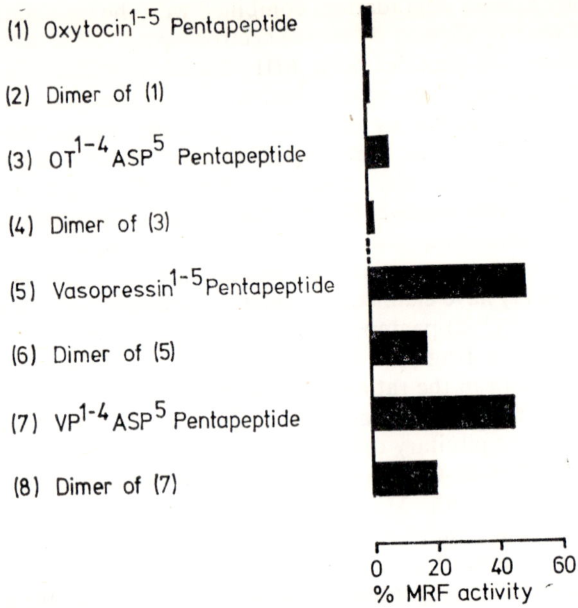

Fig. 17. MSH-releasing activity (MRF) of oxytocin-and vasopressin-fragments. MRF-activity was tested in frogs (*Rana esculenta*) by antagonism to melatonin (100 ng) (6 frogs per group, mean MRF activity in % increase over controls). Only vasopressin-fragments had significant MRF-activity.

hypothalamus, not in the posterior pituitary. Lesions of the paraventricular nucleus lead to large increases of MSH in plasma and pituitary and may indicate the site of biosynthesis or storage of MIF (Thody, 1974). Whether the factors controlling MSH-secretion are synthetised *de novo* or cleaved from preformed oxytocin, is still debated. *In vitro* experiments would argue for the possibility of cleavage *in situ* from oxytocin or vasopressin. Little is known about the secretion rate of MIF or MRF, but the fact that they can be formed from oxytocin by enzymes normally participating in the inactivation of these hormones is supported by the increase of pigmentation during pregnancy in the human.

Celis and Taleisnik (1973) have studied the enzymes and found them only in hypothalamic tissue. Both with oxytocin and vasopressin as substrates, a material with MSH-releasing activity was formed (MRF). Oestrogen promotes the formation of the enzyme producing MIF (Celis and Taleisnik, 1974), MSH itself could be formed by enzymes involved in ACTH breakdown (Landon *et al.*, 1975), thus providing an alternate regulatory mechanism.

The regulation of MSH secretion has been recently reviewed by Kastin *et al.* (1974b) (and see Chapter 12). MSH secretion is under the influence of catecholamines (Taleisnik *et al.*, 1972) and indolamines, and occurs at two levels. At the suprahypothalamic level, an α-adrenergic receptor is involved. Following noradrenaline, a depletion of hypothalamic MIF is observed. Injection of noradrenaline or dopamine into the 3rd ventricle can block pituitary MSH depletion, either by MRF or by hypertonic saline, β-adrenergic blockers are without effect on MSH depletion.

There are some difficulties with respect to the secretory pathway for MIF and MRF, due to the lack of an adequate portal vascularisation in many species. The short portal vessel system may be involved as an alternate pathway. Pituitary sensitivity to MIF and MRF can be modified by numerous pharmacological agents.

D. RECEPTOR BINDING AND MODE OF ACTION

It is doubtful in view of the high specificity of the neurophysins, whether MIF, MRF or other fragments of neurohypophysial hormones are bound to these proteins. Pituitary receptors have not been identified for either MIF or MRF, but labelled MIF is accumulated in the pituitary (see Section E). Circulating levels of MIF or MRF have not been demonstrated, the usual assays relying on measurement of pituitary MSH content. MIF and MRF activity have been demonstrated in rat blood under various conditions, but this evidence is doubtful in view of the lack of specificity of biological assays for plasma MSH.

Little is known about the mechanism of action of these peptides. The effect of MIF (Pro-Leu-$GlyNH_2$) does not seem to involve the adenyl cyclase system in the neuro-intermediate lobe of the dogfish (Deery and Jones, 1975).

A potential clinical application for MIF (Pro-Leu-$GlyNH_2$) has been investigated because of the observation that MSH aggravates the symptoms of Parkinson's disease. Very high doses of MIF, alone or in combination with L-Dopa, seem to have a beneficial effect on tremor and rigidity. In pharmacological studies designed for the detection of anti-Parkinsonian drugs, analogues of MIF have been tested for their ability to counteract oxytremorine-induced tremors in mice. The TRH/MIF hybrid Pyr-Leu $GlyNH_2$ was almost twice as active as MIF. The neuropharmacological effects of MIF and its analogues are possibly unrelated to the physiological effect of MIF, because of the extremely high doses required (Castensson *et al.*, 1974).

E. INACTIVATION, METABOLISM AND EXCRETION

(1) *MIF*

The metabolic fate of the tripeptide, Pro-Leu-$GlyNH_2$ has been studied by Redding *et al.* (1973b), Nair *et al.* (1973), Kastin *et al.* (1974a) and

Redding *et al.* (1974). Plasma elimination of Pro-[^{14}C-Leu]-GlyNH_2 and [^{3}H-Pro]-Leu-GlyNH_2 is rapid with a half-life of 3–12 minutes. The volume of distribution is larger than the plasma volume, as is the case with vasopressin, oxytocin and other hypothalamic hormones. Label was accumulated in the pituitary, pineal gland, adrenal, kidney and liver. There is no satisfactory explanation for the pineal uptake of label. A predominant metabolite was Leu-GlyNH_2 (Fig. 18). Plasma inactivation of the labelled compounds

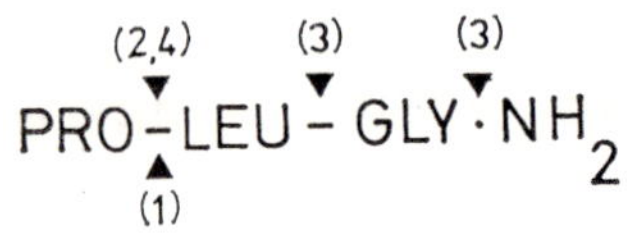

(1) Marks and Walter 1972 : brain extract, median eminence extract (arylamidases)

(2) Nair et al. 1973 : human plasma (main metabolites)

(3) Nair et al. 1973 : human plasma (trace metabolites)

(4) Redding et al. 1974 : human in vivo (main metabolites)

Fig. 18. Metabolic inactivation of MIF melanostatin. Sites of enzymatic cleavage are indicated by arrows. Main metabolite is Leu-GlyNH_2.

was rapid and showed again Leu-GlyNH_2 as the main breakdown product, as well as Pro-Leu-GlyOH and Leu-GlyOH in small quantities. All metabolites were inactive in the MIF assay in frogs (Nair *et al.*, 1973). The pituitary appears to be one of the main inactivating organs, as shown by prolonged plasma disappearance after hypophysectomy (Kastin *et al.*, 1974a). As in other instances, a constant supply of MIF to the pituitary could still lead to variable release of MSH due to changes in pituitary inactivation rate. The findings obtained in rat were essentially confirmed in human (Redding *et al.*, 1974). However, whereas only minimal amounts of the intact peptide were excreted into the urine in rats after one hour, a considerable amount was excreted in the human (17% after one hour and 23% after three hours).

(2) *MRF*

The metabolic fate of the pentapeptide derived from oxytocin has not been studied to date.

F. COMPARATIVE ASPECTS OF STRUCTURE AND FUNCTION

The evidence for the regulatory role of the tripeptide Pro-Leu-GlyNH_2 and related fragments of oxytocin and vasopressin on MSH secretion in amphibia is still controversial. It appears however, that in frogs metabolites of neurohypophysial hormones and related synthetic peptides can inhibit or stimulate MSH secretion. Extrapolation of these findings to mammals is very difficult (see Chapter 12).

NOTE ADDED IN PROOF

Recent studies have shown, that a partial dissociation of the pharmacodynamic effects of somatostatin analogues can be obtained. The ratio of insulin inhibition to glucagon inhibition is increased by lengthening of the peptide chain at the N-terminal end by Gly-Gly-Gly, by elimination of Asn^5 and/or by substitution of Ser^{13} with D-Ser. Additional substitution of Trp^8 potentiates this effect. The best split of insulin/glucagon inhibition so far obtained was seen with des-Asn^5-[D-Ser^{13}] somatostatin. On the other hand the substitutions [Ala^2] and [D-Cys^{14}] reduce insulin inhibition in favour of glucagon inhibition. Again [D-Trp^8] substitution enhances the effect. [D-Trp^8, D-Cys^{14}] ST has the highest specific glucagon inhibition of all derivatives synthetized so far. A specific effect on growth hormone inhibition is seen with the carba-analogues cyclo γ, α-des-(Ala^1, Gly^2, Cys^3)-[Glu^{14}] ST and cyclo β, γ-des-(Ala^1, Gly^2)-[4-Abu^3, Asp^{14}] ST. This effect on growth hormone inhibition is also seen with bicyclic somatostatin, where the α-carboxylgroup in Cys^{14} is linked to the α-aminogroup of Ala^1 by a peptide bond. Other substitutions which increase inhibition of growth hormone secretion are [D-Cys^{14}], [D-Trp^8] and [D-Ser^{13}]. However, if the bicyclic somatostatin analogue mentioned above is substituted with [D-Trp^8], the original spectrum of activities reappears.

ACKNOWLEDGMENTS

The authors would like to thank their colleagues Dr. Kruse and Dr. Geisen for important contributions to this review. They would also like to thank Miss G. Jerzabek and B. Krauß for preparation and revisions of the bibliography.

CHAPTER 6

Neurohypophysial Hormones—Comparative Aspects

B. T. PICKERING

I. INTRODUCTION

In recent years there have been a number of reviews of the comparative endocrinology of the vertebrate neurohypophysis. The purpose of the present chapter is to provide one link in a complete picture of the hypothalamic hormones and, for this reason, I shall not conduct an exhaustive review of the research literature but rather attempt to put the non-mammalian neurohypophysial hormones into perspective. In doing so I shall restrict literature citations to a minimum and provide a bibliography of review articles through which the interested reader can gain access to reports of the original research on which present-day ideas are based.

Broadly speaking there are two main ways in which a comparative study of an endocrine gland may be approached. Firstly, one can ask to what extent the biological properties of extracts of glands from different species are similar to each other, and leading on from this, how closely the substances responsible for these activities resemble each other chemically. The second approach is to administer the hormones isolated from the gland of one species to representatives of others, and to observe the consequent changes in the recipients' physiology. By determining the similarities and differences both of the substances produced by the glands and of the reactions to these substances of animals throughout the vertebrate subphylum, one can begin to assess the basic physiological significance (if there is one) of the organ, and the ways in which this has altered to meet the special needs of individual species.

II. The Nature of Vertebrate Neurohypophysial Hormones

(For reviews see: Sawyer, 1968, 1971; Heller and Pickering, 1970; Acher, 1974b).

The success of Kamm and his colleagues in 1928 in separating the principles responsible for the biological activities of beef posterior lobes into two fractions, one with principally oxytocic and avian vasodepressor activities and the other with principally vasopressor and antidiuretic activities, opened the door to a substantial amount of research on the neurohypophysis (see Chapter 5). The Kamm procedure was rapidly applied on a commercial scale so that Pitocin and Pitressin became available to clinicians and pharmacologists alike. Thus when, in the 1930s, Heller began the trail to the comparative biochemical studies of the neurohypophysis, he had available these partially purified preparations of the mammalian hormones for use as reference substances.

Soon after the pharmacological actions of the mammalian neurohypophysial extracts had been recognised, various workers (notably Herring and Hogben and de Beer) showed that similar effects could be elicited with extracts of the pituitaries from representatives of most vertebrate Classes. Heller began a systematic, more quantitative, pharmacological study of non-mammalian neurohypophysial extracts, determining for each species what he later came to refer to as the "pharmacological spectrum" of each extract, i.e. the relative potency of the extract to one of the mammalian hormones (usually oxytocin) in as many assay systems as possible. Significantly, one of the systems which Heller used in these early studies was the "frog water-balance" assay which measures the ability of a substance, injected into frogs kept in water, to induce an increase in their body weight by fluid uptake. He found that all of the non-mammalian posterior pituitary extracts which he studied contained far more "water-balance" activity, relative to their antidiuretic activity, than did mammalian extracts. He came to the conclusion that "The possibility must therefore be envisaged that the posterior pituitary glands of different classes . . . of vertebrates elaborate a secretion which exhibits the same activities but—by some unknown modifications of the molecule—contains them in different proportions" (Heller, 1941). As will be seen, this transpired to be amazing foresight at a time some 12 years before the chemistry of the mammalian hormones had been elucidated.

Progress in the isolation of the mammalian hormones which culminated in the determination of the structure of oxytocin, arginine vasopressin and lysine vasopressin (see Chapter 5), re-awakened interest in the non-mammalian neurohypophysial hormones. Three laboratories (Heller's in Bristol, Sawyer's in New York and Acher's in Paris) reached similar conclusions almost simultaneously: that there was present in the neurohypophysial tissue

Cys . Tyr . Ile . Gln . Asn . Cys . Pro . Leu . Gly(NH_2)

Oxytocin

Cys . Tyr . Ile . Gln . Asn . Cys . Pro . Arg . Gly(NH_2)

[Arg^8]-oxytocin, arginine vasotocin

Cys . Tyr . Phe . Gln . Asn . Cys . Pro . Arg . Gly(NH_2)

[Phe^3,Arg^8]-oxytocin, arginine vasopressin

Fig. 1. The structure of oxytocin, arginine vasotocin and arginine vasopressin showing that the non-mammailian peptide is a chemical hybrid of the two mammalian hormones.

Table. 1. Comparison of the Biological Activities of Arginine Vasotocin (AVT) with those of Arginine Vasopressin (AVP) and Oxytocin (OXY)

Activity	Potency (I.U./mg)*		
	OXY	AVT	AVP
Rat Uterotonic ($-Mg^{2+}$)	450	125	9
Rat Uterotonic ($+Mg^{2+}$)	405	238	21
Rat Vasopressor	2·7	245	400
Rat Antidiuretic	1·1	195	400
Rabbit Milk Ejection	380	275	51
Frog Water-Balance⁺	450	114,000	23

* For references see Pickering (1970).
⁺ In terms of water movement across frog bladder *in vitro*.

Cys . Tyr . Ile . Gln . Asn . Cys . Pro . Ile . Gly(NH_2)

[Ile^8]-oxytocin, mesotocin

Cys . Tyr . Ile . Ser . Asn . Cys . Pro . Ile . Gly(NH_2)

[Ser^4,Ile^8]-oxytocin, isotocin

Cys . Tyr . Ile . Ser . Asn . Cys . Pro . Gln . Gly(NH_2)

[Ser^4,Gln^8]-oxytocin, glumitocin

Cys . Tyr . Ile . Asn . Asn . Cys . Pro . Leu . Gly(NH_2)

[Asn^4]-oxytocin, aspargtocin

Cys . Tyr . Ile . Gln . Asn . Cys . Pro . Val . Gly(NH_2)

[Val^8]-oxytocin, valitocin

Fig. 2. Structures of peptides which have been characterized in neurohypophysial extracts from non-mammalian vertebrates.

of bony fishes, amphibians, reptiles and birds a peptide with a structure intermediate between that of oxytocin and arginine vasopressin. This peptide, arginine vasotocin, has the ring of oxytocin and the side chain of vasopressin (Fig. 1) and, because of its obvious pharmacological interest, had been synthesised in du Vigneaud's laboratory a year or two before its discovery in nature. The pharmacological properties of vasotocin (Table 1) are in the main just what would be predicted from its structure, i.e. it shows the activities of both oxytocin and vasopressin, with somewhat lower potency than these hormones. However, its most interesting property is that it is about 200 times more potent than the mammalian hormones when measured in the frog water-balance assay. This substance is therefore the amphibian water-balance principle which Heller had detected and possessed, as he had predicted, a structure slightly different from those of the mammalian hormones.

It soon transpired that AVT is present in the neurohypophysis of every vertebrate class except the mammals, although in all classes except Agnatha (cyclostomes and hagfish) it is accompanied by at least one other peptide. Soon after the discovery of arginine vasotocin, two other peptides were isolated from non-mammalian, neurohypophysial extracts and were shown to be structural analogues of the mammalian hormones as well (Fig. 2). [8-isoleucine]-oxytocin (mesotocin) was first isolated from amphibian extracts and later shown to occur also in the glands of reptiles and birds (see below). It differs from mammalian oxytocin by a single amino acid, having an isoleucine residue in position 8 of the chain instead of a leucine. The extracts of glands from bony fishes contained, in addition to vasotocin, another structural analogue, [4-serine, 8-isoleucine]-oxytocin (isotocin), in which there is a second amino acid difference from oxytocin: the glutamine in position 4 is replaced by serine (Fig. 2).

The remaining three peptides shown in Fig. 2 have all been isolated from cartilaginous fishes. Perks and Sawyer (1965) isolated a peptide from a ray (*Raia occellata*) and found that it had the amino acid composition of a structural analogue of oxytocin in which leucine was absent but serine present. They ruled out the possibility of a straight substitution of serine for leucine at position 8 by comparison of the "pharmacological spectrum" of the peptide with synthetic [8-serine]-oxytocin. Almost simultaneously, in Paris, Acher and his colleagues were working with a similar peptide from *R. clavata* and they were able to characterise their compound chemically and established its structure as [4-serine, 8-glutamine]-oxytocin (glumitocin). This peptide differs from oxytocin in two residues having serine substituted for glutamine in position 4 and glutamine for leucine in position 8, although as can be seen, its overall amino acid composition shows only one substitution as Perks and Sawyer had found.

For some years there was disagreement between the Bristol group and the New York group about the number of oxytocin-like peptides in the dogfishes which are members of the shark group of elasmobranch fishes (Selachii): fishes examined in the U.S.A. showed only one active region on chromatograms while in the U.K. they produced two. The problem was not resolved when Sawyer and his colleagues showed that dogfishes elaborated small amounts of vasotocin, since there could not have been enough present to account for Heller's results, and the discrepancy between one and two peptides became a discrepancy between two and three. In 1972 Acher *et al.* announced the isolation of two new peptides from the pituitary gland of the spiny dogfish (*Squalus acanthias*). Each of these differed from oxytocin by a single amino acid substitution being [4-asparagine]-oxytocin (aspargtocin) in which asparagine replaced glutamine, and [8-valine]-oxytocin (valitocin) in which valine replaced leucine.

III. Distribution of Neurohypophysial Hormones Within the Vertebrates

(For reviews see: Heller and Spickett, 1966; Geschwind, 1967; Sawyer, 1968, 1970, 1971; Perks, 1969; Heller and Pickering, 1970; Acher, 1974b; Heller, 1974).

Table 2 shows the occurrence of the various structural analogues of oxytocin in the neurohypophysis of representatives of each of the classes in the vertebrate sub-phylum. Before attempting to make some general comments on this distribution and suggesting that there may be some Class-specific peptides, it is important to emphasize the small number of individual species, let alone Orders, which have been investigated. In no Class, with the possible exception of the mammals, has a sufficient proportion of the representatives been sampled to make a statistically valid conclusion about Class-specificity.

Not withstanding this caveat, some comments are worthwhile. Arginine vasotocin has been found in every non-mammalian neurohypophysial extract which has been studied. In the cyclostomes and hagfish, AVT is the only active peptide which has been detected, while in every other animal at least one other peptide seems to be present. In recent years there has been increasing evidence that AVT persists into the mammals. Not only has the peptide been detected in mammalian pineal glands (Milcu *et al.*, 1963; Cheesman, 1970) but also Perks and Vizsolyi (1973) have isolated and characterised AVT from extracts of neurohypophyses taken from mid-term foetuses.

Oxytocin, the mammalian hormone which participates in parturition and promotes milk let-down during lactation, has also been detected in several non-mammalian vertebrates. The peptide has been chemically characterised in the modern day descendant (*Hydrolagus collei*) of the primitive cartila-

Table 2. Distribution of Neurohypophysial Peptides among the Classes of Vertebrates

PEPTIDE (As derivative of oxytocin)	VERTEBRATE CLASS											
	Cyclostomes and Hagfish	Cartilaginous Fishes			Bony Fishes				Amphibians	Reptiles	Birds	Mammals
		Chimaeroids	Sharks	Rays	Lungfishes	Polypterus	Sturgeon	Ray-finned fishes				
Oxytocin		+			+		*		(?)	(?)	(?)	+
[Arg⁸]–	+	+	+	+	+	+	+	+	+	+	+	f
[Ile⁸]–					+	(?)			+	+	+	
[Ser⁴, Ile⁸]–						+		+				
[Ser⁴, Gln⁸]–				+								
[Asn⁴]–			+									
[Val⁸]–			+									
[Phe³, Arg⁸]–												+
[Phe³, Lys⁸]–												+

* The Sturgeon contains an unknown oxytocin-like peptide (Acher *et al.*, 1973).
(?) Detected by some authors, questioned by others.
f Arginine vasotocin detected in glands of foetal mammals.

ginous fishes and hence appears to have arisen early during the phylogeny of the vertebrates. In the lungfishes there is pharmacological evidence for its occurrence in some species as there is in some amphibians and reptiles, although the major peptide apart from AVT in these species is mesotocin. In the birds, also, there has been some confusion in the literature. In 1960 Acher *et al.* isolated a peptide from chicken pituitaries which they characterised chemically as oxytocin and several groups published pharmacological data which supported their findings. However, in a rigorous re-examination of avian pituitaries from three species, Acher *et al.* (1970) could find only mesotocin in addition to AVT.

Mesotocin can be considered as the peptide of non-mammalian tetrapods as Acher has suggested. It has been chemically identified in representatives of the amphibians, reptiles and birds. Interestingly, mesotocin has also been isolated from the neurohypophysis of several species of lungfish which along with the coelocanths, are thought of as the present day descendants of the ancestor of the amphibians.

Isotocin has been found only in the bony fishes and, indeed, has been recognised in every member of this group of vertebrates which has been studied, except the lungfish. In recognition of this apparently strict Class-specificity, Heller proposed that the peptide be called ichthyotocin, but this name has never been widely accepted.

Glumitocin, aspargtocin and valitocin are all restricted to the cartilaginous fishes where there seems to be Order-specificity rather than Class-specificity, since glumitocin occurs only in that group of elamobranchs (*Batoidea*) which includes the rays, whereas aspargtocin and valitocin have been found only in the spiny dogfish which belongs to the shark group. Because of the small amounts of peptide present in elasmobranch pituitaries it has not been possible to decide whether both of these peptides occur together in single pituitaries or that there is a race of aspargtocin-producing dogfish and one which elaborates valitocin.

The vasopressins do not seem to have appeared before the emergence of the mammals but there is pharmacological evidence, at least, for the presence of arginine vasopressin (AVP) in monotremes and marsupials, as well as in most of the groups of eutherian mammals. Moreover, the peptide has been chemically identified in neurohypophysial extracts from several species (see Acher, 1974b). During the purification of the vasopressins it was soon realised that porcine vasopressin was different from that of the ox. Soon after they had established the structure of the bovine hormone as arginine vasopressin, du Vigneaud and his colleagues showed that the hormone present in the pig was lysine vasopressin (LVP), which differed from the bovine peptide by the substitution of a lysine residue for the arginine at position 8.

Lysine vasopressin seems to be widely distributed among the suborder

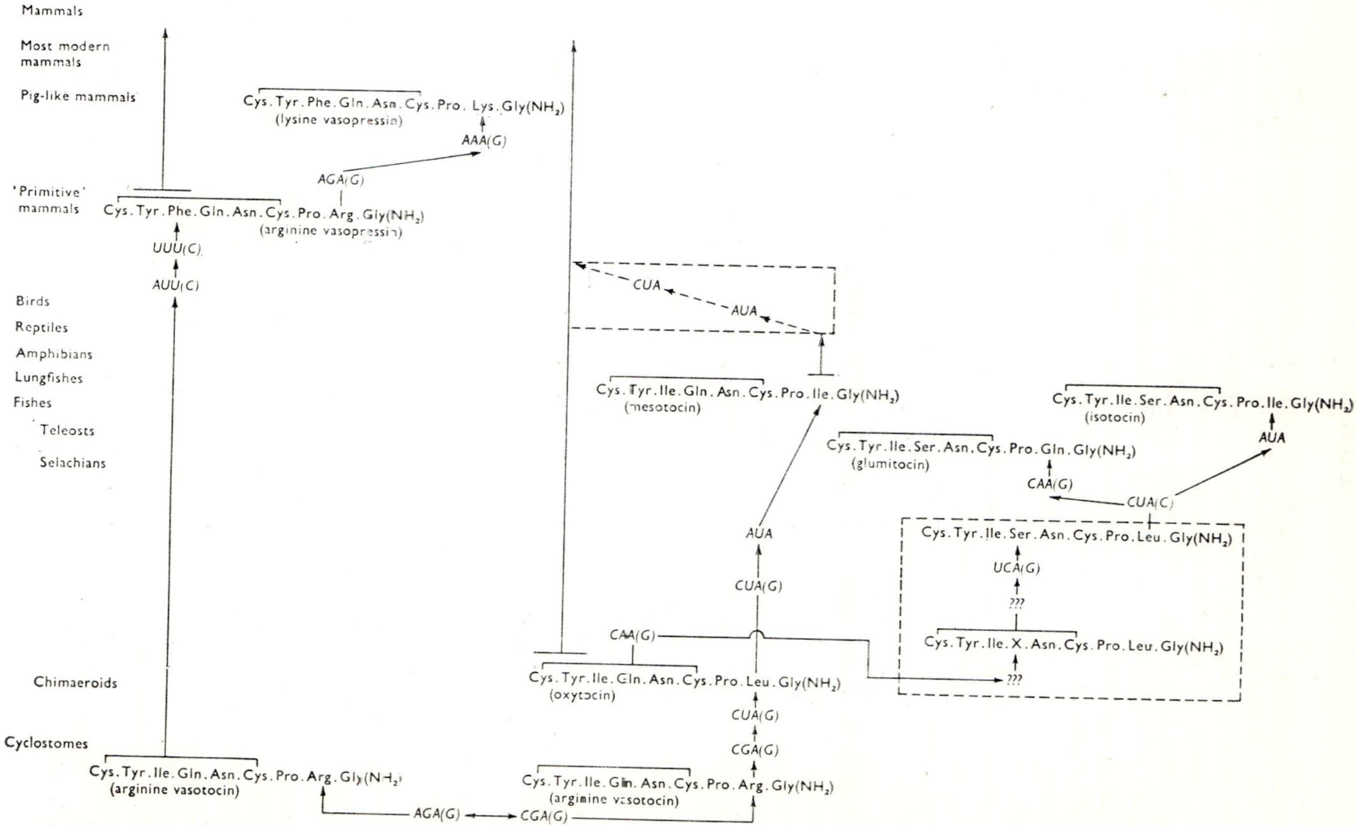

Fig. 3. A scheme for the molecular evolution of neurohypophysial hormones. Each step represents a single base mutation and the codons participating in these mutations are shown in italics. In the lower box of dotted lines are shown possible transitions involving hypothetical intermediates. The upper box illustrates the possibility that an oxytocin gene of birds and mammals may have arisen from a back mutation of the [8-isoleucine]-oxytocin gene present in reptiles and amphibians, as well as from a gene that survived unchanged from holocephalians (taken from Pickering and Heller, 1969).

of artiodactyls known as *Suina*. This group of animals includes the Old World Pigs, the peccaries (New World Pigs) and the hippopotamus: of these, only the domestic pig seems to elaborate exclusively LVP. In three reports on the glands of the hippopotamus examined in batches, Heller and Lederis (1960) found only LVP; Ferguson and Pickering (1969) found both AVP and LVP and Uttenthal and Hope (1972) found only AVP. When single glands of various members of the pig family (wild boar, bush pig, warthog, giant forest hog and peccary) were examined, it became apparent (Ferguson and Heller, 1965; Ferguson, 1969) that individuals could be homozygous for LVP, homozygous for AVP or heterozygous (containing both AVP and LVP). Indeed, when Ferguson analysed the distribution of genotypes in the warthog, he found that it obeyed the Hardly-Weinberg law for a random frequency, suggesting that none of the three genotypes had a selective advantage over the others. There is also a report that LVP occurs in the glands of one strain (Peru) of mouse (Stewart, 1971). The occurrence of polymorphism in the *Suina* suggests that we should be cautious before dismissing the reports of oxytocin along with mesotocin in some amphibians, reptiles and birds, since it is possible that a similar polymorphism occurs in these species. It will also be interesting to discover whether a similar situation exists in the sharks with respect to aspargtocin and valitocin.

The existence of a series of peptides related by single amino acid substitutions has led many authors to suggest schemes for the phylogeny of the vertebrate neurohypophysial hormones within the constraints of the genetic code of protein synthesis. In such a scheme, each peptide must be related to its immediate ancestor by a single base change in one of the codons which govern its amino acid sequence. In setting up such a scheme there is an implicit assumption that the hormones of the present-day forms of animals which arose early in the evolution of the vertebrates have survived unchanged, so that one must be cautious of over-interpretation. Nevertheless, it is possible to construct useful schemes and most of the reviews cited at the head of this section contain one. Figure 3 shows the scheme of Pickering and Heller (1969) which, in common with all others, suggests that AVT is the primitive vertebrate hormone. It is proposed that the "neurohypophysial" gene of the primitive vertebrates underwent gene-duplication from which two lines of peptides evolved: (1) a series of basic peptides consisting of AVT and the mammalian vasopressins; and (2) a series of so-called "neutral" oxytocin-like peptides. Now that it is known that mammals elaborate AVT (in the foetus), as well as AVP and oxytocin, it is necessary to postulate a further gene-duplication somewhere between the primitive vertebrates and the mammals. In this respect, it will be of interest to know whether both aspargtocin and valitocin can occur together, along with AVT, in the same dogfish, because if this is so a second duplication must have occurred rather early in phylogeny.

As can be seen from Fig. 3 there is no problem in relating AVT and the vasopressins by single base changes in the amino acid codons, but difficulty does arise in the "neutral" series. The substitutions at position 8 are all easily predictable since arginine, leucine, glutamine, isoleucine and, indeed valine, can be related as shown in the Figure. However, consideration of the genetic code shows that one cannot accomplish the substitution of serine for glutamine by a single base change. For this reason a hypothetical intermediate [X^4]-oxytocin was included in the scheme in Fig. 3. Since the preparation of this scheme aspargtocin ([Asp^4]-oxytocin) has been discovered in the dogfish but, while it is possible to explain the transition Asn → Ser in terms of a single base change (AAC → AAG), it is not possible to similarly relate glutamine and asparagine (CAA → AAC) except to note that the codon for asparagine is the mirror image of that for glutamine and to wonder whether such a small piece of nucleic acid can become reversed during replication or transcription. Geschwind (1967) pointed out that, if the transition Gln → Ser were to occur by single base changes, X would have to be proline, but pharmacological tests of relevant synthetic analogues have ruled out this possibility (for refs see Pickering and Heller, 1969). Moreover, this transition required the UCA(G) codon for serine and this would not allow a single base mutation to give asparagine (AAC).

IV. Comparative Physiology of the Neurohypophysial Hormones

(For reviews see: Jard, 1966; Morel and Jard, 1968; Sawyer, 1968, 1970, 1971; Heller, 1969, 1972; Perks, 1969; Follett, 1970; Skadhauge, 1973; Bentley, 1974; Maetz and Lahlou, 1974).

In looking for a physiological role for the neurohypophysial principles of non-mammalian vertebrates, the comparative endocrinologist begins by searching for similarities with the actions of the analogous hormones in the mammals. Here vasopressin is concerned with the maintenance of body water and oxytocin with the process of parturition and with the expulsion of milk during lactation. Most of the non-mammalian glands examined also contain two neurohypophysial peptides and it is tempting to suggest that there are two series of hormones, one with a primarily antidiuretic role, and one with a part to play in the reproductive process, as exemplified by vasopressin and oxytocin, respectively, in the mammals. However, as will be shown, such a suggestion is not supported either by the pharmacological actions of the peptides or, where it can be deduced, by their physiological effects.

A. Water Conservation and Sodium Metabolism

Arginine vasotocin has been considered as the non-mammalian antidiuretic hormone and there is considerable evidence for this, even though some of it is circumstantial. In mammals, vasopressin acts by facilitating water reabsorption from the kidney tubules, and there is evidence for a similar tubular effect of AVT in the kidneys of amphibians, reptiles and birds. In addition, AVT can depress urine flow in amphibians and reptiles by decreasing the glomerular filtration rate (GFR). This is probably achieved, not by regulating the rate at which an individual glomerulus filters, but by changing the number of active glomeruli. Some authors have shown an effect of AVT on GFR in birds but, to do so, had to use doses which would have resulted in plasma concentrations far in excess of those which might be expected physiologically.

In addition to the kidney, the urinary bladders of amphibians and reptiles, and the cloacae of amphibians, reptiles and birds are sites where water and sodium reabsorption occur. Of all these structures, only in the amphibian bladder is there clear evidence for an effect of AVT. There is one report of an action of LVP on water transport by the urinary bladder of the tortoise but this was not reproducible by other authors. The amphibian bladder, on the other hand, is extremely sensitive to neurohypophysial hormones of which AVT is by far the most potent. Both osmotically driven water transport and active transport of sodium across the bladder are augmented by the peptide. The actions of the neurohypophysial hormones on the amphibian bladder have much in common with their actions on kidney tubules and, for this reason, the tissue has been used as a model system both for the study of the cellular mechanisms involved in kidney function and for the study of structure-action relationships among the analogues of oxytocin and vasopressin. The amphibians are unique among the tetrapods in having a permeable skin and this too is a target tissue for neurohypophysial hormones in much the same way as the bladder.

It is only in the past few years that it has become apparent that AVT is an antidiuretic substance in fishes as well. In most studies on fish, neurohypophysial hormones had been found either to have no effect or to produce diuresis. However, Chester Jones *et al.* (1971) and Babiker and Rankin (1973) found that, in the freshwater-adapted eel, while AVT at doses above 10^{-8} g/kg was diuretic, a dose-dependent antidiuresis could be obtained at lower doses (10^{-13}–10^{-9} g/kg). It seems that these effects may result from a pressor action of the peptides on the renal vasculature: constriction of the afferent ateriole without an effect on the efferent arteriole would decrease GRF and hence produce diuresis; while a greater constriction of the efferent vessel could overcome and reverse this, as could recruitment of a greater number of glomeruli into action.

Neurohypophysial peptides influence the movement of sodium ions across the gills of fishes, and this is the only action on a non-mammalian vertebrate in which any analogue has a more potent effect than AVT. All the neurohypophysial peptides which show activity increase both sodium influx and outflux, with AVT having the greatest potency. When net sodium uptake is measured, however, [Ser^4, Ile^8]-oxytocin (mesotocin) is most potent, i.e. it has a greater effect on influx than on outflux.

To what extent are the effects on the water and salt metabolism described above of physiological significance rather than interesting pharmacological effects? For AVT to be a true antidiuretic hormone it must be released into the plasma at a concentration great enough to produce the required response. Studies of plasma concentrations are few and far between, although it has been shown that the plasma AVT level in amphibians increases as a result of osmotic stimulation. Similar increases could not be detected in eels stimulated in various ways, although the threshold sensitivity (10^{-11}M) of the assay used was such that an antidiuretic concentration of the peptide would not have been detected. Similar estimates have not been made in reptiles or birds, although dehydration results in a depletion of the neurohypophysial peptides in the Japanese quail with a greater effect on the vasotocin content. Thus, although it has not been completely substantiated, it does appear that AVT is the non-mammalian antidiuretic hormone, with a physiological role to play in the conservation of body water.

B. ACTIONS ON THE REPRODUCTIVE SYSTEM

One of the biological activities which aided the pharmacological characterisation of AVT in non-mammalian neurohypophysial extracts was stimulation of the chicken oviduct to contract. Moreover, Heller and his colleagues have shown that neurohypophysial extracts will elicit similar contractions in the isolated oviducts of representatives from elasmobranch and teleost fishes, amphibians and reptiles. The question which still remains is whether this represents an interesting study in comparative pharmacology or reflects a physiological role for the neurohypophysis in the delivery of eggs, or young, throughout the vertebrate sub-phylum. Certainly, as would be expected, oviposition can be induced in reptiles and birds by the injection of neurohypophysial peptides at the appropriate time. During natural oviposition in the chicken there is a decrease in the amount of AVT stored in the hen's neurohypophysis, and this can be correlated with an increase in the plasma concentration of the peptide. The problem is whether this release of AVT is a stimulus or a response to oviposition. It seems likely that AVT plays a similar role during oviposition in birds as oxytocin does during parturition in mammals (Lincoln, 1974), namely that the peptide, while not initiating, augments the intrinsic contractile activity of the muscle. There are no reports of changes in plasma AVT concentrations associated with oviposition (or parturition) in fishes, amphibians or reptiles, but the observation that sensitivity of fish and amphibian tissue to AVT increases, in some cases dramatically, prior to the expulsion of eggs or young, argues for a similar role for the peptide in these species.

Arginine vasotocin was the most active peptide in promoting contraction of oviductal/ovarian muscle from all the species which Heller studied other than the spotted dogfish (*Scyliorhinus canicula*). In this fish AVT was totally inactive, but a fraction containing the other oxytocin-like peptide(s) from the animals own neurohypophysis did stimulate contractions. It is interesting that the cartilaginous fishes differ from most other vertebrates by having very small amounts of AVT in their glands; although this may reflect their uraemic mechanism for osmoregulation (Chauvet *et al.*, 1971) rather than the sensitivity of their reproductive tract. The neurohypophysial peptides of *Scyliorhinus* have, as yet, not been characterized and it will be of interest to see if they are identical with the aspargtocin and valitocin present in *Squalus* since the latter dogfish is viviparous while the former is oviparous.

V. Conclusions

The pars nervosa of the pituitary gland of every vertebrate which has been studied contains peptides which are structural analogues of the mammalian neurohypophysial hormones oxytocin and vasopressin. In the mammals, these hormones are synthesised in the supraoptic and paraventricular nuclei of the hypothalamus in association with their specific neurophysins, and transported along the axons of the neurosecretory neurons for storage in and release from the neural lobe of the pituitary gland (see Pickering *et al.*, 1974). So far the only non-mammalian gland from which a neurophysin has been isolated is that of a bony fish, the cod, (Pickering, 1968) but it seems likely, on histological grounds, that neurophysin-like proteins are present in the hypothalamo-neurohypophysial systems of all vertebrates. Moreover, ^{3}H-tyrosine injected into the third ventricle of the goldfish (*Carassius auratus*) is incorporated into isotocin and transported along the axons of the preoptico-neurohypophysial tract at rates which are not very different from those in the mammalian system (Jones *et al.*, 1973). Thus it would appear that biosynthesis and transport in the hypothalamo-neurohypophysial systems of non-mammalian vertebrates, are comparable to these processes in mammals.

Although the evidence is by no means complete, it seems very likely that the physiological roles of the non-mammalian neurohypophysial hormones are very similar to those of their mammalian counterparts, namely, to regulate body water and electrolyte levels, and to assist the reproductive tract in the delivery of its "product". The major enigma at present is the occurrence of the second peptide alongside vasotocin in almost all glands, since, in most cases, we have yet to find a function which it (they) can fulfil better than AVT; it is only in mammals where the two hormones present can be ascribed separate and specific physiological functions. These roles cannot be finally given the seal of physiological significance in the non-mammalians until it can be shown that the gland actually functions in each group of animals, and releases its "hormones" under the appropriate conditions.

However, with the advent of more sensitive radioimmunoassays, comparative endocrinologists will soon be in a position to make the necessary measurements of changes in plasma "hormone" levels and thus settle the question.

CHAPTER 7

Regulation of Thyroid-Stimulating Hormone (TSH) Secretion

S. REICHLIN, J. B. MARTIN and I. M. D. JACKSON

I. INTRODUCTION

A. HYPOTHALAMIC-PITUITARY-THYROID AXIS

The pituitary-thyroid axis has served the endocrinologist as the example *par excellence* of a negative feedback self-regulatory system (Fig. 1). Secretion of thyroid-stimulating hormone, (thyrotrophin, TSH) is controlled by the "effective" thyroid hormone (TH) concentration in pituitary thyrotrophe cells so as to maintain a relatively constant level of TH in blood.

To the classical pituitary-thyroid feedback loop, the neuroendocrinologist has added a third component, the hypothalamic hormone, thyrotrophin

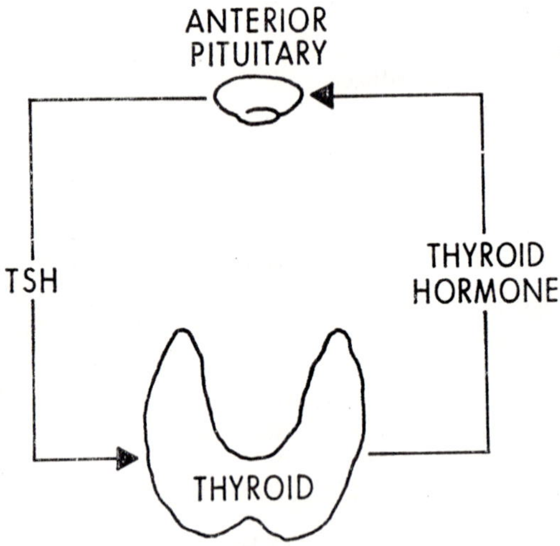

Fig. 1. Diagram of pituitary feedback system. This shows the classical negative feedback loop of control of TSH secretion by circulating thyroid hormone. High levels of thyroid hormone suppress and low levels of thyroid hormone increase secretion of TSH (see also Chapter 3). Since the rate of degradation of TSH and of thyroid hormone, the conversion of T_4 to T_3 and selective factors of uptake by T_3 in the pituitary all influence the overall TSH secretion rate, many factors enter into the regulation of TSH secretion rate but all contribute to the maintenance of the controlled variable which is the normal thyroid hormone level.

releasing hormone (TRH) which stimulates the synthesis and release of TSH (Fig. 2). In turn, TSH activates thyroid gland mechanisms for iodide uptake, hormonogenesis and release of the thyroid hormones thyroxine (T_4) and triiodothyronine (T_3). The interaction at the pituitary level between the negative feedback effects of T_4 and T_3 with the stimulatory effects of TRH appears to be the critical mechanism determining TSH secretion rate.

Although this simple model describes most of the regulatory changes in pituitary-thyroid function, other mechanisms influence the rate of TSH secretion. These include the peripheral degradation of TSH, of the thyroid hormones, of TRH and the physical state of thyroid hormones in the blood. The hypothalamic hormone, somatostatin, inhibits TSH secretion.

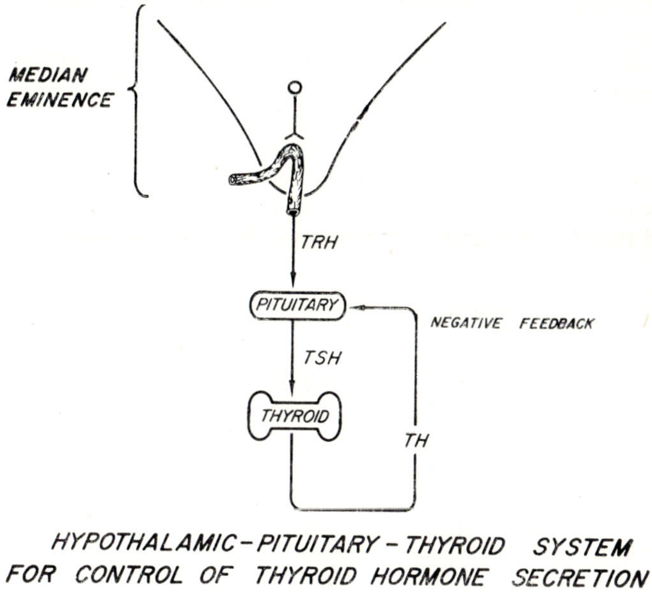

Fig. 2. Hypothalamic pituitary thyroid system for control of thyroid hormone (TH) secretion. To the classical negative feedback loop for pituitary thyroid control is added the stimulating effect of the hypothalamic hormone TRH. TRH stimulates and TH inhibits pituitary thyrotrophe cells. The TSH secretion rate is a resultant of these two interacting components. TH also may exert an effect at the hypothalamic level.

In this Chapter, we will outline these mechanisms of control and analyse the effects of many factors which influence the rate of TSH secretion. In the main, illustrations will be drawn from physiological studies in man. A number of detailed reviews provide useful background and inclusive bibliographic references (Reichlin, 1966a; Hershman and Pittman 1971; Blackwell and Guillemin, 1973; Wilber, 1973; Martin, 1974a; Sawin and Hershman, 1975).

B. TSH

Human pituitary thyrotrophin is a glycoprotein, molecular weight 28,000, which is synthesised and secreted by specific basophilic cells of the anterior pituitary distinct from those which produce the gonadotrophins. The structure of human pituitary TSH consists of two subunits termed α and β; the former is identical with the α subunit of LH and FSH. The immunological and biological determinants of the intact hormone reside in the β subunit. Circulating TSH is identical to pituitary TSH indicating that no change in structure occurs during the secretory process. Both the β subunit and the whole hormone are secreted into the blood (Kourides *et al.*, 1974).

A major methodological advance in the investigation of the physiology of TSH secretion was the development of radioimmunoassays. Such assays are now sufficiently sensitive to permit measurement of basal plasma TSH levels in most normal individuals and to study the effects of various procedures no TSH secretion. Normal levels of TSH in human serum as determined by radioimmunoassay are 1–10 μU/ml, the higher level corresponding to a concentration of pure TSH of approximately 0·3 ng/ml. The hormone circulates unbound in the blood and has a half-life of 50–60 minutes. It is estimated that 10–30% of total pituitary TSH content is secreted daily.

II. Factors Regulating TSH Secretion

A. THYROID HORMONE FEEDBACK

(1) *On the pituitary*

Administration of increasing doses of T_4 to hypothyroid individuals produces a graded suppression of plasma TSH levels. When data relating TSH to T_4 concentration are plotted to show TSH as a function of plasma T_4, a curvilinear relationship between these two variables is evident (Fig. 3). In extensive studies of this phenomenon it has been shown that plasma TSH is a function of the negative log of plasma T_4 (Cotton *et al.*, 1971). In terms of servosystems analysis, the thyroid hormone concentration in the blood can be viewed as the "controlled variable". The normal "setpoint" of pituitary-thyroid function is the resting concentration of plasma thyroid hormone which is maintained by a specific concentration of TSH. The secretion of TSH is inversely regulated by the concentration of thyroid hormones in such a way that deviations from the setpoint of control lead to appropriate changes in the rate of TSH secretion.

The precise "setpoint" of TSH regulation by thyroid hormone feedback appears to be determined by the amount of TRH secreted by the hypothalamus. Situations in which TRH secretion is reduced, such as pituitary stalk section or lesions of the hypothalamus lead to a decreased level of TSH and a consequent decline in circulating TH. In this circumstance, the secretion of TSH is exquisitely sensitive to the suppressive effects of thyroid hormones (Reichlin *et al.*, 1972). When TRH stimulation of the pituitary is increased, as, for example after the administration of synthetic TRH, the effect is blocked by prior treatment with TH. The interaction between TRH effects and TH

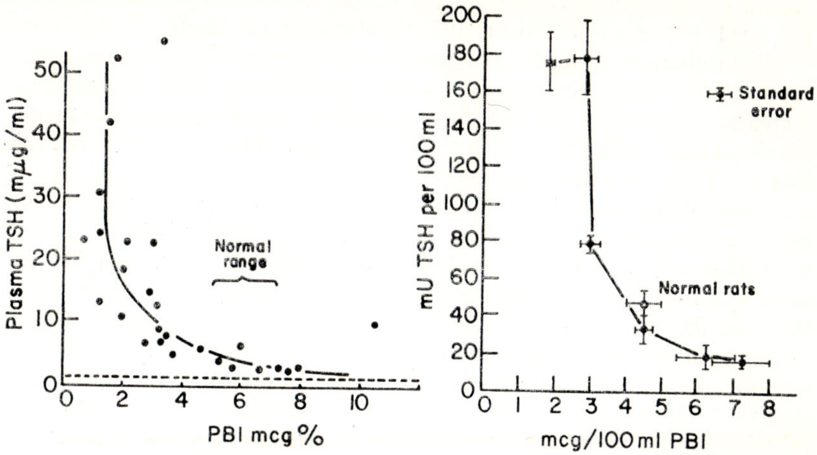

Fig. 3. The relationship between plasma TSH and thyroid hormone (as determined by plasma PBI measurements in man and in the rat). These curves illustrate in the human, left, and the rat, on the right, that plasma TSH levels are a curvilinear function of plasma thyroid hormone level. Human studies were carried out in myxoedematous patients treated with successive increments of thyroxine approximately at ten day intervals. Each point represents simultaneous measurements of plasma PBI and plasma TSH at various times in the six patients studied. The rat studies were done by treating thyroidectomised animals with various doses of thyroxine for two weeks prior to assay of plasma TSH and plasma PBI. These curves illustrate that the secretion of TSH is regulated over the entire range of thyroid hormone levels. At the normal set point for T_4, the small changes above and below the controlled level are followed by appropriate increases or decreases in plasma TSH. (From Reichlin and Utiger (1967) and Martin *et al.* (1970).)

effects is dose dependent; at the physiologically normal "setpoint" of plasma TH, pituitary responsiveness is critically altered by even minor changes in TH level. This phenomenon was well shown by Snyder and Utiger (1972) who found that the daily administration of small doses of T_4 and T_3 (15 μg T_3, and 60 μg T_4 daily for 3–4 weeks) to normal human volunteers was effective in substantially reducing plasma TSH response to intravenous TRH, even though no increase in plasma T_4 and T_3 was detectable. Conversely, minimal lowering of plasma T_4 by iodide administration sensitises the pituitary to TRH (Vagenakis *et al.*, 1974). Their observations can be taken to mean that at the normal level of plasma TH feedback regulation at the pituitary level is the most important control element.

Further evidence of the crucial importance of TH in the pituitary response to feedback control is the finding that in hypothyroidism TSH responses to TRH are enhanced, and in thyrotoxicosis they are diminished. In normal men, repeated administrations of TRH lead to a progressive decline in pituitary responsiveness, presumably due to a secondary increase in plasma TH levels (Snyder and Utiger, 1973; Rabello *et al.*, 1974).

The mechanism of feedback control at the pituitary level has been actively studied. The TSH-releasing effects of hypothalamic TRH are immediate; increases in plasma TSH levels are detectable within minutes following i.v. administration of TRH in both the rat and man. On the other hand, inhibition of TSH secretion following T_4 or T_3 administration requires prior treatment with either T_4 or T_3. Such inhibition can, however, be overcome by administration of larger doses of TRH. These studies indicate that TRH and TH interact for TSH control. It has been further shown that TH inhibition, but not TRH stimulation, is dependent upon new protein synthesis since the former, but not the latter, can be inhibited by a variety of protein synthesis inhibitors including actinomycin D, puromycin and cycloheximide (Bowers *et al.*, 1968; Vale *et al.*, 1968). These findings have led to the postulation that TH acts on the thyrotrophe by stimulating the production of an inhibitory protein which subsequently interferes with the TSH-releasing effects of TRH.

The question of the relative importance of T_4 versus T_3 in producing the feedback effects of TH on pituitary TSH secretion has been considered. Several workers report that labelled T_3 appears in the pituitary after the injection of labelled T_4, a finding interpreted to indicate the presence of an active monodeiodinating system (Reichlin *et al.*, 1966; Silva and Larsen, 1977). It has been shown by Schadlow *et al.* (1972) that the pituitary has high affinity binding sites for T_3 (as contrasted with T_4) and hence may also concentrate T_3 from circulating hormone which has been converted from T_4 elsewhere. In support of this view is the report of Galton (1975) that pituitary extracts do not convert T_4 to T_3. Because the relative biological effectiveness of T_3 and T_4 in a rat tumour model exactly parallels the relative binding of the two hormones and either hormone can displace the other from the binding sites, it has been proposed by Samuels *et al.* (1975) that both hormones act directly on the pituitary. When the conversion of T_4 to T_3 is blocked by the administration of propylthiouracil, the pituitary becomes more responsive to TRH, (Geffner *et al.*, 1975; Saberi *et al.*, 1975) but this effect may reflect peripheral changes as well as changes at the pituitary level.

(2) *On the Hypothalamus*

Although the feedback effects of TH at the pituitary level are of major importance in the control of TSH secretion, the possible effects on control of TRH secretion have also received much study but without complete resolution.

Studies in which TH has been applied directly into hypothalamic sites have been subject to criticism in the past because they did not distinguish between effects mediated at the brain level and those secondary to diffusion of the substances to the pituitary via the portal vessels. Kajihara and Kendall (1969) attempted to resolve this problem by the use of multiple pituitary transplants to the kidney capsule in hypophysectomised rats. They observed that residual pituitary-thyroid function, as indicated by thyroidal ^{131}I uptake and thyroidal radioactive hormone release rates, was obliterated in transplanted rats by bilateral lesions in the region of the paraventricular nuclei, an effect presumed due to decreased TRH reaching the transplanted pituitary via systemic blood. Implants of T_4 (0·5 and 1·5 μg) into the anterior hypothalamus, but not the cerebral cortex, were also effective in causing thyroid depression due, presumably, to inhibition of TRH release. The crux of their argument was dependent on the fact that residual radioactivity in brain sites was similar in implantations made in cortex and hypothalamus, consisting of *ca.* 46% of the administered dose 24 h after implantation. On the other hand, Averill (1970) regarded these studies as inconclusive because they do not exclude entirely the possibility that circulating T_4 from the injections may have suppressed pituitary transplants in a renal capsule site particularly since such transplanted pituitaries have been shown to be highly sensitive to parenteral T_4.

A series of interesting observations has been made by Knigge and Joseph (1971) concerning the site of feedback effects of T_4 on TSH secretion in the cat. The diffusion problem was circumvented by using hypothalamic deafferentation together with T_4 implants in the preoptic region of the hypothalamus. Both deafferented and intact cats showed compensatory thyroid enlargement following hemithyroidectomy. However, T_4 in a dose of 2·5 μg/kg/day for 2 weeks caused inhibition of thyroid function in normal cats but not in deafferented animals. Implantation of T_4 pellets into the preoptic region was shown to be effective in inhibiting thyroidectomy responses in intact cats but not in deafferented animals. The authors conclude that an inhibitory neural pathway extends from the preoptic area to the medial basal hypothalamus and that this region is sensitive to feedback actions of T_4. One may speculate that this inhibitory pathway is the somatostatinergic neurone system. In a subsequent study (Joseph and Knigge, 1972) the responses to hemithyroidectomy and T_4 treatment were examined in normal and deafferented kittens (5–21 days old). Hemithyroidectomy in 5-day-old kittens did not induce a compensatory increase in thyroid weight or ^{131}I uptake. On the other hand, deafferentation resulted in a marked increase in gland weight, as did T_4 implants into the preoptic area of the hypothalamus. They concluded that the preoptic area of the hypothalamus contains an inhibitory neural pathway in the kitten as well as in the adult, but that the neurones in the kitten respond differently to T_4 compared with those of the adult, indicating positive feedback effects in the kitten.

The elevated plasma TSH levels of thyroidectomised monkeys was decreased by microinfusion of minute amounts of T_3 into the lateral hypothalamus (Himsworth, 1975, personal communication). The time course was more rapid than seen after systemic injection, suggesting an effect not mediated by diffusion from the third ventricle.

Attempts have been made to solve this question by studying the content of TRH in the hypothalamus under varying conditions of thyroid function. It must be recognised that measurement of content is not an adequate index of function since both synthesis rate, and release rate may be influenced by thyroid status (see also Chapter 4). The first of such studies, those of Sinha and Meites (1965) suggested that bioassayable TRH levels in thyroidectomised rat hypothalamic tissue was increased after thyroidectomy and unchanged by T_4 treatment. Subsequent immunoassay studies (which are not influenced by contaminating somatostatin) have failed to show any consistent effect of prior thyroid status on TRH content (Bassiri and Utiger, 1974; Jackson *et al.*, 1974a).

The rate of biosynthesis of TRH has also been used as an indicator of the effects of thyroid status on TRH secretion. It was reported by Reichlin and collaborators that TRH was synthesised by a soluble hypothalamic extract and that the activity of this system was reduced markedly in hypothalamic tissue from hypothyroid animals and was increased slightly by thyroxine administration (Reichlin *et al.*, 1972). These observations were interpreted to indicate that thyroid hormone exerted a positive feedback effect on TRH secretion. Subsequent to these studies, it has been demonstrated that there is an active TRH degrading system in hypothalamic extracts (Bauer, 1974; Bauer and Kleinkauf, 1974; Dixon and Acres, 1975; Saperstein *et al.*, 1975). Rate of incorporation of amino acids into TRH, therefore, cannot be used as an index of synthesis rate unless effects of the degrading system have been taken into account. Also, it has been reported that the labelled peptides formed by hypothalamic fragments and which have the mobility of TRH on electrophoresis include a number of other peptides (McKelvy, 1974; Grimm-Jørgensen and McKelvy, 1974a, b). For these reasons, inferences about secretion rate from these earlier biosynthesis studies cannot be accepted without further work (see Section IV D below).

In an attempt to study the effects of thyroid status on TRH secretion more directly, several workers have utilised radioimmunoassay of TRH in urine or blood. Interpretation of the results of these studies is difficult. TRH is found in the hypothalamus and in other parts of the central nervous system as well (see Section IVC below) so that the significance of measurements of TRH in body fluids is uncertain. It is not known whether TRH present in extrahypothalamic sites is regulated in the same way as the peptide located in the hypothalamus itself. Further, as described later (see Section IV) there is some controversy regarding the authenticity of immunoreactive TRH measurements in mammalian biological fluids.

Exogenous TRH is rapidly degraded by rat blood both *in vivo* (Jackson *et al.*, 1974b) and *in vitro* (Redding and Schally, 1969a), and the rate of degradation in both instances is enhanced by thyrotoxicosis and retarded by hypothyroidism. Whether the enzymatic breakdown of TRH in blood or tissues is a physiological regulator of TRH function has not yet been determined (Griffiths, 1976). In man, however, there is no evidence that breakdown of TRH in the circulation is influenced by thyroid status (Jeffcoate and White, 1975c; Gagel *et al.*, 1976). We have attempted to measure TRH secretion in the rat (Jackson *et al.*, 1974b) and man (Gagel *et al.*, 1976). In the rat, though not in man, the secretion rate of TRH was influenced by thyroid status, being increased in thyrotoxicosis and reduced in hypothyroidism. However, further evaluation of TRH secretion rate, with more specific measurement of the endogenous immunoreactive TRH-like material present in blood and urine, is required (see Section IV).

B. COLD EXPOSURE

In laboratory and domestic animals, exposure to cold is usually followed by thyroid activation. For example, exposure of the rat to even a modest decline in ambient temperature usually causes a prompt, albeit transient, increase in plasma TSH. Chronic cold exposure is followed in most species by evidence of thyroid hypertrophy and increased secretion of TH. In man, on the other hand, it has been extremely difficult to demonstrate cold-induced thyroid activation.

Although it has been clearly demonstrated that TH is essential for survival at lowered environmental temperatures in rats, it has not been established whether changes in thyroid secretion is an important homeostatic mechanism for regulation of metabolism during cold exposure (see Gale, 1973, for review). It is appropriate, for the purposes of a consideration of this literature, to divide pituitary-thyroid responses into those which occur in response to acute (minutes to hours) versus chronic (days to weeks) exposure to lowered environmental temperature.

(1) *Acute Cold Exposure*

There is now substantial evidence that acute cold exposure in the rat causes a burst of TSH release which occurs sufficiently rapidly to imply activation by a neuroendocrine reflex. Abrupt lowering of the ambient temperature by sudden introduction of cold air into a closed cage, was shown to elicit a rise in immunoassayable plasma TSH levels within 5–10 min in unanaesthetised rats bearing chronic indwelling jugular cannulae (Martin and Reichlin, 1970) (Fig. 4). This TSH surge proved to be short-lived with a return to baseline within 40–60 min. A similar rapid increase in plasma TSH level, as measured by radioimmunoassay, was reported by Hershman *et al.* (1970) in rats transferred to a 2–4°C room.

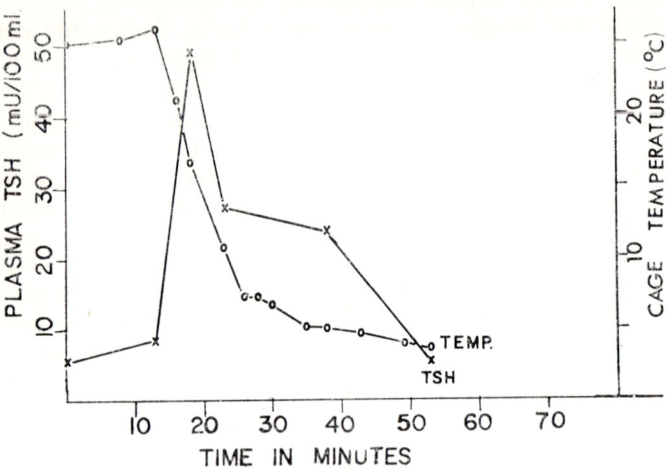

Fig. 4. Effect of cold exposure on plasma TSH in the rat. In this experiment a rat bearing a chronic cannula in right atrium was exposed suddenly to a lowered environmental temperature which was accomplished by connecting the uptake to the external environment. External temperature fell rapidly from 25 to 4°C and plasma TSH levels promptly rose and then gradually declined. This reflex has been shown to be due to cold stimulation of both internal and external cold receptors (from Martin and Reichlin, 1970).

That acute cold-induced TSH release is probably triggered by pheripheral cold receptors is suggested by the finding that during acute cold exposure, hypothalamic core temperature either shows no change or a slight rise. On the other hand, TSH release can also be induced by lowering of core temperature, indicating that both peripheral and central thermoreceptor inputs may trigger TSH secretion (Martin and Reichlin, 1970; Reichlin *et al.*, 1972). These observations confirm that experiments showing activation of TH secretion following hypothalamic cooling in the goat (Anderson, *et al.*, 1962), rat (Reichlin, 1964) and baboon (Gale, *et al.*, 1970) are probably mediated by enhanced pituitary TSH secretion, although one recent report failed to show this effect in the rat (Jobin *et al.*, 1971). The effect of stress in this preparation must be considered as a possible cause of the failure to observe increased TSH secretion.

Recently it has been reported by Montoya *et al.* (1975) that acute cold exposure in the rat leads to an increase in peripheral *plasma* TRH levels, a finding not confirmed by Emerson and Utiger (1975), Harris *et al.* (1977) or ourselves (Mueller *et al.*, 1977), though a TSH rise comparable to that of Montoya was obtained by the latter groups. Passive immunisation with anti-TRH blocks the TSH rise to acute cold (Szabo and Frohman, 1976;

Mueller *et al.*, 1977; Harris *et al.*, 1977) suggesting that the TSH rise due to acute cold results from TRH secretion.

(2) *Chronic Cold Exposure*

There is much less certainty concerning the role of the hypothalamic-pituitary-thyroid system in homeostatic responses to prolonged cold exposure (Galton and Nisula, 1969; Reichlin *et al.*, 1972; Gale, 1973). The bulk of experiments performed in a number of species, including man, have, in general, failed to show any change in plasma T_4 levels (total or free) during chronic cold exposure and, in fact, a fall in plasma TH level has been noted in some studies. The observation that TSH and TH secretion are increased, however, during prolonged cold exposure has been demonstrated repeatedly in the rat. This response has been attributed by Galton and Nisula (1969) to increased utilisation, excretion and/or degradation of T_4. In this species, increased faecal excretion of T_4 secondary to enhanced food intake is an important component of the response to cold exposure. Activation of the pituitary-thyroid axis may thus not involve hypothalamic mechanisms, and may be entirely secondary to a decrease in negative feedback effects on TSH secretion resulting from falling plasma TH levels.

Exposure of rats to 5°C for 32 days was shown by Fortier *et al.* (1970) to result in no detectable change in plasma TSH levels and a 50% reduction in plasma T_4 levels. However, calculated TSH secretory rate showed an increase in cold-exposed animals from 236 ± 25 to 452 ± 95 μU min^{-1} associated with an increased metabolic clearance rate. The secretory rate of T_4, calculated from disappearance rates of labelled T_4 from blood, showed no change but the metabolic clearance rate of T_4 was increased. These data are interpreted to mean that TSH release is stimulated initially during cold exposure by hypothalamic mechanisms, but this is not reflected by an elevation of plasma TSH levels because of increased metabolic clearance of TSH. During prolonged cold exposure, enhanced clearance of T_4 from the blood leads to an increased secretion of TSH which, because of its more rapid clearance from the blood, is not reflected by any absolute increase in plasma TSH level. Continued hypothalamic activation of the pituitary-thyroid axis was not thought to be a necessary component of the response. In man, most (e.g. Berg *et al.*, 1966; Hershman *et al.*, 1970) but not all (Golstein-Golaire *et al.*, 1970) find no effect on plasma TSH.

The pituitary-thyroid component of chronic cold adaptation is but one aspect of the larger homeostatic question as to how TH interact with the catecholamines in the regulation of thermogenesis, (c.f. Himms-Hagen, 1967) and the role of core temperature in determination of this response. Thyroid deficiency in cold-exposed animals elicits a compensatory increase in catecholamine secretion in both the rat (Johnson *et al.*, 1965) and the baboon (Gale *et al.*, 1969), and T_4 administration decreases catecholamine secretion. We have proposed that a deficit in autonomic mechanisms might be expected to elicit a compensatory increase in the pituitary-thyroid component (Reichlin *et al.*, 1972; Martin, 1972).

To test this hypothesis, pituitary-thyroid responses were assessed in immunosympathectomised-adrenomedullated rats exposed to 6°C for 2 weeks. Such animals were shown to have a chronic core hypothermia of about 10°C (Martin, 1972). Although enhanced TSH secretion was indicated by increased thyroid weight and plasma TSH levels, there were no changes in either total or free (diffusible) T_4 levels in plasma. These experiments thus also failed to show any change in setpoint of regulation as far as T_4 levels were concerned.

Another aspect to this question has been suggested by the observations of Nejad *et al.* (1972) and Jobin *et al.* (1973) which indicate that plasma T_3 levels are increased in cold-exposed rats (4 weeks at 6°C) as is kidney T_3 concentration. In cold-exposed rats. T_3-binding proteins in parenchymal tissues are increased (Balsam and Leppo, 1974) as is T_4 degradation rate. Whether the rise in plasma T_3 level is due to increased T_3 secretion by the thyroid, to increased peripheral deiodination of T_4 to T_3 or to reduced plasma clearance of T_3, is unknown. These observations provide an alternative explanation for the numerous experimental failures to demonstrate a change in hypothalamic-pituitary-thyroid setpoint during cold exposure.

Final proof of hypothalamic-activated TSH release during chronic cold exposure requires demonstration that TRH synthesis and secretion are enhanced. Redding and Schally (1969b) have reported increased TRH levels as measured by an *in vitro* bioassay for TRH in the peripheral plasma of hypophysectomised cold-exposed rats. By immunoassay, a slight fall in hypothalamic TRH content has been reported in chronically cold-exposed rats (Jackson and Reichlin, 1973; Martin and Jackson, 1975). Such findings require confirmation.

C. DIURNAL RHYTHMS

Plasma TH levels vary only slightly throughout the twenty-four hours. In part, this change is due to alterations in water concentration in the blood related to altered fluid dynamics in the erect as compared with the supine position. More important perhaps, is the diurnal variation in TSH and thyroid hormone secretion. By means of a sensitive double labelling method, it is shown that thyroid gland release of iodide reaches its highest value during early morning hours (Nicoloff *et al.*, 1970). This appears to be a reflection of the corresponding changes in plasma TSH (Patel *et al.*, 1972), although not all authors have been able to confirm a diurnal change in TSH secretion (e.g. Nicoloff *et al.*, 1970). It is not known with certainty whether intrinsic rhythm in the secretion of TRH is responsible for this effect or, that the diurnal shift in TSH secretion is in turn dependent upon changes in secretion of cortisol, which is well known to follow a diurnal pattern of release (Nicoloff *et al.*, 1970). This viewpoint is supported by the finding that the administration of cortisol is followed by the suppression of thyroid gland secretion as manifested by a change in the pattern of urinary iodide excretion, and by a decline in plasma T_3. Recently Greer *et al.* (1975) found persistence of circadian rhythm in the rat even after adrenalectomy.

D. STRESS

The effects of stress on pituitary-thyroid function have intrigued many generations of investigators. This concern is based in part on the traditional view that Graves' disease is a physical response to psychic stress, and on the more recent studies of endocrine responses, particularly those of the pituitary-adrenal system to non-specific stress. The study of stress-induced changes in pituitary-thyroid function has led to ambiguous results which do not make up a clear-cut pattern.

Stressful factors may alter pituitary-thyroid function at many levels of controls; these include altered TRH secretion, changes in responsiveness of the pituitary to TRH, altered metabolism of TSH, altered metabolism of TH in peripheral tissue and alterations in the physical state of TH-binding proteins in blood. Other hormones such as those of the adrenal gland may also influence TSH secretion. Moreover, the concept of non-specificity does not apply in all situations to all the stressful conditions commonly studied. For example, cold exposure generally increases TSH secretion in small animals, but if the intensity of the cold exposure is great enough, inhibition of TSH secretion is observed, a finding generally attributed to the "non-specific" component of cold stress. Bacterial infection has also been regarded as a non-specific stress, but recent work indicates that during a severe infection, such as pneumonia in man, there is an increase in the rate of peripheral disappearance of TH from plasma, together with an increase in the rate of degradation of T_4 by leukocytes (see DeRubertis and Kosch, 1975 for review and critique). The finding that plasma TH levels remain unchanged despite these changes in TH clearance suggests a compensatory increase in pituitary-thyroid secretion, possibly related to the normal operation of the negative feedback control system. In rats, on the other hand, bacterial pneumonia leads to thyroid inhibition. The administration of adrenaline, which causes thyroid gland inhibition in the rat and rabbit, may cause thyroid activation in species such as the mouse which has a rich innervation of thyroid secretory cells by catecholaminergic secretomotor nerves. In the dog, adrenaline promotes the release of inorganic iodide from the thyroid gland. Thus, adrenaline can hardly be termed a non-specific stress.

Another example of the problem in attributing pituitary-thyroid responses to stress to primary alterations in TSH secretion, is that seen in patients suffering from chronic illness. Many patients with chronic wasting disease have a decrease in the blood TH-binding proteins, together with an increased concentration of free T_4 (Sterling and Hegedus, 1962; Oppenheimer *et al.*, 1963; Ingbar *et al.*, 1965). Such patients have been shown by Sullivan and collaborators (1972) to have a T_3 deficiency, believed to be due to a defect in the peripheral conversion of T_4 to T_3, because the tissues of such patients have low T_3 concentration. In diseases of the liver or kidney, similar findings are reported, and certain of these individuals may actually have slightly elevated plasma TSH concentration, suggesting that the defect of tissue conversion has led to compensatory activation of pituitary-thyroid function.

More recent studies of peripheral thyroid hormone metabolism show that T_4 is converted to both T_3 and to reverse T_3 (rT_3), and that in stressed individuals conversion in the direction of rT_3, a metabolically inactive isomer, is favoured. Associated starvation, particularly of carbohydrate may be the most important factor in the genesis of stress-induced T_3 deficiency. Whether these peripheral hormone changes lead to functional thyroid hormone deficiency is unclear. Plasma TSH levels and TSH responses to TRH are generally normal in the sick euthyroid suggesting that "effective" tissue thyroid hormone levels are normal.

Important new data indicate that somatostatin secretion may play a modulating role in TSH regulation, at least in the rat in whom stress characteristically suppresses TSH secretion. TSH responses to both cold exposure (Gordin *et al.*, 1976) and to TRH (Arimura and Schally, 1976) are enhanced by prior treatment with an anti-somatostatin antiserum.

Bearing in mind the reservations recorded above, certain observations about the effects of stress can be summarised. In general, most studies have indicated that TSH secretion (inferred from direct measurement of plasma TSH or indirect measures of thyroid function) is inhibited during exposure to severe stress (see Gregerman, 1971, for review). A few studies have shown that TH may be increased after acute stress in the sheep or man or after aversive conditioning in the rhesus monkey (Mason and Mougey, 1972). In all such studies, critical changes in blood water concentration, TH-binding, direct effects on catecholamines on thyroid gland function, or changes in other hormones have not been fully excluded.

In both rat and rabbit, the acute stress of tibial fracture causes a decline in TSH secretion. In the rat, exposure to a noisy laboratory setting also caused a decrease in plasma TSH. It was reported by Ducommon *et al.* (1967) that repeated exposure to noise in the rat was followed by an adaptive change so that increase in TSH results from noisy exposure. We were unable to confirm this finding.

Insulin-induced hypoglycaemia has also been reported to cause release of TSH in men with GH deficiency (Guansing *et al.*, 1975) and normal rats

(Leung *et al.*, 1974), although surgery and electro-shock therapy are without effect on pituitary-thyroid function other than the acute change in concentration attributable to blood volume shifts.

A few isolated reports have claimed that emotional stress in man is followed by increased pituitary-thyroid function. Even if it could be shown that any form of stressful exposure could increase pituitary-thyroid function, it would still be largely irrelevant to the problem of Graves' disease. With the exception of a few cases (see Section G below), all patients with thyrotoxicosis have low, or unmeasurable, plasma levels of TSH. This finding has been confirmed using the specific immunoassay for the β chain of TSH (Kourides *et al.*, 1973). Thus if stress factors play any role in the pathogenesis of Graves' disease, they do not do so through an increase of TSH secretion.

E. HORMONES

A number of hormones other than the thyroid hormones are known to interact with the secretion of TSH in the human. Work dealing with the pre-immunoassay era has been thoroughly reviewed (Reichlin, 1966a).

(1) *Adrenal Corticoids*

Perhaps the most important non-thyroidal hormones involved in TSH regulation are glucocorticoids. The earlier literature dealing with the effects of the adrenal cortex on thyroid function indicates that one or more measures of pituitary-thyroid secretion were inhibited by cortisol or other adrenal corticoids. The advent of immunoassay has now made it possible to show that either a brief or chronic exposure to excessively high levels of cortisol will inhibit TSH secretion in man or in the rat (Wilber and Utiger, 1969). The residual pituitary-thyroid activity observed in patients maximally suppressed with exogenous T_3 is further reduced by treatment with high doses of corticoids (Nicoloff *et al.*, 1970). Corticoids apparently act at two loci in bringing about this response: they appear to reduce the sensitivity of the pituitary to TRH presumably by acting at the pituitary level, and under some circumstances probably act to reduce the secretion of TRH as indicated in some experiments by the finding of unchanged pituitary sensitivity to TRH in cortisol-suppressed glands. Corticoids, also reduce peripheral conversion of T_4 to T_3 (Burr *et al.*, 1976).

Stress-induced TSH suppression (see above) is not mediated by concomitant activation of ACTH-adrenal secretion, since it is observed even in animals adrenalectomised and maintained on a constant maintenance dose of glucocorticoid.

(2) *Oestrogens*

The high incidence of thyroid disease in women as compared with men (3 : 1 for multinodular goitre, 7 : 1 for Graves' disease, 9 : 1 for myxoedema) has provoked inquiry as to differences in TSH secretory dynamics in relation to sex hormone status. The early literature, reviewed previously (Reichlin, 1966a), is complex and contradictory, partly because oestrogens modify peripheral metabolism of TH, and partly because in the widely studied rodent model doses of oestrogen usually administered cause inanition with secondary effects on thyroid function. Basal levels of TSH are similar in women and prepubertal children. On the other hand, pituitary TSH secretory responses to exogenous TRH is greater in women than in men (Jacobs *et al.*, 1973). That this effect is due to oestrogen sensitisation is suggested by the finding that pituitary responsiveness is greatest in that phase of the menstrual cycle associated with the highest levels of oestradiol (late follicular phase), (Sanchez-Franco *et al.*, 1973), and that the response of the pituitary in men is enhanced by prior treatment with oestrogen (Carlson *et al.*, 1973). This is probably due to a direct effect of oestrogen on thyrotrophe cells analogous to the sensitising effects on the growth hormone and prolactin-releasing effects of hypothalamic extracts, and on the LH responses to LHRH in women (see Chapter 8). The precise nature of the oestrogen effect on the pituitary is unknown, nor is it necessarily related to the increased incidence of thyroid disorder in women. Little is known about the effects of progestagens and testosterone on TSH secretion.

(3) *Growth Hormone*

Although it is well-known that the synthesis and secretion of growth hormone is controlled to some degree by TH, an effect of growth hormone on the secretion of TSH has only recently been recognised. In patients with hypopituitarism presumed to be of hypothalamic origin, the administration of growth hormone in therapeutic amounts has been shown to decrease the pituitary TSH response to TRH (Root *et al.*, 1973; Lippe *et al.*, 1975). This phenomenon may be in part responsible for the reduced pituitary TSH secretory responsiveness to TRH observed in patients with acromegaly. The mechanism underlying this response has not been established. It has been proposed that GH administration may bring about the release of somatostatin through a short loop feedback mechanism. Somatostatin would, in turn, inhibit pituitary responsiveness to TRH.

F. DRUGS

Many drugs influence pituitary thyroid function, most by altering peripheral metabolism of TH or by changing the TH-binding characteristics of plasma; secondary changes may then occur through activation of normal pituitary-thyroid feedback control. Examples of such phenomena are seen after treatment with phenobarbital or dilantin, drugs which induce microsomal changes in the liver leading to an increase in the rate of breakdown of thyroxine. Dilantin, aspirin, and 2,3,-dinitrophenol compete with thyroxine for binding sites on thyroxine-binding globulin and cause an increase in percent free T_4 in plasma. Compensatory decrease in TSH secretion is then required to maintain constancy of the free TH level. The conversion of T_4 to T_3 in peripheral tissue is partially blocked by propylthiouracil (Bernal and Escobar del Rey, 1974), presumably by affecting a peripheral tissue monodeidoinase. Under carefully controlled conditions, in hypothyroid patients maintained on constant T_4 replacement, this slight lowering of plasma T_3 is capable of increasing TSH secretion (Saberi *et al.*, 1975; Geffner *et al.*, 1975).

Most drugs which alter pituitary-thyroid function do so by changing elements in the peripheral control system as noted above. Certain drugs may act more directly on hypothalamic-TSH control. Morphine is reported to decrease TSH secretion in man. In experimental animals, phenobarbital in anaesthetic doses will cause a lowering of plamsa TSH levels. Reserpine has also been reported to cause a decrease in plasma TSH, a response attributed to interference with the central monoaminergic control of TRH excretion (see Section III F).

G. HYPOTHALAMIC-PITUITARY-THYROID DISEASE

Modifications in TSH-secretion observed in clinical disease of the pituitary and thyroid generally follow classical formulations of pituitary-thyroid negative feedback regulation with a few paradoxical exceptions.

(1) *Primary Hypothyroidism*

Effects of long-standing thyroid insufficiency on the pituitary were recognised long before the concept of pituitary-thyroid feedback was developed, or even the understanding of the function of the thyroid gland. Thus, Niépce, in 1847, reported the finding of an enlarged pituitary gland in patients dying of cretinisim (Reichlin, 1966). More recently, a number of authors have drawn attention to the fact that enlargement of the pituitary fossa may occur in long-standing hypothyroidism, so much so that a patient may appear on superficial study to have tumour of the pituitary with secondary thyroid failure (Jackson, 1970). It has been proposed that a pituitary TSH-secreting tumour may rarely arise in man secondary to prolonged hypothyroidism. Such a phenomenon is frequently observed in the hypothyroid mouse (Furth thyrotrophic tumour).

Primary disease of the thyroid gland with thyroid failure usually causes an increase in plasma TSH levels; this is true after thyroidectomy, thyroiditis, and treatment with radioactive iodine. Latent thyroid failure, for example, Hashimoto's disease, or post-radiation thyroiditis, may be followed by an increase in plasma TSH when thyroid secretion is reduced by concomitant administration of therapuetic doses of inorganic iodide. Even in normal individuals, the slight lowering of plasma TH caused by iodide administration may sensitise the pituitary to TRH. In patients with marginal thyroid secretory defects, plasma TSH levels may be disproportionately elevated in relationship to the TH level, suggesting a compensatory stage of thyroid failure in which adequate thyroid function is maintained only by TSH hypersecretion. This phenomenon may account for the disproportionate elevation in TSH levels observed in thyrotoxic patients after treatment.

(2) *Secondary Hypothyroidism*

Failure of TSH secretion causes secondary hypothyroidism. As expected, most such patients have low or unmeasureable TSH levels in blood. Paradoxically, a few such patients have been found to have TSH in the blood within the normal or even high range (Patel and Burger, 1973; Sawin and Hershman, 1975). A number of explanations of this phenomenon have been offered but the mechanism is still obscure.

(3) *Hypothalamic Hypothyroidism* (*Tertiary Hypothyroidism*)

The introduction of TRH into clinical medicine made it possible to evaluate pituitary secretory reserve in hypothyroid patients with low plasma TSH levels, commonly considered to be due to pituitary insufficiency. A number of patients with secondary hypothyroidism ("thyrotrophic failure") have now been shown to have normal or even supranormal responses to exogenous TRH (see Wilber, 1973; Sawin and Hershman, 1975 for reviews). Since the pituitary is capable of normal function in these individuals, it has been inferred that the defect is due to failure of stimulation by the hypothalamus. For reasons as yet unexplained, anencephalic infants have almost normal plasma TH levels in the face of hypothalamic agenesis (Hayek *et al.*, 1973; Allen *et al.*, 1974).

The importance of hypothalamic hypothyroidism as a cause of thyrotrophic failure depends upon the population studied. The majority of children suffering from "idiopathic" partial or total hypopituitarism have TRH-responsive pituitaries. This important observation indicates that the idiopathic dwarfism is due in most instances to an intrinsic hypothalamic disorder. In spontaneously occurring "hypothyrotrophic" hypothyroidism, hypothalamic failure may be due to hypothalamic tumours, inflammation, pituitary stalk section, rarely in Sheehan's syndrome (then attributed to hypothalamic infarction), and also occasionally in idiopathic cases. Autopsy findings in the latter group of patients have not been reported. It may be speculated that a group of such cases will prove to have hypophysiotrophic neurone failure analogous to the atrophy of the supraoptico-hypophysial neurones in idiopathic diabetes insipidus.

(4) *Hyperthyroidism*

That thyrotoxicosis is due to TSH hypersecretion is excluded by the finding that plasma immunoassayable TSH levels are absent or barely detectable in patients with Graves' disease and that they do not respond to the injection of TRH. This observation supports the view that pituitary secretion is suppressed and is in fact an "innocent bystander" to the thyroid stimulation caused by thyroid stimulating immunoglobulin, a gamma globulin directed against thyroid cell membranes. However, in recent years, a few cases have been recognised with paradoxically elevated levels of immunoassayable TSH in the face of elevated TH concentrations. The occurrence of such cases and their frequency has been disputed, but there now appear to be several patients in whom thyrotoxicosis is accompanied by normal, or even high TSH levels. Several forms of this syndrome have been reported (see Gershengorn and Weintraub, 1975 for review). In two or three cases, there appears to be an altered "setpoint" for pituitary-thyroid feedback control, since increasing levels of T_3 were able to bring about partial or complete suppression of thyroid function. Because prolonged TRH stimulation has not been reported to produce thyrotoxicosis in man, it is proposed that these cases most likely represent a defect in the receptor function for TH at the pituitary level, although primary TRH hypersecretion, has not been excluded. The defect may be analogous to that of Cushing's disease in which there is an altered set-point of ACTH secretion.

III. Neural Regulation of TSH Secretion

Most work on the neural control of TSH secretion has been carried out in experimental animals, and the results extrapolated to the human. These studies will be summarised here. Several detailed reviews of this material have been published (Reichlin, 1966a; Martin, 1974a; Brown and Martin, 1973); Martin and Jackson, 1975).

A. ABLATIVE STUDIES (see also Chapter 2)

The importance of intact hypothalamic pituitary relationships for the maintenance of basal pituitary-thyroid function has been documented by a number of experimental manipulations. Anatomical disconnection of the pituitary from the hypothalamus (by either stalk section or transplantation) decreases basal pituitary TSH secretion resulting in a hypothyroid state. These studies imply a tonic drive by the hypothalamus which is necessary for the maintenance of basal TSH secretion. Only the "hypophysiotrophic area" is needed for maintenance of basal TSH secretion. Nearly normal TSH secretion is observed in rats with hypothalamic isolation carried out by the Halász method and even small conical hypothalamic islands, containing only the ventromedial and arcuate nuclei attached to the pituitary stalk, are capable of maintaining normal thyroid size and radioiodide uptake in the rat and the cat. These studies suggest that hypothalamic-pituitary islands can maintain basal pituitary-thyroid functional capacity without input from other parts of the brain.

Lesions placed in the vicinity of the paraventricular nuclei result in a decrease in pituitary-thyroid function without any definite direct effects on other pituitary trophic hormones. Anterior hypothalamic lesions in the paraventricular region cause a reduction in plasma PBI levels and a fall in both plasma and pituitary TSH levels. In one study of the effects of such lesions on feedback regulation of TSH secretion, it was shown that the lesions resulted in a fall in plasma PBI from $3{\cdot}0 \pm 1{\cdot}0$ to $1{\cdot}7 \pm 0{\cdot}2$ μg/100 ml and plasma TSH levels, as measured by radioimmunoassay, became undetectable (Martin *et al.*, 1970). Pituitary TSH levels were reduced to one-third of control levels. Although TSH levels were undetectable in the plasma of hypothalamic-lesioned rats, such animals were capable of increased TSH secretion following a further reduction in plasma TH after thyroidectomy. Hypothalamic-lesioned rats were also shown to be more sensitive than intact controls to the feedback effects of small doses of T_4. These studies were interpreted to indicate that anterior hypothalamic lesions lower basal TSH secretion and modify, but do not prevent, qualitatively normal secretory responses to alterations in plasma TH levels. The hypothalamus thus acts to determine the sensitivity of the pituitary to feedback inhibition, presumably by reducing the quantity of TRH which reaches the pituitary. Additional evidence for the importance of an anterior hypothalamic region in TSH control has been provided by the segmental deafferentation experiments of Halász *et al.* (1967a). These investigators observed that a coronal section through the midhypothalamus anterior to the arcuate nucleus and posterior to the paraventricular nuclei, caused a decrease in plasma PBI levels from $4{\cdot}0 \pm 0{\cdot}2$ to $1{\cdot}6 \pm 0{\cdot}2$ μg/100 ml.

New immunoassay methods have been used to define the extent of the TRH hypophysiotrophic area, and the effects of hypothalamic lesions on TRH secretion (Jackson and Reichlin, 1977a). In the rat a small amount of TRH is present throughout most of the brain (see Section IV C below). Within the hypothalamus, there is a concentration gradient from dorsal hypothalamus, ventral hypothalamus and stalk median eminence region

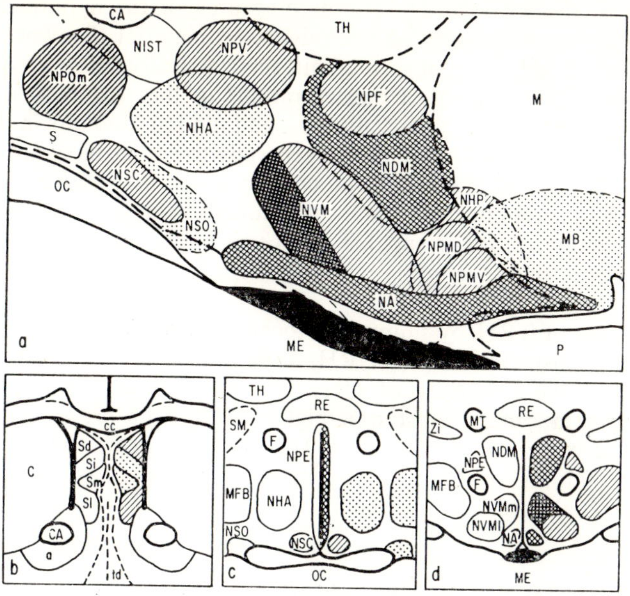

Fig. 5. Localisation of TRH in the hypothalamus, septum and preoptic area. Drawing (a) is of a parasagittal section through the rat hypothalamus and (b) to (d) are of frontal sections. Drawing (b) depicts the septal region, (c) the anterior hypothalamus, and (d) the tuberal region. Abbreviations: C, nucleus caudatus; CA, comissura anterior; CC, corpus callosum; F, fornix; M, mesencephalon; MT, tractus mammillothalamicus; NIST, nucleus interstitialis striae terminalis; OC, chiasma opticum; P, pituitary; RE, nucleus reuniens thalami; S, nucleus preopticus suprachiasmaticus; SM, stria medullaris; TH, thalamus; Zi, zona incerta; a, nucleus accumbens; and td, nucleus tractus diagonalis. Shaded areas refer to concentration of TRH. Solid area, 38 ng/mg protein, dark cross hatched, 9 ng/mg, light cross hatched 3·5 to 4·5 ng/mg, hatched 1–3 ng/mg, stippled, less than 1 ng/mg. (From Brownstein *et al.*, 1974.)

of 49, 64 and 3570 pg/mg respectively. Lesions of the hypothalamus characteristically associated with decreased thyroid function and "goitre block" reduce total hypothalamic content by about two-thirds, and the TRH content of the SME by the same amount (Jackson and Reichlin, 1977a). In terms of traditional questions about hypothalamic-pituitary control, it appears that there is residual TRH secretion, which may in part account for the residual TSH secretion, hitherto attributed to the "intrinsic autonomy" of pituitary TSH secretion. It had previously been shown that the autonomous TSH secretion of pituitary transplants is due in large measure to TRH-like substances reaching the transplant by way of the circulating blood.

Detailed analysis of the TRH-containing area of the hypothalamus has been made by immunoassay of microdissected material (Brownstein *et al.*, 1974) (Fig. 5). TRH is found throughout the classically defined hypophysiotrophic area with the greatest concentrations in the periventricular areas. Little is found in the mammillary nucleus, and modest amounts in the medial preoptic area. These findings conform generally to earlier bioassay studies of Mess *et al.* (1970), and of Krulich and collaborators (1968). With the possible exception of the preoptic area, the localisation also comforms to the distribution of hypothalamic tissue capable of incorporating precursor amino acids into peptides with the chromatographic mobility of TRH (Reichlin and Mitnick, 1973a).

Localisation of TRH by immunohistochemical methods has not been as readily accomplished as with LHRH, somatostatin and neurophysin; this may be due to the diffuseness of TRH distribution throughout cells, lack of complexing to a carrier substance, high solubility or presence outside of granules. However, Hökfelt *et al.* (1975c, d) have demonstrated immunofluorescent staining of TRH in the median eminence, extrahypothalamic brain and spinal cord.

B. EFFECTS OF ELECTRICAL STIMULATION

Over the years, several attempts have been made to define the hypophysiotrophic region of the hypothalamus by utilising electrical stimulation techniques (see Chapter 2). Prior to development of radioimmunoassay methods, most of such studies used thyroid activation as evidence of TSH release, although a few reports appeared in which TSH was measured by bioassay. Elevated levels of TSH have now been reported following electrical stimulation in the rat, dog and rabbit. In stimulation studies in the rat (see Martin, 1974a for review), radioimmunoassayable plasma TSH levels rose rapidly (in pentobarbital-anaesthetised rats) after hypothalamic stimulation, (Fig. 6) whereas in sham-stimulated animals, plasma TSH levels fell during the period of observation (Martin and Reichlin, 1972). The area from which positive responses were obtained was extensive, including the anterior hypothalamus and preoptic area, the paraventricular nuclei and the dorsomedial, ventromedial and arcuate nuclei (Fig. 7). A similar distribution of effective stimulation sites for TSH release was found by Averill and Salaman (1967) in the rabbit using a TSH bioassay. The largest responses in the rat occurred with stimulation of the anterior hypothalamus in the vicinity of the paraventricular nuclei. Negative responses occurred with stimulation of the lateral hypothalamus, thalamus and mammillary bodies.

As the currents used in these experiments were large, the possibility was considered that the localisations were non-specific. Arguing against this fact was the observation that parallel studies of plasma GH levels in the same rats indicated that GH responses were strictly limited to the ventromedial and arcuate regions. These results would seem to be of valid localising value since TRH is present in a distribution similar to that defined by electrical stimulation (see previous section), whereas GRF is restricted to the ventromedial nuclei (Krulich *et al.*, 1972).

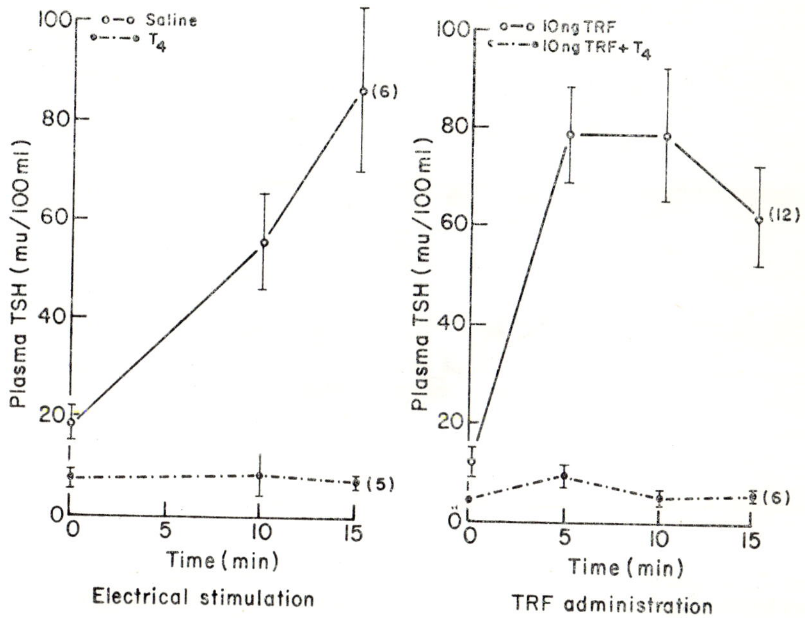

Fig. 6. This figure shows the effect of electrical stimulation of the hypothalamus on TSH release in the rat, the effect of prior administration of thyroxine on this response which is to inhibit the effect, and the effect of TRH administration on plasma TSH which is also blocked by thyroxine (from Martin and Reichlin, 1970).

The time course of electrically-induced TSH release in the rat was compared with that which followed intravenous administration of synthetic TRH. In each case, plasma TSH levels were significantly elevated within 5 min; the electrically-induced response was slightly more sustained, probably because stimulation was continued for 5–10 min, whereas TRH was administered in a single injection. Although the effects of incremental doses of TH on electrically-induced TSH release were not examined, it was found that T_4 in a dose of 50 μg per 100 g body weight administered 2 h before the experiment was effective in blocking TSH release to both electrical stimulation and synthetic TRH. As will be considered below, it is likely that this effect was due to a blockade of the TRH effect at the pituitary level.

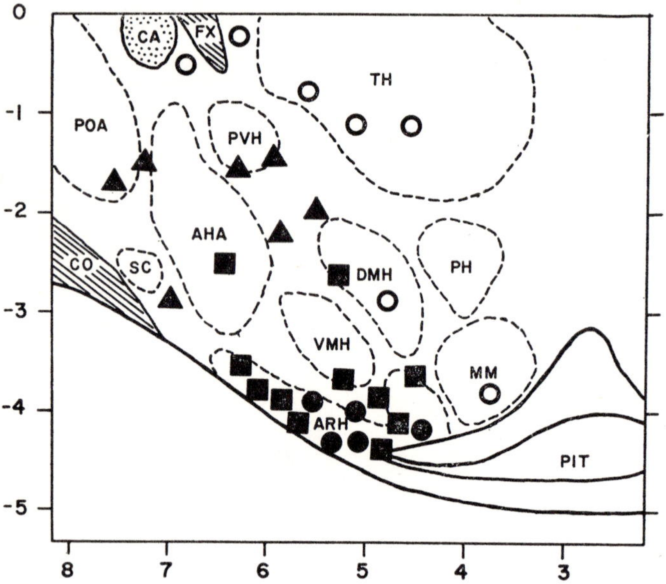

Fig. 7. Localisation of regions of the hypothalamus which responded to electrical stimulation by the release of TSH. This figure illustrates the places in the hypothalamus of the rat which can be electrically stimulated to produce increase in plasma TSH level. The solid symbols indicate positive stimulating areas, the open circles, negative. Sites outside the hypothalamus were ineffective. The distribution of these stimulating sites should be compared with the microdissection studies (Fig. 5) since they are roughly comparable (from Martin and Reichlin, 1972).

The similarity of the time course of TSH release following electrical stimulation and TRH administration and its prevention by T_4 pretreatment, suggest that hypothalamic electrical stimulation is effective by causing release of stored TRH. Wilber and Porter (1970) have reported that bioassayable TRH activity appears in the portal blood after electrical stimulation of the anterior hypothalamus. Prior T_4 administration had no effect on this response.

C. HYPOTHALAMIC INHIBITION OF TSH SECRETION

The observation that certain stimuli, such as stress in the rat and rabbit, cause a rapid reduction in plasma TSH levels, raises the question whether such an effect is mediated by suppression of tonic TRH secretion or by release of a specific thyrotrophin-inhibiting factor (TIF). Lomax *et al.* (1970) have studied the effects of morphine-induced TSH suppression in the rat. Both systemic administration and local intrahypothalamic injection of morphine were shown to depress the release of radioiodine from the thyroid gland of rats. Effective hypothalamic sites for such inhibition included the suprachiasmatic region and supramammillary and posterior hypothalamic regions within 1 mm of the midline. The latter placements were similar in location to the sites which were found to be effective in blocking the inhibitory effects of systemically-administered morphine. The possibility that the posterior hypothalamus contains an inhibitory regulatory region for thyroid control, was also suggested by the electrical stimulation studies of Vertes *et al.* (1965).

Evidence derived from experiments in the goldfish have indicated that hypothalamic inhibition of TSH secretion is the primary regulatory mechanism in this species. Stalk secretion in the goldfish causes increased TSH secretion and pituitaries from the species incubated *in vitro* show increased secretory activity resembling that observed in prolactin-producing cells of the mammalian pituitary explant. Peter (1970) has postulated that the hypothalamic inhibition is mediated by a TIF.

In the light of these speculations based on physiological experiments, recent studies with somatostatin, are particularly relevent. This hypothalamic hormone, identified in 1973 by Guillemin and collaborators on the basis of its ability to inhibit GH secretion both *in vitro* and *in vivo* in a variety of assay preparations (see Chapters 5 and 9), has also been found to inhibit pituitary TSH secretory responses to TRH administration: this is true for pituitary cells grown in culture, for studies in the whole rat, and in the human (see Siler *et al.*, 1974 for review). Anti-somatostatin serum increases TSH responses to TRH, and enhances cold-induced TSH release in the rat (Arimura and Schally, 1976; Gordin *et al.*, 1976). It is probable that there is a dual hypothalamic control for TSH regulation similar to that which exists for prolactin and for growth hormone regulation. It may be that somatostatin effects are observed after only certain forms of stress, and that there may be a species difference in the importance of this hormone in TSH regulation.

D. EXTRAHYPOTHALAMIC CONTROL OF TSH SECRETION

Considering the number of observations which have been made concerning the role of extrahypothalamic structures in ACTH and gonadotrophin regulation, it is somewhat surprising that so little information is available with respect to the function of these structures in TSH control. Although the medial basal hypothalamus appears capable of maintaining basal TSH secretion, phasic release of TSH, such as occurs in diurnal or circadian rhythms, stress-induced inhibition or cold-induced release are probably mediated via extra-hypothalamic structures and inputs. In this section, evidence suggesting a role of extrahypothalamic structures in TSH secretion will be considered. Detailed references will be found in Reichlin (1966a) and Martin (1974a).

(1) *Habenula and Pineal*

Of extrahypothalamic structures, the function of the habenula in thyroid control has received the most attention, particularly by Szentágothai *et al.* (1972). Based on the effects of habenular lesions on thyroid responses to cold exposure (utilising a TSH bioassay in which TSH effects were assessed by changes in karyometrics of thyroid follicle cells), these workers postulated that the habenula functions in an indirect and primarily inhibitory manner, outside the primary TH feedback loop, to integrate neural homoestatic responses with hormonal needs. They proposed that these effects are mediated via the anterior hypothalamic thyrotrophic area, since lesions of this region of the hypothalamus blocked supranormal thyroid responses which occurred in habenula-lesioned rats. Unfortunately, these observations were based upon bioassay determinations of TSH and do not entirely exclude the possibility of non-TSH mediated effects. Earlier reports also implicated inhibitory effects of the habenula in thyroid control but there have been no recent reports to either confirm or deny the relevance of these observations. It is of interest that the major efferent pathway from the habenula is the fasciculus retroflexus which terminates in the medial mesencephalic region (interpeduncular nucleus), an area which forms part of the midbrain limbic system as defined by Nauta and Haymaker (1969) and which has important relays to the hypothalamus via the medial forebrain bundle. Thus, significant anatomical inputs do exist to connect the habenula with the hypothalamus.

Of equal uncertainty is the role of the pineal in TSH and thyroid control. Observations which have employed various parameters of thyroid secretion have suggested a pineal influence, predominantly inhibitory, and perhaps mediated by melatonin. Thus melatonin decreases thyroid weight and ^{131}I uptake and causes a reduction in thyroid secretory rate. On the other hand, pinealectomy is reported to cause thyroid hyperactivity in the tadpole and an enhanced thyroid secretory rate in the rat. The question as to whether these effects are mediated by changes in TSH secretion has been examined. Rowe *et al.* (1970) found no differences in plasma TSH levels, as measured by radioimmunoassay, 27–28 days after pinealectomy under varied conditions of illumination. On the other hand, pinealectomy in 21-day-old rats is said to cause a transient period of increased TSH secretion which is detectable on the 4th but not the 8th day after pinealectomy (Relkin, 1972). The mechanism of this transient elevation in TSH levels has not been demonstrated but it would appear to be of minimal physiological significance.

(2) *Extrapyramidal and Limbic Structures*

In 1962, Lupulescu and his co-workers reported that bilateral lesions of the globus pallidus and septal area cause increased thyroid activity as indicated by increased thyroid weight, ^{131}I uptake and enhanced goitrogenic response to propylthiouracil. Cheifetz (1969) confirmed this finding and suggested that tonic inhibitory influences may arise from extrapyramidal structures.

The role of the "limbic system" in TSH control has not been adequately defined. Shizumi *et al.* (1962) reported that electrical stimulation of the hippocampal formation in thiopental-anaesthetised dogs caused thyroid activation and increased TSH blood levels (within 30 min) as measured by the McKenzie bioassay. Stimulation of the orbital frontal cortex of the amygdaloid nuclear complex did not result in a significant change in thyroid secretion. Eleftheriou and Zolovick (1968) found that corticomedial amygdaloid lesions in the deermouse caused thyroid atrophy and increased bioassayable TSH levels in the pituitary and decreased levels in plasma. These effects, which were interpreted to indicate a decrease in TSH secretion, persisted for as long as 16 days. Kovács *et al.* (1965) failed to detect changes in thyroid activity in rats with amygdaloid lesions, but reported that low frequency stimulation (15 Hz) produced an elevation in ^{131}I uptake.

The hippocampus has also been implicated in inhibition of ACTH. Because of the reciprocal effects of stress on ACTH and TSH (causing release of the former and inhibition of the latter), Dupont *et al.* (1972) assessed the effects of hippocampal stimulation on TSH and ACTH responses to simultaneous cold exposure and mild psychic stress. Control rats exposed to cold showed an increase in both plasma corticosterone and TSH levels (McKenzie bioassay). The presence of an observer in the room during exposure to cold caused an enhanced corticosterone response but prevented TSH release. Stimulation of the hippocampus in unanaesthetised rats was shown to inhibit corticosterone responses and to reinstate the TSH response to cold. It was concluded from these experiments that hippocampal activation may induce these reciprocal responses, causing inhibition of ACTH secretion and activation of TSH release.

It is evident at this point that data on the role of extrahypothalamic structures in TSH control are fragmentary and inconclusive. It is interesting that both inhibitory and excitatory inputs have been suggested. However, until the precise role of these structures in pituitary-thyroid control has been more adequately defined, and until it has been clearly shown that none of the effects are secondary to changes in adrenal or other hormonal alterations, definitive conclusions concerning their role must remain tentative.

E. TRANSPORT FUNCTION OF THE MEDIAN EMINENCE

According to the classical formulation of the portal-vessel chemo-transmitter hypothesis, secretions of the hypophysiotrophic neurones are liberated at nerve endings in the median eminence and upper pituitary stalk in relationship to the primary portal vessel capillaries. A second, or supplemental view of this system, has developed from the work of Lofgren, and elaborated by Knigge *et al.* (1972). These workers propose that some portion of the releasing hormone reaches the primary portal plexus by trans-median eminence transport. It is postulated that TRH is secreted into the ventricular system, is then taken up by the lumenal processes of the tanycytes of the median eminence, and thence actively transported for release at the capillary end of the cell. The tanycyte is a specialised ependymal cell which runs from the lumenal side through to the external zone. The trans-median eminence transport hypothesis was originally based on the paradoxical finding that isolation of the median eminence itself by Halász-type section did not result in hypothyroidism, despite the fact that essentially all neurones connecting with this region had been severed. Although the possibility was considered that some regeneration of neurones might take place, it was proposed that the residual TSH function in such animals was due to transport of TRH from the third ventricle. In support of this postulation is the report of TRH being present in fluid removed from the third ventricle of the rat (Knigge and Joseph, 1974), and that TRH injected into the third ventricle, or lateral ventricle, is followed by release of TSH (measured either directly or inferred from changes in thyroid function) with a somewhat more prolonged time course (Kendall *et al.*, 1971; Gordon *et al.*, 1972). Also in support of the trans-median eminence hypothesis is the demonstration of active transport of other substances such as labelled amino acids, cortisol, from the third ventricle to the pituitary, and that median eminence tissue *in vitro* can concentrate labelled TRH (Joseph *et al.*, 1973) and T_4 (Silverman *et al.*, 1972) from incubation media. The relative importance of the trans-median eminence route, as compared with the classical release route, has not been determined.

F. MONOAMINES IN THE REGULATION OF TSH SECRETION

Despite extensive experimental observations which support the hypothesis that central monoamines (in particular noradrenaline, dopamine, and serotonin) have important modulating effects on hypothalamic secretion of hypophysiotrophic hormones for regulation of LH, FSH, prolactin and GH (see Chapters 8, 9 and 10) the effects of these substances on TRH release have been little studied. Fuxe and Hökfelt (1969) found no alteration in dopamine fluorescence of the median eminence or arcuate nucleus following thiouracil or T_4 administration. On the other hand, Lichtensteiger (1969) reported an increase in fluorescent intensity of tuberoinfundibular neurones in reponse to acute cold exposure, an effect which was prevented by prior T_4 administration. Brown *et al.* (1972) found a significant decrease in dopamine levels in the anterior and middle hypothalamus after daily T_4 adminstration in a dose sufficient to elevate plasma PBI levels (50 μg T_4 per day orally). There were no effects on noradrenaline levels, however.

A more direct approach to the problem of the role of monoamines in hypothalamic TRH release has been provided by Grimm-Jørgensen and Reichlin (1973). Hypothalamic fragments from mice were pulse-labelled with [^{3}H]-histidine and release of chromatographically separated "TRH" followed. Noradrenaline (10^{-4} M) caused an increase in "TRH" release from incubated hypothalamic tissue. Dopamine also had a stimulatory effect but this appeared to result from conversion of dopamine to noradrenaline, since pharmacologic blockade of this enzymatic step prevented dopamine-induced TRH release. Conversely, serotonin consistently inhibited TRH release, whereas acetylcholine had no effect. It was concluded from these experiments that catecholaminergic neurones may function in a regulatory manner to control synthesis and/or release of TRH in a manner analogous to the role of noradrenaline in induction of pineal synthesis of melatonin. Their findings are further supported by data that the release of TRH from sheep median eminence synaptosomes is increased by dopamine and reduced by serotonin (Bennett *et al.*, 1975).

L-DOPA administration, which in man causes acute GH release, is not associated with any change in plasma TSH levels; on the contrary, chronic L-DOPA administration is reported to interfere with TSH release induced by TRH, suggesting an effect of L-DOPA at the pituitary level (Spaulding *et al.*, 1972). This is of interest because L-DOPA causes inhibition of prolactin secretion *in vitro* in the rat by direct effects on pituitary prolactin-producing cells. In man, neither the administration of phentolamine (an α-adrenergic blocking agent) nor propranolol (a β-adrenergic blocker) has any effect on TRH-induced TSH secretion (Woolf *et al.*, 1972). In the rat, Tuomisto *et al.* (1975) report that the TSH response to acute cold exposure is decreased by drugs which interfere with noradrenaline synthesis or action L-DOPA decreased TSH response to cold as did apomorphine, a drug which stimulates dopaminergic receptors.

IV. Thyrotrophin Releasing Hormone (TRH)

A. INTRODUCTION, HISTORY

As the first of the hypothalamic hormones to be chemically identified, synthesised and administered to the human, the identification of TRH has been a landmark in neuroendocrinology and has, in a sense, legitimised the entire field (see Wilber, 1973 for review). That a thyrotrophic hormone releasing hormone existed was predicted by the studies of G. W. Harris and collaborators in the early 1950s, and the name TRF applied by Shibusawa *et al.* (1959) to material extracted from dog hypothalamus and human urine. Unfortunately, the initial claims of the Shibusawa group could not be substantiated, and proof of the TRH activity of hypothalamic extracts had to await the development of sensitive and specific assays, the first successful one being the preparation of Yamazaki and Guillemin. Others, developed shortly thereafter by Reichlin, by Schally and collaborators and by Schreiber led to the proof that the hypothalamus contained materials capable of releasing TSH from the pituitary, and that the effect was blocked by the administration of thyroid hormone. The laboratories of Guillemin and of Schally, finally identified this material. The small amount of compound available, even after the extraction of literally hundreds of thousands of hypothalamic fragments, plus the unique structure of the compound, led to a number of false starts (see Chapter 5); but, finally, the compound was found to be a tripeptide amide with a protected amino terminus (pyro-glutamic acid), and a terminal amide group (Fig 3 in Chapter 5) (Bøler *et al.*, 1969; Bowers *et al.*, 1970; Burgus *et al.*, 1970).

B. CHEMISTRY, STRUCTURE-ACTIVITY RELATIONSHIPS

TRH is a tripeptide amide, with a molecular weight of 362. A number of analogues have been synthesised in an effort to delineate the structural requirements for activity (see Chapter 5), and to devise a molecule which would either have longer acting properties, or block TRH effects. The presence of the pyroglutamic acid terminus, and the terminal amide are essential for the activity of the compound and groups added to either the N-terminus or the C-terminus markedly reduce activity. As summarised by Wilber (1973), " . . . the central histidine residue is a critical determinant of TRH activity. The TRH diasteroisomeric analogue, in which the histidine residue is optically reversed to the D form, has only 3% of the native potency, and photooxidation of the imidazole ring abolishes TRH activity completely." All substituted derivatives of TRH have less activity than the parent compound except for the unique compound methylated on the 3 position of histidine which has an 8-fold higher TSH-releasing potency than native TRH (see Fig. 3 of Chapter 5). Several analogues have recently been reported which block the effect of TRH. The most potent of these is cyclopentylcarbonyl-histidylpyrolidine. Although it is reasonable to suppose that resistance to degradation might be a factor in longer activity of a given analogue, it appears that differences in activity are related primarily to differences in binding of the analogue to specific binding sites in pituitary cells.

C. EXTRAHYPOTHALAMIC DISTRIBUTION AND COMPARATIVE ASPECTS

On the basis of early physiological study and bioassays, it was reasonable to assume that TRH would be demonstrated in the hypothalamus. This distribution has been confirmed by more recently developed immunoassay methods (see Section III A above). Beyond the hypothalamic distribution, the introduction of immunoassay has permitted the study of extrahypothalamic TRH, and the phylogenetic aspects of TRH secretion.

(1) *Extrahypothalamic Distribution of TRH*

Though low in concentration as compared with the hypothalamus (Jackson and Reichlin, 1974a), over 70% of total brain TRH lies outside the hypothalamus (Oliver *et al.*, 1974a; Winokur and Utiger, 1974). Substantial quantities of TRH are found in the spinal cord (Jeffcoate *et al.*, 1976b; Kardon *et al.*, 1977), and immunohistochemical examination has localised TRH around the motoneurones of the spinal cord and many cranial nerve nuclei (Hökfelt *et al.*, 1975d). Classical "thyrotrophic area" lesions that bring about a reduction in hypothalamic TRH content by two thirds and cause marked hypothyroidism do not affect the quantity of TRH present in the extrahypothalamic brain suggesting that the TRH located there is synthesised *in situ* (Jackson and Reichlin, 1977a). That TRH may alter brain function is shown by behavioural and neurophysiological studies (see IV I below). It has recently been reported to alter the turnover of noradrenaline in cerebral tissue, and to be a potent hypothermic agent.

(2) *Phylogenetic Distribution of TRH*

When studied from the phylogenetic point of view, TRH proved to be widely distributed throughout the vertebrates (Jackson and Reichlin, 1974b) and has even been found in prevertebrates, including the amphioxus and an invertebrate, the snail (Grimm-Jørgensen *et al.*, 1975). TRH has been found in the hypothalami of chicks, snakes, frogs, tadpoles and salmon, the values being particularly high in amphibia where the concentration may be as high as 3,620 pg/mg tissue, as compared with whole rat hypothalamus which contains 300 pg/mg. Substantial amounts of TRH are found in the neurohypophysis of rat and chicken, and is greater than the amount in the anterior pituitary. The "pituitary complex" of lower vertebrates such as snake and frog, believed to be analogous to the median eminence and/or neurohypophysis, contains large concentrations of TRH (865 pg/mg in snake, and over 5,000 pg/mg in the frog). In the larval lamprey, the pituitary contains at least 700 pg/mg compared with the whole brain tissue which contains 38 pg/mg. TRH has been found in the neural ganglia of the snail (Grimm-Jørgensen *et al.*, 1975). That this material is probably identical with pyro Glu-His-Pro-NH_2 is shown by parallel inhibition curves in immunoassay, and by the demonstration of biological TSH-releasing activity in material extracts from frog brain and salamander brain.

The functional significance of TRH in lower forms is a matter of conjecture. It is found in the lamprey, which is believed to lack TSH, in the amphioxus, which has no pituitary, and in the snail ganglia which lacks any endocrine systems homologous with the vertebrate. It has been proposed by Jackson and Reichlin (1974b) that TRH is a primitive molecule, which has been "co-opted" by the pituitary for TSH secretory control. The function of TRH has been well shown in mammals, and is also capable of stimulating TSH release in birds. However, in the poikilotherms there is no evidence that TRH is a TSH-stimulating substance. TRH does not induce metamorphosis in frog tadpoles (which are sensitive to increased TH levels), nor in the Mexican axolotl, and does not stimulate thyroid function in the lungfish. In view of its widespread neural distribution (and lacking a TSH-stimulating effect), it seems most reasonable to propose that TRH plays some role in regulation of brain function, perhaps as a neurotransmitter (see Chapter 4). Of interest is the finding that the concentration of TRH in frog pineal varies with conditions of illumination and at different seasons (Jackson *et al.*, 1977). Recently TRH has been shown to be present outside the nervous system altogether and to occur in high concentration in the skin and retina of the frog (Jackson and Reichlin, 1977b).

D. BIOSYNTHESIS AND DEGRADATION IN HYPOTHALAMUS

(1) *Synthesis*

The isolation and chemical synthesis of TRH made it possible to study the mechanism of biosynthesis of this hormone. In light of current views of peptide biosynthesis, two possibilities for the route of synthesis may be considered from the theoretical point of view, one by breakdown of a large prohormone synthesised by ribosomes by classical mechanisms such as is the case for insulin and vasopressin, and the other by a non-ribosomal enzymic process analogous to the biosynthesis of adrenaline, acetylcholine and the thyroid hormones (see Reichlin and Mitnick, 1973a, b). Evidence favouring the latter mechanism has been presented by several investigators. Rat hypothalamic fragments, incubated *in vitro* in the presence of labelled amino acid precursors of TRH were found to incorporate radioactivity into peptides which had chromatographic and electrophoretic behaviour similar to authentic TRH (Mitnick and Reichlin, 1971; Reichlin *et al.*, 1972). This incorporation process was not blocked by prior treatment with puromycin or cycloheximide, despite the presence of concentrations of inhibitor sufficient to reduce total protein synthesis by 90–95%. These experiments were criticised by McKelvy (1974) on the basis that proof of identity of the labelled product was not sufficiently rigorous and that only a fraction of the radioactivity moving with TRH by electrophoresis was authentic TRH. Using guinea-pig hypothalamic organ culture, this worker showed by extremely rigorous isolation procedures, including gas chromatography of N-trifluoroacetyl n-butyl ester derivatives, that ^{3}H-proline was incorporated into a

TRH-like compound. Similar findings were reported by Grimm-Jørgensen and McKelvy (1974a, b) using a newt. Mouse hypothalamic fragments were reported to incorporate ^{14}C-histidine into a peptide with chromatographic identity with TRH and were co-chromatographed to constant specific activity with authentic ^{3}H-TRH (Grimm-Jørgensen and Reichlin, 1973).

Studies of biosynthesis (Mitnick and Reichlin, 1972) have also been carried out using a soluble, non-particulate extract of porcine and rat hypothalamic tissue in the presence of amino acid precursors of TRH, together with Mg^{++} and ATP. A peptide with the electrophoretic and chromatographic behaviour of TRH was identified in these studies, and it was shown that puromycin did not interfere with the reaction. These observations led to the proposal that TRH was synthesised by non-ribosomal enzymatic system which was given the name "TRH synthetase". It was also determined that prior thyroid status of the animal determined the activity of the incorporating activity, hypothyroidism reducing, and thyroxine administration increasing the number of counts incorporated into a peptide with the electrophoretic behaviour of TRH (Reichlin *et al.*, 1972). From these observations it was proposed that TRH synthesis (and possibly secretion) was controlled by a "positive feedback" by thyroid hormone.

Subsequently several groups have attempted to repeat these findings. The identity of the labelled product as TRH has been questioned on the basis of more elaborate separation procedures (McKelvy, 1974; Dixon and Acres, 1975), and several groups have shown that the enzyme preparations used to study biosynthesis have active TRH degrading systems (Bauer, 1974; Bauer and Kleinkauf, 1974; Dixon and Acres, 1975; Saperstein *et al.*, 1975). Grimm-Jørgensen and McKelvy (1974a) have found that a cell-free extract of newt brain incubated with appropriate co-factors is capable of incorporating ^{3}H-proline into a TRH-like substance, the identity of which was established by rigorous physicochemical methods. The incorporation was not blocked by ribosome synthesis inhibitors, including the highly potent purified diphtheria toxin, but was blocked by pancreatic RNAase, leading to the proposal that TRH synthesis was mediated by a non-ribosomal, RNA-dependent enzymic mechanism.

In view of the presence of the TRH degrading enzyme, conclusions about the effects of prior thyroid status on synthesising activity should now be reserved until the relative effects on synthesis and degradation are evaluated. Also, the present conflicting claims of the identity of the TRH product in soluble systems will have to be resolved before one can be completely certain of the existence or nature of a TRH synthetase.

(2) *Degradation*

TRH is degraded within the hypothalamus, extrahypothalamic brain by certain tissues, and by peripheral blood (Bauer, 1974; Bauer and Kleinkauf,

1974; Reichlin *et al.*, 1976 for review). On the basis of the finding that hypothalamic and extrahypothalamic TRH breakdown is blocked by a large variety of amidated, and non-amidated peptides, it has been proposed by Saperstein *et al.* (1975) that TRH breaks down by action of one or more endopeptidases with activity directed at both peptide bonds and at the amide grouping. Deamidases and endopeptidases have also been suggested to occur in hypothalamus on the basis of identification of the de-amido form of TRH as well as other split products. That there may be a regulatory role for peptide-degrading enzymes, has been proposed as a general mechanism of peptide hormone secretion control, but the relevance of this mechanism to TRH secretion is unknown. Since the activity is widely distributed in both particulate fractions and cell sap, and is found in a number of ammonium sulphate fractions, it will be difficult to assign a regulatory role to any one of these enzymes.

In the blood, deamidation and peptidolysis (Vale *et al.*, 1971; Nair *et al.*, 1971b) is exceedingly rapid. The nature of the degrading system(s) has not been fully characterised. On the basis of blocking experiments, it has been shown that benzamidine interferes with breakdown, and that a non-amidated analogue pyro Glu-His-O methyl ester is capable of preventing breakdown. Since this non-amide substrate may block deamidation, it may be that in blood, as in hypothalamus, the peptidase(s) can attack either peptide or amide bonds.

E. BLOOD TRH, URINE TRH, CEREBROSPINAL FLUID TRH

(1) *Blood*

TRH was initially detected in hypophysial-portal blood by bioassay (Wilber and Porter, 1970). More recently TRH has been detectable by radioimmunoassay in the peripheral blood of the rat (Oliver *et al.*, 1973: Montoya *et al.*, 1975: Emerson and Utiger, 1975; Jackson and Reichlin, 1974a), in man (Oliver *et al.*, 1974a) and in several lower forms including frog (Jackson and Reichlin 1977b), chicken and newt (Jackson, unpublished). Most work has been done with man and rat, and there are several conflicting estimates of TRH level in plasma of these species. In part this is due to the rapid inactivation of TRH in the blood, and also to the low concentrations found. Estimates have been given ranging from unmeasurable to values as high as 700 pg/ml. Extimates based on indirect criteria (production rate and urinary secretion) suggest normal values in both the human and the rat to be less than 5 pg/ml (Jackson *et al.*, 1974b, 1975). There is controversy about the effects of thyroid status (Oliver *et al.*, 1973) and prior cold exposure (Montoya *et al.*, 1975; Emerson and Utiger, 1975) on plasma TRH and the general caution has been raised that plasma levels should not be taken as a measure of TRH secretion rate in view of the wide alterations of peripheral turnover which may be observed in rats exposed to different thyroid function changes, being very rapid in hyperthyroid rats, and low in the hypothyroid animal (Jackson *et al.*, 1974b).

In constrast to the situation in mammals where circulating TRH levels are quite low, we have found very high values in chicken plasma (up to 300 pg/ml) and in the leopard frog (but not bullfrog), extraordinary high levels of up to 750 ng/ml have been found (Jackson and Reichlin, 1977b). This material releases TSH from rat pituitaries *in vivo*.

(2) *Urine*

The first reports of a TRH-like material being present in the urine (as well as in the blood) of mammals was that by Shibusawa *et al.* (1959). However, subsequent studies by other workers using bioassays failed to confirm these workers' findings (c.f. Reichlin, 1966a).

TRH appears in urine after injection of exogenous hormone in man (Bassiri and Utiger, 1973; Jackson *et al.*, 1975; Jeffcoate and White, 1975d; Gagel *et al.*, 1976) in quantities ranging from 5·5–14·5% dose administered, and in the rat where 15% dose is excreted (Jackson *et al.*, 1974b). TRH when added to urine, unlike blood, shows little or no breakdown. Endogenous IR-TRH like material is present in rat (Montoya *et al.*, 1973; Jackson and Reichlin, 1974b) and human (Jackson *et al.*, 1974a, 1975; Oliver *et al.*, 1974b; Gagel *et al.*, 1976) urine. There have been reports that excretion of endogenous urine IR-TRH is increased in T_4 treated rats but reduced in thyroidectomised animals (Jackson *et al.*, 1974b) and enhanced in response to acute cold exposure (Montoya *et al.*, 1973). However, there is some controversy over the specificity of the immunoreactive material in urine (Jeffcoate and White, 1975d; Vagenakis *et al.*, 1975; Emerson *et al.*, 1977) though it is probable that some of the IR-TRH is the authentic tripeptide (Leppaluoto *et al.*, 1976; Jackson *et al.*, 1976). Until the nature of this substance has been clarified with improved methods of purification, conclusions should be withheld regarding its physiologic significance.

It must be emphasised that TRH measured in the blood or urine probably comes from both hypothalamic and extrahypothalamic brain, so that it may not be possible to make any implication as to hypothalamic-TRH secretion rate from measurements of TRH in blood and urine.

(3) *Cerebrospinal Fluid* (*CSF*)

TRH has been reported in rat and human CSF. Using a bioassay, Knigge and Joseph (1974) reported the third ventricle CSF of normal rats to contain TRH levels of 18·5 ng/ml. By immunoassay Oliver *et al.*, 1974c) found TRH levels in human CSF to range between 64 and 290 pg/ml. Where such TRH comes from, and its significance, is uncertain.

F. DEVELOPMENTAL ASPECTS

The brain content of TRH rises from 16 pg at 14 days to 380 pg at term (Eskay *et al.*, 1974), and over the first ten days following birth, rises sharply to adult levels (Shambaugh and Wilber, 1974). Eskay *et al.* (1974) also reported that serum TRH rose markedly between the 10th and 35th day of post-natal life. In the human foetus TRH ontogenesis occurs as early as 4½ weeks after conception; the highest levels in human foetuses were in the hypothalamus, though significant levels were also noted in the cerebellum. It appears that TRH is present in the human foetus and may be involved in early pituitary differentiation (Winters *et al.*, 1974).

G. HORMONAL SPECIFICITY OF TRH EFFECTS

When first introduced into clinical investigation, the specificity of TRH was established by the demonstration that infusions of this releasing hormone has no effect on plasma levels of GH, FSH, LH or cortisol in normal man. Subsequent work, however, has forced a re-evaluation of this conclusion. Following the demonstration by Tashjian *et al.* (1971) that TRH stimulated

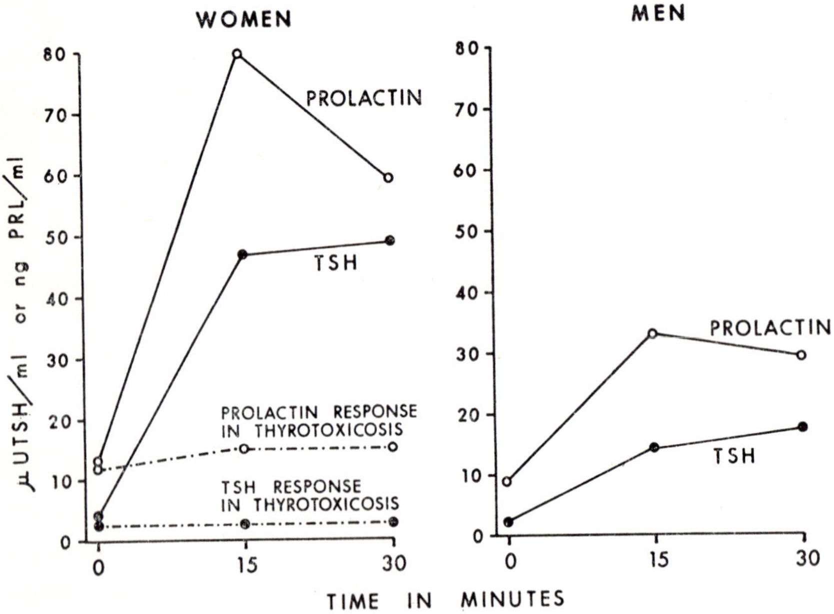

Fig. 8. The effect of TRH on plasma prolactin and plasma TSH in men and women and in relationship to thyrotoxicosis. TRH was shown to release both prolactin and TSH, the effect was greater in women and the effect in women was blocked by the presence of excessive amounts of thyroid hormone (from Reichlin, 1973a).

prolactin secretion by isolated rat pituitary tumour cell cultures, it was shown that TRH stimulated prolactin release in man and that the effect was blocked by T_4 (Bowers *et al.*, 1971, 1973b; Jacobs *et al.*, 1971, 1973; Dussault and Coulombe, 1974) (Fig. 8). Careful dose-response comparisons have been made between TSH and prolactin responses, and it appears that thresholds for response of the two hormones are quite similar. These findings suggest that in the human, basal TRH secretion may have a significant effect on basal secretion of prolactin as well as TSH. The effects of TRH on either hormonal response are blocked by exogenous TH administration and potentiated by prior oestrogen administration. In the rat, the threshold of the prolactin response to TRH is much higher than that to TRH so that in a number of studies, TRH has been reported to have no effect on prolactin release.

The possible role of TRH as a prolactin-releasing factor is relevant to the problem of hypothalamic control of prolactin secretion (see Chapter 8). The bulk of work indicates that prolactin secretion is tonically suppressed by the hypothalamus, presumably through the action of a prolactin inhibitory factor (PIF). There are a number of circumstances in which prolactin secretion rises sharply over a few minutes, as in stress, or after suckling. This finding suggests that there may be separable prolactin-releasing factors. In support of this view is the finding that crude hypothalamic extracts stimulate prolactin release both *in vivo* and *in vitro* (Nicoll *et al.*, 1970; Valverde *et al.*, 1972; Dular *et al.*, 1974; Boyd and Jackson, 1975; Frohman and Szabo, 1975). Since these extracts contain TRH, a known PRF, it is important to determine whether there is releasing hormone distinct from TRH as well. Several groups have now found that the prolactin-releasing activity can be separated from TRH (Valverde *et al.*, 1972; Machlin *et al.*, 1974).

Under certain circumstances, TRH can also act as a growth hormone releasing factor. This is true in a proportion of patients with GH-secreting tumours of the pituitary (Faglia *et al.*, 1973), and when applied to the rat pituitary in high concentrations *in vitro* (Carlson *et al.*, 1974).

H. MECHANISM OF TRH ACTION ON THE PITUITARY

The effects of intravenous TRH are extremely rapid with measurable increases in plasma TSH levels occurring within 1–2 min in both man and animals. This is in contrast to the negative feedback effects of TH on pituitary TSH secretion which require at least 15 min to develop *in vitro* and an even longer period *in vivo*. The potency of TRH is indicated by the fact that estimations of the multiplication factor of TRH-induced TSH release, on a molar ratio, are of the order of 1 : 100,000; i.e. a single molecule of TRH induces release of more than 100,000 molecules of TSH (Martin and Reichlin, 1970).

TRH stimulatory effects appear to be the result of activation of cellular secretory processes initiated by attachment of the TRH molecule to specific membrane receptor sites on the thyrotrope cell (Grant *et al.*, 1972; Labrie *et al.*, 1972; Eddy *et al.*, 1973; Wilber and Seibel, 1973b). Labrie and his coworkers (1972) have shown that *in vitro* binding of tritiated TRH to bovine pituitary cell membrane occurs very rapidly (within 5 min) and is dissociable by the addition of unlabelled hormone. The binding of TRH was not altered by administration of MSH-release inhibiting factor (MIF), lysine vasopressin, ACTH, GH, prolactin, luteinising hormone, insulin, glucagon, or significantly, the thyroid hormones T_4 and T_3 (see also Schally *et al.*, 1973a for review).

The sequence of events which are set into motion by the attachment of TRH to the receptor site appear to involve activation of an adenyl cyclase-cyclic AMP system. This effect was first reported by Wilber and Utiger in 1968, and has since been confirmed by others. Prolonged pituitary stimulation by TRH apparently leads to increased TSH synthesis as well as release (Wilber, 1971). TSH release *in vitro* can also be stimulated by addition of excess K^+ and both TRH and excess K^+ require the presence of Ca^{++} (Vale and Guillemin, 1967). Milligan and Kraicer (1971) have shown that addition of K^+ results in a reversal of transmembrane potential in most rat anterior pituitary cells incubated *in vitro* accompanied by increased cellular ^{45}Ca uptake. The precise role of calcium in the secretory response to TRH remains unidentified. Recent evidence suggests that a protein kinase present in anterior pituitary tissue may be involved in the cyclic AMP activation of hormone secretion (Barden and Labrie, 1973b). The action of TRH in inducing acute TSH release is not affected by pretreatment with protein synthesis inhibitors such as puromycin or cycloheximide.

I. TRH EFFECTS ON THE BRAIN

A number of mild side effects are observed in volunteers given TRH after i.v. bolus injection. These include nausea, flushed feeling, an urge to urinate, taste in mouth, lightheadedness, headache, and urge to defecate (Anderson *et al.*, 1971). The site of action of the hormone in producing these effects is unknown, but in a single case, it has been reported that the urge to urinate was present in a patient with paraplegia, anaesthetic from the neck down. It would appear therefore that TRH can act on the central neurones subserving urinary bladder sensations.

Chronic effects of TRH have been reported in patients with depression, a number of investigators claiming a relief of clinical symptomatology (Kastin *et al.*, 1972; Prange *et al.*, 1972), but others report that TRH given either by injection or by mouth is without effect on depressive symptoms (Coppen *et al.*, 1974; Ehrensing *et al.*, 1974; Hollister *et al.*, 1974; Mountjoy *et al.*, 1974). In the light of conflicting claims, this question cannot be resolved now. One unexplained aspect of studies of TRH in depression is that many patients have subnormal TSH secretory responses; this can be interpreted as indicating a pituitary abnormality associated with or due to the depressed state, the nature of which is still unknown. The possibility that this is due to effects of a neurotransmitter is suggested by the finding that in normals, L-DOPA (a precursor of dopamine) inhibits pituitary responses to exogenous TRH (Noel *et al.*, 1973; Rapoport *et al.*, 1973; Refetoff *et al.*, 1974).

The finding that TRH is widely distributed in brain, and evidence for its primitive phylogenetic appearance in the animal kingdom, has given rise to the question of whether TRH has functional significance in brain control. Several preliminary lines of evidence suggest that TRH has behavioural and electrophysiologic effects not explainable on the basis of its pituitary stimulating effects. The most extensive studies of the behavioural effects of TRH in experimental animals have been carried out by Plotnikoff *et al.* (1972, 1974). Administration of pargyline, a monoamine oxidase inhibitor, in combination with L-DOPA causes marked behaviour activation in the mouse. This response which includes jumping, running, fighting, piloerection, salivation and marked irritability is suppressed by the administration of TRH. The TRH effects persist in thyroidectomised and hypophysectomised animals, suggesting that TRH has direct depressive effects on brain. Large doses of TRH are reported to be tremorogenic, and this response is also mediated by direct effects on the brain, independent of pituitary (Schenkel-Hulliger *et al.*, 1974). Since these experiments utilized large doses it is difficult to assess the physiologic role of TRH in brain function. TRH in large doses was reported to enhance cerebral noradenaline turnover (Keller *et al.*, 1974) and to cause hypothermia (Metcalf, 1974).

Recent electrophysiologic studies, moreover, indicate that TRH may have potent effects on electrical activity of single neurones. Microiontophoretic application of TRH on single hypothalamic neurones results in depression of firing in a certain population of cells (Dyer and Dyball, 1974; Renaud and Martin, 1975; Renaud *et al.*, 1975). The response in the hypothalamus which was observed with small ejection currents, was closely related to the application of TRH, and normal cellular firing was restored within a few seconds after termination of TRH application.

These observations are suggestive of, but do not as yet provide definitive evidence for, a role of such small molecular weight peptides in neuronal function, perhaps acting as neurotransmitters. Final proof for such an hypothesis will require demonstration that these substances are contained in and released from nerve terminals in the brain regions where they have been shown to be localised, that there are specific binding sites, and that there is a system for degradation (see Chapter 4). The latter criterion has already been met, and immunoassayable TRH has been shown to be localised to the "synaptosome" fraction of hypothalamic extracts (Barnea *et al.*, 1975; Winokur *et al.*, 1977).

V. Conclusions

Initially conceived as a sluggishly reactive system concerned primarily with long-term metabolic adjustments, the pituitary secretion of TSH has been shown to be rapidly responsive to changes in blood levels of thyroid hormone, to changes in the external environment such as temperature, and to alterations in brain function such as hypothalamic stimulation or hypothalamic lesions. The evidence is overwhelming that the primary factor determining the rate of TSH secretion is the circulating level of thyroid hormone, but that the actual "set point" of TSH regulation is controlled by the hypothalamic secretion of releasing hormones (simple peptides) involving both stimulatory (TRH) and inhibitory (somatostatin) factors, with the former playing the dominant role—at least in higher animals. The hypothalamic hormones are released at nerve endings in the median eminence, in relationship to the primary portal vessel capillaries, and thence transported to the anterior pituitary. The possibility that part of the secretion of hypophysiotrophic neurones may be directly released into the CSF and actively transported to the median eminence by ependymal tanycytes lining the third ventricle, remains unproven.

There is evidence that the peptidergic secretion of the hypophysiotrophic neurones are influenced by the secretion of mono-aminergic neurones, and preliminary data suggest that noradrenaline may stimulate, and serotonin inhibit, TRH secretion. The effect of thyroid hormone on the hypothalamic secretion of TRH is uncertain, there being evidence reported from different laboratories for a positive feedback, a negative feedback and absence of any effect.

The development of sensitive assays for TRH has demonstrated that this hormone has a wide phylogenetic and anatomic distribution suggesting a role in neuronal function outside the pituitary-thyroid axis. Whether such extrahypothalamic hormone is affected by the same factors that may play a role in the regulation of TRH secretion from the hypothalamus such as cold stress or altered states of thyroid function, is uncertain. The evidence at present indicates that extrahypothalamic brain TRH (in the rat) arises *de novo*. In lower species such as amphibia and fishes, TRH is plentiful throughout neuronal tissue but has no known thyroid function.

Further refinement of the radioimmunoassay for TRH and improvement in the methodology, especially with respect to the specificity of the measurement in body fluids, should permit elucidation of the factors involved in the physiologic control of its secretion from both hypothalamic and extrahypothalamic sites. Further progress in neurophysiologic and molecular biologic techniques will lead to a greater understanding of the role played by TRH in brain function, including its possible action as a neurotransmitter. Such knowledge might bring with it the development of new TRH analogues and/or antagonists—agents that may have a therapeutic potential in brain disease.

Acknowledgement

This review was prepared with the support in Part of U.S.P.H.S. Grant AM 16684, AM 05166, and a grant from the Canadian National Research Council.

CHAPTER 8

Control of Reproduction

P. J. SHARP and H. M. FRASER

I. Introduction

Reproduction is influenced by many different environmental stimuli acting through the central nervous system. Ovulation is induced by coitus in several mammalian species (e.g. rabbit, cat, llama, ferret) and seasonal breeding, while depending ultimately on food supply, is controlled by "proximal" stimuli such as changing day-length and temperature (Baker, 1938; Amoroso and Marshall, 1960). Psychological stress such as that experienced by wild animals held in captivity or by inmates of concentration camps depresses reproductive activity.

Early experiments directly showing that the brain controls reproduction involved physical interference with the hypothalamus and its vascular connections with the pituitary (Harris, 1955, 1961; Chapter 1). Electrical stimulation of the hypothalamus but not of the pituitary, in the oestrous rabbit caused ovulation, while transplantation of the pituitary away from contact with the hypothalamus in the rat, resulted in gonadal atrophy. Destruction of the hypothalamus leaving the pituitary and its vascular supply intact, was found in various mammals and the duck, to result in gonadal regression or to prevent gonadal development.

The discoveries of neurosecretion (Bargmann, 1949; Scharrer and Scharrer, 1954) and of the functional significance of the hypophysial portal vasculature (Green and Harris, 1949; Chapter 1) formed the basis of current ideas about the hypothalamic control of reproduction. Three of the trophic hormones produced in the pars distalis, follicle-stimulating hormone (FSH), luteinising hormone (LH) and prolactin, are concerned with reproduction. FSH and LH are glycoproteins and are produced in the basophil cells. The traditional view that LH and FSH are produced by different cell types has been challenged by immunohistochemical observations showing that these hormones might be produced in the same cell (Nakane, 1970). The production of prolactin in cells which stain with acidic dyes has been confirmed using immunohistochemical techniques.

The target tissues for FSH, LH and prolactin are different components of the gonads. FSH in the presence of LH stimulates gametogenesis, termed oogenesis in the female and spermatogenesis in the male. LH in the presence of FSH stimulates the release of steroids from the endocrine component of the gonads, consisting of Leydig cells in the testes, and of theca and granulosa cells in the ovary. LH is also the ovulation-inducing hormone causing the rupture of the mature ovarian follicle. Generalisations cannot be made about prolactin, but in some mammals it is involved in the maintenance of the corpus luteum, the structure formed from the post-ovulatory follicle. Prolactin is also involved in another aspect of reproduction, lactation, which is discussed in Chapter 9.

Gonadotrophins are secreted by the pituitary in relation to the function of the gonads. The quantity and types of steroids secreted by the gonads act as "feedback signals" (Chapter 3) which are detected by means of specific steroid binding proteins, or steroid receptors present in certain hypothalamic neurones and in pituitary cells. The steroids bound by these receptors influence FSH and LH secretion and by this means a state of dynamic equilibrium is maintained between the secretion of hypothalamic gonadotrophin releasing hormones, pituitary gonadotrophins and gonadal steroids. This system of nterdependent endocrine organs forms the hypothalamo-pituitary-gonadal axis (HPGA). After the onset of puberty the feedback equilibrium between FSH, LH and gonadal steroids in the male does not vary a great deal, but in the female a switch in the hypothalamo-pituitary response to steroid feedback plays a key role in the induction of ovulation. Instead of exerting an inhibitory or negative feedback effect on LH release, steroids secreted by the pre-ovulatory follicle exert a stimulatory or positive feedback effect to trigger the pre-ovulatory release of LH.

An increase in the hypothalamic threshold of response to the negative feedback action of steroids initiates the onset of puberty and probably, seasonal breeding. Animals tend to reproduce only at times which favour the survival of the young and gonadal growth and regression are controlled by changes in the internal or external environment. The hypothalamus monitors these changes which are detected as either neural or endocrine inputs, and increases or decreases the activity of the HPGA.

Hypothalamic hormones, other than gonadotrophin releasing factors, concerned directly or indirectly with reproduction, are discussed in other chapters in this book. Oxytocin, which is stored in the neural lobe of the pituitary, is important in parturition, causing the uterus to contract and to expel the conceptus. Prolactin is involved in lactation and, in some species, in the maintenance of the corpus luteum; its release is controlled by a hypothalamic inhibitory factor (Chapter 9).

The terms LRF, FRF and GnRF, are used to describe "factors" in the hypothalamus which have not been chemically identified, but which stimulate the release from the pars distalis of respectively, LH, FSH and gonadotrophins. LHRH is used to describe the chemically defined decapeptide hormone (Section II) which causes both FSH and LH to be released from the pituitary.

II. Hypothalamic Gonadotrophin Releasing Factors

Knowledge that the release of pituitary gonadotrophins is controlled by hypothalamic neurosecretion led to attempts to demonstrate gonadotrophin releasing factors in hypothalamic extracts. A luteinising hormone releasing factor (LRF) was demonstrated by showing that hypothalamic extracts infused into the pituitary of the rabbit at oestrus or of the pentobarbital blocked rat at pro-oestrus, would induce ovulation in doses which had no effect when given systemically (Harris, 1961). Another approach was used by McCann *et al.* (1960) who measured LH using the rat ovarian ascorbic acid depletion (OAAD) bioassay, and showed that rat pituitary stalk-median eminence extract caused LH release when injected into gonadotrophin-pretreated immature rats. Subsequently, a more sensitive test model, the oestrogen-progesterone primed ovariectomised (O-P-OVX) rat, was developed (Ramirez and McCann, 1963) which continues to be used. LRF was also detected using *in vitro* techniques. Rat hypothalamic extracts released LH into the medium when incubated with pituitaries from O-P-OVX rats (Ramirez and McCann, 1963).

A follicle-stimulating hormone releasing factor (FRF) in the rat hypothalamus, was indicated by a rise in plasma FSH in O-P-OVX rats following injection of hypothalamic extracts (Igarashi and McCann, 1964). FSH was measured using the mouse uterine weight augmentation assay. The more specific, but less sensitive Steelman-Pohley rat ovarian augmentation assay for FSH has been used to detect FRF, by measuring the depletion of pituitary FSH after injection of hypothalamic extracts in castrated or steroid-blocked rats (Mess *et al.*, 1970).

By the early 1970s measurement of FSH and LH by bioassay was complemented by the development of radioimmunoassays for these gonadotrophins, making it easier to measure FRF and LRF using *in vivo* or *in vitro* bioassay techniques. It should be noted that the demonstration of FRF and LRF activity in hypothalamic extracts was not evidence that these factors existed as different chemical entities.

The development of methods to detect LRF and FRF was linked to attempts to isolate them and determine their chemical structure (Chapter 5). Two groups of workers, one headed by Schally using porcine hypothalamus, and the other by Guillemin using ovine hypothalamus, isolated a decapeptide with both LRF and FRF activity with the sequence, p-Glu-His-Trp-Ser-Tyr-Gly-Leu-Arg-Pro-Gly-NH_2. Neither group were able to separate LRF and FRF activity. The decapeptide called luteinising hormone-releasing hormone (LHRH) was synthesised and found to release FSH and LH as effectively as the natural product (Schally *et al.*, 1971b).

A. LHRH: THE SINGLE RELEASING FACTOR FOR LH AND FSH?

Since LHRH releases LH more readily than FSH it has been suggested that there may be a second gonadotrophin releasing hormone which conversely, releases FSH more readily than LH (Igarashi *et al.*, 1974). Alternatively there may as originally believed (McCann and Porter, 1969), be a separate FRF, specifically releasing FSH. The evidence most often advanced to support the case for releasing factor distinct from LHRH and releasing predominantly FSH, is as follows. (1) Localised electrical stimulation of the anterior hypothalamic area in the rat was observed to cause a selective rise in plasma FSH concentration in two cases (McCann, 1974). (2) Stimulation of LH but not of FSH release by progesterone, is blocked in the rat by lesions in the suprachiasmatic area of the hypothalamus (Bishop *et al.*, 1972). (3) Lesions in the paraventricular area of the rat hypothalamus cause a drop in hypothalamic FRF but not LRF activity (Motta *et al.*, 1971a). (4) An extract of bovine testes infused into the lateral ventricle of a castrated male rat was observed to cause a fall in FRF but not LRF activity in the hypothalamus (Lugaro *et al.*, 1974). (5) A highly purified porcine hypothalamic fraction free of LHRH, with essentially FRF activity has been described by Bowers *et al.* (1973a). (6) Changes occurring in plasma FSH : LH ratio at the onset of puberty, after castration (Section III D) and between the follicular and luteal phases of the menstrual and oestrous cycle (Section V), are readily explained by the existence of two gonadotrophin releasing factors.

The evidence in favour of a FRF separate from LRF has not been confirmed generally (points 1–5) or could be explained in other ways (point 6) (Arimura and Schally, 1974).

A major criticism of points (3) and (4) is that hypothalamic FRF activity was measured using the pituitary "depletion" method. This method can not be accepted as satisfactory because it is impossible to control the initial content of pituitary LH and FSH.

The claim for a chemically purified FSHRH (point 5) has not been confirmed, despite rigorous attempts to do so (Schally *et al.*, 1976). Claims for an FSHRH enriched preparation may arise because of the contamination of such fractions with metallic ions, vasopressin, cyclic AMP etc., which may synergise with LHRH to stimulate FSH release to a greater extent than does LHRH alone.

Changes in plasma FSH: LH ratio (point 6) can be explained partly as due to a differential response of FSH- and LH-producing cells to LHRH. This response is modulated by the levels of steroids and inhibin to which the cells are exposed (Section III D). The half life of FSH is longer than that of LH and in some situations this may shift the FSH:LH ratio. When the pituitary is exposed for prolonged periods to LHRH, pituitary FSH secretion is preferentially stimulated, leading to a shift in the FSH:LH ratio.

The postulate that LHRH controls the secretion of both FSH and LH is also supported by immunological studies showing that, in rats immunised against synthetic LHRH, the gonads are atretic and concentrations of FSH and LH in the pituitary and blood are depressed (Fraser *et al.*, 1974, 1975; Fig. 1).

In conclusion, there can be no doubt that LHRH is a naturally occurring hypothalamic hormone capable of releasing both gonadotrophins and the balance of evidence suggests that it is the *only* GnRF.

B. LOCALISATION AND TRANSPORT TO THE PITUITARY

There is good evidence for the presence of LHRH-containing nerve terminals in the median eminence. Using either bio- or radio-immunoassay to measure LHRH in different regions of the hypothalamus, activity is higher in the median eminence than in any other region including the arcuate nuclei and preoptic area (Palkovits *et al.*, 1974a). Material cross reacting with synthetic LHRH has been localised, using an immunohistological technique, in nerve terminals in the median eminence of mammals (Barry *et al.*,

1973a, b, 1975; Naik, 1975; Sétáló *et al.*, 1975, 1976) and in birds (Sharp *et al.*, 1975) (see Fig. 3 in Chapter 2).

Classical silver impregnation studies show that the axon endings in the rat median eminence are derived from the tubero-infundibular tract and

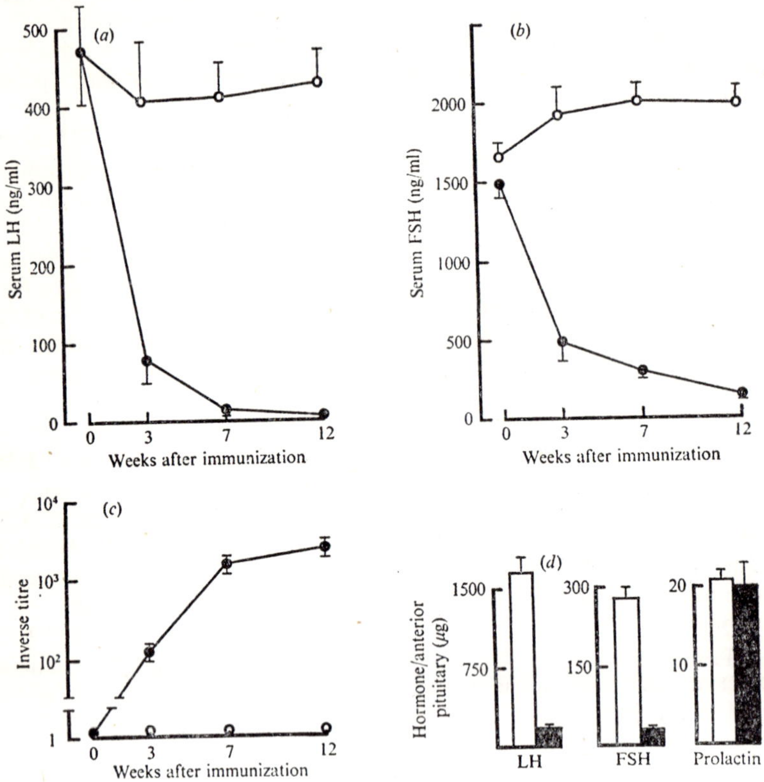

Fig. 1. Levels of (a) LH; (b) FSH; (c) LH-releasing hormone (LHRH) antibody titres in ovariectomised rats immunised with LHRH conjugated to bovine serum albumin (BSA) (●) or with BSA alone (○); (d) anterior pituitary LH, FSH and prolactin content 12 weeks after immunisation in LHRH-immunised (black columns) and BSA-immunised rats (white columns). Means ± S.E.M. (From Fraser *et al.*, 1975.)

that at least a portion of the fibres in this tract can be traced back to cell bodies in the arcuate and anterior periventricular nuclei (Szentágothai, 1962). It is likely that many of these fibres contain GnRF activity, because pituitary grafts in the arcuate nuclei and retrochiasmatic area of the hypothalamus contain basophilic gonadotrophes that enlarge after castration and maintain gonadal activity after hypophysectomy (Szentágothai, 1962). It is assumed that gonadotrophic activity is maintained in the grafts because they are in direct contact with GnRF-containing neurones. The part of the hypothalamus capable of supporting pituitary grafts in this way is termed the hypophysiotrophic area (Chapter 1). It excludes the pre-optic and septal areas of the hypothalamus.

FRF and LRF have been measured by bioassay in the rat hypothalamus, after surgically isolating the arcuate-ventromedial region from the rest of the brain or after lesioning the suprachiasmatic area, the paraventricular area, the arcuate-ventromedial nuclei or the median eminence (Mess *et al.*, 1970). It was concluded that FRF is synthesised exclusively in the paraventricular area and LRF is synthesised in both the suprachiasmatic and arcuate-ventromedial areas. As mentioned previously (Section II A), the chief criticism of this work is that the pituitary depletion bioassay used for measuring FSH is not reliable.

Experiments involving the measurement of LH released from pituitaries incubated *in vitro* with extracts of different parts of the hypothalamus, also support the view that both the suprachiasmatic and arcuate-median eminence regions are sites of LRF synthesis (McCann, 1974). Palkovits *et al.* (1974a) measured LHRH directly in 16 different hypothalamic nuclei, including the suprachiasmatic nucleus, and found in contrast, that activity was restricted almost exclusively to the median eminence and arcuate nucleus.

Using an immunohistological technique in the guinea-pig (Barry *et al.*, 1973a), rat (Naik, 1975; Sétáló *et al.*, 1975) and mouse (Naik, 1975), nerve cell bodies containing LHRH have been found chiefly in the preoptic-septal area. If these observations are valid, a large proportion of LHRH-producing cell bodies occur outside, and anterior to, the hypophysiotrophic area. Barry *et al.* (1975) observed that in primates, in contrast with rodents, immunoreactive LHRH-positive cell bodies were located mainly in the medio-basal hypothalamus. The implications of this observation are discussed in Section III A.

Direct evidence that the capillary plexus covering the median eminence serves to transport LHRH to the pars distalis was provided by Porter *et al.* (1971) who took blood directly from the hypophysial portal vessels in the rat and showed that it released LH in an *in vitro* assay.

The route by which GnRF enters the hypophysial portal vasculature has been studied with reference to the possible role of the modified ependymal cells called tanycytes, which form a major portion of the median eminence. The tanycyte cell bodies are in contact with the cerebrospinal fluid in the third ventricle of the brain and their long basilar processes terminate as "end feet" in apposition to the capillary plexus covering the superficial surface of the median eminence. Various proteins injected into the third ventricle have been observed to be taken up by the tanycytes (Kobayashi *et al.*, 1972) or to appear rapidly in the hypophysial portal vasculature (Ondo *et al.*, 1973). LHRH is rapidly taken up by the tanycytes (Scott *et al.*, 1974b) and LHRH injected into the third ventricle results in a rapid and prolonged release of LH (Ben-Jonathan *et al.*, 1974). There are however, no direct data to support the view that GnRF is normally secreted into the third ventricle. Cramer and Barraclough (1975) using a radioimmunoassay for LHRH, failed to detect it in the cerebrospinal fluid of rats which had elevated plasma LH concentrations as a result of electrochemical stimulation of the medial pre-optic area of the hypothalamus (see also Chapter 2).

C. EFFECTS ON GONADOTROPHIN SECRETION

The LH and FSH responses to LHRH depend on the dose and the route of administration (Schally *et al.*, 1972). It is effective when given intravenously, intramuscularly, subcutaneously or intranasally. A dose-response relationship between LH released and LHRH injected intravenously can usually be established with doses of between 0·01 and 1 μg LHRH/kg. In this dose range, the maximum increase in plasma LH concentration occurs between 5 and 25 min after injection and basal levels are regained after 60 to 90 min. After single injections of higher doses of LHRH the duration of LH release increases until a plateau is reached between 2 and 3 h after injection. When LHRH is continuously infused intravenously, the plasma LH concentration tends to rise for between 2 and 3 h and thereafter may start to fall despite the continued infusion of LHRH. It appears that the pituitary has only a limited releasable pool of LH.

LHRH does not release FSH as readily as it releases LH except in prepubertal subjects (Franchimont *et al.*, 1974) and in certain endocrine disorders, such as primary hypogonadism (Donald and Espiner, 1974). The maximum incremental change in plasma concentration is always less than that for LH. After an injection of LHRH, plasma FSH levels rise more slowly than do LH levels but remain elevated for longer.

Pituitary sensitivity to LHRH depends in part, on the extent to which the gland has previously been exposed to LHRH (Fink *et al.*, 1975; Edwardson and Gilbert, 1976). This priming effect can be demonstrated by injecting the same dose of LHRH separated by an appropriate time interval; the second injection causes a larger release of LH than the first (Fig. 2). This may be important in situations in which there is a progressive increase in plasma LH concentration as for example, during the induction of a preovulatory LH surge or at the onset of puberty.

LHRH also stimulates the synthesis of gonadotrophins. The pituitary LH and FSH concentrations in rats actively immunised against LHRH are low, suggesting that gonadotrophin synthesis is impaired when LHRH activity is inhibited (Fraser *et al.*, 1974; Fig. 1). Addition of LHRH to a rat pituitary culture causes an increase in the concentration of LH and

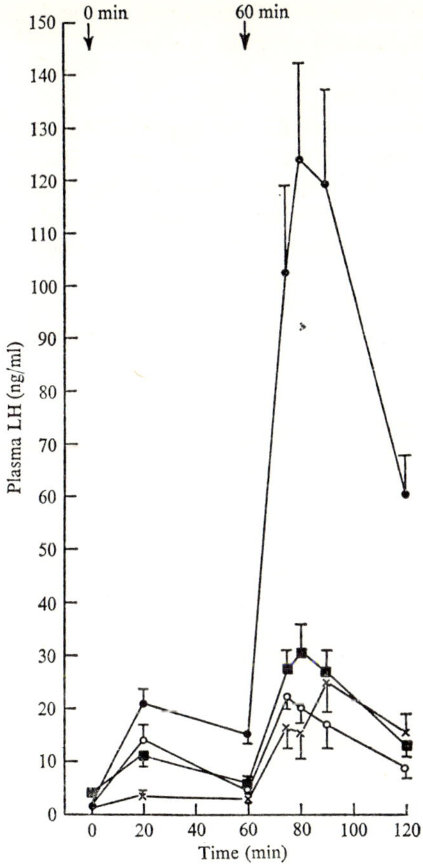

Fig. 2. Mean plasma luteinising hormone concentrations after two successive i.v. injections of 50 ng LHRH/100g body wt, 60 min apart in female rats. The first dose of LHRH was injected 30–60 min after the administration of sodium pentobarbitone at 13.30 h on each day of the cycle. Vertical bars indicate ± S.E.M., ●, pro-oestrus; ○, oestrus; ■, metoestrus; X, dioestrus. Nine animals studied at pro-oestrus; six animals in other groups. (From: Aiyer *et al.*, 1974a.)

FSH both in the culture tissue and medium (Redding *et al.*, 1972). It was also found that if ^{3}H-glucosamine was added to the culture, addition of LH RH caused it to be incorporated into FSH and LH. In hypophysectomised rats bearing pituitary grafts under the kidney capsule, treatment with LHRH for 2 months stimulated spermatogenesis in males and ovarian development in females (Debeljuk *et al.*, 1973b). Since the cells producing gonadotrophin normally regress in pituitaries transplanted to the kidney capsule, LHRH must have stimulated gonadotrophin synthesis.

D. COMPARATIVE ASPECTS

There is extensive evidence for the hypothalamic control of reproduction in submammalian species (Dodd *et al.*, 1971). Neurosecretory nerve terminals are found in the postoptic floor of the hypothalamus in a region corresponding to the median eminence and lie in close proximity to the cells of the pars distalis in all vertebrate classes. In teleost fish, neurosecretory nerve terminals invade and interdigitate with the cells of the pars distalis while in amphibians, reptiles and birds, there is a hypophysial portal vascular system. In elasmobranch fish, the part of the pars distalis which contains

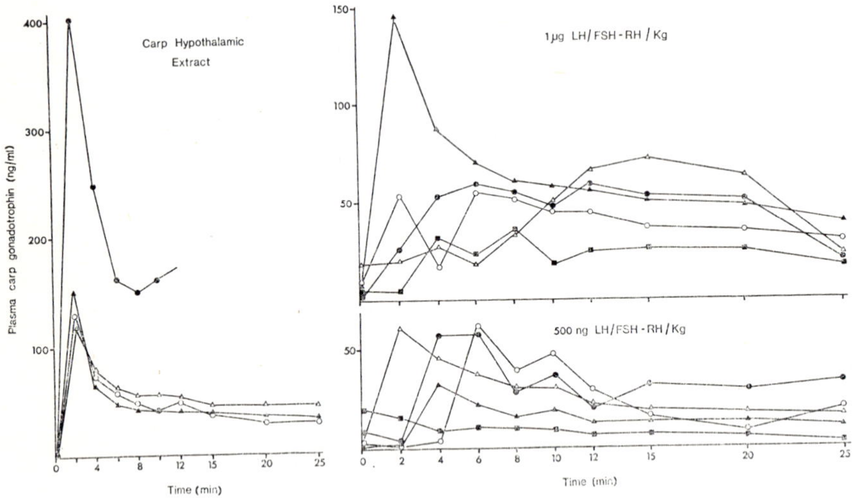

Fig. 3. Effects of carp hypothalamic extract and synthetic LHRH on the secretion of carp gonadotrophin *in vivo*. (From: Breton and Weil, 1973.)

gonadotrophic activity has no direct vascular or neural connection with the hypothalamus.

The effects of hypothalamic ablation on reproduction have been studied in teleosts, amphibians and birds: in all cases the chief effect is to block gonadal growth or cause gonadal regression.

Synthetic LHRH stimulates gonadotrophin secretion in fish, amphibians and birds. However, the rate of release of gonadotrophin in carp (*Cyprinus carpio* L) after the injection of carp hypothalamic extract, was not the same as that after seen injecting LHRH (Fig. 3). Further, in RIA systems for synthetic LHRH, immunological activity was not detected in hypothalamic extracts from fish, but was present in hypothalamic extracts from amphibians (Deery, 1974) and birds (Jeffcoate *et al.*, 1974c). These findings suggest that though GnRF may not have changed markedly during the course of evolution there are differences between fish GnRF and synthetic LHRH.

III. Hypothalamic and Extrahypothalamic Control Mechanisms

The secretion of GnRF is influenced by a variety of complex mechanisms. Different neuronal circuits, many of which have components outside the hypothalamus, using different neurotransmitters, come into synaptic contact with GnRF synthesising neurones. The activities of these circuits are controlled by stimuli originating externally or from the different components of the HPGA.

A. Levels of Control

(1) *Basal and Cyclic Gonadotrophin Secretion*

Base-line concentrations of plasma FSH and LH maintain the basic integrity of the gonads and are regulated by the negative feedback effects of gonadal steroids (Section I). As discussed later (Section III E), basal gonadotrophin concentrations are maintained by a series of episodic discharges. In the breeding female, cyclical or pre-ovulatory release of gonadotrophins is triggered by the positive feedback action of steroids secreted by the pre-ovulatory follicle(s) (Section I).

Extensive studies on the rat have resulted in the "dual control" hypothesis according to which two regions of the hypothalamus, the medial basal region and the suprachiasmatic pre-optic area, control respectively, the basal and cyclic secretion of gonadotrophins (Flerkó, 1966; Barraclough, 1967; Everett, 1969; Halász, 1969; Gorski, 1971).

The medial basal hypothalamus encompasses nearly all of the "hypophysiotrophic" area; it includes part of the anterior hypothalamic area, the ventromedial and arcuate nuclei and the median eminence. FSH and LH release is increased if it is stimulated electrically and blocked if it is lesioned. The medial basal hypothalamus is also concerned with the negative feedback action of gonadal steroids, since oestrogen implanted into it in amounts which have no effect when given systemically, inhibit gonadotrophin secretion and cause gonadal atrophy.

The suprachiasmatic-preoptic area has extensive connections with the forebrain, and more specifically, with the various components of the limbic system (septum, amygdala, hippocampus). It is not regarded so much as the region controlling the cyclic release of gonadotrophins, but more as an integration centre for the different neuronal circuits outside the medial basal hypothalamus which modulate GnRF release. If the suprachiasmatic-preoptic area is lesioned, the ovaries fill with large follicles which secrete large amounts of oestrogen. When rats with such lesions are ovariectomised plasma FSH and LH concentrations rise, but not to the same extent as in unlesioned controls: this observation supports the view that many of the neurones concerned with steroid-mediated negative feedback are located in the medial basal hypothalamus.

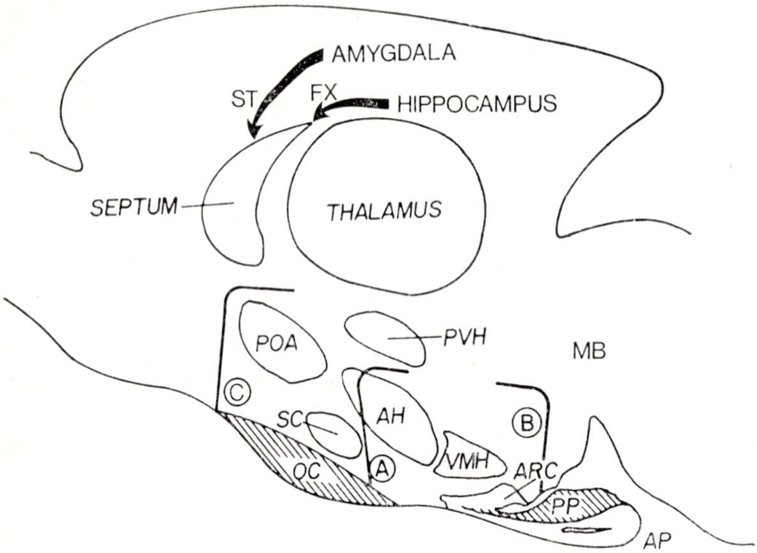

Fig. 4. Mid-sagittal section of rat brain showing the position of deafferentation cuts made with a Halász knife (see Chapters 1 and 2). A, cut in the retrochiasmatic region; B, cut posterior to the hypophysiotrophic area; C, cut between the preoptic area and the forebrain.

Abbreviations: AH, anterior hypothalamic area; AP, adenohypophysis; ARC, arcuate nucleus; FX, fornix; OC, optic chiasma; MB, mid-brain; PP, neurohypophysis; POA, preoptic area, PVH, paraventricular nucleus; SC, suprachiasmatic nucleus; ST, stria terminalis; VMH, ventromedial nucleus. (From: Rogers, 1975.)

The suprachiasmatic-preoptic area appears to contain neurones concerned with the positive feedback action of ovarian steriods on gonadotrophin release; thus, implants of small amounts of oestrogen into the preoptic area in immature female rats induce premature onset of cyclic gonadotrophin release, while the LH surge which can be induced with progesterone in oestrogen-primed female rats, cannot be induced if the suprachiasmatic area is lesioned (McCann, 1974). Further evidence that the suprachiasmatic-preoptic area contains oestrogen-sensitive neurones comes from the neurophysiological studies of Yagi and Sawaki (1973) showing that cells in this area have long axons reaching to the arcuate nucleus and that the electrical activity of these cells changes in response to oestrogen.

After surgical separation of the hypophysiotrophic area from the rest of the brain in the female rat (Halász, 1969; Chapter 1), basal secretion of gonadotrophins is maintained but ovulation ceases and the ovaries become polycystic. The same effect is observed if a single cut is placed in the retrochiasmatic region of the hypothalamus (at A, Fig. 4); this procedure interrupts the neural connections between the suprachiasmatic-preoptic area and the median eminence-arcuate nucleus (ME-ARC). If a single cut is made posterior to the hypophysiotrophic area (at B, Fig 4), ovulation is not disrupted. Following anterior deafferentation of the hypophysiotrophic area (cut A, Fig. 4) concentrations of LH, prolactin and oestradiol-17β in the blood are increased but that of FSH remains unchanged (Blake *et al.*, 1973a). Plasma LH levels rise after ovariectomy in rats with anterior hypothalamic deafferentation or with suprachiasmatic-preoptic lesions, indicating that the negative feedback system is intact. This system seems to lie principally in the retrochiasmatic area rostral to the ME-ARC; thus, in animals with cuts made caudal to that shown at A (Fig. 4), the extent to which LH levels rise after ovariectomy depends on how much of the retrochiasmatic area remains connected to the ME-ARC (Hayashi *et al.*, 1974). If the retrochiasmatic area of the hypothalamus is entirely disconnected from the ME-ARC, plasma LH levels remain low after ovariectomy but are higher than in intact dioestrous rats.

The effects of anterior deafferentation of the hypophysiotrophic area have also been studied in the male rat (Blake *et al.*, 1973b). Spermatogenesis is not impaired and plasma LH and FSH levels are not depressed.

The view that the medial basal hypothalamus and the suprachiasmatic-preoptic areas contain oestrogen receptors and by implication gonadal steroid sensitive neurones, has been supported by studies using autoradiography (Stumpf, 1968; Pfaff and Keiner, 1973) or tissue extraction techniques (Kato, 1973) following the injection of ^{3}H-oestradiol into immature female, or ovariectomised rats. These studies show that oestrogen uptake is greater in the medial preoptic nucleus and the arcuate nucleus than in other areas of the hypothalamus.

There are two variants of the "dual control" hypothesis. In one, the suprachiasmatic area is thought to contain GnRF synthesising neurones with long axons terminating in the median eminence (McCann, 1974). These are believed to be involved in the cyclic release of gonadotrophins while others in the medial basal hypothalamus are thought to be concerned with the basal release of gonadotrophins (McCann, 1974). This hypothesis is supported by the observation that a small part of the suprachiasmatic area, the anterior periventricular area, contains immunoreactive LHRH and neurones, identified by antidromic stimulation, with axons terminating in the median eminence (Kawakami *et al.*, 1975). The other variant of the "dual control" hypothesis holds that the suprachiasmatic-preoptic area contains no GnRF-synthesising neurones, and that the cyclic release of gonadotrophins is brought about by the modulating influence of neuronal activity in the suprachiasmatic-preoptic area on the GnRF neurones that occur exclusively in the hypophysiotrophic area (Halász, 1969). This concept has been supported by ultra-structural observations showing that if the hypophysiotrophic area is disconnected from the test of the brain, degenerating nerve endings do not occur in the surface zone of the median eminence (Rêthelyi and Halász, 1970).

Whichever variant of the "dual control" hypothesis is correct, both propose that the suprachiasmatic area is involved in triggering the pre-ovulatory LH surge. However this concept requires re-evaluation in the light of reports in the guinea-pig and rat, that the cell bodies producing GnRF may be located in the preoptic-septal area (Section II B). The loss of cyclicity after anterior deafferentation of the hypophysiotrophic area could be due to a reduction in the number of terminals containing GnRF in the median eminence, which results in insufficient GnRF being available to stimulate a preovulatory surge of LH. In support of this interpretation, immunoreactive LHRH activity is reduced in the rat median eminence after deafferentation of the medial basal hypothalamus (Weiner *et al.*, 1975). As mentioned previously (Section II B), it appears that in primates, in contrast with rodents, immunoreactive LHRH-positive neurones are found mainly in the medial basal hypothalamus; this observation is of interest because in the monkey, spontaneous and oestrogen-induced ovulation can still occur after the medial-basal hypothalamus is surgically isolated from the rest of the brain (Knobil, 1974). This finding lends further support to the view that loss of cyclicity after retrochiasmatic deafferentation in the rat could be due to a reduction in the number of GnRF terminals in the median eminence.

However, the rat oestrous and monkey menstrual cycles differ in two important respects. The first is that the rat brain, unlike that of the monkey, is sexually differentiated (Section IV C). Secondly, unlike the monkey, the timing of the preovulatory LH surge in the rat is controlled by a circadian rhythm (Section V B). It is therefore to be anticipated that the rat brain contains neurones concerned with the circadian control of the cyclic release of LH and others which are sexually differentiated.

A sexually differentiated region has been located in the rat hypothalamus in the mid-dorsal part of the preoptic area which is in synaptic contact with fibres from the amygdala (Raisman, 1974). Certain axons whose origin is unknown terminate in this area and the number of synaptic contacts they make are twice as numerous in females as in males. Ovulation is not blocked by lesioning this sexually dimorphic region, but is prevented by lesions in the suprachiasmatic nuclei. Raisman (1974) suggests that the suprachiasmatic nuclei may be concerned with the timing of the pre-ovulatory LH surge.

(2) *Extrahypothalamic Control*

The extensive connections between the preoptic area and the limbic system in the rat, suggest that the cyclic control of gonadotrophin release could be outside the hypothalamus (Kawakami and Terasawa, 1974). Since the stria terminalis, amygdala and hippocampus, which are components of the limbic system, have been shown by autoradiography to bind oestrogen specifically (Pfaff and Keiner, 1973), they could well form part of the oestrogen-mediated positive feedback mechanism. It should be noted however, that ovulation continues, albeit at a reduced rate, in rats in which the connections between the preoptic area and the forebrain have been severed (Cut C, Fig. 4) (Köves and Halász, 1970).

The amygdala is directly connected to the pre-optic area by the stria terminalis and indirectly by the septum; it is also connected to the medial basal hypothalamus by the ventral amygdalofugal tract. The hippocampus is connected to the pre-optic area by the fornix and septum and to the medial basal hypothalamus by the medial cortico-hypothalamic tract. Electrochemical stimulation of the amygdala and hippocampus in rats produces opposing effects on gonadotrophin secretion (Velasco and Taleisnik, 1969). Stimulation of the amygdala induces ovulation in animals in which constant oestrus has been induced by exposing them to continuous light: this effect is prevented by cutting the stria terminalis. In contrast, spontaneous ovulation is blocked by stimulating the hippocampus just before the critical period for LH release; this effect is partially prevented by transecting the cortico-hypothalamic tract.

Neuronal circuits in the mid-brain also modulate GnRF release. Depending upon which part of the midbrain is stimulated, gonadotrophin release can be either facilitated or inhibited (Carrer and Taleisnik, 1970). These effects are mediated by components of a complex system of interconnected fibre tracts (the medial forebrain bundle) running rostrocaudally through the brain connecting the limbic system, the preoptic area, the medial basal hypothalamus and the midbrain. The complexity of the extra-hypothalamic neuronal systems, exerting both inhibiting and facilitating effects on gonadotrophin secretion, indicate the fine neural control involved in GnRF release mechanisms.

The pineal is another extra-hypothalamic structure which influences gonadotrophin secretion and is particularly important in the regulation of circadian and seasonal rhythms (Reiter, 1974). Its possible role in the control of gonadotrophin secretion is discussed below (Section III F).

B. NEUROTRANSMITTERS

Sawyer and his co-workers were the first to demonstrate that ovulation can be induced by adrenergic and cholinergic neuro-transmitters and inhibited by drugs which block their activity (Sawyer *et al.*, 1949). Subsequently, it was shown that hypothalamic catecholamine content or turnover, and hypothalamic monoamine oxidase activity are related to stages of the ovulatory cycle or to levels of gonadal steroids in the blood (Anton-Tay and Wurtman, 1971). The presence of noradrenergic and 5HT (5-hydroxy-tryptamine) terminals in the suprachiasmatic-preoptic area makes it likely that they are involved in modulating the cyclic release of gonadotrophins, while a direct effect of noradrenaline and dopamine on GnRF release is suggested by the presence of these neurotransmitters in the median eminence (Chapters 1, 3 and 4). Noradrenaline stimulates ovulation when injected into the third ventricle of the oestrogen-primed rabbit (Sawyer *et al.*, 1974), or into rats made anovulatory by electrolytic lesions in the anterior hypothalamus (Tima and Flerkó, 1974). *In vitro*, noradrenaline does not stimulate gonadotrophin release when incubated with rat anterior pituitaries alone, but does so in the presence of hypothalamic tissue (McCann, 1974).

The view that noradrenergic terminals in the preoptic area are involved in stimulating the pre-ovulatory surge of LH, is supported by the observation that, after the hypothalamus is depleted of noradrenaline by drug treatment, LH release cannot be induced by electrical stimulation of the preoptic area or by steroid injections, but can be induced by electrochemical stimulation of the ME-ARC (McCann, 1974). Noradrenaline may also be involved in maintaining the secretion of gonadotrophins after gonadectomy, since pharmacological blockade of noradrenaline synthesis prevents the rise in plasma LH (Cocchi *et al.*, 1974).

Dopamine, injected into the third ventricle of the brain of the proestrous or oestrogen-primed ovariectomised rat stimulates LH secretion and *in vitro*, it stimulates gonadotrophin release from rat anterior pituitaries co-incubated with hypothalamic tissue (McCann, 1974). In contrast, intraventricular injection of dopamine in the female rabbit, not only fails to stimulate LH release, but blocks the stimulatory effect of noradrenaline (Sawyer *et al.*, 1974). Dopamine, but not noradrenaline, stimulates the *in vitro* release of LHRH from nerve endings (synaptosomes) isolated from sheep and rat hypothalami (Bennett *et al.*, 1975).

Fuxe and Hökfelt (1970) have studied the turn-over rate of dopamine in axon terminals in the median eminence of rats in various reproductive conditions. Turnover is high when gonadotrophin secretion is low, as in the lactating female rat or where sex steroids are exerting an inhibitory feedback effect. Conversely, dopamine turnover is low when gonadotrophin secretion is high, as in castrates, or during the preovulatory LH surge. These observations suggest that dopaminergic neurones inhibit the release of gonadotrophins, although other explanations are possible. The situation is complicated, since the dopaminergic axon terminals in the median eminence are involved in controlling the release of other pituitary hormones (Chapter 3).

Elevation of hypothalamic 5HT concentration by drug treatment in rats blocks both spontaneous ovulation (Labhsetwar, 1971) and superovulation in gonadotrophin-primed immature females (Kordon *et al.*, 1968). Intraventricular injection of 5HT suppresses both cyclic and tonic release of gonadotrophins in rats (Kamberi, 1973).

It has been proposed that the hypothalamus exerts a dual control over ovulation—its inhibitory influence being transmitted via a 5HT pathway and a stimulatory one through a catecholaminergic pathway (Labhsetwar, 1971). A pre-ovulatory release of gonadotrophins is induced if the equilibrium between these systems changes in favour of the catecholaminergic pathways: this change could be brought about by the stimulatory effects of ovarian steroids acting on the catecholaminergic neurones. Evidence in favour of progesterone acting in this way has been obtained, by showing that this steroid fails to induce ovulation in the gonadotrophin-primed immature rat pre-treated with an α-adrenergic bolcker (Zolovick and Labhsetwar, 1973).

Further evidence that the feedback effects of steroids on gonadotrophin release involve catecholaminergic pathways has been reviewed by McCann (1974). Oestradiol blocks the stimulatory effect of dopamine on GnRF release both *in vitro* and *in vivo*. This effect can be reversed *in vitro* by inhibitors of protein synthesis, suggesting that oestrogen stimulates the synthesis of a protein which then inhibits the response of GnRF-synthesising neurones to dopamine. Studies showing that intraventricular injections of atropine will block the secretion of FSH and LH, indicate that cholinergic systems may also be involved in the control of gonadotrophin secretion (Libertun and McCann, 1973).

There is no doubt that monoamines and acetylcholine regulate LHRH secretion. Generally, noradrenaline and dopamine tend to stimulate LHRH release while 5HT inhibits it. The exact roles of noradrenaline and dopamine are not clear. It is likely, as suggested by Fuxe and Hökfelt (1970), that central noradrenergic neurones regulate LHRF release through the tubero-infundibular dopaminergic neurones.

C. STEROID FEEDBACK MECHANISMS

It was first recognised in the 1930s that plasma gonadotrophins and gonadal steroids are maintained in a state of dynamic equilibrium. This was deduced from experiments showing that castration resulted in increased gonadotrophin secretion and that injection of large amounts of oestrogen caused gonadal atrophy.

From the time of these early experiments there has been uncertainty as to whether the site of steroid feedback is at the level of the pituitary, as originally suggested by Moore and Price (1932) or of the brain, as proposed by Hohlweg and Junkmann (1932). In addition to causing gonadal atrophy, oestrogen was also found to induce ovulation. This effect was first shown by Hohlweg in 1934 when he observed that corpora lutea were formed in prepubertal rats after injecting oestrogen. In the late 1940s Everett firmly established the concept that oestrogen induces the release of LH to cause ovulation (Everett, 1961). He showed that, in pregnant rats or rats in the mid-dioestrous phase of their ovulatory cycles, ovulation can be induced by an injection of oestrogen.

The development of methods to measure plasma gonadotrophins and steroids in the late 1960s and the more recent general availability of synthetic LHRH, has made it possible to study these steroid feedback mechanisms in detail.

(1) *Negative Feedback*

Gonadal steroids regulate the basal secretion of gonadotrophins by a negative feedback mechanism; thus, if the gonads become non-functional or are surgically removed, there is an increase in gonadotrophin secretion. After gonadectomy in the rat, the rates of increase of FSH and LH levels in the blood vary independently and according to age and sex (McCann, 1974; Eldridge *et al.*, 1974a). LH rises in the adult male within a day, whereas in the adult female the rise is delayed by several days. The sex difference in the LH response to gonadectomy probably reflects differences in the negative feedback capacities of the gonadal steroids in each sex. Oestrogen in the female may have a higher affinity for negative feedback receptors in the hypothalamus and pituitary than testosterone in the male and thus be effective for longer following gonadectomy.

Plasma FSH also rises rapidly after gonadectomy at all ages with levels rising faster in the male than in the female. The mechanism is different from that for LH however, since in the adult rat FSH levels rise faster after ovariectomy, than LH (Ramirez and Sawyer, 1974).

The ability of administered gonadal steroids to suppress gonadotrophin release decreases with time following gonadectomy. Thus, in the prepubertal rat, physiological doses of gonadal steroids are less effective in suppressing gonadotrophin secretion 5 days after gonadectomy than when given immediately after surgery (Negro-Vilar *et al.*, 1973). Similarly, within a few days of castration of the adult male rat there is a decrease in the ability of oestrogen to suppress the secretion of FSH but not of LH (Swerdloff and Walsh, 1973). This decline in the sensitivity of the negative feedback mechansim suggests that the integrity of steroid feedback receptors is maintained by the continuous presence of steroids.

The principal steroids concerned with the negative feedback control of gonadotrophins are oestrogens and androgens; the former are more potent than the latter (McCann, 1974). Because oestrogens are principally produced by the ovary and androgens by the testis, the negative feedback steroids are probably androgens in the male and oestrogens in the female. Physiological doses of progesterone given to gonadectomised animals do not usually suppress gonadotrophin secretion, although pharmacological doses may do so (McCann, 1974).

There is evidence that androgens may exert a greater negative feedback effect in the male than in the female, since in the rat, lower doses of testosterone propionate were needed to suppress LH secretion after castration than after ovariectomy (McCann, 1974). Unphysiological doses of testosterone propionate are needed to suppress FSH release in the castrated male rat and it is possible that androgens normally act synergistically with oestrogens to control FSH release. Such synergism has been observed after administration of low doses of testosterone propionate and oestradiol benzoate to the castrated rat (McCann, 1974). A similar synergism has been found between progesterone and oestrogen in the female rhesus monkey (Karsch *et al.*, 1973a). A dose of oestradiol given after ovariectomy which was sufficient to maintain plasma oestradiol at pre-operative levels, did not prevent an increase in pituitary gonadotrophin secretion. If progesterone was also given so that pre-operative levels of both oestradiol and progesterone were achieved, the rise in plasma gonadotrophin concentrations following ovariectomy was prevented. It is possible that in the male, testosterone itself does not exert a negative feedback control over LH release, but must first be converted to oestrogen in the liver or at the site of feedback (Jeffcoate, 1975b). A further possibility is that testosterone is converted to 5α-reduced metabolites at the site of feedback (Jeffcoate, 1975b).

Although, as mentioned above, FSH release in the male can be suppressed by oestrogen and testosterone acting synergistically, there is doubt whether FSH release is exclusively controlled by steroid feedback. The existence of a non-steroidal inhibitor of FSH secretion produced by the germinal epithelium was proposed in the 1930s and called inhibin. A protein with the properties of inhibin has been partially characterised (Franchimont *et al.*, 1975a, b, Section III F).

(2) *Positive Feedback*

Cyclic release of gonadotrophins is triggered by the positive feedback action of steroids secreted by ovarian follicles during the final phase of rapid growth prior to ovulation. In most mammals so far studied (including women, monkey, horse, sow, pig, dog, rat, guinea-pig and hamster) the pre-ovulatory LH surge is preceded by increased oestrogen levels in the blood. In all cases where it has been investigated, an injection of oestrogen given when progesterone levels are low or falling, will induce an ovulation or an LH surge.

Perhaps the most elegant demonstration that oestrogen exerts a positive feedback effect on LH release, comes from the observation that the preovulatory LH surge can be blocked in the monkey and rat antibodies to 17β-oestradiol (Ferin *et al.*, 1969, 1974).

Special mention should be made of the ewe, in which oestrogen levels may not always rise before a pre-ovulatory LH surge (Fig. 14, Section V B). In this species during the breeding season, the ovary always contains waves of rapidly growing follicles which secrete sufficient oestrogen to induce an LH surge (Baird *et al.*, 1975). This action of oestrogen is antagonised by progesterone secreted by the corpus luteum. As soon as the corpus luteum regresses and progesterone levels fall, the levels of oestrogen in the circulation are sufficient to trigger an LH surge.

The hen is the only species so far studied in which oestrogen does not induce a pre-ovulatory LH surge. LH release is not stimulated when oestrogen is given at any stage of the ovulatory cycle or when administered to oestrogen-progesterone-primed ovariectomised birds (Wilson and Sharp, 1976a, b). Oestrogen is essential, however, together with progesterone to *prime* the positive feedback mechanism to enable progesterone to induce LH release (Wilson and Sharp, 1976a). Progesterone will induce LH release when given at any stage of the hen's ovulatory cycle except when pre-ovulatory LH levels are falling (Wilson and Sharp, 1975a).

A single injection of progesterone will induce LH release in the rat in the presence of high levels of oestrogen (Brown-Grant and Naftolin, 1972) and in the hen in the presence of oestrogen and progesterone (Wilson and Sharp, 1976a). In both species, pre-ovulatory LH release is associated with a pre-ovulatory rise in progesterone secretion. This implies that in both the rat and hen, progesterone is physiologically important in facilitating the pre-ovulatory LH surge. Progesterone in the hen may indeed be the major steroid of physiological significance, because ovulation is blocked by injection of antibodies to progesterone (Furr and Smith, 1975). In contrast, in the rat, injection of progesterone antibodies does not block ovulation (Ferin *et al.*, 1969).

In the rat and the hen progesterone is involved in priming the positive feedback mechanism. Thus, an injection of oestrogen failed to stimulate LH secretion in ovariectomised rats after adrenalectomy but was able to do so if the animals were pretreated with progesterone (Fig. 5). The adrenals are an important source of progesterone in the rat (Shaikh and Shaikh, 1975). Ovariectomised hens must be primed with progesterone and oestrogen before a single injection of progesterone can stimulate LH release (Wilson and Sharp, 1976a).

While it is certain that progesterone is involved in facilitating the pre-ovulatory LH surge in the rat and hen, its role in primates is less clear. There is a small increase in progesterone secretion at the same time as the pre-

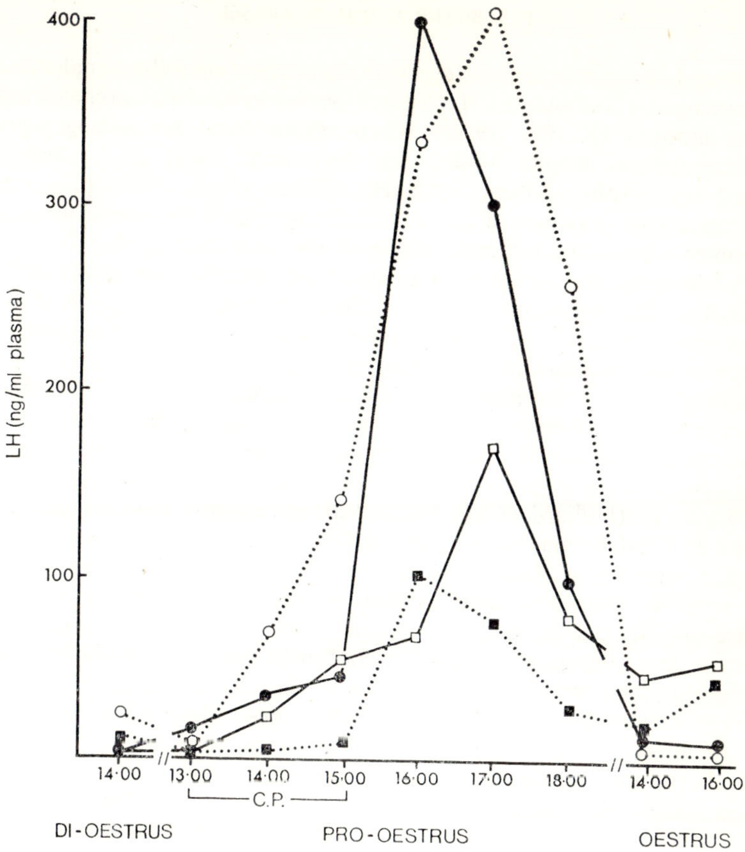

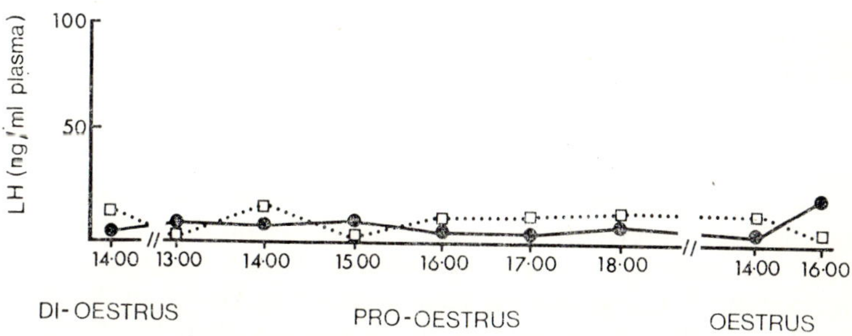

Fig. 5. *Upper panel:* plasma LH concentrations on pro-oestrus in rats ovariectomised at 10.00 h of dioestrus day 2 and injected with 10 μg oestradiol 17*β*/100 g body wt immediately following surgery.

Lower panel: plasma LH concentrations in rats ovariectomised and adrenalectomised at 10.00 h of dioestrus day 2 and immediately injected with 10 μg oestradiol 17*β*/100 g body wt C.P., critical period. (From: Mann and Barraclough 1973.)

ovulatory LH surge, but in contrast with the ovariectomised-adrenalectomised rat, oestrogen will stimulate LH release in the ovariectomised-adrenalectomised rhesus monkey (Knobil, 1974). When administered to oestrogen-primed post-menopausal women (Odell and Swerdloff, 1968) or to oestrogen-primed ovariectomised rhesus monkeys (Clifton *et al.*, 1975), progesterone will trigger a surge release of LH. It will also stimulate LH release when given to women during the follicular phase of the menstrual cycle (Shaw *et al.*, 1975), but in contrast when given to the rhesus monkey during the follicular phase, it inhibits ovulation and blocks the positive feedback action of oestrogen (Clifton *et al.*, 1975). These observations indicate that in primates progesterone plays a minor role in triggering the pre-ovulatory release of LH.

Elevated levels of plasma 17α-hydroxyprogesterone (17α-OH) and 20α-hydroxypregn-4-en-3-one (20α-OH) occur respectively, in women and rabbits in association with the pre-ovulatory LH surge. In rabbits, 20α-OH synergises with oestrogen to prolong the copulation induced pre-ovulatory LH discharge (Hilliard *et al.*, 1967). Various authors have suggested that in women 17α-OH may also synergise with oestrogen to prolong pre-ovulatory LH release (e.g., Gual *et al.*, 1975).

A specific positive feedback effect of androgen on the pre-ovulatory release of FSH has been shown in rats. Intravenous injections of an anti-serum to testosterone prevented the continued rise in the FSH level which normally occurs during the early hours of the day of oestrus (see Fig. 13). The anti-serum did not prevent the initial pre-ovulatory rise in the FSH level or affect the LH surge (Gay and Tomacari, 1974).

In all mammalian species, including rodents and primates, progesterone acts in various ways to prepare the animal for pregnancy. During the luteal phase of the cycle increased levels of circulating progesterone antagonise the positive feedback action of oestrogen on LH release: this ensures that ovulation does not occur when there is a possibility that the uterus may contain a fertilised egg.

(3) *Sites of Feedback*

The lesioning, deafferentation, electro-physiological and autoradiographic evidence discussed previously (Section III A) leave no doubt that the hypothalamus is a major site for steroid feedback. Direct evidence comes from the observation that gonadectomy leads to a rise in GnRF in the hypophysial portal blood (Ben-Jonathan *et al.*, 1973) and to a rise in the incorporation of ^{3}H-labelled tyrosine into the hypothalamic peptide fraction containing LHRH (Moguilevsky *et al.*, 1975).

Bogdanove (1963) refocussed attention on the pituitary as an alternative site of steroid feedback in his formulation of the "Implantation Paradox". He pointed out that oestrogen implanted in the hypothalamus could diffuse to the median eminence and from there be distributed via the hypophysial portal vessels to the pars distalis. An intrahypothalamic oestrogen implant could thus inhibit gonadotrophin release directly at the pars distalis, in addition to reducing the hypothalamic secretion of GnRF. That oestrogen directly inhibits gonadotrophin release from the pars distalis was demonstrated by implanting oestrogen into the pars distalis of castrated rats and showing that the development of "castration cells" was reversed in the region contiguous with the implant (Bogdanove, 1963). Intra-pituitary implants of testosterone propionate and dihydrotestosterone benzoate were also observed to inhibit the formation of castration cells (Kingsley and Bogdanove, 1973).

Steroids have also been shown to affect the pituitary response to GnRF or LHRF *in vitro* but this type of research has yielded inconsistent results and further work will be needed before firm conclusions can be drawn. Tang and Spies (1975) found that addition of 17β-oestradiol to a monolayer culture of rat pituitary cells inhibited LHRH-induced secretion of LH but not of FSH. Progesterone did not affect LHRH-induced gonadotrophin release although a major metabolite of this steroid, 20α-hydroxypregn-4-en-3-one, inhibited the basal secretion of FSH and augmented LH secretion. Testosterone increased the basal secretion of LH in the absence of LHRH and augmented the FSH response to LHRH. In another study in which rat pituitaries were superfused *in vitro* with the test steroid and/or hypothalamic extract, testosterone, 3β-androstane-3α,17β-diol and dihydrotestosterone were found initially to facilitate and subsequently to inhibit GnRF-induced LH secretion (Kao and Weisz, 1975). Dihydrotestosterone, but not the other androgens, initially facilitated GnRF-induced release of both LH and FSH. In contrast, oestrogen was found initially to inhibit and subsequently to facilitate GnRF-induced LH secretion.

The observation that, after administration of oestrogen, the pituitary response to LHRH is first inhibited and then facilitated, has been demonstrated *in vivo* in women, the rhesus monkey, sheep, cow and rat. Sexual differences have been demonstrated in the rat (Debeljuk *et al.*, 1973a; van Dieten *et al.*, 1974). Immediately after the administration of oestrogen to intact males and females, the pituitary becomes less sensitive to LHRH; this is followed, in the females only, by a phase of increased sensitivity. The absence of a hypersensitive phase in the male is ascribed to the antagonistic effect of testosterone, since a hypersensitive phase develops in long term but not short term castrated males after oestrogen injection (Debeljuk *et al.*, 1973a).

By measuring changes in basal gonadotrophin secretion and pituitary sensitivity to LHRH after steroid administration, it is possible to distinguish between central nervous and pituitary sites of steroid feedback (see also Chapter 4). Thus, in the rat, oestrogen exerts a negative feedback effect at the hypothalamic level, and a positive feedback effect at the pituitary level, since

Table 1

Effect of a Single Injection of an LHRF/FSHRF Preparation on Plasma LH in Diestrous Rats, and in Saline or Oestrogen-Treated, Ovariectomised Rats.

Group	No. of rats	Plasma LH (ng/ml)		
		Initial	Increment after injection at: 10 min	30 min
Dioestrus (day 1)	8	1·6 ± 0·4[a]	6·6 ± 0·7**	3·7 ± 0·6
Ovariectomised				
Saline control	10	15·6 ± 2·4	8·7 ± 1·7*,****	4·0 ± 1·3
3 h after oestrogen	10	5·6 ± 1·1	2·3 ± 0·4*,***	1·2 ± 0·3
6 h after oestrogen	10	6·8 ± 0·9	7·3 ± 1·4**	4·4 ± 1·0
9 h after oestrogen	10	7·2 ± 1·2	15·3 ± 1·9**,****	8·5 ± 1·7

[a] Mean ± S.E.M.
*$P<0{\cdot}05$ for LHRF/FSHRF evoked release.
**$P<0{\cdot}01$ for LHRF/FSHRF evoked release.
***$P<0{\cdot}01$ compared to all other groups at 10 min.
****$P<0{\cdot}025$ for saline control v oestrogen treated at 9 h. (From: Cooper *et. al.*, 1974)

ovariectomised rats treated with oestrogen showed an increased pituitary sensitivity to LHRH and lowered basal plasma LH concentrations (Table 1). Smith and Davidson (1974) extended this experimental approach by implanting steroids in the rat hypothalamus and observing changes in basal LH secretion and pituitary sensitivity to LHRH (Fig. 6). Oestrogen implants lowered plasma LH levels and in the female, increased pituitary responsiveness to LHRH.

Intrahypothalamic implants of progesterone in gonadectomised rats depressed basal LH levels without changing pituitary sensitivity to LHRH (Fig. 6). However, in breeding female rats, intrahypothalamic implants of progesterone in the medial basal hypothalamus but not the suprachiasmatic area facilitated LH release and if the neural connections between these two areas were severed at the retrochiasmatic level (cut A, Fig. 4), this facilitatory effect did not occur (Kobayashi, 1974). From these observations it is concluded that, in the rat, progesterone exerts a feedback effect on gonado-

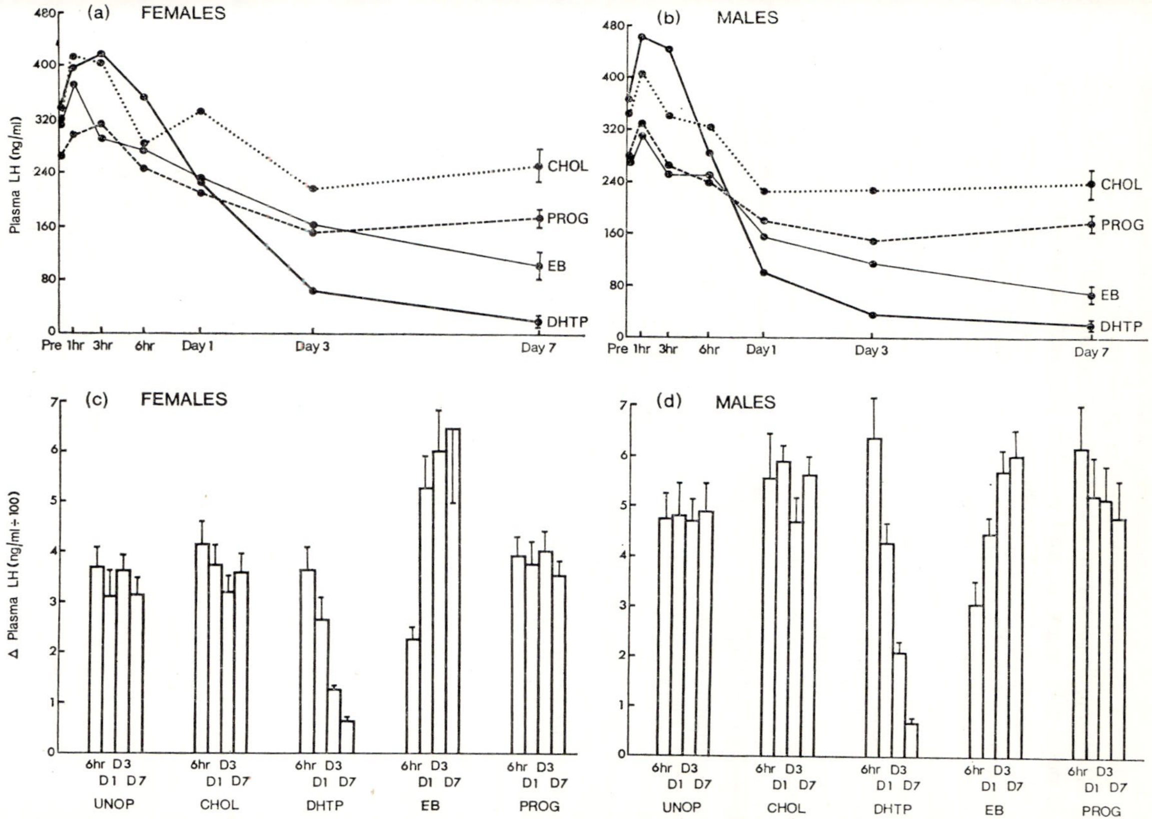

Fig. 6. Location of steroid feedback receptors in the hypothalamo-hypophysial complex of gonadectomised rats. (a) and (b), effects of intrahypothalamic steroid implants on plasma LH. (c) and (d), effects of intrahypothalamic steroid implants on the responsiveness of the pituitary to an injection of 10 ng synthetic LHRH. Abbreviations and number of rats per group: cholesterol (CHOL, 12); progesterone (PROG, 8); oestradiol benzoate (EB, 7); and dihydrotestosterone propionate (DHTP, 9). UNOP = unoperated controls; D1, D3, D7 refer to days after the steroids were implanted. (Modified from Smith and Davidson, 1974.)

trophin release at the hypothalamic level. Presumably, the concentration of progesterone reaching the hypothalamus and the level of circulating oestrogen determines whether this effect is negative or positive.

Intrahypothalamic implants of dihydrotestosterone (DHT) resulted in a decrease in basal LH secretion and a fall in pituitary sensitivity to LHRH (Fig. 6). This implies that DHT could exert a negative feedback effect at hypothalamic or pituitary levels. The *in vivo* studies and intrapituitary implantation studies mentioned previously suggest that the pituitary is a major feedback site for androgens.

In summary, steroid regulation of gonadotrophin secretion is complex. Negative and positive feedback control is exerted at both the level of the hypothalamus and of the pituitary and may involve synergism between different steroids and their metabolites. Studies *in vivo* and *in vitro* have done much to unravel the complexities of feedback mechanisms but both approaches have their drawbacks (see also Chapter 4). In the studies *in vitro* and implantation studies it is difficult to distinguish between physiological and pharmacological effects. In the studies *in vivo*, difficulties arise because steroid-induced release of GnRF may result in GnRF potentiating its own action; this could be misinterpreted as a direct effect of the steroid on the pituitary. Assuming that there is only one GnRF, the question of how the two gonadotrophins can be differentially released from the pars distalis has not been fully answered. It is established however, that the steroid environment does modify the ratio of FSH: LH secreted by the pituitary. Generally, in a low oestrogen environment, GnRF tends to stimulate a greater release of FSH and in a high oestrogen environment to stimulate a greater release of LH. The FSH response to GnRF also appears to be facilitated by androgens. It seems likely that steroid metabolites play an important role in modulating the pituitary response to GnRF.

D. SHORT AND ULTRASHORT FEEDBACK

Gonadotrophins may exert an influence on their own secretion by acting directly on the pituitary or on the hypothalamus to regulate the release of GnRF (short-loop feedback) (Chapter 3). In support of this suggestion, chronic injection of gonadotrophin or implants of FSH/LH in the median eminence reduced pituitary gonadotrophin content, while hypophysectomy induced secretion of GnRF (Motta *et al.*, 1969). Further, the electrical activity of some hypothalamic neurones can be changed by the iontophoretic application of LH and FSH (Kawakami and Sakuma, 1974; Chapter 2).

There are also data which do not support the short-loop feedback hypothesis. Coppings and Malvan (1975) infused ovine LH into the ovariectomised sheep and found that, despite the elevated levels, plasma LH continued to show the same pattern of episodic discharge as was observed in the controls. If LH does regulate its own release, constant infusion of LH would be expected to abolish the characteristic episodic pattern of endogenous LH secretion (Section III E). In another experiment, plasma FSH and LH were found to be further elevated in castrated rats after bilateral nephrectomy (Gay, 1974). This operation prolongs the half-life of LH and FSH and it might be anticipated, that if a short-feedback mechanism exists, there would be a compensatory reduction in gonadotrophin secretion to maintain the pre-operation levels.

There is evidence that GnRF can influence its own release by an ultrashort feedback loop (Chapter 3). Kawakami and Sakuma (1974) found that the basal hypothalamus contains neurones whose electrical activity can be either enhanced or reduced by the iontophoretic application of synthetic LHRH (see Chapter 2).

E. EPISODIC RELEASE

Increased plasma gonadotrophin concentrations observed in gonadectomised animals are maintained by regular episodic discharges each lasting about ten minutes and occurring at approximately hourly intervals. Episodic LH release was first demonstrated in the ovariectomised rhesus monkey (Knobil, 1974) and subsequently in man, sheep, rats and chickens (Fig. 7). Generally, LH is also released episodically in intact animals, although the frequency and/or the amplitude of the discharges are less (Fig. 7). Episodic discharges of LH in men, bulls and rams, are associated with pulsatile discharges of testosterone (e.g. Sanford *et al.*, 1974).

It has not always been possible to demonstrate episodic LH release, notably in the female rhesus monkey (Knobil, 1974) and the laying hen (Wilson and Sharp, 1975b) which suggests that small amounts of LH can be released non-episodically. This is supported by the observation that low, stable plasma LH levels were found in the ovariectomised rhesus monkey after episodic release of the hormone had been blocked pharmacologically (Knobil, 1974).

Episodic release of LH in the ovariectomised monkey can be inhibited with α- but not β-adrenergic blocking agents, suggesting that it is controlled by a central α-adrenergic nervous mechanism (Knobil, 1974). This agrees with the finding in the rat, that α- but not β-adrenergic blockers prevented dopamine-induced release of GnRF from hypothalamic fragments *in vitro* (McCann, 1974; Section III B). Control of episodic LH release may reside in the basal hypothalamus, because in both the monkey (Knobil, 1974) and the rat (Blake and Sawyer, 1974) LH continues to be secreted episodically after deafferentation of the medial based hypothalamus. There is also evidence that episodic LH release is caused by a pulsatile release of GnRF and not by rhythmic pituitary insensitivity to GnRF. Thus, bioassayable circulating GnRF was found to show pulsatile fluctuation in postmenopausal woman (Seyler and Reichlin, 1973), while LH release from superfused rat pituitaries

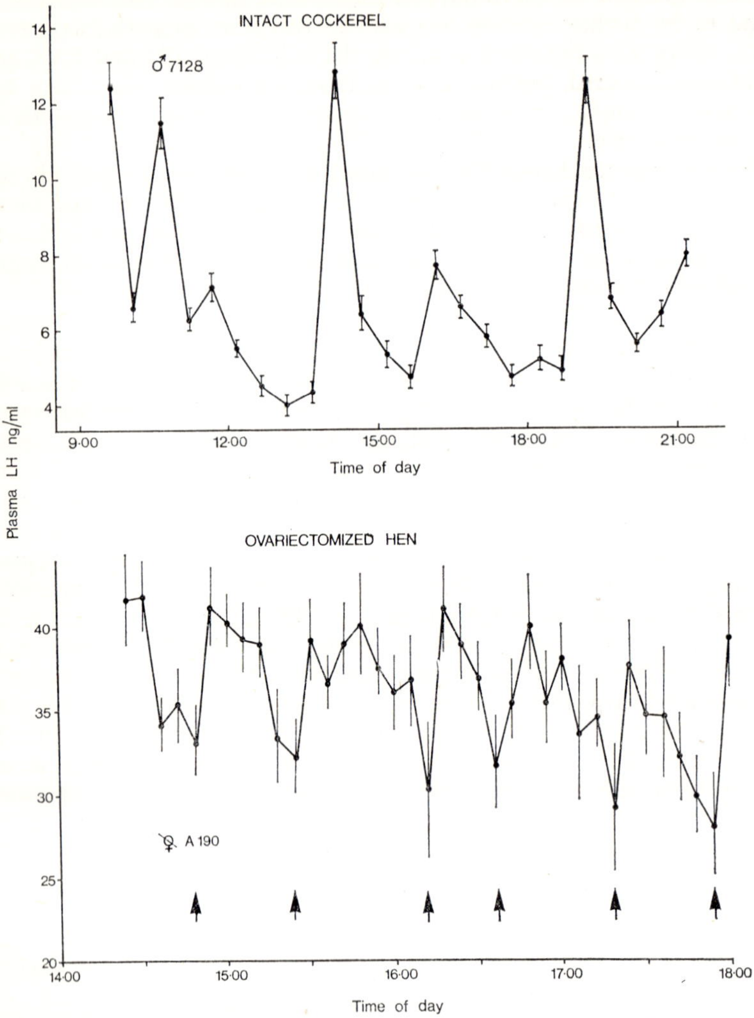

Fig. 7. Episodic discharge of LH in an intact cockerel (*upper panel*) and an ovariectomised hen (*lower panel*). Arrows indicate pulsatile discharges of LH. Vertical lines represent 95% confidence limits. (From: Wilson and Sharp, 1975b.)

was continuous when LHRH was administered continuously (Osland *et al.*, 1975). The possibility that the pulsatile release mechanism is caused by a short feedback effect of LH can also be discounted, since as mentioned earlier, episodic LH release was not abolished in castrated sheep infused with exogenous LH (Coppings and Malven, 1975).

F. INTERACTION WITH HORMONES OTHER THAN GONADAL STEROIDS

Perhaps the most important non-steroidal hormone thought to be involved in the regulation of gonadotrophin release is inhibin. The existence of this substance was first proposed by McCullagh (1932) to explain how FSH secretion is controlled in the male, and it is believed to be produced by the germinal epithelium. Part of the evidence supporting the existence of inhibin is that in patients or animals with reduced spermatogenesis and normal Leydig cells, the level of FSH in the blood tends to be elevated while that of LH remains in the normal range. In such cases, there is an exaggerated FSH response to exogenous LHRH suggesting that inhibin acts on the pituitary to modulate the FSH response to GnRF (Jeffcoate, 1975b). A protein fraction isolated from bull semen has been shown to have the properties of inhibin (Franchimont *et al.*, 1975a).

The adrenal steroids may be involved in the control of reproduction. Increasing levels of plasma androstenedione and oestrone, secreted by the adrenals, have been implicated in decreasing the sensitivity of the negative feedback mechanism to initiate the onset of puberty (Collu and Ducharme, 1975). Also the circadian rhythm of adrenal progesterone secretion, which is entrained by the daily photoperiod, may regulate the ovulatory cycle in the rat (Rogers *et al.*, 1974). Exposure of rats to constant light results in a loss of the circadian rhythm of adrenal steroid secretion (Krieger, 1973) and this might be connected with the development of persistent oestrus in rats kept in continuous light (Everett, 1961). During the human menstrual cycle, there are pre-ovulatory peaks of ACTH and cortisol secretion suggesting that the hypothalamo-pituitary-adrenal axis may be involved in the induction of ovulation (Genazzani *et al.*, 1975).

The thyroid exerts an influence on the hypothalamo-pituitary-ovarian axis but the mechanism(s) is not understood and experimental observations are contradictory. After thyroidectomy in the rat, ovarian weight decreases and compensatory ovarian hypertrophy is suppressed (Evans *et al.*, 1964; Florsheim *et al.*, 1967). In contrast, studies on ovariectomised rats show that FSH and LH secretion is increased after thyroidectomy (LaRochelle and Freeman, 1974). Serum TSH and triiodothyronine levels do not vary during the human menstrual cycle (Weeke and Hansen, 1975).

Pineal extracts were shown in the 1930s to possess anti-gonadotrophic properties in the mouse, rat and rabbit and this has been confirmed in more recent years (Reiter, 1974). Pineal hormones may act on all components of the HPGA via the systemic circulation but could also be secreted into the cerebrospinal fluid and there directly influence the GnRF-secreting neurones. Attempts to identify pineal hormones have produced evidence for both pro- and anti-gonadotrophic activities. Indolamines are important pineal "hormones", one of which, melatonin, exerts an anti gonadotrophic influence in the rat and weasel (Reiter, 1974). In contrast, melatonin has also been shown to exert a gonadotrophic effect, since when it was implanted in blinded male hamsters, it prevented the testicular regression normally induced by blinding (Reiter and Vaughan, 1975). Melatonin is only one of a number of pineal hormones which can influence gonadotrophin release. Other pineal hormones include small proteins which have been fractionated by gel filtration and assayed for pro- and anti-gonadotrophic activity (Ebels *et al.*, 1975). A confusing picture emerged, the physiological significance of which is difficult to assess with some fractions possessing pro-, and others anti-, gonadotrophic properties. Some fractions acted directly in the hypothalamus and others on the pars distalis. It is concluded that the pineal contains a complex of hormones capable of modulating gonadotrophin secretion.

The pineal is activated by darkness; animals exposed to continuous lighting or to long daily photoperiods are virtually physiologically pinealectomised. Consequently, pinealectomy tends to reverse the effects of short daily photoperiods, or total light deprivation, on sexual functions (Reiter, 1974). Thus, the delay in sexual maturation normally observed in blinded rats is prevented by pinealectomy. Similarly, pinealectomy prevents gonadal regression which would otherwise occur in sexually mature hamsters when they are deprived of light by decreasing the photoperiod or by blinding.

G. INFLUENCE OF THE ENVIRONMENT

Light is a major factor controlling reproduction and acts in two ways. First, the pattern of alternating light and dark during 24 h entrains various circadian rhythms which, in the rat and chicken, result in the pre-ovulatory release of LH occurring during a restricted period of the day (Section V B). Secondly, the length of the daily light period regulates the induction of seasonal gonadotrophin release (Section V D). Increasing day-lengths in many birds and small mammals stimulate gonadotrophin secretion, while decreasing day-lengths exert the same effect in large mammals such as sheep and deer. If the daily pattern of light and dark is interrupted and animals are kept in continuous light or dark, this treatment may lead to a loss or desynchronisation of circadian rhythms and a disruption of cyclic reproductive activity. Thus, if female rats are exposed to continuous light they stop ovulating and develop the condition known as persistent oestrus (Everett, 1961).

Decreased food supply results in a decline in reproductive activity

(Leathem, 1958) and direct measurement of FSH and LH in starved rats confirms that the pituitary release of gonadotrophins is reduced (Howland and Skinner, 1973).

Various stressful stimuli affect the functioning of the hypothalamo-pituitary-gonadal axis. Plasma gonadotrophins tend to rise in response to ether and bleeding in rats (Ajika *et al.*, 1972), while surgery in the female rat results in a release of progesterone from the adrenal which affects the ovulatory cycle. Depending upon the stage of the cycle at which surgery is performed, the stress-induced release of progesterone either advances or delays ovulation (Nequin *et al.*, 1975). Surgical stress in human male patients causes a transitory rise in plasma LH followed by a fall, and a sustained fall in plasma testosterone (Aono *et al.*, 1972). In contrast with mammalian studies, the stress caused by repeated handling in the chicken results in an immediate fall in plasma LH (Wilson and Sharp, 1975b).

Sexual stimuli of various kinds have an important influence on HPGA activity. Anticipation of sexual activity increased the rate of beard growth in a human male (Anon, 1970) and in the bull, stimulated pituitary release of LH (Katangole *et al.*, 1971).

Olfactory, visual and tactile stimuli can all play a part depending on the species, in regulating hypothalamo-pituitary activity. Pheromones are important in many species and have been extensively studied in rodents (Whitten and Champlin, 1974). Thus, reproductive activity in female mice is influenced by pheromones produced by the male. Female mice kept together without a male have irregular cycles, but as soon as a male is housed with them, cycles become synchronised. Pheromones produced by an alien male mouse, other than the successful mate, causes the pregnant mouse to abort or absorb the developing embryos.

The above discussion gives only an indication of the diverse environmental factors which influence reproductive activity and serves to emphasise the importance of the central nervous system in ensuring that young are produced when their chances of survival are the greatest.

IV. Development and Sexual Differentiation of Hypothalamic Control Mechanisms

The pituitary is potentially under hypothalamic control before birth, since GnRF activity is present in the foetal hypothalamus and gonadotrophins are found in the foetal pituitary. The vascular connection between the hypothalamus and pituitary develops during the later stages of embryogenesis. The components of the foetal HPGA are influenced by maternal gonadotrophins and steroids; consequently, the HPGA is not a self-contained regulatory system until after birth. The HPGA in the juvenile is functional and characterised by a high sensitivity of the negative feedback mechanism to gonadal steroids: this sensitivity decreases at the onset of puberty. Development of the positive feedback mechanism can occur before, or at the onset of puberty, depending on the species. Further, in some species the neural pathways for the positive feedback mechanism may be present in both sexes, and in other species, only in the female.

A. FOETAL AND NEONATAL FEEDBACK RELATIONSHIPS

During the early stages of foetal development, the median eminence becomes progressively innervated with GnRF-containing terminals. The GnRF released stimulates a gradual increase in gonadotrophin synthesis and secretion which is not subject to negative feedback control by steroids. The neural basis for the negative feedback control mechanism becomes established during the later stages of foetal development, when maternal steroids chiefly from the placenta, suppress foetal gonadotrophin secretion. After birth, the negative feedback influence of these steroids is lifted and gonadotrophin secretion increases: this stimulates steroidogenesis in the neonatal gonad, and the increase in circulating steroids eventually suppresses gonadotrophin secretion to the levels found in the juvenile. The details of the events described above vary from species to species and in many they are not fully known.

In the human female foetus plasma FSH rises to a peak between 10 and 30 weeks of gestation with values equal to those in castrated adults (Faiman *et al.*, 1974). Since these high FSH levels occur in the presence of high plasma concentrations of oestrogen of maternal origin, it seems that foetal gonadotrophin secretion is not under negative feedback control until after 20–30 weeks of gestation. In contrast, a similar increase in plasma FSH is not found in the human male foetus at the same stage of development, although the oestrogens are as high as in the female. It is concluded that negative feedback control of gonadotrophin in the male develops earlier than in the female. This difference could be a result of the high plasma testosterone levels found in the male foetus between 10 and 30 weeks of gestation. The testosterone is derived from the foetal testis and is secreted in response to the high levels of plasma chorionic gonadotrophin which are still present at this stage of pregnancy (Faiman *et al.*, 1974).

In the guinea-pig (Donovan *et al.*, 1974), FSH and LH levels in the blood rise to maxima during the last few days of gestation (gestation period 68 days) and fall abruptly on the day of birth. Negative feedback control does not develop until the very final stage of foetal life, since oestrogen given at 65, but not at 50, days of gestation suppresses gonadotrophin secretion. Patterns of FSH and LH secretion are sexually differentiated during the last week of intrauterine life: plasma FSH in the male does not rise as sharply as it does in the female, while plasma LH tends to be higher in the male than in the female.

In the sheep (Foster, 1974), plasma LH rises to a peak between 90 and 100 days of gestation and then falls during the remaining 50 and 60 days of intrauterine life. The foetal pituitary is sensitive to LHRH, although GnRF activity is undetectable in the near term foetal hypothalamus. In view of the latter observation it is not clear whether the sheep pituitary comes under hypothalamic control before or after birth.

After birth, circulating sex steroids of maternal origin begin to disappear: in human neonates, this takes between 5 and 7 days. The lifting of the negative feedback effects of maternal steroids results in an increase in gonadotrophin secretion after birth in man, the guinea-pig and the ewe, and after hatching, in the chicken.

LH levels in the blood of male and female human babies rise during the few weeks after birth and return to a low level at about 4 months of age. LH levels do not rise as much in females as they do in males (Faiman *et al.*, 1974). FSH levels also rise and fall during the first 4 months of life in males, but in females they continue to rise to reach a peak at between 2 and 5 months of age; thereafter, they gradually fall over the next 4 years. The reason for this sex difference in FSH secretion is not known. A similar increase in LH levels occurs after birth in the guinea-pig with levels being higher in the male than in the female (Donovan *et al.*, 1974). FSH also rises after birth in the male but not in the female guinea-pig. LH and FSH levels are high in rats two days after birth, but in contrast with the guinea-pig, concentrations of both hormones are higher in females than in males (Döhler and Wuttke, 1974). In the chicken, LH levels rise during the first week after hatch in both sexes (Sharp, 1975).

Plasma gonadotrophins remain low during the 4 weeks after birth in sheep, probably because at this time there is a pituitary deficiency of readily releasable gonadotrophin (Foster, 1974; Lee *et al.*, 1976). Between 5 and 7 weeks after birth, gonadotrophin secretion starts to increase. Gonadotrophin secretion appears to come under negative feedback control earlier in the male than in the female, since in sheep gonadectomised at birth, plasma LH starts to rise above the levels in intact controls at 7 days of age, in the male and after about 6 weeks of age, in the female (Foster, 1974).

Generally, levels of circulating gonadal steroids are lower in juvenile animals than in adults and in the juvenile, exert a strong negative feedback effect on gonadotrophin release. Gonadotrophin levels in the juvenile may be higher, the same as, or lower than in the adult, depending on sex and species. Plasma gonadotrophins in primates of both sexes (Faiman *et al.*, 1974; McArthur, 1972) and LH in male chickens (Sharp, 1975) are lower before the onset of puberty than in the adult. LH levels are higher in 7 to 9 week-old ewes than between pre-ovulatory surges in adults and are maintained by episodic discharges. FSH levels in the ewe at this stage are stable and within the adult range (Foster, 1974). In 5 to 7 week-old rams plasma FSH is higher than in adults, while the plasma LH concentration is similar to that at maturity (Lee *et al.*, 1976).

FSH, LH and sex steroid concentrations in the blood have been extensively studied in pre-pubertal rats. In the female rat, FSH and oestrogen levels start to rise between 5 and 10 days after birth, reach a peak at around 15 days and fall steeply to base-line levels between 20 and 25 days of age (Meijs-Roelofs *et al.*, 1973). The steroid negative feedback mechanism is fully operational in rats between 5 and 10 days after birth, since if they are ovariectomised, plasma gonadotrophin levels increase. The negative feedback effect of the high levels of plasma oestrogen on FSH secretion is limited, because the steroid is bound to a specific oestradiol binding protein present in the plasma of 5 to 10 day-old rats. Concentrations of this protein gradually fall between 15 and 20 days of age and increased free plasma oestrogen, in combination with rising levels of progesterone, is sufficient to suppress FSH release to basal levels (Meijs-Roelofs *et al.*, 1975). Brown-Grant (1974) suggests that the neonatal plasma FSH peak may be necessary for the later development of ovarian follicles or for moderating the rate of follicular atresia which normally occurs in 10 to 15 day-old rats. There are conflicting reports on LH levels in juvenile female rats but generally they have been reported to be elevated. Döhler and Wuttke (1974) found very high plasma LH in 10–15% of female rats killed between 10 and 20 days of age.

Plasma FSH and LH levels also tend to be elevated between 10 and 20 days of age in juvenile male rats, but this elevation is not as marked as in the females (Döhler and Wuttke, 1974).

B. CONTROL OF THE ONSET OF PUBERTY

The onset of puberty is controlled by a change in the sensitivity of the pituitary to GnRF, in the response of the gonads to gonadotrophins (Odell and Swerdloff, 1974), and in a decrease in hypothalamic sensitivity to the negative feedback effects of steroids. The changes are probably interrelated and triggered by a decrease in the sensitivity of the negative feedback mechanism (Ramirez, 1973; (Section III C)).

Pituitary responsiveness to LHRH increases at the onset of puberty in humans (Franchimont *et al.*, 1974; Grumbach *et al.*, 1974) and rats (Debeljuk *et al.*, 1972a, b). In view of the trophic effect of LHRH on pituitary gonadotrophin synthesis (Section II C), the increase in pituitary response to LHRH at the onset of puberty may be a consequence of an increase in pituitary gonadotrophin synthesis stimulated by a pre-pubertal increase in LHRH secretion. Although pituitary responsiveness to LHRH may increase prior to the onset of puberty, the rising levels of steriods in the blood secreted by the gonads as puberty approaches, tend to exert an increasing negative feedback effect directly on the pituitary. This effect is particularly well illustrated in the female domestic chicken, where pituitary LH responsive-

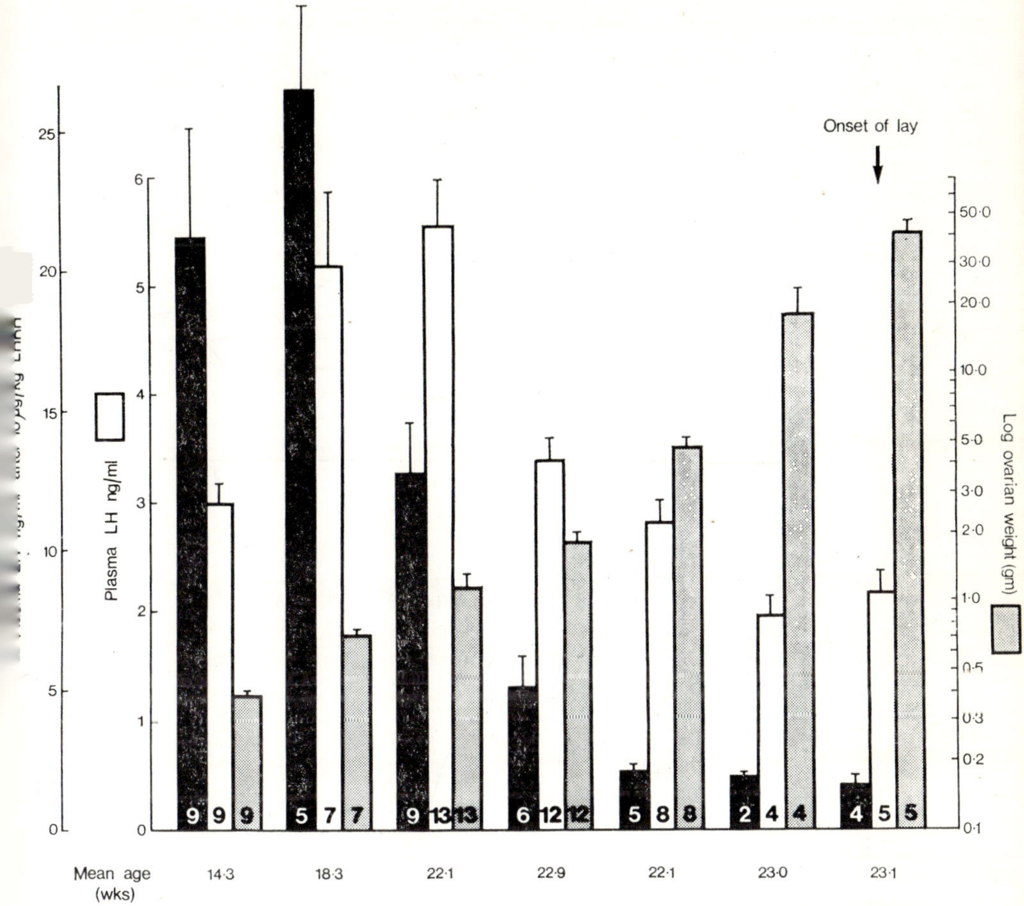

Fig. 8. Changes in plasma LH levels and pituitary responsiveness to synthetic LHRH in hens during the period immediately before the onset of puberty. The birds are categorised into seven stages of sexual development according to the weights of their ovaries: the mean ages shown beneath the figures relates to these groups. The figure at the bottom of each column is the number of hens studied. Vertical lines, = S.E.(S. C. Wilson and P. J. Sharp, unpublished data.)

ness to LHRH is much less in the adult than in the juvenile (Fig. 8). Such a dramatic fall in pituitary sensitivity is not observed in the adults of other species, although a fall has been noted in the rat (Debeljuk *et al.*, 1972b) and man (Franchimont *et al.*, 1974), particularly with respect to the FSH response to LHRH.

Hohlweg and Dohrn (1932) were the first so suggest that the onset of

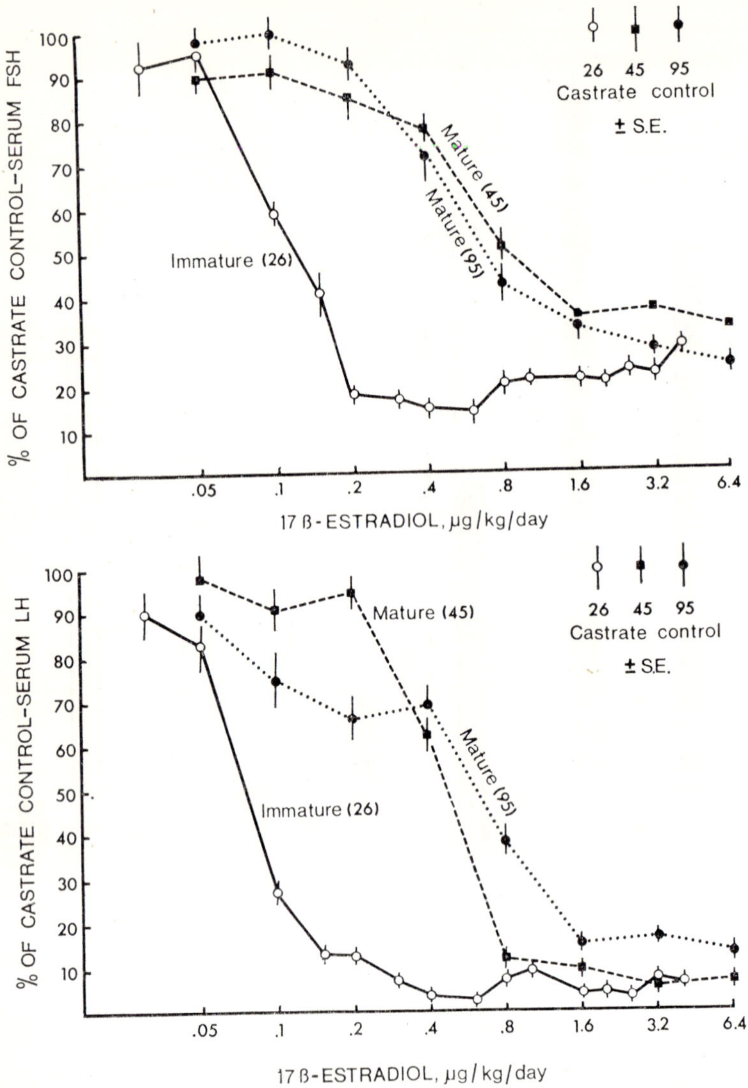

Fig. 9. A comparison of the ability of different doses of oestrogen to suppress gonadotrophin secretion in pre- and post-pubertal female rats. The animals were ovariectomised at either 26, 45 or 95 days of age and immediately injected with various doses of 17 β-oestradiol for 5 days. The post-ovariectomy rise in serum concentration of both gonadotrophins (FSH, *upper panel*; LH, *lower panel*) is suppressed by a lower dose of oestrogen in the immature rats than in the adults. Each point represents the mean ($\pm$ S.E.) of not less than 6 animals. The figures in brackets refer to the age of the animals. (From: Eldridge *et al.*, 1974).

puberty is initiated by a decrease in hypothalamic sensitivity to the negative feedback effects of steroids and later studies were carried out by Donovan and van der Werff ten Bosch (1965). Recent evidence in support of this suggestion comes from studies in man on the effect of clomiphene citrate on gonadotrophin release (Kulin and Reiter, 1974). The drug depresses circulating gonadotrophin levels in prepubertal subjects and stimulates gonadotrophin secretion in adults. The weak oestrogenic properties of the drug appear to be sufficient to exert a negative feedback effect in prepubertal subjects, but not in adults. The rise in LH levels observed after clomiphene treatment in the adult may be due to competition with endogenous steroids for negative feedback sites in the hypothalamus and pituitary. A decrease in the sensitivity of the negative feedback mechanism in humans at the onset of puberty, is also suggested by the observation that gonadotrophin secretion is suppressed by lower doses of ethinyl oestradiol in prepubertal boys than in adult men (Grumbach *et al.*, 1974).

A decrease in hypothalamic sensitivity to the negative feedback action of steroids during the onset of puberty has been demonstrated in rats of both sexes by comparing, in pre-pubertal and adult animals, the minimum dose of testosterone or oestrogen needed to prevent a rise in plasma gonadotrophin levels after gonadectomy (Fig. 9). Greater amounts of steroid are required to prevent a rise in plasma gonadotrophin after gonadectomy at 45 days of age (after the onset of puberty) than after gonadectomy at 26 days of age (before the onset of puberty).

There are two explanations for the decrease in the sensitivity of the hypothalamic negative feedback mechanism at the onset of puberty. First, assuming that GnRF secretion is stimulated by catecholamines and that steroid negative feedback is mediated by a catecholaminergic mechanism (McCann, 1974), it is suggested that at the onset of puberty, there is a progressive increase in the numbers of stimulatory catecholaminergic terminals making synaptic contact with GnRF-producing neurones (Ruf, 1973): this would result in increased GnRF release and a decrease in the negative feedback action of gonadal steroids. Secondly, the decreased hypothalamic sensitivity to negative feedback could be brought about by the increase in adrenal steroid secretion, the "adrenarche", which occurs before the onset of puberty (Collu and Ducharme, 1975; Ramaley, 1974). Increased secretion of adrenal oestrone and androstenedione occurs at adrenarche and direct evidence that these steroids can desensitise the negative feedback mechanism in the pre-pubertal animal was obtained by Döcke and Dörner (1974), who observed this effect in juvenile rats treated with oestrogen. These authors also showed that the prepubertal decrease in hypothalamic sensitivity to steroid feedback occurs in the absence of the gonads. A similar decrease in hypothalamic sensitivity is suggested by the increase on gonadotrophin secretion which occurs at the usual age for the onset of puberty in agonadal children

(Winter and Faiman, 1972) and gonadectomised chickens (Sharp, 1973) All these observations would be in accord with a role for the adrenal in the induction of puberty.

Both mechanisms mentioned above could involve functional changes in the forebrain, since gonadotrophin secretion appears to be neurally inhibited in the pre-pubertal animal. Thus, damage to the hypothalamus or forebrain of the pre-pubertal rat can advance the onset of puberty (Donovan and van der Werff ten Bosch, 1965; Donovan, 1971). Lesions in the dorsal and ventral areas of the rat rostral hypothalamus are especially effective in advancing puberty (Sherwood and Timiras, 1974), and it is thought that these lesions interrupt a diffuse inhibitory pathway from the forebrain. This pathway may originate in the amygdala or the hippocampus (Donovan, 1971).

A positive feedback response to ovarian steroids is not observed in the rhesus monkey (Knobil, 1974) or the domestic hen (Wilson and Sharp, 1975c) before the onset of puberty. Ovulation and corpus luteum formation can be induced in immature female rats with oestrogen after they are about 25 days old and is termed the "Hohlweg effect" (Hohlweg, 1934). Spontaneous ovulation does not normally occur in rats until they are about 36 days old. In the sheep, oestrogen can induce LH release in pre-pubertal females (Foster and Karsch, 1975).

C. SEXUAL DIFFERENTIATION OF THE BRAIN

The first evidence of a sexual differentiation of the gonadotrophin release mechanism was obtained in the rat by Pfeiffer (1936). He showed that ovarian grafts transplanted into adult females underwent cyclical activity with the periodic formation of corpora lutea, while ovarian grafts transplanted into adult males became polycystic and failed to ovulate. If the grafts were placed in males which had been castrated at birth, they showed cyclic activity. Pfeiffer suggested that androgen secreted by the juvenile testes masculinised the pituitary gland. It was later shown that it is the brain, not the pituitary that is sexually differentiated because pituitaries from male rats, transplanted next to the median eminence in hypophysectomised female rats, were able to support cyclical ovarian activity (Harris and Jacobsohn, 1952). Barraclough (1967) confirmed Pfeiffer's hypothesis regarding androgenisation by showing that a single injection of testosterone given to a female rat during a "critical period" of between 2 to 6 days after birth, inhibits the development of a sex steroid positive feedback mechanism and cyclic ovarian activity. The mode of action of testosterone may be to cause a degeneration of neurones which respond to the positive feedback action of oestrogen. This suggestion is supported by anatomical studies showing that the sexually differentiated neuropil in the preoptic area in androgenised females, is the same as that in males and contains fewer synapses than in females (Raisman, 1974).

Studies in other species suggest that the brain is not sexually differentiated in man (Stearns *et al.*, 1973), the rhesus monkey (Knobil, 1974) and chicken (P. J. Sharp, unpublished) but is differentiated in the sheep (Short, 1974; Karsch and Foster, 1975). Administration of oestrogen to anoestrous or ovariectomised ewes induces a surge release of LH about 12 h later: a similar response is not observed in intact or castrated rams (Fig. 10; Karsch and Foster, 1975). The "critical period" in sheep during which androgens block the development of the steroid positive feedback mechanism is between 20 and 60 days of gestation (Short, 1974).

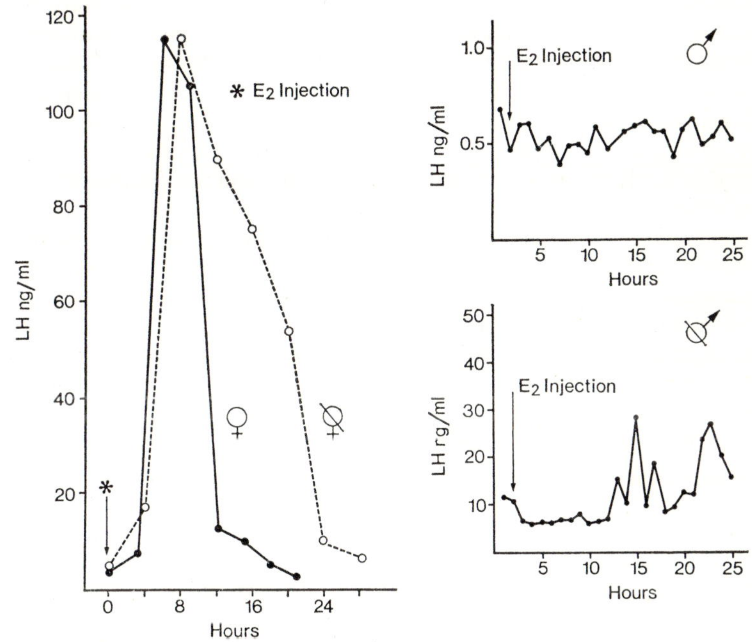

Fig. 10. Comparison of the effects of a single injection of oestrogen (E_2) on LH secretion in adult gonadectomised or intact sheep of both sexes. *Left panel*: females. (From Pelletier and Signoret, 1969.) *Right panel*: males. (From: Short, 1974.)

A single injection of oestrogen will not evoke an LH surge in the intact male or ovariectomised female monkey (Knobil, 1974). For a positive feedback response to occur the brain must be exposed to a priming dose of oestrogen. Thus, in both castrated and ovariectomised rhesus monkeys LH secretion is stimulated by an injection of oestrogen, if the animals are first implanted with oestrogen to bring plasma LH levels down to those found during the early follicular phase of the menstrual cycle (Knobil, 1974). A positive feedback response to progesterone has been demonstrated in castrated men pre-treated with oestrogen (Stearns *et al.*, 1973).

A positive feedback response to a single injection of progesterone can be evoked in adult intact or gonadectomised chickens of either sex, provided they have been primed with oestrogen and progesterone (P. J. Sharp, unpublished).

D. AGEING

A decline in reproductive activity with advancing age occurs in man and domesticated animals. Relatively little is known about wild animals, but it is likely that few mammals survive to advanced old age. The loss of reproductive capacity with advancing age has been most studied in women, the Macaque monkey, the rat and the mouse (Jones, 1975). It has been found in all cases, that with age, there is a progressive loss of oocytes from the ovary caused partly by ovulation but mainly by atresia. This loss means that there are progressively fewer oocytes available for ovulation and because the oocytes are a major source of oestrogen, there is a decline in baseline circulating oestrogen levels. This decline affects positive and negative feedback relationships and leads to changes in the ratios and amounts of FSH and LH in the circulation.

Menstrual cycles cease entirely in women and in the Macaque monkey at the menopause (Jones, 1975). In the Macaque, the menopause occurs shortly before the end of the life, whereas in women, the menopause occurs at about two-thirds of the total life span. The menopause is due to ovarian failure; there is no impairment of the competency of the hypothalamo-pituitary component of the HPGA. The decrease in oestrogen secretion in post-menopausal women leads to a lifting of the negative feedback effect of this steroid and to an increase on gonadotrophin secretion. The positive feedback mechanism remains potentially functional in post-menopausal women, as does the capacity of the pituitary to respond to LHRH (Franchimont *et al.*, 1974).

There is no equivalent to the menopause in rodents. Loss of reproductive activity with age is a gradual process; first oestrous cycles become irregular, then there is a period of persistent oestrus followed by a series of pseudopregnancies and finally, just prior to the end of the total life span, a period of anoestrus. An exception to this is the CBA strain of mouse which becomes completely anoestrous, due to a drastic decline in the number of oocytes in the ovary at about one-third of the total life span (Jones, 1975). Generally, in rodents, the initial decline in cyclical activity with advanced age can be ascribed to a failure of hypothalamo-pituitary function rather than to a loss of ovarian competency: this has been shown by transplanting ovaries from young rats into old rats and *vice versa* and observing cyclic ovarian activity (Peng and Huang, 1972). If ovaries from old rats, showing irregular cycles are transplanted into ovariectomised young hosts, they start showing regular cycles, characteristic of the host. Conversely, irregular cycles persist in old hosts with transplanted ovaries from young donors. Similar experiments with transplanted pituitaries suggest that, although there may be some loss in pituitary function in old age, the absence of cyclical activity in old rats is due to failure at the level of the central nervous system (hypothalamus).

V. Cycles and Rhythms

Most animals show rhythmic variations in behaviour (e.g. sleeping, eating) and in physiological activities including the secretion of hormones, which are synchronised by the daily pattern of light and dark; some of these rhythms disappear in continuous light or dark while others persist. The latter are endogenous rhythms driven by an internal biological clock which is believed to be located in the central nervous system but the nature of which is not fully understood (Bünning, 1967). The daily changes in light and dark serve to "set the hands" of the biological clock and thereby synchronise endogenous rhythms with the external lighting pattern.

A second kind of "biological clock" originates in the ovary and is a function of the cyclic growth and development of ovarian follicles and formation of corpora lutea.

Changes in the general activity of the hypothalamo-pituitary-gonadal axis are profoundly influenced by diurnal rhythms. They are involved in the control of the onset of puberty, seasonal breeding and, in laboratory rodents and the hen, in the timing of the pre-ovulatory LH surge. The ovary acts as the chief biological clock regulating the oestrous/menstrual cycle in other animals.

A. Diurnal Rhythms

There are two problems associated with attempts to demonstrate diurnal rhythms of gonadotrophin or gonadal steroid secretion. First, both types of hormone are released episodically (Section III E). This means that many blood samples must be taken to obtain even an approximate picture of the mean level of a circulating hormone. The second problem is that stress caused by frequent handling or by withdrawing a large volume of blood, can affect the rate of hormone release (Section III G). It is therefore not surprising that there have been conflicting reports on the existence of diurnal rhythms of gonadotrophin and gonadal steroid secretion.

Although a diurnal rhythm in LH secretion is difficult to demonstrate in adult men or young boys, one is observed during the early stages of the onset of puberty (Kapen *et al.*, 1974). It is characterised by an increase in the frequency and amplitude of episodic release during sleep. Initially this increased LH secretion occurs only at the onset of sleep but as sexual maturation proceeds, it continues for longer periods each night until at sexual maturity, LH is secreted at an enhanced rate during waking hours. Circulating LH levels rise at night in boys at mid-puberty, even when they were deprived of sleep, implying that there is an endogenous rhythm controlling LH release. However, this rhythm is not entirely endogenous since LH secretion increases in boys going to sleep during normal waking hours (Kapen *et al.*, 1974).

Numerous studies on adult men and male domesticated animals, excluding the rat, show that there is no clear diurnal rhythm in LH secretion (Rowe *et al.*, 1975; Ellendorf *et al.*, 1975). Testosterone however, shows a diurnal pattern of release in all species studied, with levels generally rising during the latter part of the dark period to peak values after dawn or "lights on". Levels then fall gradually during the rest of the day and/or early night.

Circulating LH levels show a diurnal rhythm in adult rats with peak levels in the male occurring 6 h earlier than those in the female (ter Haar *et al.*, 1974). Peak levels occur in males in the morning and in females, in the afternoon. Sexual differentiation of the LH diurnal rhythm is reflected in a sexually differentiated diurnal rhythm of ^{35}S-methionine incorporation into protein in various regions of the brain (amygdala, thalamus, preoptic area, putamen, median eminence) and pars distalis. The ^{35}S-methionine incorporation rhythm is dependent on gonadal steroids, since it disappears after gonadectomy and assumes the male pattern in androgenised females (ter Haar *et al.*, 1974).

There is no agreement on a diurnal rhythm of FSH secretion. Faiman and Winter (1971) found a diurnal rhythm for FSH in men which was similar to that for testosterone but Krieger *et al.* (1972) found no evidence for this. FSH levels were found to show diurnal rhythmicity in male and androgenised female rats by Dunn (1974) but not by ter Haar *et al.* (1974).

B. OESTROUS, MENSTRUAL AND OVULATORY CYCLES

Follicular development, ovulation and formation of corpora lutea are the first stages of the breeding cycle which ends with the behavioural and physiological changes associated with the care of young. In some species (e.g. rabbit, cat, ferret), ovulation does not occur unless a stimulus is provided by coitus; instead the mature follicles become atretic and do not form corpora lutea. Long successions of oestrous or menstrual cycles can be regarded as abortive attempts to initiate breeding cycles and are usually artifacts of domestication: they are rare in nature and normally pregnancy follows the formation of corpora lutea. In birds, corpora lutea are not formed and in wild species ovulatory cycles cease once a clutch of eggs is laid.

(1) *Environmental Control*

Seasonal changes in day-length regulate the occurrence of oestrous cycles in wild animals and domesticated species, such as the sheep and the horse (Hoffman, 1973; Section V D). Some domestic species such as the cow, the dog, and the laboratory rat have lived in association with man so long that they are not influenced by seasonal changes in day-length and ovulate at any time of year. The menstrual cycles of primates including man, are not greatly influenced by seasonal factors.

In rodents and hens, pre-ovulatory LH surges are initiated only during a restricted period of the day. This period, called the "critical period" in rodents and the "open period" in hens, is determined (entrained) by the alternating rhythm of light and dark during 24 h. If the lighting pattern is reversed or shifted, the period for LH release becomes re-entrained (rat, Colombo *et al.*, 1974; hen, P. J. Sharp, unpublished).

The "critical period" occurs between 14.00 and 16.00 or 17.00 h in rats and hamsters, kept on the lighting schedule used by most workers (lights on at 05.00 h and off at 19.00 h). In the mouse, the critical period starts at about 14.00 but lasts longer and ends between 18.00 and 21.00 h. The "open period" in hens lasts 8–9 h and starts soon after the onset of darkness. A hen lays eggs at daily intervals in sequence, and as a sequence proceeds each egg is laid later in the day than its predecessor. This pattern of lay is caused by successive pre-ovulatory LH surges occurring later in successive "open periods". No LH discharge occurs on the day following a pre-ovulatory surge initiated at the end of the "open period". On the next day, the LH release mechanism is reset and a pre-ovulatory release of LH is initiated at the beginning of the "open period" to start a new sequence.

It is not known what controls the "open period", but the basis of the "critical period" is a diurnal rhythm in the responsiveness of the positive feedback mechanism. This was shown by the early studies of Everett and co-workers (Everett, 1961), who found that in the rat an injection of pentobarbital at 14.00 h on the day of pro-oestrus delays ovulation by 24 h. The fact that a central nervous system depressant delayed ovulation suggested that the positive feedback rhythm has a neural component.

Ovarian steroids can only exert a positive feedback effect on LH release when the positive feedback rhythm is in its responsive phase, i.e. during the "critical period". In the cycling rat, the underlying diurnal nature of the "critical period" is obscured because the plasma steroid environment (high oestrogen, low progesterone) needed to trigger LH release only occurs every 4 or 5 days, on the day of pro-oestrus (Section III C). The diurnal nature of the "critical period" can be shown experimentally by injecting long-term ovariectomised rats with oestradiol benzoate (Fig. 11). A single injection causes a sustained suppression of basal LH secretion, which is interrupted by a large increase in LH levels on each of the following 4 days at 17.00 h: this is the time when pre-ovulatory levels of LH reach a peak in intact rats (see Fig. 13). This daily discharge of LH can be blocked by administration of sodium pentobarbital confirming that the positive feedback rhythm has a neural component.

A major difference between the rodent "critical period" and the hen "open period" is that a positive feedback response to an ovarian steroid can be induced in the hen, but not in the rat, at any time of day (Wilson and Sharp, 1975a). The "open period" is not controlled by the ovary because on the day between two sequences of ovulations, the ovary contains a follicle which matures and becomes capable of ovulation after the end of the "open period". Ovulation of the follicle is delayed until after the beginning of the next "open period".

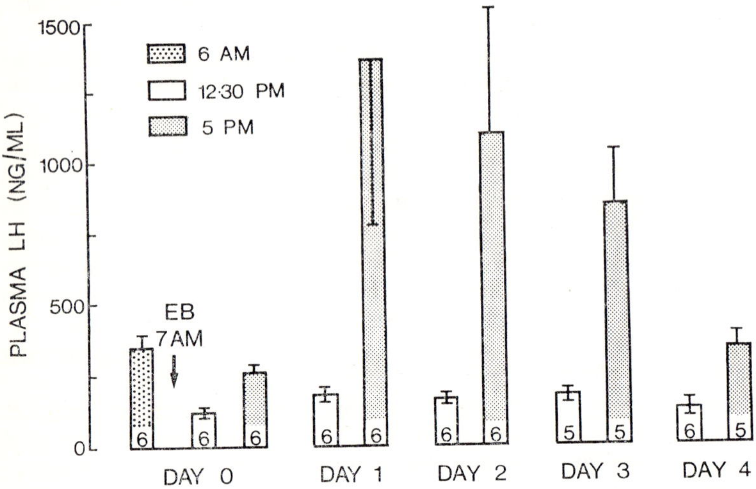

Fig. 11. Induction of a daily LH surge by administration of oestradiol benzoate (EB) on the 15th day after-ovariectomy (day 0). Each bar depicts mean ± S.E. of plasma LH concentrations. Numbers at the base of the bars represent the number of observations. (From: Legan *et al.*, 1975.)

Another difference between the role of light in controlling the pre-ovulatory LH surge in the rat and in the hen, is that in the rat, but not the hen, exposure to constant light leads to a cessation of oestrous cycles and the development of persistent oestrus. A surge release of LH and ovulation can be induced by copulation (Davidson *et al.*, 1973). In a flock of hens exposed to constant light, egg laying continues but is no longer restricted to a limited period during 24 h.

There is no "open period" or "critical period" in species other than rodents and birds; the "clock" which times ovulation is located in the ovary rather than the brain. Thus, the LH response to oestradiol-17β is not influenced by the time of day in the anoestrous ewe (Jackson and Thurmon, 1974) or the ovariectomised rhesus monkey (Knobil, 1974).

(2) *Follicular and Luteal Phases*

The ovary contains a graded series of maturing follicles of which only a few are destined to ovulate; the majority become atretic. Follicular atresia occurs at any stage of development but is particularly common in mammalian cycles during oestrus or the period immediately prior to ovulation (Greenwald 1974). Since large ovarian follicles become atretic after hypophysectomy, atresia is caused by a withdrawal of hormonal support.

The dependency of follicular growth on the pituitary has been mostly studied in rodents. In 1930, Smith reported that in the rat after hypophysectomy, development of primordial follicles continued up until the stage when an antrum (cavity) started to form. Both FSH and LH are needed to stimulate development into the rapid growth phase and must always be present in low concentrations if the follicles are to become mature. Even a temporary absence of gonadotrophins rapidly causes follicular atresia (Schwartz, 1974).

A basic rhythm of follicular maturation and regression underlies ovarian activity in all species and is particularly obvious in species like the sheep and the rat, in which the rhythm is not affected by the presence of active corpora lutea. In other species (e.g. guinea-pig, pig, cow, primates), active corpora lutea slow down, to various degrees, the rate of follicular maturation (Greenwald, 1974). The regression of the corpora lutea lifts this inhibitory effect and initiates the final growth of the follicle(s) next due to ovulate. It is assumed that steroids secreted by the corpora lutea, particularly progesterone and oestrogen, exert a negative feedback effect on gonadotrophin secretion. Plasma gonadotrophin levels would then be too low during the luteal phase to sustain follicular development; this could be the case in women and the rhesus monkey, since plasma FSH and LH levels are lower during the luteal phase than during the follicular phase (Fig. 12). It is not clear why the corpus luteum in sheep does not suppress follicular growth; it could be connected with the fact that it secretes only progesterone and no oestradiol (Baird *et al*., 1975).

Plasma FSH levels start to rise at the end of the luteal phase of the menstrual cycle in primates at a time when the corpora lutea are regressing and plasma progesterone levels are falling (Fig. 12). This rise may seem to be due to the removal of negative feedback, but the situation is not simple because in the rhesus monkey FSH continues to rise when progesterone levels are artificially maintained (Resko *et al*., 1974). FSH levels remain high during the early follicular phase of the menstrual cycle but start to fall as the rapidly growing follicles begin to secrete increasing amounts of oestradiol (Fig. 12). The fall is due to the negative feedback action of oestradiol-17β.

The high levels of plasma FSH observed in the rat during late proestrus and on the day of oestrus (Fig. 13) may synchronise waves of follicular growth or more probably, stimulate the final maturation of follicles. Attempts

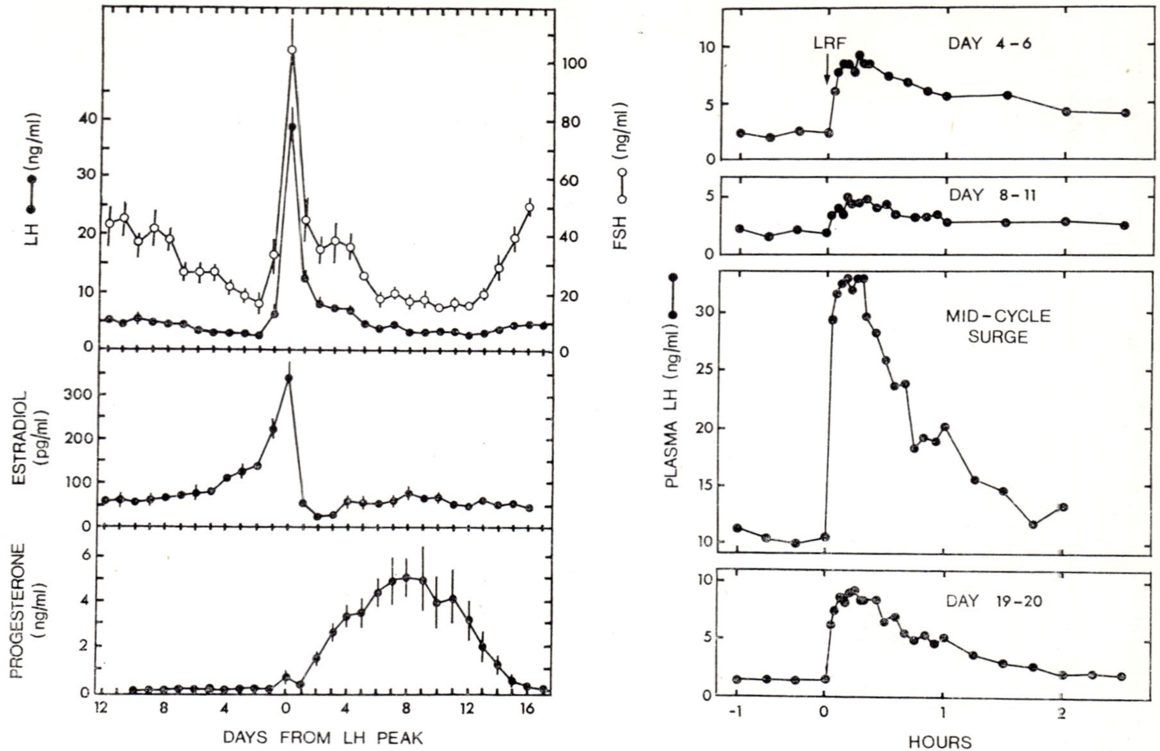

Fig. 12. *Left panel*: concentrations of plasma LH, FSH, oestradiol and progesterone throughout the normal cycle of the rhesus monkey normalised to the day of the mid-cycle peak (day 0). *Right panel*: four individual responses to single intravenous injections of 10 μg LHRH (arrow) during the early follicular (days 4–6), late follicular (days 8–11), mid-cycle surge and mid-luteal (days 19–20) phases of the cycle. (From: Krey *et al.*, 1973b.)

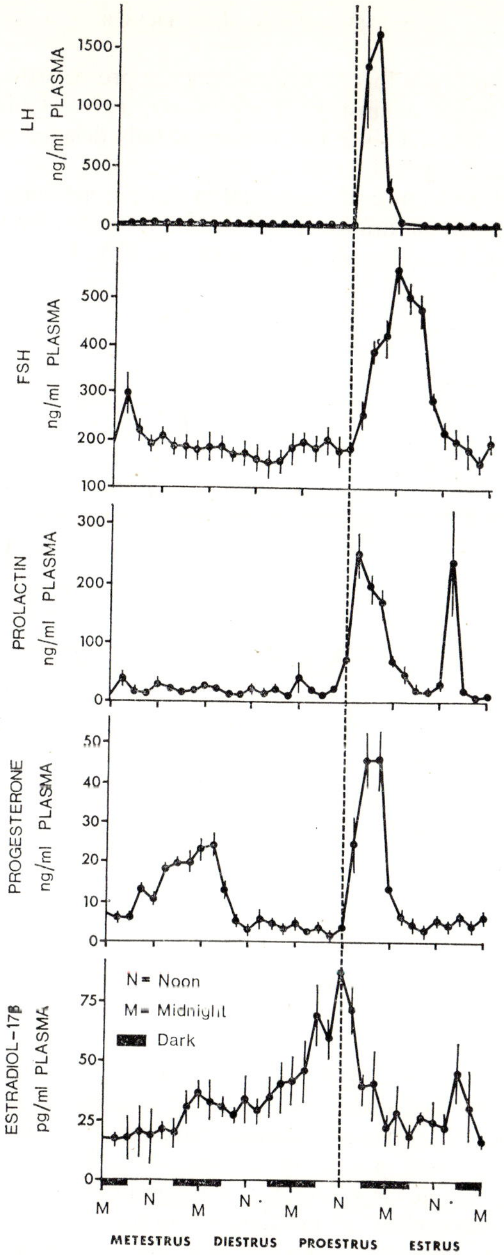

Fig. 13. Mean concentrations (± S.E.) of plasma LH, FSH, prolactin, progesterone and oestradiol-17β during the 4-day oestrous cycle of the rat. Plasma samples were taken at 3-h intervals beginning at 0300 h of met-oestrus with 5 animals sampled at each point. (From: Butcher *et al.*, 1974.)

have been made to prove this by neutralising the pro-oestrous/oestrous FSH surge with antibodies against FSH (Schwartz, 1974). Unfortunately, the results were not clear-cut since this treatment only delayed the next vaginal oestrus at the most, by 2–4 days.

The lengths of oestrous and menstrual cycles are not only determined by the rate of follicular maturation but more importantly, by the life-span of the corpora lutea. The oestrous cycles of the rat (Fig. 13) the mouse and the

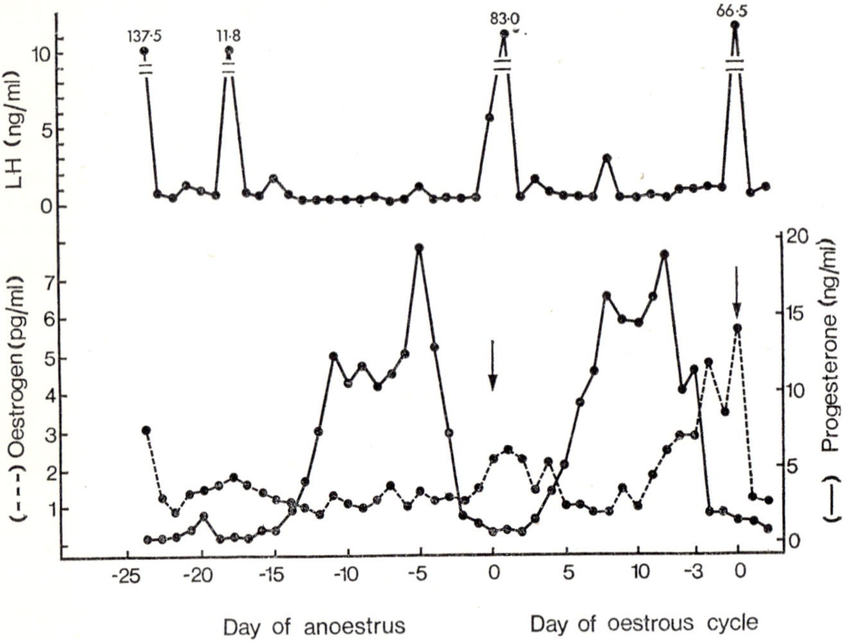

Fig. 14. Daily peripheral levels of serum LH (*upper line*), oestrogen (- - - - -) and progesterone (—) in a ewe during late anoestrus and the first oestrus of the breeding season. Oestrus is indicated by arrows. (From: Yuthasastrakosol *et al.*, 1975.)

hamster are short (4–6 days) because the corpora lutea start to regress as soon as they form and secrete only a small amount of progesterone. The corpora lutea in other species including man, rhesus monkey (Fig. 12) cow, sheep (Fig. 14), pig and guinea-pig are maintained for 10–15 days after they are formed and secrete large amounts of progesterone. Hypophysial support is required for the corpora lutea of the cycle in some species during the entire luteal phase (e.g. primates, sheep), or only during the first few days after ovulation (e.g. guinea-pig), but in others (e.g. pigs), it is not required at all (Nalbandov, 1973).

(3) *Ovulation*

The pre-ovulatory release of LH involves positive feedback by steroids secreted by the pre-ovulatory follicle(s) in both reflex ovulators like the rabbit, and spontaneous ovulators like the rat, the sheep and man. The role of oestrogen and progestins in controlling the pre-ovulatory LH surge is discussed in Section III C.

An LH surge comparable with that at mid-cycle can be induced in the rhesus monkey during the early follicular phase of the menstrual cycle if the concentration of plasma oestradiol-17β is maintained experimentally at 200–400 pg/ml for 36 h or 100–200 pg/ml for 24 h (Karsch *et al.*, 1973b). These concentrations correspond with the lower and upper range of levels observed prior to a spontaneous LH surge (Fig. 12). The triggering of an LH surge is, therefore, not only dependent upon achieving a threshold concentration of oestradiol-17β but also on the time of exposure to this steroid. Similar studies in women (Yen *et al.*, 1975), the ewe (Goding *et al.*, 1969) the rat and the hamster (Everett, 1964) have shown that physiological amounts of oestradiol-17β can advance the LH surge.

There are a number of possible components of the steroid mediated positive feedback mechanism. Ovarian steroids may act at the level of the hypothalamus to alter the output of LHRH; act directly on the anterior pituitary to release gonadotrophin by changing membrane permeability; and, as discussed in Section III C, increase the sensitivity of the gonadotrophes to LHRH. In addition, repeated exposure of the pituitary to pulses of LHRH may have a priming effect (Section II C).

Although it is likely that the hypothalamic signal for the LH surge is mediated by LHRH, it has proved difficult to determine whether there is a surge of LHRH secretion prior to the pre-ovulatory release of LH. However, a rise in the concentration of LHRH in pituitary stalk blood has been detected on the afternoon of pro-oestrous in the rat (Sarkar *et al.*, 1976). LRF has also been detected by bioassay in woman at mid-cycle (Malacara *et al.*, 1972), but an increase in LHRH activity at this time was not found using a radioimmunoassay (Jeffcoate *et al.*, 1975). Hypothalamic LRF concentrations, measured by bioassay, have been shown to fall in the morning (Chowers and McCann, 1965) or the afternoon (Ramirez and Sawyer, 1965) of pro-oestrous in the rat.

An enhanced pituitary response to LHRH which is observed after administration of oestrogen (Section III C), has also been observed during the pre-ovulatory phase of the menstrual (Fig. 12) or oestrous cycle when plasma oestrogen levels are high (women, Yen *et al.*, 1972; ewe, Hooley *et al.*, 1974; rat and the hamster, (Arimura *et al.*, 1972; Aiyer *et al.*, 1974b).

It is concluded that the principal factors causing the pre-ovulatory gonadotrophin surge are an enhanced release of LHRH, an increased sensitivity of the pituitary to LHRH and the self-potentiating effect of LHRH on its own release.

C. PSEUDOPREGNANCY, PREGNANCY AND LACTATION

Following copulation the life span of the corpus luteum of the oestrous or menstrual cycle is extended to support the early stages of pregnancy. An exception to this generalisation is found in the marsupials in which the gestation period is very short and about the same length as the oestrous cycle. The factors controlling the "rescue" of the corpus luteum from degeneration in early pregnancy vary from one species to the next and do not all involve the pituitary.

In the rat, the condition of pseudopregnancy which can be induced by mechanical stimulation of the uterine cervix or by sterile mating, has been exploited to study the hormonal control of the corpus luteum (Smith *et al.*, 1975). Following cervical stimulation at 19.00 h on the day of pro-oestrus, the increased plasma progesterone levels which are normally seen on the day of metoestrus (Fig. 13) are maintained for about 12 days; this is associated with an increased secretion of prolactin which is released in two surges each day and confirms the studies carried out in the late 1930s and early 1940s showing that the corpus luteum of the rat is maintained at least in part, by prolactin (Nalbandov, 1973).

Study of the hypothalamic control of pseudopregnancy, has shown that electrical stimulation of the medial hypothalamus at sites ranging from the anterior hypothalamic area and paraventricular nuclei caudally into the premammillary complex and dorsally to the dorsomedial nucleus, induces pseudopregnancy but not ovulation (Everett and Quinn, 1966). Transection of the preoptic area-hypothalamic connections inhibit the establishment of pseudopregnancy following vaginal-cervical stimulation, and it seems that the neural influence supporting prolactin release originates from the preoptic area or forebrain (Arai and Yamanouchi, 1975).

The pituitary is indispensable for prolonging the life of the corpus luteum and maintaining pregnancy in the rabbit, the ferret, the pig and the goat, since hypophysectomy causes abortion. In the sheep, the rat and the guinea-pig, the corpus luteum is only dependent on the pituitary during the early stages of pregnancy. In primates, the corpus luteum of pregnancy is not dependent on the pituitary but is supported by hormones secreted by the conceptus (Nalbandov, 1973).

The hormones required to maintain luteal function during pregnancy vary from one species to another and may change during different stages of pregnancy. Prolactin and LH are needed in the rat and sheep, while in the pig, prolactin is required towards the end of pregnancy and LH is needed in mid-pregnancy. Luteal function is maintained by prolactin, FSH and low levels of LH in the rabbit (Heap *et al.*, 1973).

There are mature ovarian follicles present in the rat during pregnancy and ovulation occurs within 48 h of parturition (Ying *et al.*, 1973). It seems that during pregnancy the high levels of circulating progesterone inhibit hypothalamic activity. When the concentration of progesterone declines at the end of pregnancy, positive feedback by oestrogen leads to an LH surge. Levels of circulating LH and FSH remain very low, however, in the ewe and women for some time after parturition, and menstrual cycles are suppressed during lactation in women. The first cycle after parturition occurs earlier in women if prolactin release and lactation are suppressed by bromocriptine (Rolland *et al.*, 1975a, b).

During the course of pregnancy there is a progressive decrease in the response of the maternal pituitary to LHRH in the ewe (Chamley *et al.*, 1974) and in women (Jeppsson *et al.*, 1974). There is little or no response to LHRH during the first few days after parturition, but a gradual recovery of sensitivity, first for FSH and then for LH, occurs during the following 5 weeks.

The suppression of menstrual cycles during lactation may involve not only a depression of hypothalamo-pituitary function, but also the development of ovarian refractoriness to gonadotrophin stimulation (Zarate *et al.*, 1972).

D. ANNUAL CYCLES

Most wild animals breed seasonally, thus ensuring that young are produced at a time of year when the food supply and other environmental factors favour their survival. Seasonal growth of the gonads is initiated in response to environmental cues, such as changes in rainfall, temperature and day-length. The importance of day-length was established in pioneer studies on the migratory bird, *Junco hymalis* (Rowan, 1925), the drake (Benoit, 1934) and ferret (Bissonette, 1932). In these species, out of season gonadal growth could be induced by artificially increasing the day-length. More recent studies have focussed on the possibility that such annual rhythms of reproductive activity are endogenous and entrained by the seasonal changes in day-length (Pengelly, 1974). An important feature of the breeding cycle of many birds is the development of photorefractoriness during the summer, when the gonads regress despite the long day-lengths. The development of photorefractoriness ensures that young are hatched in the spring and have enough time to grow large enough to face the rigours of winter.

A number of models involving circadian rhythms have been proposed to explain seasonal breeding. The simplest and most popular is the "external coincidence" model (Follett, 1973) in which a circadian rhythm of photosensitivity is thought to be entrained by the daily pattern of light and dark. Each day there is a "photosensitive period" lasting a few hours during which the hypothalamus can respond to light and increase the secretion of GnRF. Increased GnRF secretion stimulates gonadotrophin release and consequently, an accelerated rate of gonadal growth. The "photosensitive period"

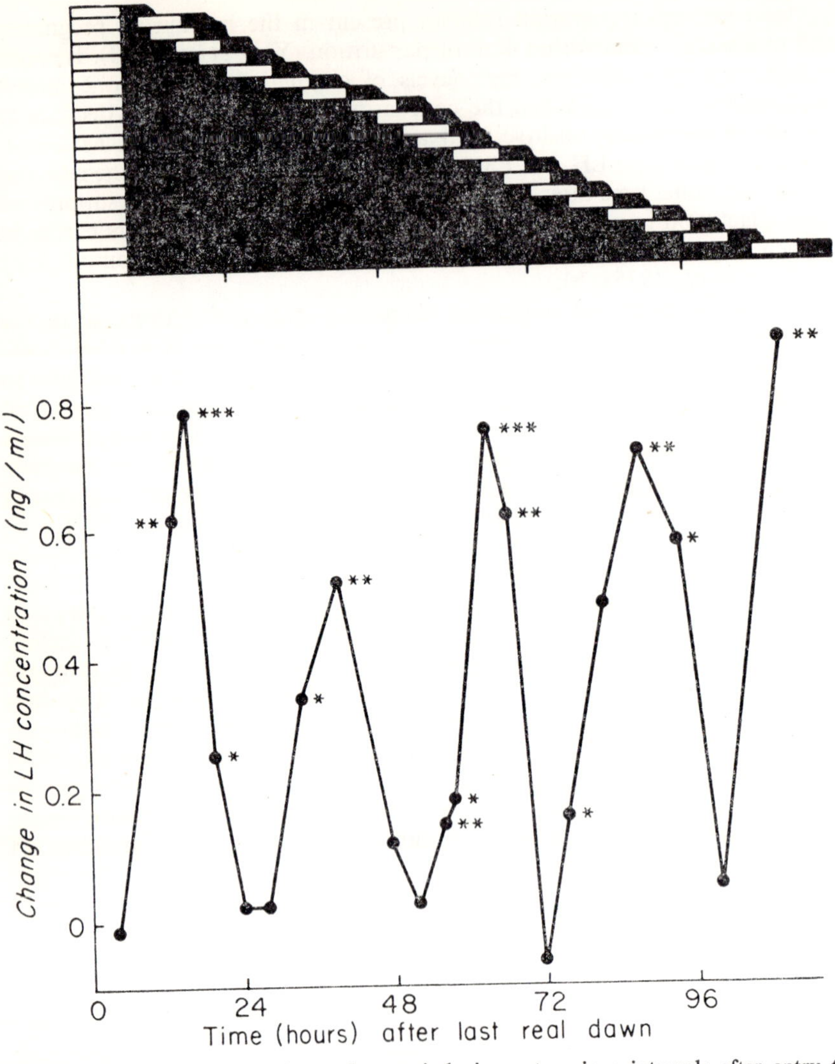

Fig. 15. The effect of an 8-hour photoperiod given at various intervals after entry to darkness on plasma LH concentration in White-crowned Sparrows. The white and black bars at the top illustrate periods of light and darkness respectively. Birds were previously maintained as a daily photoschedule of 8 h light and 16 h darkness and a pre-experimental blood sample was taken for all birds early in the last 8 h light period. The post-experimental sample was taken 7–16 h after the end of the test photoperiod. The ordinate shows the change in plasma LH concentration between these two samples that resulted for a particular treatment. The time after the last real dawn is shown on the abscissa. ***P< 0·001, **P<0·01, *P<0·05. (From: Follett *et al.*, 1974.)

starts at least 11 h after dawn or "lights on" so that light can only induce gonadotrophin release if there are more than 11 h of light in a day. If there are less than 11 h, the "photosensitive period" occurs when it is dark and gonadotrophin secretion is not induced. There is a similarity between this "photosensitive period" and the "critical period" of the rat oestrous cycle (Section V B), since both are entrained by the daily photoperiod. The external coincidence model can be tested experimentally by exposing sexually quiescent birds to cycles of light and dark, in which the length of the light period is fixed (6–8 h) and that of the dark period is variable. If there is an endogenous circadian rhythm of photosensitivity, gonadotrophin release should only occur when the interval of darkness between two successive light periods is such that the photosensitive phase occurs during a light period. An example of such a "resonance experiment" using the White-crowned sparrow (*Zonotrichia leucophrys gambelii*) is shown in Fig. 15. These data show that there is a circadian rhythm in photosensitivity with photoinducible phases occurring at 24 h intervals.

A more complex explanation of seasonal breeding involving circadian rhythms is provided by the so-called "internal coincidence" model (Meier, 1974). This model (see Figs. 4 and 5 in Chapter 9) was developed as a result of observations made on the effects of injections of corticosterone and prolactin on testicular growth, fattening and migratory behaviour in sparrows In one such experiment, photo-refractory house sparrows (*Passer domesticus*) and white-throated sparrows (*Zonotrichia albicollis*) were kept in continuous light and injected daily with prolactin, 0, 4, 8, 12, 16 or 20 h after an injection of corticosterone. Testicular growth was induced in house sparrows when the prolactin and corticosterone injections were separated by 4 or 8 h, and in white-throated sparrows when the injections were separated by 12 h. Gonadal growth was not stimulated by injections of prolactin and corticosterone separated by other time intervals. Testicular growth was not induced by any temporal arrangement of prolactin and corticosterone injections in sparrows held in continuous darkness. Meier (1974) suggested that seasonal breeding could be brought about by alterations in the phase relationships of diurnal rhythms of corticosterone and prolactin secretion (Fig. 5 in Chapter 9). As the daily pattern of light and dark changes throughout the year the two rhythms alter their phase relationships. The hypothalamus becomes photosensitive or photorefractory depending on specific phase relationships between the two rhythms. Thus, in the house sparrow, the hypothalamus becomes photorefractory as lengthening day-lengths progressively re-entrain the rhythms of corticosterone and prolactin secretion, so that the difference between their maxima shifts from 8 to 12 h.

In the circadian models proposed by both Murton (1975) and Meier (1974) it is suggested that there may be separate photosensitive periods for the induction of FSH and LH release. A consequence of this would be that, depending on the entraining properties of the lighting pattern, it should be possible for LH or FSH to be released independently of each other; this mechanism would explain why many non-migrating birds such as the rook, the robin and the duck, show sexual behaviour in the autumn (Marshall, 1952). The autumn lighting pattern may be such that the LH photosensitive period occurs during daylight, but that the FSH photosensitive period occurs at night. The increased LH secretion would stimulate testosterone secretion and, in turn, sexual behaviour, but not gonadal development. Increased LH secretion has been observed in the male mallard in the autumn (Haase *et al.*, 1975).

Less emphasis has been placed on the role of circadian rhythms in seasonal breeding in mammals and instead, it is believed that seasonal breeding is due to changes in the sensitivity of the negative feedback mechanism controlling gonadotrophin secretion (Millar and Glover, 1973; Pelletier and Ortavant, 1975). A decrease in sensitivity would lead to increased gonadotrophin secretion at the beginning of the breeding season, while an increase in sensitivity would inhibit gonadotrophin release at the end of the breeding season. Pituitary LH content has been shown to increase during the breeding season in various male mammals (e.g. rock hydrax, sheep, roebuck and red deer), but in all these species the LH content is depressed when testicular or plasma testosterone concentrations are at a maximum. The decreased pituitary content of LH probably reflects a high rate of release of the hormone: since this high rate of release occurs when testosterone levels are maximal, it seems that the negative feedback mechanism is relatively insensitive at this time (Brüggemann *et al.*, 1965; Lincoln, 1971; Millar and Glover, 1973). A seasonal change in the sensitivity of the negative feedback mechanism in the sheep, is suggested by the observation that testosterone suppresses LH secretion after castration more effectively in rams held on long daily photoperiods than in animals held on short daily photoperiods (Pelletier and Ortavant, 1975). In the ewe, it appears that the sensitivity of the positive feedback mechanism to oestrogen decreases during the non-breeding season and increases during the breeding season (Land *et al.*, 1976).

Changes in hypothalamic sensitivity to steroid feedback may also underlie seasonal breeding in birds. Cusick and Wilson (1972) found that in the tree sparrow (*Spizella arborea*) the development of photorefractoriness can be prevented by intra-hypothalamic implants of an anti-androgen, cyproterone. It seems that cyproterone prevents endogenous androgens from exerting a negative feedback effect on gonadotrophin release, by competing with them for negative feedback receptors in the hypothalamus. Further support for the view that photorefractoriness may be caused by an increased sensitivity to steroid negative feedback, comes from the observation that in male red grouse (*Lagopus scoticus*) the plasma LH concentration increases rapidly after castration in photorefractory birds (Sharp and Moss, 1977). Other data do not support the view that seasonal breeding is caused only by changes in the sensitivity of the negative feedback mechanism. Nicholls and Storey (1976) observed that in contrast with grouse, castration of photorefractory canaries does not result in an increase in LH release. Similarly, Davis and Meyer (1973) observed that while pituitary LH content increased in snowshoe hares after castration in the breeding season, no such increase could be detected after castration in the non-breeding season.

The episodic release of LH and the change in pituitary sensitivity to LHRH have been studied during the cycle of testis growth and regression in the Soay sheep (Lincoln, 1976). LH is released episodically in breeding and non-breeding animals, but the frequency and amplitude of each discharge is the lowest when the testes are regressed and the highest when they are developing. The amplitude of the pituitary response to LHRH increases as the testes develop. When the testes are developed the amplitude of the pituitary response to LHRH is less than when the testes are regressed but the duration of the induced LH surge is longer.

Seasonal breeding has been extensively studied in reptiles, amphibians and fish. As in birds and mammals, increasing day-lengths stimulate seasonal reproductive activity, but in all poikilotherms temperature also plays an important role (reptiles, Licht, 1972; amphibians, Jørgensen, 1970; fish, de Vlaming, 1972 and Dodd, 1975). Plasma gonadotrophin concentration is low in various species of trout and salmon during the early phases of gametogenesis in both sexes, but increases during the breeding season (Crim *et al.*, 1975). In the carp (*Cyprinus carpio* L), pituitary responsiveness to LHRH or carp hypothalamic extract is higher during the breeding season than during the non-breeding season (Weil *et al.*, 1975).

VI. Clinical Aspects

In patients presenting with disorders of the hypothalamo-pituitary-gonadal axis, the primary lesion may reside in any of three sites. The gonads may lack the ability to respond to gonadotrophin, the anterior pituitary may be unable to respond to GnRF or there may be an abnormality in the output of GnRF from the hypothalamus, caused by a lesion in the hypothalamus or higher neural areas. A primary gonadal disorder can be diagnosed from high circulating gonadotrophin levels, but it is difficult to distinguish between hypothalamic and pituitary causes of hypogonadotrophic hypogonadism. Differences in the rate of secretion of GnRF may be present but these cannot be satisfactorily detected. A possible way of distinguishing between hypothalamic and pituitary causes of hypogonadotrophic hypogonadism, is to assess the capacity of the pituitary gland to respond to exogenous LHRH. A standard LHRH test comprises an intravenous injection of 100 μg LHRH and measurement of plasma LH and FSH immediately before and 20 and 60 min after the injection (Mortimer *et al.*, 1973).

The LHRH test can be used to assist the diagnosis of gonadotrophin deficiency which may be caused by various pituitary defects such as chromophobe tumours. In such cases, there is a complete absence of response to LHRH.

The heterogenous group of conditions termed hypogonadotrophic hypogonadism occurs in both adults and juveniles in whom it can be a cause of delayed puberty. One form of the condition is isolated pituitary gonadotrophin deficiency, in which plasma LH levels may be very low while FSH levels are in the normal range. The gonadotrophic response to LHRH in patients with this condition is generally subnormal but is not necessarily due to a defect of the pituitary (Marshall *et al.*, 1972; Mortimer *et al.*, 1973). It could be secondary to deficient LHRH secretion, since LHRH is needed to maintain the synthesising capacity of pituitary gonadotrophes (see Section II C).

Precocious puberty of neurogenic origin is a rare disease in which the hypothalamo-pituitary-gonadal axis is activated at an early age. Gonadotrophin secretion appears to be neurally inhibited prior to the onset of puberty (Section V B) and in precocious puberty the inhibitory neural pathways are damaged by tumours or some other lesion (Reichlin, 1974).

Hypogonadism of a hypothalamic origin is not as often observed in adult men as in women, probably because testicular function, but not full ovarian activity, can be maintained by low gonadotrophin levels. One established organic disorder of the hypothalamus in men is Kallman's syndrome, in which hypogonadism is associated with anosomia due to agenesis of the olfactory lobes and maldevelopment of the hypothalamus (Reichlin, 1974). Secondary amenorrhoea in women can be due to pituitary disease but commonly it is of hypothalamic origin and caused by environmental or emotional factors. In pseudocyesis a non-pregnant woman believes she is pregnant, and physiological changes associated with pregnancy occur. The induction of pseudocyesis seems to be the result of a fear of being, or a desire to be, pregnant; it is associated with an absence of an ovulatory LH surge, persistence of the corpus luteum and increased prolactin secretion (Reichlin, 1974). Different degrees of stress or depression without any other influences can lead to "psychogenic" amenorrhoea in which pituitary gonadotrophin and probably GnRF secretion are suppressed. Amenorrhoea is often present in anorexia nervosa in which psychological factors are involved arising out of the desire to slim to the extent of starvation. Occasionally amenorrhoea is due to a hypothalamic lesion, as in the Chiari-Frommel syndrome in which amenorrhoea and galactorrhoea persist after the puerperium. It is thought that in this condition secretion of PIF is inhibited (Reichlin, 1974). Although an LHRH test may not distinguish the site of secondary amenorrhoea, it is used to assist in the diagnosis of the different types of this condition (Mortimer *et al.*, 1973).

LHRH has been used therapeutically to stimulate gonadotrophin secretion in cases of anorexia nervosa, galactorrhoea and certain types of male hypogonadism (Keller, 1973; Mortimer *et al.*, 1974b). Although the initial response is impaired, continued LHRH administration leads to enhanced gonadotrophin secretion; this suggests that these diseases are associated with a deficiency of GnRF (Mortimer *et al.*, 1974b).

It is anticipated that LHRH therapy will be successful in inducing ovulation in women with amenorrhoea originating from hypothalamic dysfunction. Current treatment with gonadotrophin preparations to stimulate follicular development and induce ovulation are effective, but have to be carefully monitored to prevent multiple ovulations. By stimulating endogenous pituitary gonadotrophin release by LHRH, a more physiological response should be generated so that treatment would require less supervision. LHRH must be administered in repeated doses to try and mimic gonadotrophin release during the follicular phase of the cycle to induce follicular maturation, followed by a large dose to induce ovulation. The mimicking of gonadotrophin release during the follicular phase is difficult, since LHRH induces preferential release of LH, which is opposite to the pattern of release by the pituitary during the follicular phase. Although such treatment has resulted in the induction of ovulation, the success rate is low (Keller, 1973; Zanartu *et al.*, 1974). In future, long acting super-active LHRH analogues are likely to be more successful; their use could achieve more prolonged FSH secretion to improve the rate of follicular maturation (Schally *et al.*, 1975).

LHRH has been used in the veterinary field to try and induce ovulation outside the breeding season. If given to the anoestrous ewe, ovulation can be induced but the corpus luteum functions for a short time only and more work is required to determine its usefulness, and that of long-acting analogues (Crighton *et al.*, 1975).

A potentially wider application of hypothalamic control of fertility is in contraception. An early suggestion was that administration of LHRH could be used in the normal cycle to induce an LH surge and to pinpoint ovulation, but this possible application has not been developed. A more likely alternative is to inhibit ovulation using substances which prevent the action of LHRH. Synthetic peptides related to the LHRH molecule are being developed which compete with LHRH at pituitary binding sites and can inhibit the preovulatory LH surge in laboratory rodents (Vilchez-Martinez *et al.*, 1974c; Corbin and Beattie, 1975). Such substances might be successful in inhibiting GnRF during the late follicular phase of the cycle in women and thus prevent ovulation. The main problem is likely to be dosage of antagonist, since very high doses would have to be used if taken orally.

Another approach is to inhibit LHRH by antibodies. Passive transfer of antibodies, although effective in inhibiting ovulation in the rat (Fraser and Gunn, 1973; Koch *et al.*, 1973a), is impracticable since large quantities of antibody would be required and the injection of foreign protein could lead to antibody formation. A better method would be to inhibit LHRH by active immunisation which has been shown to prevent ovulation for long periods in the rat (Fraser, 1976).

VII. Summary and Conclusions

Reproduction is controlled by the hypothalamus which regulates the secretion of LH and FSH by means of hypothalamic hormones. Spermatogenesis and oogenesis are stimulated and maintained by the basal secretion of both gonadotrophins while ovulation is controlled by the cyclic release of LH. Basal and cyclic secretions of gonadotrophin are regulated by the negative and positive feedback action of gonadal steroids respectively. One of the sites of steroid feedback is the hypothalamus and its projections to the forebrain, the other is the pars distalis. At the level of the hypothalamus, changes in the internal and external environment influence neural activity which in turn, directly or indirectly regulates GnRF secretion. Indirect regulation is achieved by an alteration in the "threshold" of response of the steroid feedback systems (Section I).

Only one gonadotrophin releasing hormone, LHRH, has been extracted from hypothalamic tissue and chemically identified. LHRH is a decapeptide with the sequence, (pyro) Glu-His-Trp-Ser-Tyr-Gly-Leu-Arg-Pro-Gly(NH_2). After injection of LHRH, plasma FSH rises more slowly than LH, but remains elevated for longer. Except in certain endocrine disorders and in pubertal subjects, LHRH does not release FSH as readily as LH. Claims for the existence of hypothalamic factors, releasing exclusively or predominantly FSH, have not been confirmed. LHRH activity has been located in the hypothalamus using immunological and bioassay techniques. High concentrations occur in the median eminence and basal hypothalamus in a region corresponding to the "hypophysiotrophic" area. Smaller concentrations occur in the preoptic area of rodents and it is likely that a substantial proportion of LHRH containing cell bodies are located anterior to the classical hypophysiotrophic area (Section II).

The concept that basal and cyclic gonadotrophin secretion are regulated at two levels by the medial basal and suprachiasmatic-preoptic area respectively, requires re-evaluation. First, it may only apply to rodents, since in the monkey the medial basal hypothalamus is capable of initiating the cyclic release of LH. Secondly, the loss of cyclicity after anterior deafferentation of the hypophysiotrophic area in the rat, could be due to a reduction in the number of LHRH containing terminals in the median eminence, resulting in insufficient LHRH being available to stimulate a preovulatory LH surge. Finally, extrahypothalamic areas, especially in the limbic system, appear to play an important role in the cyclic release of LH (Section III A).

The role of monoamines in the control of gonadotrophin secretion has been extensively studied. Noradrenaline and dopamine tend to stimulate LHRH release, while 5-hydroxytryptamine inhibits it. The exact roles of noradrenaline and dopamine are controversial but it is likely that central noradrenergic neurones regulate LHRH release through the dopamine containing neuronal system in the basal hypothalamus. Catecholamines have also been shown to mediate the feedback actions of gonadal steroids on gonadotrophin release (Section III B).

Oestrogens and androgens are the principal negative feedback steroids (Section III C). Acting alone in experimental conditions they suppress gonadotrophin release but under physiological conditions they probably synergise with each other or with progesterone to maintain basal gonadotrophin secretion. In addition to being controlled by steroids, the release of FSH in the male may also be regulated by a protein, inhibin, originating from the germinal epithelium.

Synthetic LHRH has been used to distinguish pituitary and hypothalamic sites of feedback. The ability of oestrogen to facilitate the LHRH induced release of gonadotrophins has been demonstrated in women, the rhesus monkey, the sheep, the cow and the rat. Testosterone and oestrogen given chronically, exert a negative feedback effect on gonadotrophin release at both pituitary and hypothalamic levels. Progesterone may exert both positive and negative feedback effects exclusively at the hypothalamic level.

Oestrogen and progesterone regulate the cyclic discharge of gonadotrophins. Oestrogen triggers the pre-ovulatory release of LH in all mammals studied, while in the domestic chicken it "primes" (together with progesterone) the positive feedback mechanism. Oestrogen induces the pre-ovulatory release of LH in mammals by stimulating the secretion of LHRH and increasing the sensitivity of the pituitary to LHRH. The increased LHRH secretion potentiates its own action and thereby further increases the magnitude of the LH surge.

Progesterone exerts a positive feedback effect on LH release in the domestic chicken, rodents and primates, but not in other mammals. Rising levels of plasma progesterone are essential to trigger the pre-ovulatory LH surge in the domestic chicken but not in rodents or primates. Progesterone may also "prime" the positive feedback mechanism in the rat and the domestic chicken.

In all mammals, including rodents and primates, the increased level of circulating progesterone which occurs during the luteal phase of the cycle antagonises the positive feedback of oestrogen on LH release: this ensures that ovulation does not occur when there is a chance that the uterus contains a fertilised egg.

Assuming that LHRH is the only gonadotrophin releasing factor, it is necessary to explain how changes in the plasma FSH : LH ratio can occur at the onset of puberty, after castration and between the follicular and luteal phases of the menstrual cycle. Part of the explanation lies in the differential response of FSH and LH producing cells to LHRH after exposure to different levels of steroids, steroid metabolites, and inhibin. High levels of oestrogen tend to favour LH release while low levels favour FSH release. LHRH induced FSH release in the rat, is facilitated by androgens. The half-life of FSH is longer than that of LH and in acute situations this shifts the FSH: LH ratio. FSH cells are predominantly stimulated when the pituitary is exposed for prolonged periods to LHRH, this also leads to a shift in the FSH : LH ratio.

There is neurophysiological evidence that gonadotrophins and LHRH influence their own secretion by short- and ultrashort feedback loops respectively. It is doubtful however, if these mechanisms have any physiological significance (Section III D).

Basal gonadotrophin secretion is maintained by a series of episodic discharges, the amplitude and frequency of which are determined by levels of circulating steroids (Section III E).

The roles of the adrenal, thyroid and pineal in relation to reproduction have been studied. A role for the thyroid is not established but steroids secreted by the adrenals may be involved in the initiation of puberty, and in rodents, in timing the pre-ovulatory surge of LH. The pineal is activated by darkness and consequently pinealectomy tends to reverse the effects of short daily photoperiods or total light deprivation, on sexual functions (Section III F).

Behavioural interactions and other environmental factors such as changes in the lighting pattern, stressful stimuli or food availability, exert an important influence on the activity of the hypothalamo-pituitary-gonadal axis (Section III G).

Maturation of the negative feedback mechanism begins in mid- to late-foetal life, but the hypothalamo-pituitary-gonadal axis does not become a self-contained regulating system until after birth because of the influence of maternal gonadotrophins and steroids. In most species studied, the positive feedback mechanism matures at the onset of puberty. In the ewe, however, the positive feedback mechanism is functional in the juvenile (Section IV A).

The onset of puberty is characterised by a decrease in the sensitivity of the negative feedback mechanism, a change in the sensitivity of the pituitary to LHRH and an increase in the sensitivity of the gonads to gonadotrophins. It is not known what triggers the onset of puberty, but it is likely to involve functional changes in diffuse inhibitory pathways originating in the forebrain and terminating in the hypothalamus. Other factors may include the adrenarche which occurs before the onset of puberty. Increased secretion of adrenal steroids may decrease the sensitivity of the negative feedback mechanism.

There may also be a progressive increase in the number of stimulatory catecholaminergic terminals making synaptic contact with LHRH producing neurones. Since catecholamines mediate steroid feedback, this increased catecholaminergic innervation could lead to a decreased hypothalamic sensitivity to steroid feedback (Section IV B).

The neural basis for the positive feedback action of steroids on LH release is sexually differentiated in the rat and the sheep but not in man, the rhesus monkey and the domestic chicken (Section IV C).

Loss of reproductive capacity with age is due to ovarian failure in women (the menopause) but in animals, such as the rat, it is due to failure at the level of the central nervous system (Section IV D).

Changes in the general activity of the hypothalamo-pituitary-gonadal-axis are profoundly influenced by diurnal rhythms which are synchronised by the daily pattern of light and dark (Section V A). Such rhythms are involved in the control of the onset of puberty, seasonal breeding, and in rodents and the domestic chicken, in the timing of the pre-ovulatory LH surge. In mammals other than rodents, the ovary may act as the chief biological clock regulating the oestrous or menstrual cycles.

A diurnal rhythm in LH secretion is observed at the onset of puberty in man but not in juveniles or adults. LH levels show a diurnal rhythm in adult rats with peak values in the male occurring 6 h earlier than those in the female. However, diurnal rhythms in gonadotrophin release have not been generally observed in farm mammals.

The mechanism(s) controlling seasonal breeding are not well understood (Section V D). Particular attention has been given in birds to the role of light entrained circadian rhythms of photosensitivity. More limited studies in mammals have tended to emphasise seasonal changes in the sensitivity of the hypothalamus to the negative feedback action of steroids; this implies that there may be a common mechanism controlling the onset of puberty and of seasonal breeding.

The lengths of oestrous, menstrual and ovulatory cycles (Section V B) are determined by the rate at which ovarian follicles mature and in mammals, by the life span of the corpora lutea. The corpora lutea in the rat start to degenerate as soon as they are formed, consequently the luteal phase of the cycle is short. The corpora lutea in primates and farm animals are maintained by a complex of hormones which usually includes LH and prolactin.

Functional corpora lutea are a characteristic feature of pregnancy or pseudopregnancy (Section V C). The mechanisms involved in the "rescue" of the corpus luteum vary from one species to another and do not always involve the pituitary. The corpus luteum of pregnancy requires pituitary support, particularly in the early stages of pregnancy in many species, including the sheep, and the rat, but is not required in primates.

In sheep and primates, the high levels of circulating progesterone and oestrogen present during pregnancy depresses hypothalamic and pituitary activity. The maternal pituitary contains reduced quantities of gonadotrophins and is insensitive to LHRH.

LHRH is used to assist in the diagnosis of pituitary tumours and disorders of the hypothalamo-pituitary-gonadal-axis which cause such conditions as delayed puberty, secondary amenorrhoea and hypogonadotrophic hypogonadism. It has not proved possible to differentiate between hypothalamic and pituitary causes of hypogonadism. LHRH has been used therapeutically in the treatment of male and female infertility, but its short half-life and expense limits its usefulness: the problem may be overcome by the use of long-acting analogues of LHRH.

Further studies on LHRH may lead to the development of new methods of birth control. One method would be to inhibit LHRH by active immunisation; another, more attractive method, would involve the use of inhibitory analogues of LHRH. The first generation of inhibitory analogues has been synthesised and the chance of producing more potent LHRH inhibitors seems promising (Section VI).

CHAPTER 9

Control of Prolactin Secretion

J. S. TINDAL

There has been a wealth of published information in the prolactin field in recent years and various aspects of this vast literature have been the subject of numerous comprehensive reviews (Cowie and Tindal, 1971; Meites, 1972, 1973; Meites and Clemens, 1972; Meites *et al.*, 1972; Nicoll and Bern, 1972; Nicoll, 1973, 1974; Grosvenor and Mena, 1974; Neill, 1974; Tindal, 1974). The problem facing the reviewer is the diversity of function of prolactin and its numerous release-stimuli. In order to try and set the subject in its correct perspective, therefore, it would be appropriate to conduct a brief survey of the role of prolactin, before proceeding with a more detailed exploration of the control of prolactin secretion.

I. Functions of Prolactin

C. S. Nicoll has suggested that prolactin would have been better named "versatilin", and not without good reason. A total of 85 different actions have now been claimed for prolactin, whose influence extends over the whole range of vertebrates and a bewildering array of mechanisms. These can only be dealt with in brief in the present chapter and the reader is referred to the masterly review by Nicoll (1974) for a more detailed and fully documented account.

A. OSMOREGULATION

Prolactin is intimately concerned with electrolyte and water balance in vertebrates and indeed this type of control mechanism accounts for approximately a quarter of the postulated actions of prolactin (Nicoll, 1974). The marine teleost excretes excess ions by an active transport mechanism across the gills and concentrates the urine to a high degree, while in the freshwater teleost the permeability of the gills to outward diffusion of ions is reduced, there is an active uptake of ions by the gills and the kidneys excrete a large volume of very dilute urine from which most of the ions have been reabsorbed. In certain, though by no means all, teleosts which have been investigated, prolactin plays an important role in these transport and absorption processes. Thus, many euryhaline species cannot survive in freshwater after hypophysectomy because of loss of salts from the gills and kidneys, unless given injections of prolactin. Prolactin appears to act primarily on reduction of Na^+ loss through the gills rather than on Na^+ uptake, which is under the control of adrenal corticosteroids. Prolactin can also act on the kidney of freshwater teleosts to assist in the elimination of excess water, and on the gut, urinary bladder and skin of euryhaline fishes to help in the adaptation to freshwater, although as Nicoll (1974) pointed out, it does not necessarily act on all these organs in a particular species.

The function of prolactin in teleosts which live in seawater is not known at present. However, this is not the case for certain birds which are able to drink seawater, in which specialised nasal glands excrete the excess salts. The control mechanism of these glands is complex, involving a cholinergic innervation and a synergism between corticosteroids and prolactin. In mammals, including man, there is now evidence that prolactin may play an important role in osmoregulation, particularly in preventing renal sodium and water loss. However, since blood prolactin levels also appear to follow plasma osmolality or Na^+ level (Buckman and Peake, 1973; Relkin, 1974) the situation appears complex and prolactin may well synergize with many other osmoregulatory mechanisms (Horrobin *et al.*, 1973; Nicoll, 1974).

B. GROWTH AND METAMORPHOSIS

Mammalian prolactin has been reported to cause growth of various body tissues in all classes of vertebrates, in fact about half of the reported actions of prolactin fall into this classification (Nicoll, 1974). Prolactin also plays an important role in amphibians where it promotes growth of the tadpole stage and opposes metamorphosis in both anurans and urodeles, but facilitates the second metamorphosis of urodeles (Etkin, 1970).

C. BEHAVIOUR AND PARENTAL CARE

Some species of teleosts display complex parental behaviour, such as "water conditioning" the eggs by fanning a constant stream of water through the nest, and prolactin has been shown to increase such activity, to suppress fighting and to promote secretion of cutaneous mucus which is used in some species for nest building and in others for feeding the young. Prolactin is also responsible for the "water drive" of the red eft stage of salamanders which actively seek out, and migrate to, water.

Prolactin plays an important role in maternal behaviour and parental care in birds. It is believed to be responsible for incubation behaviour ("broodiness") in several species which have been investigated, although its role in pigeons and doves may be complicated by an interaction with progesterone. Prolactin synergises with gonadal steroids in the development of the brood patch in birds, where the abdominal skin becomes defeathered, highly vascularised and oedematous to facilitate incubation. The well-known crop-sac bioassay for prolactin is based on the fact that in a few species of birds prolactin stimulates the lining of the crop to proliferate and desquamate a fat-rich "crop milk" on which the young are fed.

In mammals, there is now evidence based on pup retrieval by the rat and nest building by the rabbit, that both prolactin and ovarian steroids are essential for the display of maternal behaviour and it should be noted that these are the same hormones which participate in mammary growth. It may be recalled that MacLean (1972) suggested that the corpus striatum may be a neural repository for genetically constituted forms of behaviour and it is not inconceivable that some of the actions of prolactin, acting either alone or in conjunction with steroids, may be explained on the basis of a lowering of the threshold of the striatal complex for the release of such behaviour.

D. LACTATION

(1) *Mammary Growth*

Prolactin, together with the ovarian steroids, appears to be an essential member of the complex of hormones required to stimulate mammary growth, at least under experimental conditions (see Cowie and Tindal, 1971). However, prolactin levels are, in general, low during pregnancy in the mouse, rat, sheep and goat, but not in man where the rising prolactin levels throughout pregnancy are possibly related to the high circulating levels of oestrogens, or the hamster, in which prolactin levels are raised during the first half of pregnancy (Bast and Greenwald, 1974; Hart, 1974; Karg and Schams, 1974; Ôta *et al.*, 1974; Sinha *et al.*, 1974). It is now also becoming clear that in some species at least placental lactogen may be of greater significance than prolactin for mammary growth.

(2) *Milk Secretion*

Prolactin appears to be essential for the initiation of milk secretion in the ruminant and may govern the peak and total milk yield of that particular lactation. Once lactation is established, however, even though the amount of prolactin released at milking can be correlated with the milk yield as lactation advances and then declines (Hart, 1974), nevertheless, the physiological role of prolactin at this time is not clear (Karg and Schams, 1974). In the non-ruminant mammal, prolactin appears to be essential at all stages of lactation for milk secretion. In the rabbit, prolactin appears to act virtually unaided in this respect, but in the rat and mouse prolactin and ACTH are both essential for the induction and maintenance of milk secretion, although in certain strains of mouse growth hormone may be substituted for prolactin (Cowie and Tindal, 1971).

The seasonal variations in basal prolactin secretion which occur in ruminants and which appear to be related to day-length have been discussed elsewhere in this chapter and it has also been shown that an innate seasonal oscillation in the rate of milk secretion occurs in the non-pregnant lactating goat (Linzell, 1973) which parallels the changes in basal prolactin levels. Hart (1974) pointed out that early in the year pregnancy may inhibit the otherwise spontaneous increase in the ability of the animal to release larger quantities of prolactin in response to increasing day-length, with the inference that this is held back until parturition after which the increased prolactin levels facilitate the rapid initial rise of the lactation curve. Later in the year during the declining phase of lactation, the amount of prolactin released at milking and the milk yield of the goat both fall progressively. However, the two events are not directly linked, since if milking-induced prolactin release is maintained at a high level in the autumn by exposing the animal to summer lighting conditions, the milk yield nevertheless declines in the normal manner (Hart, 1975b).

E. LUTEAL FUNCTION

Perhaps in no other field of reproductive physiology has there been so much disagreement and uncertainty as the control of the corpus luteum. All that will be attempted here is a brief summary of the status of prolactin as a luteotrophic hormone since, for references and a more detailed discussion, several excellent reviews are available (Greenwald and Rothchild, 1968; Kann and Denamur, 1974; Nalbandov, 1973; Nicoll, 1974). The evidence at present is that prolactin is luteotrophic (i.e. maintains the structure of, and promotes progesterone production by, the corpus luteum) in the rat, mouse, hamster and sheep. In the rat, LH may also be necessary, in the mouse FSH is also necessary, the hamster requires FSH, LH and prolactin, while in the sheep both prolactin and LH are necessary for corpus luteum maintenance.

In view of the conflicting opinions which have arisen in the past as to the luteotrophic role of prolactin, three points should be noted. First, since different species have solved this problem in different ways, it is more helpful to think in terms of a luteotrophic complex of hormones, rather than a specific hormone; second it appears that in some species prolactin may be more concerned with ensuring that the corpus luteum has a sufficient pool of cholesterol esters and preventing induction of enzymes which catabolise progesterone, while small quantities of LH may control the actual synthesis of progesterone; and third, that a continuous high level of circulating prolactin is not essential for maintenance of the corpus luteum in those species where prolactin is luteotrophic. Finally, in addition to its luteotrophic function in some species prolactin can also destroy corpora lutea (i.e. is luteolytic) in the rat. When a newly-formed corpus luteum is exposed to prolactin, the effect of the hormone is luteotrophic, but after the corpus luteum is over a certain critical age prolactin becomes luteolytic. It is now thought that the surge of prolactin release which accompanies the release of LH and FSH at pro-oestrus in the rat and which is not essential for ovulation (see Tindal, 1974) is responsible for destroying the old set of corpora lutea (see Nicoll, 1974).

II. Stimuli Associated with Prolactin Release

A. LACTATIONAL STIMULI

(1) *Suckling and Milking*

The teat, or nipple, is one of the most highly innervated tissues in the body and the most effective stimulus for the release of prolactin is the activation of its receptors by suckling or milking (Tindal and Knaggs, 1970; Cowie and Tindal, 1971). It has been known for nearly 40 years that suckling causes a fall in adenohypophysial prolactin content and the advent of radioimmunoassay procedures has made it possible to confirm the release of prolactin in response to suckling or milking in the goat, cow, sheep, rat and man, and the rise in circulating prolactin level after suckling in the rat has also been correlated with discharge of granules from the lactotrophe cells of the adenohypophysis (see Tindal, 1974; Lamming *et al.*, 1974).

Although there is probably a common release mechanism for prolactin in response to suckling or milking, there are variations among species in the patterns of such releases. In the rat, for instance, studies by Grosvenor and Whitworth (1974) have shown that after an 8 h period of non-suckling, suckling by a hungry litter causes an immediate and sustained increase in the circulating level of prolactin which then falls after approximately 30 min, when the pups fall asleep. If, however, the pups are replaced by fresh pups, the high circulating level of prolactin is maintained and this pattern of circulating prolactin can be mimicked accurately by constant infusion of prolactin. Grosvenor and Whitworth (1974) conclude that in rats at mid-lactation, prolactin is not released during suckling in a large amount over a short time but rather is released continuously in relatively small quantities during suckling: this could, perhaps, be considered as an adaptation to the particular pattern of suckling behaviour in the rat, where the pups remain attached to the teat for long periods of time and milk is ejected to them at intervals of about 5–15 min (Lincoln *et al.*, 1973).

In the goat, the sheep and the cow, where discrete episodes of suckling or milking occur, the rise of circulating prolactin within 1–2 min of the start of milking or suckling is rapid, reaching a peak in some 5–10 min. In the goat, prolactin release may then continue after cessation of the milking stimulus (Hart, 1973, 1974). In lactating goats from which blood samples were taken every 30 min for 24 h at the height of lactation, the rate of decline of circulating prolactin levels after both morning and afternoon milking was much slower than could be accounted for by the half-life of circulating prolactin, with the inference that the initial stimulus is capable of triggering a slow, sustained release of prolactin, possibly by prolonged after-discharge of teat receptors. This would ensure that the mammary gland of the lactating goat is exposed, for 14–16 h of the day, to far higher levels of prolactin than are found in virgin females, castrated males or lactating female goats which have not been exposed to the milking stimulus. Unlike the rat, however, no correlation was found between the duration of milking and the quantity of prolactin released into the circulation by milking in the lactating goat (Hart, 1975a).

There is conflicting evidence regarding the effectiveness of suckling in depleting pituitary prolactin content in the rat. On the one hand it was reported by Grosvenor and his group that 30 min of suckling caused a rapid discharge of prolactin from the pituitary, while on the other, it was claimed by Meites and his colleagues that there was a gradual fall during 3 h of suckling. These differing points of view were to some extent reconciled by Nicoll (Tindal, 1974, for references) who obtained values intermediate between these two extremes and who noted that prolactin appears to be present in the rat pituitary in at least three fractions, or pools, of which one can be rapidly discharged, one is more slowly releasable and one is essentially non-releasable. He suggested that different proportions of the three pools may be present in different strains of rat and also computed that the total prolactin content of the pituitary in the rats used by Meites's laboratory was only about one quarter of that found in the rats used by Grosvenor's group.

(2) *Other Stimuli*

In addition to the suckling stimulus, other tactile stimuli may play a minor role in the release of prolactin. Thus, self-licking of the nipples by the pregnant rat leads to a greater degree of mammary development than occurs if licking is prevented, which would suggest a licking-induced release of prolactin, while burns or surgery or damage to the chest wall in man can lead to spontaneous galactorrhoea. Irritation of central ascending pathways has also been shown to cause galactorrhoea in man and lactogenesis in the pseudopregnant rabbit (Tindal and Knaggs, 1972; Tindal, 1974). Contact between a lactating rat and her pups after removal of her teats leads to inhibition of ovarian activity, although this latter effect might be the result of tactile triggering of impulses by deformation through the skin of the stumps of peripheral nerves (see Tindal, 1974).

It must be borne in mind that the release of prolactin in the lactating rat is not governed merely by physical contact between mother and young, since certain environmental stimuli play specific roles at particular stages of lactation. A whole series of elegant studies have stemmed from the original observation of Grosvenor (Grosvenor and Mena, 1974) that merely placing a lactating rat close to, but not in contact with, her pups led to a fall in pituitary prolactin concentration in the absence of suckling. Subsequent studies showed that this phenomenon developed in primiparous rats between the 7th and 14th day of lactation and that this conditioned-release mechanism could be maintained over an ensuing pregnancy and was operative when tested on the 7th day of the second lactation. The stimulus for this release is olfactory, although visual stimuli are effective in the absence of olfaction. It is not known whether this olfactory stimulus is merely the general odour of the litter or whether a specific pheromone (see Chapter 3) is involved. Later still in lactation, the rat's prolactin-releasing mechanism becomes sensitive to a wider range of stimuli and by the 21st day of lactation prolactin is released in response to general stimuli emanating from lactating rats and their litters in the same room. This result recalls the earlier findings of Grosvenor and Turner that suckling could not, apparently, deplete pituitary prolactin concentration on the 21st day of lactation, although it could do so earlier in lactation. It is now clear that the suckling stimulus was not ineffective but merely that environmental stimuli had already triggered the release of prolactin before the pups were allowed to suckle. In the mouse too, odours can affect prolactin release, although here, the odour of a strange male can terminate pregnancy by inhibiting prolactin release (see Tindal, 1974).

In view of the effectiveness of stimuli other than suckling in the rat to trigger release of prolactin at certain stages of lactation it might be thought that release of prolactin in the ruminant is subject to the same influences. However, such is not the case, at least in the goat, where conditioned or other environmental stimuli were eliminated by milking goats under cyclopropane-oxygen anaesthesia (Hart, 1974). It was found that increases in circulating prolactin which occurred under anaesthesia did not differ significantly from those obtained at control milkings in the conscious animal two days earlier. It is known that the milk-ejection reflex (i.e. reflex release of oxytocin in response to milking) is blocked by cyclopropane-oxygen anaesthesia during milking in the goat (Cowie and Tindal, 1971) and that milk yield is unaffected when this regime is continued for up to two weeks. Hart's studies on the goat make it clear that whereas oxytocin release may be blocked by the anaesthetic, prolactin release is definitely not; this may explain why milk yield was similarly unaffected by anaesthesia during suckling in the rabbit, even though milk ejection was blocked (Cowie and Tindal, 1971).

In studies in collaboration with J. L. Linzell, Hart also showed that the milking stimulus was activated by tactile sensory pathways and was not causing release of prolactin by some unknown metabolic stimulus secondary to the removal of milk. Lactating goats were used in which half of the udder (i.e. one mammary gland) had been transplanted to the neck. The transplants had a patent vascular system, but no nervous connection to the central nervous system, and it was found that a rise in circulating prolactin occurred only in response to milking the intact gland (Hart, 1974).

B. NON-LACTATIONAL STIMULI

As mentioned earlier (see Section 1) prolactin is involved in the control of many physiological mechanisms, yet little is known about the stimuli which evoke its release at the appropriate time. However, in the case of osmoregulation, recent work suggests that in at least some euryhaline species of teleost the pituitary can release prolactin in direct response to changing plasma osmotic pressure (Ingleton *et al.*, 1973) and control of the pituitary prolactin secreting cells appears to be mediated by an interaction between the direct catecholaminergic innervation of the teleost adenohypophysis and plasma osmolality (Zambrano *et al.*, 1973/74). Other stimuli which are known to elicit prolactin release range from painful stimuli to non-specific stresses as mild as handling an animal. Almost any stimulus which arouses, alerts or frightens an animal, may evoke release of prolactin and in many cases several sensory modalities must be involved.

C. RHYTHM OF PROLACTIN RELEASE

In some mammalian species there is a free-running rhythm of prolactin release with a period of about 24 h (see Chapter 3). In the rat, circadian variations in adenohypophysial content and plasma concentration of prolactin have been reported (Tindal, 1974) although some of the earlier data may be misleading because of infrequent sampling. In female rats there are reports of maximum circulating levels in the late afternoon or early evening, and in male rats at about midnight. However, this type of information should be treated with caution, since Dunn *et al.* (1972) found that the maximum level in male rats occurred at 23.00 h in unstressed animals and at 17.00 h in stressed animals, which suggests that the experimental sampling technique may itself modify any underlying cyclic pattern. In another study, also in the male rat, the time of the nocturnal surge of plasma prolactin was found to be 3.30 h–5.00 h. This surge is abolished by pinealectomy (Rønnekliev *et al.*, 1973) which suggests that the light/dark cycle may govern this cyclic release via a pineal principle (see Chapter 3).

In man there is a marked increase in circulating prolactin at night (Nokin *et al.*, 1972) which is entrained by sleep and appears to be related to the onset of rapid eye movement (REM) sleep episodes (Sassin, *et al.*, 1972; Parker *et al.*, 1973, 1974). The release pattern shifts immediately and completely with shifts of sleep onset of 3, 6 and 12 h and hence is not based on an inherent rhythm related to the time of day (Sassin *et al.*, 1973).

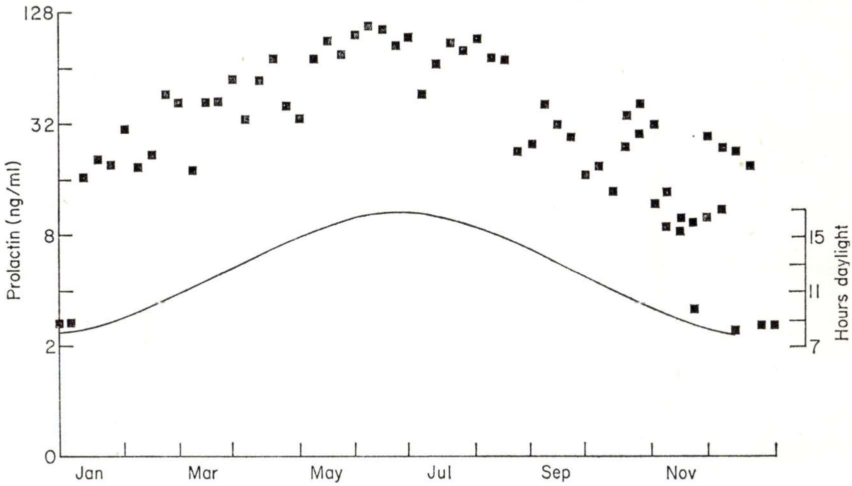

Fig. 1. The concentrations of circulating prolactin in three intact male goats over the period January to December 1971. Each point is the mean of three samples, one from each goat. Note that prolactin levels follow the seasonal change in day-length (——). (Dr. H. L. Buttle, unpublished.)

In ruminants, although a circadian pattern of prolactin release was reported in the cow, the data are not entirely convincing (see Tindal, 1974) and a comprehensive study in lactating, non-lactating virgin female and castrated male goats showed that whereas levels may rise and fall slowly within a 24 h period, there was considerable variation between animals and nothing to suggest a circadian pattern of release, except the bimodal pattern induced by twice-daily milking of the lactating animal (Hart, 1973).

In the longer term, however, ruminants do have a seasonal photoperiodic response, with higher basal circulating prolactin levels associated with long day-length and low levels with short day-length in the ram (Pelletier, 1973), sheep (Driver *et al.*, 1974), goat (Buttle, 1974) and the bull and non-lactating

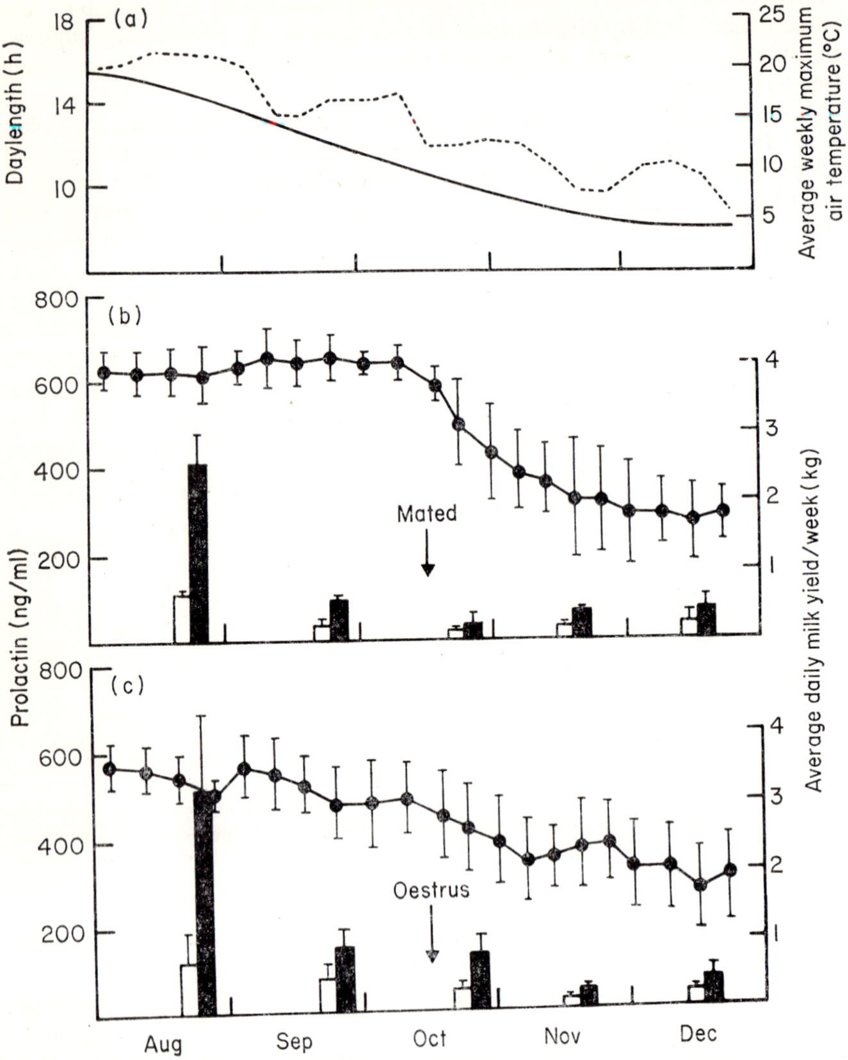

Fig. 2. (a) Changes in day-length (——) and average weekly maximum air temperature (- - - -) occurring between August and December 1972. (b, c) The average daily milk yield/week (●——●) and average concentration of prolactin released at milking (■) by three mated goats (b) and three non-mated goats (c) between August and December 1972. □, Pre-milking controls. Vertical lines indicate ± S.E.M. Note that average concentration of prolactin released at milking declined in the autumn in both mated and non-mated goats. (From Hart, 1975b.)

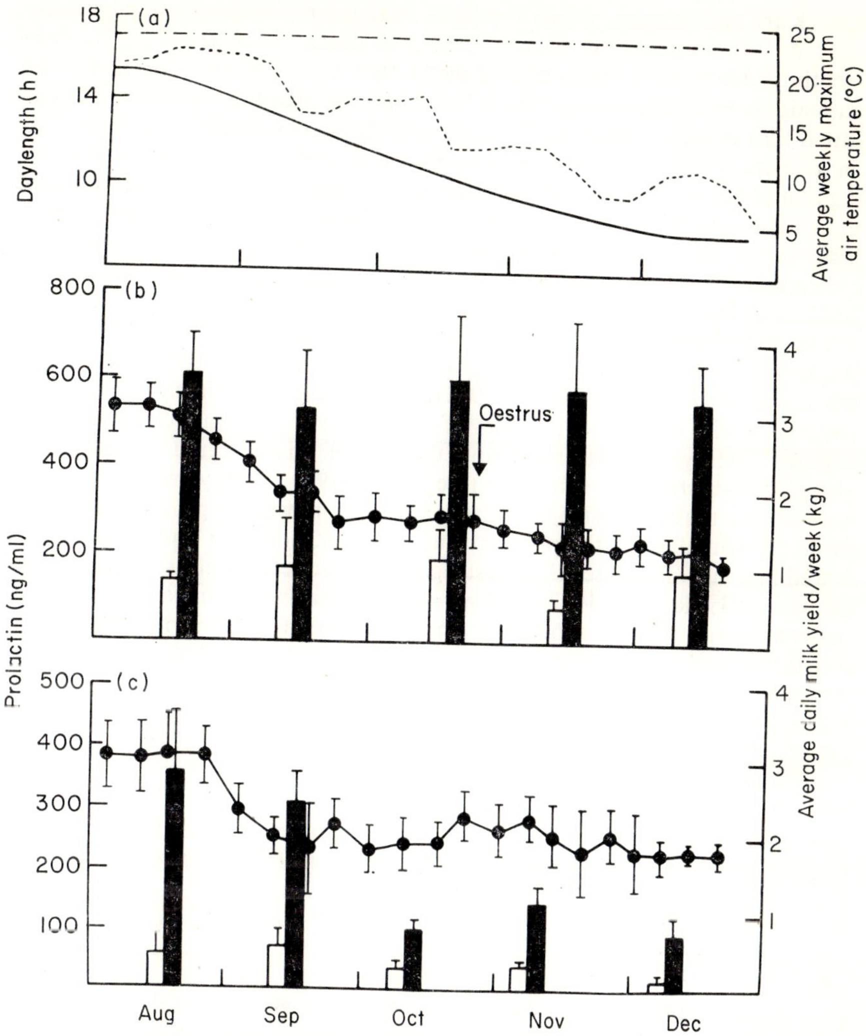

Fig. 3. (a) Artificially maintained day-length (—·—·). natural changes in day-length (——) and average weekly maximum air temperature (- - - -) occurring between August and December 1972. (b, c) The average daily milk yield/week (●——●) and average concentration of prolactin released at milking (■) by four goats maintained in summer lighting conditions (b) and three ovariectomised goats (c). □, Premilking controls. Vertical lines indicate ± S.E.M. Note that average concentration of prolactin released at milking did not decline in the autumn in those goats maintained in summer lighting conditions although milk yield declined as in the other groups in this and Fig. 2. (From Hart, 1975b.)

cow (Karg and Schams, 1974). Similarly, both the pre-milking and post-milking levels of circulating prolactin in the goat decreased with decreasing day-length. Subsequent experiments showed that this was not caused by the animals entering a breeding state or becoming pregnant in the autumn. It was found that exposing the goats to summer lighting conditions (17 h light: 7 h dark) from August to December maintained the release of prolactin at milking at a consistently high level throughout this time (Hart, 1975a, b). It seems quite likely, therefore, that the pineal may influence seasonal rhythms of prolactin in the ruminant and, in addition, environmental temperature may also have a role to play in the modulation of basal prolactin levels (Wettemann and Tucker, 1974).

The well-documented pre-ovulatory surge of LH (see Chapter 8) which occurs on the afternoon of pro-oestrus in the cycling rat is accompanied by a release of prolactin (see Tindal, 1974); this release of prolactin is not itself concerned with ovulation. A series of studies involving ovariectomy with or without oestrogen replacement and the use of an antiserum to oestradiol and an anti-oestrogenic substance (see Tindal, 1974; Yokoyama *et al.*, 1974) have shown that the oestrogen surge on the day preceding pro-oestrus in the rat is the stimulus for the pro-oestrous surges of both LH and prolactin. It is well known, of course, that oestrogen can evoke the release of prolactin (Meites and Clemens, 1972) and a release of prolactin similar to the pro-oestrous surge can be evoked by the injection of oestrogen in the adult ovariectomised rat, but not in the male castrated when adult or in androgen-sterilised females ovariectomised when adult. The oestrogen-triggered release of prolactin can be prevented by a retrochiasmatic cut between the anterior hypothalamus and median eminence (see Tindal, 1974).

Although pseudopregnancy has been reviewed in Chapter 8, mention should be made of the pattern of prolactin release during this state in the rat. Originally, pseudopregnancy in this species was thought to be dependent on a continuous secretion of prolactin (Freeman and Neill, 1972). However, not only were serum prolactin values found to be elevated for only 2–3 days after induction of pseudopregnancy, after which they declined to low levels, but it was also observed that pseudopregnancy could be induced in the rat even when the initial rise in prolactin was prevented by injection of ergocornine (see Tindal, 1974). It was then reported that after the initial elevation of circulating prolactin in the 24 h following cervical stimulation, there is an early morning surge of prolactin release, beginning at 1.00 h, reaching a peak at 3.00 to 5.00 h and ending at 9.00 to 11.00 h (Freeman and Neill, 1972). In later work (Freeman *et al.*, 1974) it was found that the stress of sampling cannulated rats was distorting the pattern of prolactin release and, that if samples were taken after decapitation, there were two daily surges of prolactin during pseudo-pregnancy, one at 17.00 to 19.00 h and one at 3.00 to 7.00 h. Hypothalamic retrochiasmatic cuts abolished both surges. The evening surge is attributed to ovarian influence, while the early morning surge is explained by postulating a "mnemonic system" which "remembers" the initial cervical stimulation each night. However, there could be a simpler explanation. As pointed out previously (Tindal, 1974) present knowledge of the effects of 5-HT and of melatonin on prolactin release, together with the fact that in the male rat a nocturnal surge also occurs and is abolished by

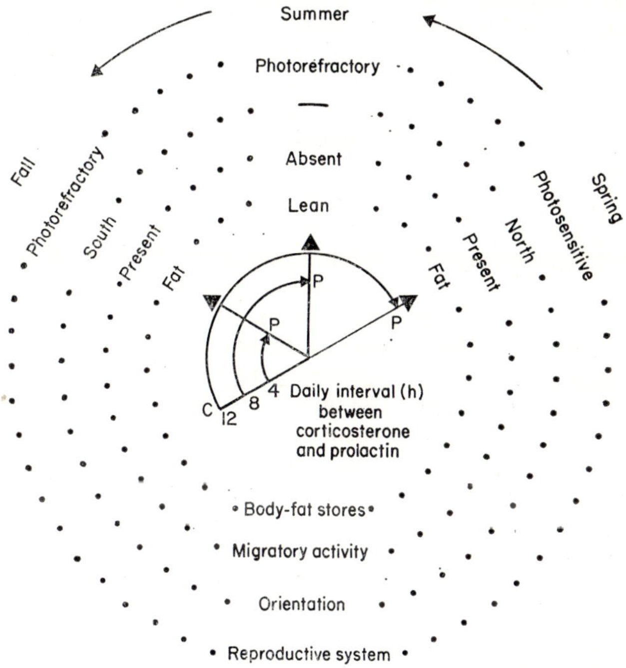

Fig. 4. The responses of the white-throated sparrow *Zonotrichia albicollis* to exogenous corticosterone (C) and prolactin (P) vary, depending on the temporal relation of the injections. Daily injections of prolactin administered 12 h after corticosterone stimulate spring conditions of migration and reproduction. Similarly, an 8 h pattern of hormone injections induces summer photorefractory conditions, and a 4 h pattern of injections induces autumnal conditions. (From Meier, 1973.)

pinealectomy, at least raises the possibility of a circadian rhythm of prolactin secretion and release which is entrained by pineal principles and which may normally be masked by other endocrine events in the cycling rat.

Recent work by A. H. Meier and his colleagues has revealed that the time of day at which a particular circulating hormone level rises, or a hormone injection is given, may be as important as the absolute amount of hormone involved (Meier, 1973; Meier and Dusseau, 1973; Nicoll, 1973). Most of these studies have been carried out on a migratory bird, the white-throated sparrow *Zonotrichia albicollis*.

Meier's previous work in D. S. Farner's laboratory had shown that prolactin injections caused fattening and pre-migratory restlessness in a related species, the white-crowned sparrow *Zonotrichia leucophrys gambelii*, yet when the experiment was repeated later in the white-throated sparrow the effect could not be confirmed. The first significant finding was that if lean birds in summer were injected with prolactin daily in the morning, fat stores were depleted, whereas if they were injected in the afternoon there was an explosive increase in fat deposits which reached the peak level seen during migration after only a week of injections, and, in addition, these afternoon injections induced nocturnal pre-migratory restlessness (this bird is a nocturnal migrant). Corticosterone and prolactin are both released spontaneously in this species each day, although the actual time of day at which the release occurs varies throughout the year, as does the interval between the releases of the two hormones. To investigate the possible significance of this relationship, Meier utilised an experimental technique whereby putting the birds in continuous light suppressed these endogenous hormone rhythms, allowing the effects of exogenous hormones to be interpreted. In an elegant series of experiments it was found that a 12 h interval, corresponding to natural events in spring, between corticosterone and prolactin injections was associated with a dramatic increase in fat deposits, development of gonads and a north-orientated pre-migratory restlessness. An 8 h interval, corresponding to the non-breeding state in midsummer, was associated with reduced fat deposits, absence of gonadal stimulation and no nocturnal restlessness, while a 4 h interval (and to some extent a 0 h interval, i.e. simultaneous injections) corresponding to the autumnal migratory period, was associated with increased fat deposits, no gonadal stimulation and a south-orientated pre-migratory restlessness.

Hence, the effects seen under artifical conditions of continuous light and exogenous hormone injections mimicked those which occur naturally as the year progresses. Corticosterone appears to phase periods of sensitivity and insensitivity in certain tissues to the action of prolactin and the photoperiod appears to play the vital role in triggering release of corticosterone and prolactin, yet how it does so, and the mechanism of the seasonal shift in the actual and relative times of release of the two hormones, is not understood at present. The pineal could well have a role to play here. A temporal relationship between daily rhythms of corticosteroids and prolactin has also been demonstrated for the control of fat deposits in fish, amphibia, lizards and other birds and different time-relationships exist in the pigeon for two actions of prolactin, for maximal and minimal stimulation of fat deposits and intestinal weight on the one hand and of the crop sac on the other (see Meier, 1973; Nicoll, 1973). In the white-throated sparrow prolactin can act directly on the gonads to render them refractory to the action of gonadotrophins but only when raised prolactin levels are in a particular time-relationship with raised corticosterone levels. In their later work, Meier and Dusseau (1973) put forward a scheme to attempt an explanation of the seasonal photoperiodic control of gonadal function in the white-throated sparrow, which involves an ongoing sequence of events, initiated each day, but requiring almost 48 h for completion. It is envisaged that the beginning of the photoperiod (dawn) entrains a daily rhythm of plasma corticosterone. In photo-stimulated birds in May this

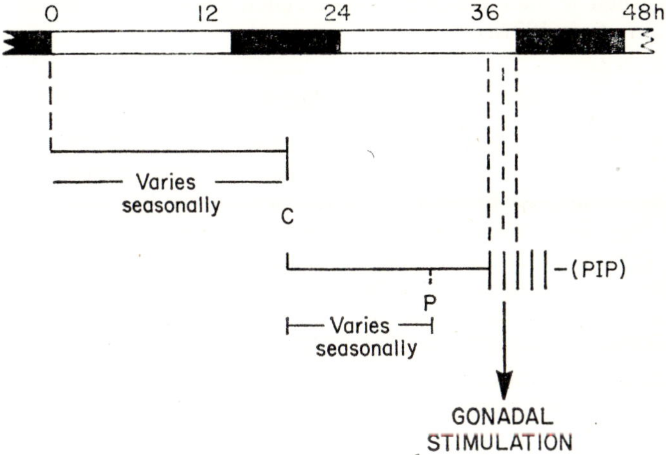

Fig. 5. A model of the photostimulatory effect in the white-throated sparrow *Zonotrichia albicollis*. The beginning of the photoperiod entrains a daily rhythm in the concentration of plasma corticosterone. In photo-stimulated birds in May, the increase in corticosterone (C) occurs after midnight, about 20 h after dawn. The rise of plasma corticosterone, in turn, entrains a photo-inducible phase (PIP) (for FSH?) about 17 to 22 h later. A PIP for LH may occur several hours earlier than the FSH PIP. If light is present during the photoinducible phase(s), i.e. if the daily photoperiod is longer than 13 h, it induces gonadal growth. Photosensitivity and photo-refractoriness during the annual cycle are determined by the temporal relation of the daily increase in levels of plasma corticosterone (C) and the daily release of prolactin (P). In photosensitive birds in May the interval between increases in plasma levels of the two hormones is about 12 h. (From Meier and Dusseau, 1973.)

increase occurs about 20 h after dawn and this rise, in turn, entrains photo-inducible phases for LH and FSH release some 17–22 h later. If light is present during these photo-inducible phases (i.e. if day-length is sufficiently long) gonadal growth will be induced, provided that the interval between the daily release of corticosterone and prolactin is 12 h (which it is in May). Other time-relationships between these two hormones result in inhibition of gonadal growth, even though the gonadotrophins may have been released. Meier and Dusseau (1973) point out, however, that even if other birds do utilise this mechanism the details of timing will probably show considerable variation between species.

D. CLINICAL ASPECTS

Originally, it was claimed that the actions of prolactin in man could be attributed to the lactogenic activity of hGH and hCS and the existence of a discrete human prolactin was hotly disputed. However, this problem disappeared when human prolactin was isolated by Friesen's group in 1970 (see Kaplan *et al.*, 1972a).

(1) *Control of Prolactin Secretion*

There is no evidence that control of prolactin secretion in man differs from the hypothalamic-inhibitory mechanism which has been demonstrated in experimental animals (see Section III). It has been shown by many investigators that thyrotrophin releasing hormone (TRH) has the ability to release prolactin as well as TSH in man and its use has been suggested, where appropriate, for the augmentation and prolongation of lactation, with a concomitant prolongation of the period of postpartum infertility, thought to be caused by the raised circulating prolactin levels (Tyson *et al.*, 1972). TRH has also been claimed as a physiological control mechanism for prolactin release (Bowers *et al.*, 1973b). However, the situation is far from simple, since the most effective stimulus for prolactin release, postpartum suckling, fails to release TSH (Gautvik *et al.*, 1974) and on the basis of the lack of feedback effect of thyroid hormones on the prolactin-releasing effect of TRH, L'Hermite *et al.* (1974) concluded that the action of TRH on prolactin secretion was purely pharmacological. However, whether or not TRH is a physiological releaser of prolactin in situations other than suckling, the TRH stimulation test serves as an extremely valuable tool for clinical testing of prolactin release to distinguish between primary hypopituitarism, where the fault lies with the pituitary, and secondary hypopituitarism where there is a disturbance of hypothalamic function (Kaplan *et al.*, 1972a; Besser, 1974).

(2) *Prolactin and Reproduction*

Prolactin levels remain relatively constant during the menstrual cycle (McNeilly *et al.*, 1973) and, unlike non-primates, prolactin levels rise progressively during pregnancy in man to reach a peak at term, and then decline (McNeilly, 1974). After parturition, prolactin is released in response to suckling (Tyson *et al.*, 1972; Jaffe *et al.*, 1973; Gautvik *et al.*, 1974; Noel *et al.*, 1974) and the absence of lactation during pregnancy, despite high circulating prolactin levels, must be attributed to the concomitant high circulating levels of steroids, an inhibition which is removed by parturition. Prolactin is essential for the initiation and maintenance of milk secretion in man. Although L-Dopa can reduce circulating prolactin levels and inhibit lactation (Frantz *et al.*, 1973) the effect of each dose is short-lived and L-Dopa is not without side-effects. Far superior clinical therapy is now possible with bromocriptine, a synthetic ergot compound, which has a prolonged period of action, no undesirable side-effects and has also superseded the use of steroid mixtures, which probably act directly on the mammary gland, to suppress lactation. Inhibition of human prolactin secretion by bromocriptine results in inhibition of both natural post-puerperal lactation and spontaneous galactorrhoea (del Pozo and Flückiger, 1973), although in the case of spontaneous galactorrhoea, stopping bromocriptine therapy may result in the return of the inappropiate lactation (Lloyd *et al.*, 1975).

Although prolactin is a member of the luteotrophic complex in a number of mammals (see Section I), most evidence is against prolactin having a significant luteotrophic role in man. A low level of prolactin in follicular fluid does, however, appear to facilitate, and a high level to inhibit progesterone synthesis. It has been suggested that this leads to a reduced positive feedback to trigger the ovulatory surge of LHRH and of LH and FSH (McNatty *et al.*, 1974; see also Chapter 7). The inhibition of progesterone synthesis by high levels of prolactin in follicular fluid could explain the inhibition of ovulation during normal post-parturient lactation and also affords an explanation of the hormonal mechanism underlying galactorrhoea and its accompanying amenorrhoea (McNatty *et al.*, 1974; McNeilly, 1974). Three types of syndrome have been described: the Chiari-Frommel, in which persistent lactation follows a normal pregnancy, and the Argonz-del Castillo and the Forbes-Albright syndromes in which lactation occurs spontaneously, all accompanied by amenorrhoea. Treatment with bromocriptine not only reduces circulating prolactin levels and abolishes galactorrhoea, but also leads to the return of ovulation and menstrual cycles, presumably in response to the lowered prolactin level in the follicular fluid (McNatty *et al.*, 1974; Lloyd *et al.*, 1975).

III. Role of Releasing and Release-inhibiting Mechanisms

A. Hypothalamic Factors and Hormones

It has become apparent during the past twenty years (see Chapter 1) that in mammals the secretion of prolactin is normally restrained by the brain, that severance of the link between adenohypophysis and brain results in increased prolactin release and that, in general, hypothalamic extracts inhibit the release of prolactin from the adenohypophysis. In birds, however, the brain appears to facilitate and not to inhibit release of prolactin (Meites *et al.*, 1972). The concept became established (Meites and Clemens, 1972; Tindal, 1974) that in mammals, a substance of hypothalamic origin called prolactin-inhibiting factor (PIF) restrains prolactin release and that another factor, prolactin-releasing factor (PRF) probably plays a minor role.

B. MECHANISM OF ACTION OF PIF AND PRF

It is believed that the adenohypophysial trophic cell membrane is electrically charged and that it can be depolarised or hyperpolarised. Depolarisation is thought to be associated with hormone release and hyperpolarisation with inhibition of release. However, whereas most trophic cells require a releasing factor to trigger depolarisation and hormone release, the lactotrophe cell appears to depolarise spontaneously and release prolactin, unless it is maintained in a state of hyperpolarisation by PIF. The presence of Ca^{++} is necessary for prolactin release and Ca^{++} must enter the lactotroph cell before prolactin granules can leave. The current view is that releasing factors act on specific receptor sites on the trophic cell membrane to cause a change in permeability, which then allows Ca^{++} to enter the cell and hormone granules to leave. In the case of PIF it is thought that there is a PIF-membrane interaction which renders the cell membrane impermeable to Ca^{++} and prevents the release of prolactin; this effect is thought to be mediated by activation of the cyclic AMP system and inhibited by an increase in intracellular prostaglandin E_1 concentration (Ojeda *et al.*, 1974b) (for a more detailed discussion see Tindal, 1974).

As regards the function of PIF and PRF in the living animal, it seems that they may be affected by different mechanisms. Thus, in the rat, a minor stress such as ether causes a rapid but relatively small release of prolactin, while the suckling stimulus results in a slightly delayed but much larger release. Moreover, the suckling-induced release, but not the ether-induced release, can be blocked by nicotine (Terkel *et al.*, 1972; Blake and Sawyer, 1972b). These workers suggested that two separate mechanisms exist for prolactin release. A specific PRF (Valverde *et al.*, 1972) may be involved in the relatively minor immediate release induced by stress, while the delayed but much larger release triggered by suckling may be due to a decrease of PIF activity. In this connection, it is of interest that Dular *et al.*, (1974) have reported the isolation of a fraction containing specific PRF activity from an extract of bovine neurohypophysis, stalk and median eminence.

It appears that a constant trickle of PIF prevents prolactin release, that a stimulus such as suckling temporarily arrests the release of PIF into the portal vessels, thus allowing a burst of prolactin release to occur, while a minor stress evokes release of a PRF which over-rides the action of PIF and induces prolactin release. In addition to these two factors, thyrotrophin-releasing hormone (TRH) also releases prolactin in man (see Section II), sheep, goat, cow and rat (Mueller *et al.*, 1973; see Tindal, 1974). Whether or not TRH functions as a physiological PRF is still in dispute.

C. HYPOTHALAMIC AMINERGIC AND CHOLINERGIC MECHANISMS AND THE IDENTITY OF PIF AND PRF

Hypothalamic monoamines play a vital role in the control of adenohypophysial function. These monoamines are concentrated in nerve terminals which converge on the basal hypothalamus and, in particular, on the median eminence and primary plexus of the pituitary portal system (see Chapters 1, 3 and 5; Tindal, 1974). Although there are still minor differences of opinion on the details of the distribution of monoaminergic terminals in the hypothalamus, it appears that dopaminergic terminals in the median eminence arise from cell bodies in the neighbouring arcuate nucleus and anterior periventricular hypothalamic nucleus, that noradrenergic terminals in the median eminence and throughout the hypothalamus and preoptic area in general arise from cell bodies within the hypothalamus and further away in the pons and medulla. A serotoninergic (5-hydroxytryptamine, 5-HT) pathway also ascends from the mesencephalon, in association with the ventral noradrenergic path in the medial forebrain bundle, and terminates diffusely in the hypothalamus, with a denser innervation of the suprachiasmatic nuclei, medial retrochiasmatic area and anterior median eminence. Both the 5-HT and noradrenergic systems have a large network of collaterals (see Tindal, 1974).

Serotoninergic mechanisms play a role in the release of prolactin, although the detailed mechanism of action is not clear at present. Thus, injection of 5-HT or a related amine, melatonin, into the third ventricle of the rat caused the release of prolactin (Kamberi *et al.*, 1971a) as did systemic injection of precursors of 5-HT (tryptophan or 5-hydroxytryptophan) (Lu and Meites, 1973), presumably by causing increased synthesis of 5-HT (Fernstrom and Wurtman, 1972). Blockade of 5-HT synthesis by parachlorophenylalanine (PCPA) blocked the release of prolactin in response to suckling in the rat, an effect which could be overcome by administration of 5-hydroxytryptophan. However, PCPA only blocked prolactin release in response to a neural stimulus such as suckling and did not affect basal levels of the hormone (Kordon *et al.*, 1973/74).

It has been known for some time that catecholamines play a major role in the control of prolactin release. In the rat, intraventricular dopamine led to increased PIF activity in pituitary stalk plasma and reduced prolactin levels in peripheral plasma; injection of dopamine directly into the portal vessels had no effect (Kamberi *et al.*, 1970, 1971a). It was also known that administration of the precursor of dopamine, L-Dopa, which, unlike dopamine, can pass the blood-brain barrier, led to increased hypothalamic PIF content and the appearance of prolactin-inhibiting activity in the serum of the rat, caused remission of galactorrhoea in man and a drop in serum prolactin levels in both species (Lu and Meites, 1972; Turkington, 1972). In addition, other studies indicate the presence of a cholinergic mechanism involved in the inhibition of prolactin release which appears to precede the dopaminergic link (Libertun and McCann, 1974; Grandison *et al.*, 1974; Lawson and Gala, 1975).

Although the work of van Maanen and Smelik (1968), McLeod (1969) and Birge *et al.* (1970) had suggested that catecholamines could act directly on the adenohypophysis, this view was discounted for various reasons. However, in 1974 evidence from studies in the rat completely changed the situation. Apomorphine, a specific dopamine receptor stimulant (i.e. a dopamine agonist) was shown to prevent the suckling-induced release of prolactin and to inhibit the release of prolactin *in vitro* (Smalstig *et al.*, 1974; MacLeod and Lehmeyer, 1974). Dopamine itself was found to inhibit the release of prolactin *in vitro* at concentrations as low as 10^{-9}M (Shaar and Clemens, 1974) and this effect could be antagonised by TRH at the level of the dopaminergic receptors of the lactotrophe cells (Hill-Samli and MacLeod, 1974). L-Dopa was shown to cause a decrease in prolactin secretion by acting directly on pituitary grafts, an effect which was antagonised by L-α methyldopa hydrazide which inhibits peripheral dopa decarboxylase (Donoso *et al.*, 1974). Furthermore, pimozide, a specific dopamine antagonist, reduced the inhibitory action of apomorphine *in vitro* (Smalstig *et al.*, 1974) and when implanted locally in the intact animal, pimozide was effective in raising prolactin levels when present either in the median eminence or in the anterior pituitary (Ojeda *et al.*, 1974a).

The final conclusive proof that catecholamines can act directly on the pituitary *in vivo* came from two studies with hypothalamic extracts. In one of them, a fraction of purified porcine stalk-median eminence (SME) extract which contained high PIF activity was also shown to be rich in the catecholamines noradrenaline and dopamine. It was found that when this PIF preparation, as well as dopamine and noradrenaline, were dissolved in freshly made 5% glucose solution to protect them from oxidation, they each had the ability to inhibit prolactin secretion when infused into a pituitary portal vessel of the rat and the PIF activity of the SME extract could be fully accounted for by its catecholamine content (Schally *et al.*, 1974; Takahara *et al.*, 1974a). Since other fractions of this porcine SME extract also possess PIF activity, it remains to be seen whether there is a discrete PIF in addition to the catecholamines, but in view of the other study by Shaar and Clemens (1974), this seems unlikely. These workers showed that rat hypothalamic extracts, dopamine, noradrenaline and adrenaline each inhibited prolactin release by rat pituitaries *in vitro* and that when the hypothalamic extract was pre-incubated with monoamine oxidase or was subjected to alumina catacholamine adsorption to remove catecholamines, it lost its ability to inhibit prolactin release. Furthermore, the fact that pepsin treatment of the extract was without effect on its prolactin release-inhibiting activity was evidence against PIF being a peptide. The dopamine and noradrenaline content of the hypothalamic extract was measured fluorometrically and, in addition, the catecholamine-containing acid eluate from the initial alumina adsorption procedure was found to inhibit prolactin release to the same degree as the original untreated hypothalamic extract. These results indicated that dopamine and noradrenaline, in amounts less than that found in the hypothalamus, can act directly on the rat adenohypophysis to inhibit prolactin release and that hypothalamic PIF activity can be totally accounted for by amounts of endogenous hypothalamic catecholamines.

While there is now convincing evidence for the identity of hypothalamic prolactin-inhibiting activity, the same cannot be said of prolactin-releasing activity. Stated simply, it is a question of whether the inherent prolactin-releasing activity of TRH represents PRF, or whether there is another specific PRF. Some claim that TRH only acts as a pharmacological releaser, while others believe that TRH is a physiological regulator of prolactin secretion although, as Meites (1973) pointed out, the stimuli which are associated with prolactin release do not usually trigger the release of TSH as well.

D. FEEDBACK MECHANISMS

Prolactin does not have a single target organ and hence there is no simple target organ hormone "long" feedback (see Chapter 3). However, prolactin has been shown to exert a negative feedback on its own secretion ("short" feedback) at least under experimental conditions. The site of action appears to be the median eminence and the mechanism involves the activation of dopaminergic terminals in the median eminence (Meites and Clemens, 1972; Tindal, 1974).

E. CENTRAL PATHWAYS

Although a general account of pathways from higher brain areas to the hypothalamus appears in Chapters 1 and 3, a more detailed account of the ascending pathways involved in prolactin release would be appropriate here (for references see Tindal, 1974). The first studies in this field were carried out before the availability of radioimmunoassays for prolactin and involved the use of a relatively crude bioassay. Discrete electrical stimulation was applied twice daily for 11 days through chronically implanted electrodes in pseudopregnant rabbits. If the stimulation resulted in the release of prolactin, this was assessed by the lactogenic response of the mammary glands at autopsy. It was found that a pathway for prolactin release ascended in the lateral tegmentum of the midbrain, entered the posterior hypothalamus in the region between the third ventricle and the mammillo-thalamic tract and, from there, passed rostrally in the medio-dorsal hypothalamus and then laterally to enter the far-lateral hypothalamus. It then passed forward within the medial forebrain bundle to the lateral preoptic area, medially into the medial preoptic area and then turned caudally and appeared to terminate in the medial anterior hypothalamic area, close to the third ventricle. This probably represents the ascending pathway which is triggered by suckling, since it has much in common at the mesencephalic level with the ascending oxytocin-release pathway of the milk-ejection reflex (Tindal and Knaggs, 1970; Cowie and Tindal, 1971, Tindal, 1974). In addition to this direct ascending pathway it was also

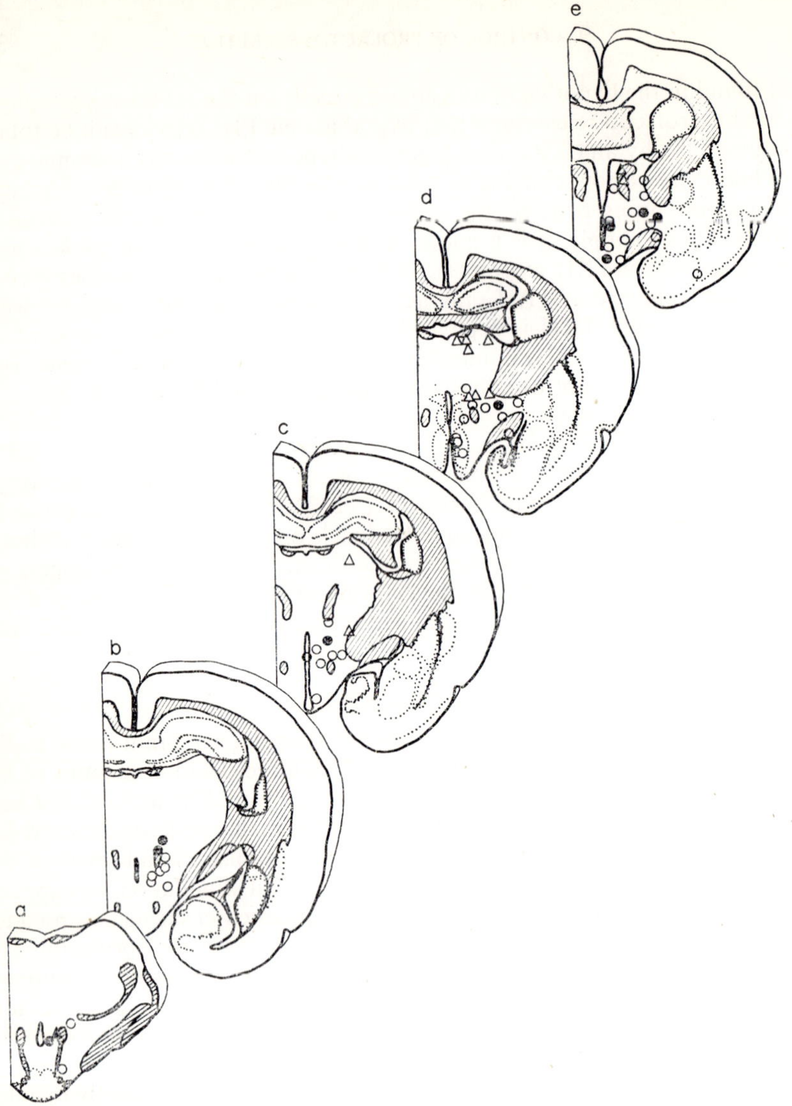

Fig. 6. Pathways for release of prolactin in the forebrain of the rabbit showing sites where twice-daily electrical stimulation during 11 days of pseudopregnancy caused the release of prolactin and lactogenesis. It was found that control electrode tips, if situated in a pathway for prolactin release, could also evoke release of prolactin by causing irritation of neurones. Stereotaxic planes pass rostrally at 1 mm intervals from Fig. 6a through to Fig. 7f. Note stimulation sites associated with lactogenesis close to third ventricle in posterior hypothalamus (a) and further rostrally in medio-dorsal (b, c) and then in lateral hypothalamus (d, e) in association with medial forebrain bundle, also two sites near the third ventricle in the anterior hypothalamus (e). ●, Stimulation followed by lactogenesis; ▲, control electrode, no stimulation followed by lactogenesis; ○, stimulation, no lactogenesis; △, control electrode, no stimulation, no lactogenesis. (From Tindal and Knaggs, 1972.)

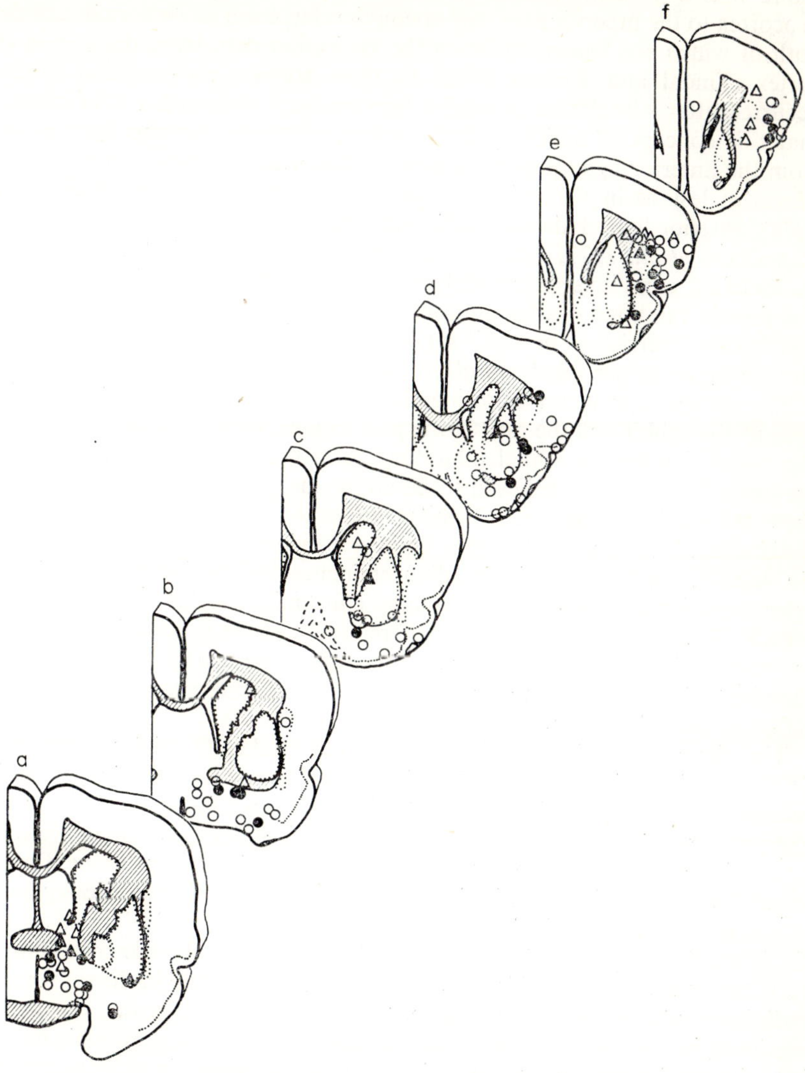

Fig. 7. Pathways for release of prolactin in the forebrain of the rabbit (see Fig. 6). Note stimulation sites associated with lactogenesis in lateral and medial preoptic area (a), and further forward in, and ventral to, the external capsule (b, c, d), in the claustrum (d, e) and in the adjacent orbito-frontal cortex (e, f). (From Tindal and Knaggs, 1972.)

found that a prolactin-release pathway descended from the orbito-frontal neocortex to the preoptic area, traversing the claustrum and external capsule, both of which are known to lie on the projection path from this region of cortex (Tindal and Knaggs, 1972). In these studies, no prolactin release occurred after stimulation of medial hypothalamic sites and it is of interest that this concept of major prolactin-releasing stimuli exerting their action from the anterior hypothalamic region is supported by studies involving the effects of lesions in the goldfish, *Carassius auratus* (Peter and McKeown, 1974), anterior hypothalamic deafferentation in the rat (Blake *et al.*, 1973b; Kawakami *et al.*, 1973; Arai and Yamanouchi, 1975) and electrical stimulation of the conscious sheep (Malven, 1975). Interpretation of the picture in the female rat after deafferentation, however, can be complicated by the consequent "persistant oestrus" syndrome and its associated high levels of circulating oestrogen (Blake *et al.*, 1973b).

The release of prolactin following placement of lesions in the median eminence in the rabbit and the rat (Cowie and Tindal, 1971; Tindal, 1974) led to the conclusion that the final inhibitory control over prolactin release was mediated by PIF (now believed to be dopamine, see Section III C) released from the median eminence. In similar studies on the sheep (Dománski and Polkowska, 1973) posterior median eminence lesions caused increased prolactin release, whereas anterior median eminence lesions inhibited prolactin release and it was suggested that the latter effect might be due to destruction of a PRF. However, since complete deafferentation of the hypothalamus results in greatly reduced circulating levels of prolactin in the rat (Blake *et al.*, 1973b), the effect of anterior median eminence lesions in the sheep could well be attributed to loss of sensory input.

The choice of the conscious rabbit for our studies (Tindal and Knaggs, 1972) was dictated by the necessity of using a bioassay for assessing prolactin release. However, in the light of recent evidence it was a fortunate choice, since the problem of studying the release of prolactin under anaesthesia are only now being appreciated (see Tindal, 1974). Ether and pentobarbitone, for instance, will both trigger an initial release, but pentobarbitone subsequently blocks release of prolactin from the pituitary, and even the widely-used anaesthetic urethane has been shown to block LHRH release. In our own laboratory, the studies in the rabbit were repeated in greater detail in the rat when a radioimmunoassay became available. Initially, these were carried out under urethane, but when it was suspected that the anaesthetic was probably blocking release of prolactin, the decorticate animal, prepared initially under the short-acting barbiturate methohexitone sodium, was used and it was then possible to confirm in the rat the findings from the rabbit (Tindal and Knaggs, 1977).

There have been many conflicting reports on the effects of electrical or electrochemical stimulation of the preoptic area and medial basal hypothalamus, some claiming inhibition, and some claiming facilitation, of prolactin release (see Tindal, 1974). However, our studies in the rabbit and rat make it clear that one cannot emphasise too strongly that the pathways for prolactin release are essentially discrete, and that one cannot stimulate indiscriminately in, say, the preoptic area, and expect to evoke release of prolactin consistently.

Olfactory information could influence prolactin release via at least three pathways to the hypothalamus. It can travel within the medial forebrain bundle direct from the olfactory bulb and tubercle to the lateral preoptic and lateral hypothalamic area, or via pyriform cortex, amygdala and ventral amygdalofugal path to the lateral preoptic area (see Tindal, 1974) or, in the light of recent work, by a pathway from the amygdala to the dorsomedial thalamic nucleus and orbitofrontal cortex (Motokizawa, 1974; Tanabe *et al.*, 1974) and thence to the preoptic area and hypothalamus (Tindal and Knaggs, 1972).

F. CURRENT VIEWS

It must be stated at the outset that the general picture of how and why certain incoming stimuli evoke the release of prolactin is still not understood. Various segments of the prolactin-release mechanism are, however, becoming clearer and, as mentioned earlier in the chapter, minor releases appear to be mediated by a PRF, while major releases caused, for instance, by suckling, are achieved by inhibition of catecholamine release from the median eminence. The apparent termination of major prolactin-releasing pathways some distance from the median eminence has proved a major stumbling block. In order to explain this apparent paradox, it has been suggested that there may be a change of transmitter at this point (see Tindal and Knaggs, 1972; Tindal, 1974). It is envisaged that incoming stimuli achieve prolactin release by inhibiting a subsequent neurone or chain of neurones which, in the light of current evidence, could well be dopaminergic. Although it is well-known that dopaminergic neurones are found in the arcuate nucleus and send short axons to the median eminence (see Chapters 1 and 3), another hypothalamic group of dopaminergic neurones has been described, extending from the rostral hypothalamic/preoptic area in the midline, caudally to the rostral median eminence (Björklund and Nobin, 1973). It would be feasible, therefore, for an appropriate stimulus to inhibit these dopaminergic neurones and hence to inhibit the release of dopamine from the median eminence. The lactotrophe cells, released from dopamine inhibition, would then respond by releasing prolactin. It must be stressed, of course, that this is only a tentative suggestion, and indeed there are other mechanisms by which this system could be influenced. For instance, the region surrounding the ventromedial hypothalamic nucleus is known to contain a massive network of dendrites, many of which are long (see Tindal, 1974). This raises the possibility that neurones in the anterior hypothalamic and preoptic areas, whose stimulation evoked release of prolactin, may have been exerting an effect on the tips of dendrites whose cell bodies lay some distance away in the ventromedial-arcuate region. Also, although oestrogen can act directly at the level of the adenohypophysis or the median eminence to cause release of prolactin (see Cowie and Tindal, 1971), the limbic system may also be involved in modulation of the prolactin release mechanism. Release of prolactin occurs after local placement of oestrogen in specific regions of the amygdala that were shown later to be preferential sites for uptake of labelled oestrogen. Available evidence at the time (Tindal *et al.*, 1967) suggested that these oestrogen-sensitive neurones exerted their effect on the hypothalamus by means of the discrete stria terminalis path, and not by the diffuse ventral amygdalofugal path.

Recent elegant studies (De Olmos and Ingram, 1972) have described the origin of the different components of the stria terminalis, the relative positions of these components within the stria terminalis and the termination of these components within the hypothalamus from which it is evident that the sites of origin of two of the components in the amygdala coincide with the oestrogen-sensitive sites concerned with prolactin release. One of them terminates in a shell-like manner surrounding the ventromedial hypothalamic nucleus (VMH), while the other penetrates the core of this nucleus. Within the VMH two different types of neurones have been reported which can be activated by stimulation of the amygdala (Murphy, 1972). One type is bipolar and occurs at the periphery of the nucleus, while the other is multipolar and is found in the medial and basal regions of the nucleus. The bipolar cells are considered to be inhibitory interneurones which can be activated either by stimulation of the ventral amygdalofugal path or of the stria terminalis. In either case, a facilitatory signal into the bipolar cells would cause inhibition of the neighbouring multipolar cell or cells, the system as a whole acting as a type of polarity switch. The VMH itself is devoid of catecholamines in the cat (Cowchock, *et al.*, 1974), while high levels of dopamine are present in the medial posterior subdivision of this nucleus in the rat (Palkovits *et al.*, 1974b), so that axons of the VMH multipolar neurones might establish contact with dopaminergic neurones either in the posterior medial VMH or in the adjacent arcuate nucleus and, perhaps, facilitate the release of dopamine from the median eminence into the pituitary portal blood vessels. It has been known from the early studies of C. H. Sawyer and his colleagues that basal hypothalamic lesions in the rabbit which destroy the arcuate and basal portion of the VMH nuclei, cause release of prolactin. With this in mind Tindal (1974) suggested that the ability of oestrogen implants in specific regions of the amygdala to evoke release of prolactin might possibly be explained in terms of VMH inhibitory interneurones "switching off" PIF release when they were activated by the stria terminalis fibres, or by a known projection from the anterior hypothalamus which is itself associated with inputs from another stria component (Chi, 1970).

In the light of present knowledge, the last sentence could be re-written to express the view that the inhibitory interneurones "switch off" neighbouring dopaminergic neurones, restrain dopamine release into the portal system and hence allow the lactotrophe cells to release prolactin. As mentioned in section III E, the ascending pathway believed to be associated with the suckling stimu-

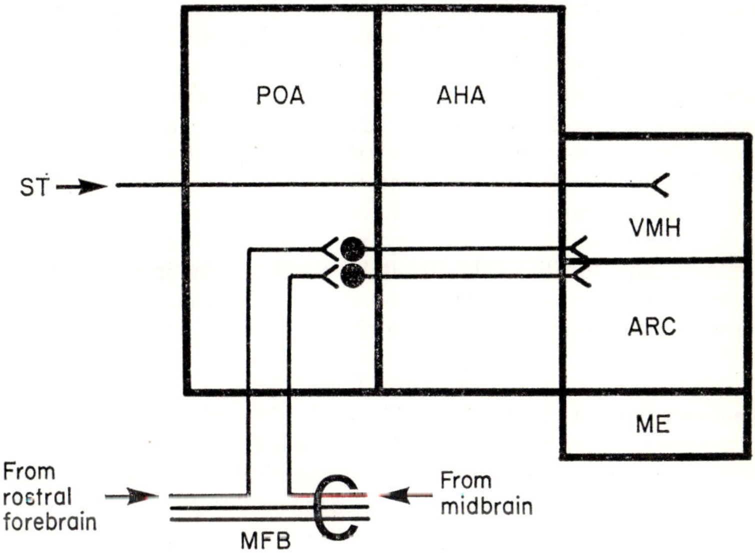

Fig. 8. Possible mechanism for major releases of prolactin. Impulses from prolactin-releasing stimuli ascend via spinothalamic tract and medial forebrain bundle or descend from orbital cortex, claustrum, external capsule and part of striatum to enter the preoptic area (Tindal and Knaggs, 1972). Stimuli could act at the preoptic level on a subsequent neurone or chain of neurones involved in tonic release of PIF (i.e. catecholamine) by the basal hypothalamus. It is possible that while the somas of such neurones could lie in the arcuate-ventromedial region their dendrites might extend rostrally to the preoptic-anterior hypothalamic region (see Tindal, 1974). In either case it is proposed that transmission in these final neurones would be inhibited by prolactin-releasing stimuli, causing a temporary inhibition of catecholamine release and hence a surge of prolactin release. The mechanism could also be modulated by a specific region of the amygdala acting via the stria terminalis on the ventromedial nucleus (see Tindal, 1974). AHA, anterior hypothalamic area; ARC, arcuate nucleus; ME, median eminence; MFB, medial forebrain bundle; POA, preoptic area; ST, stria terminalis; VMH, ventromedial nucleus. (From Tindal, 1974.)

lus and the descending pathway from the neocortex converge and appear to terminate in the region of the preoptic/anterior hypothalamic boundary. This final segment of the pathway is not yet resolved, although several possibilities exist (see Tindal, 1974). The pathway could pass caudally with the projection described by Chi (1970), or it may link up with the far-reaching dendritic network of the VMH, or it may relay directly with the newly described dopaminergic pathway from anterior hypothalamus to anterior median eminence (Björklund and Nobin, 1973). Whichever proves to be the case, incoming sensory stimuli destined to release prolactin could be visualised as inhibiting a dopaminergic system, either directly at the level of the anterior hypothalamus, or indirectly via intervening neurones, which might be cholinergic, and a stimulus such as suckling could then achieve release of prolactin by removing, for a short time, the constant trickle of dopamine which normally inhibits prolactin release from the lactotrophe cells. Once the effect of the stimulus had worn off, dopaminergic activity would be resumed and prolactin release would again be suppressed. The events occurring at the preoptic level, and higher, present a considerable challenge, since little is known of the integrative mechanisms which "decide" whether a particular stimulus will release prolactin, even though the appropriate pathway may have been stimulated. Even for the apparently straightforward tactile pathway of suckling this mechanism is not yet understood, and for the orbitofrontal-preoptic prolactin-release path the difficulty of the problem increases enormously, since much of the forebrain and limbic system may be called into play.

IV. Conclusions

The diversity of function exhibited by prolactin in all classes of vertebrates entails many different release stimuli. Some originate from the environment while others arise as a consequence of physiological processes. Knowledge of the release mechanism is largely confined to the mammal where PIF is released from the median eminence into the portal vessels. The prolactin-inhibiting activity of the hypothalamus can be totally accounted for by its catecholamine content and there is compelling evidence that PIF is dopamine. Major prolactin-release stimuli probably inhibit the release of dopamine. Minor stresses may trigger prolactin release by releasing a specific PRF from the median eminence which over-rides the action of dopamine on the lactotrophe cell. Central nervous pathways which convey prolactin-release stimuli appear to terminate in the anterior hypothalamic-preoptic area. Future work will involve investigation of (1) the precise role of diencephalic serotoninergic and cholinergic pathways in the release mechanism; (2) whether a chain of neurones linking anterior hypothalamic-preoptic area with median eminence is "switched off" by incoming stimuli for major releases of prolactin; (3) the integration centre, possibly in the preoptic area, which "decides" whether or not prolactin will be released and (4) the role of higher nervous centres in arriving at such a decision.

CHAPTER 10

Control of Growth Hormone Secretion

A. PECILE AND V. R. OLGIATI

I. Introduction

The study of the regulatory physiology and neuroendocrinology of growth hormone (GH) secretion has been extensively advanced in recent years. Comprehensive reviews of GH regulation (Pecile and Müller, 1966; Reichlin, 1966b; Müller, 1973; Reichlin, 1973b; Martin, 1973) are available and in this Chapter some of the most interesting and stimulating topics will be discussed referring for more exhaustive information to these reviews. Only a few of the original references are cited here.

The difficulties of elucidating the problems of the control of GH secretion were, in the "pre-immunoassay" era, the necessity of an evaluation of pituitary GH control using bioassays and, in the "immunoassay" era, the uncertainty that the changes in GH activity as measured by bioassay that occur in a number of physiological conditions, including metabolic adjustments, are parallelled to changes in immunoreactive GH. The absence of a single target gland for GH action is a further drawback in the research into factors regulating its secretion.

II. Evidence for Hypothalamic Control of Growth Hormone Secretion

A. Pituitary Stalk Section

Surgical transection of the pituitary stalk (see Chapter 1) induces metabolic disturbances and growth retardation indicating lack of GH which will be temporary if regeneration of the portal vessels is not prevented (e.g. by insertion of a plate).

B. Pituitary Transplants

Transplantation to sites remote from the sella turcica are not associated with a maintenance of growth despite good "takes" and rich vascularisation of the transplanted tissue even in animals with more than one pituitary transplanted. However, growth is retarded when the pituitary is removed and then placed back in the region of the median eminence and pituitary stalk. When pituitary tissue is transplanted into the evacuated sella turcica the body weight gain of the recipient animals is similar when the donated pituitary tissue is obtained from old animals, compared with young animals, which suggests that, if there is any difference in activity between pituitary glands from young or old animals in the "basal" secretion of GH, this difference is imposed from the median eminence.

C. PITUITARY EXPLANTS

Pituitary explants incubated *in vitro* also indicate that GH secretion by pituitary tissue is modifiable. The release of GH into the medium may be increased by addition of an acid extract of hypothalamic tissue.

D. BRAIN LESIONS

Massive lesions in the hypothalamus of the rat, particularly those involving the anterior hypothalamus, impair growth: this is observed also in experiments in which interference in the secretion of hormones, other than GH, has been avoided and pair fed controls have been used to exclude the effect of differences in food intake by the animals with lesions. Normal growth rates could be restored only when GH was administered.

Electrical lesions in the rat confined to the ventromedial nucleus (VMN) of the hypothalamus and sparing the median eminence (ME) cause a reduction of both pituitary and plasma GH. With increasing lesion size in and around the VMN, a progressive decrease in pituitary weight, pituitary and plasma GH and growth is found.

A hypothalamic control over GH secretion is suggested also (in mice) by the bilateral destruction of the VMN by parenterally administered goldthioglucose, a toxic glucose analogue, which is trapped by discrete hypothalamic areas.

Facilitatory influences on GH secretion exerted by the central nervous system have also been suggested by the technique of brain lesions in primates. Small lesions in either the anterior or posterior median eminence, or in the region of mammillary bodies in the squirrel monkey, block the GH response to the emotional stimulus of capture.

Inhibitory influences on GH secretion are also reported. Recent findings indicate that in the rat surgical isolation of the medial basal hypothalamus causes increased linear growth associated with a marked elevation in plasma GH.

No firm conclusions can be derived from studies of pathological conditions in humans. Situations considered similar to those induced experimentally by hypothalamic lesions in laboratory animals (suprasellar craniopharyngioma or other tumours) may be so divergent and, of necessity, so poorly controlled, that the observations may lead to misleading interpretations.

E. BRAIN STIMULATION

In addition to the evidence for the nervous control of GH secretion derived from experiments involving destructive hypothalamic lesions, the results of brain stimulation also deserve a mention. Electrical stimulation within the hypothalamus (paraventricular nuclei of the weanling kitten, ventromedial or median eminence in the rat), or in extrahypothalamic sites (dorsal hippocampus, basolateral amygdala and mesencephalic interpeduncular nucleus) induces acceleration of growth (as measured by body weight and tibia length in the weanling kitten) and rises in plasma GH. Stimulation of the corticomedial amygdala, however, inhibits GH secretion in rats.

These findings suggest that both excitatory and inhibitory inputs to the hypothalamus from several regions of the limbic system may act to modulate GH secretion. In the rhesus monkey the state of cerebral glucoprivation obtained by central administration of a glucose analogue, 2-deoxy-d-glucose, an agent that interferes with cellular glucose metabolism, indicates the existence within the lateral hypothalamus of chemo-receptors sensitive to a lack of metabolisable glucose.

F. CONCLUSIONS

Considerable information has accumulated from these different experimental approaches concerning the central control of GH secretion. The information is sometimes equivocal, because many of the approaches have clear limitations. No convincing evidence is available, for instance, on the precise site(s) in the brain concerned with GH control, because lesions or stimulations, may have been placed in one of the steps along the neural pathway that originate from cells beyond the nucleus. However, the conclusion may be drawn that neural mechanisms implicated in GH secretion control and dysfunction at the neural level, may result in the pathophysiologic alteration of GH secretion in many clinical disorders.

III. Growth Hormone Releasing and Growth Hormone Inhibiting Factors

A. GROWTH HORMONE-RELEASING HORMONE (GHRH)

Evidence that the control of GH secretion depends on the hypothalamic and extrahypothalamic brain structures has stimulated the search for the hypothalamic factors concerned with this regulation, either by stimulation or inhibition of release.

During the so-called "pre-immunoassay" era (Reichlin, 1973b) bioassay techniques were used to measure the release of GH *in vitro* from cultured rat adenohypophyses (halved or quartered), or *in vivo* from the adenohypophyses of variously treated rats. Release of GH was demonstrated in rat pituitaries incubated with extracts of porcine or bovine hypothalamus (usually SME). Evidence for the presence of GH-releasing activity in the hypothalamic extracts from mammals (rat, mouse, cat, dog, sheep, cow, monkey and man), birds, and amphibia was obtained *in vivo* by several authors, who measured the depletion of pituitary GH after administration of hypothalamic extracts presumed to contain the growth hormone releasing factor (GRF). Increases of GH-like activity in plasma, at the same time as the pituitary depletion and extrusion of GH-containing granules from adenohypophysial somatotrophs, were also observed. The bioassay methods for GHRH were used for the purification and isolation of the hypothalamic substance which would release GH and a decapeptide was isolated. However, this peptide did not release radioimmunoassayable GH (RIA-GH) in the rat, sheep, pig, monkey and man. This unexpected result questions the possibility that the decapeptide represents the hormone of physiological importance in GH regulation (see Chapter 5).

In addition to this discrepancy between bioassayable and RIA-GH data there has also been a lack of correlation, in the rat, between the reduced pituitary GH as determined by bioassay in several conditions, (cold exposure, insulin hypoglycaemia, prolonged fasting), and the unchanged or even reduced RIA-GH in plasma. This problem remains unresolved.

Release of bioassayable GH from the rat pituitary induced by the above mentioned conditions has been confirmed in primates where plasma RIA-GH increases. However, it has been reported that the plasma GH, as measured by radioimmunoassay, may measure only a small fraction of the total growth potency which is detectable by the tibial assay in hypophysectomised rat; the two assays apparently yield values which are inversely related (Grindeland *et al.*, 1974).

As for rat RIA-GH, the report of Wilber *et al.* (1971) of a releasing material from pig hypothalami that is able to stimulate the release of RIA-GH from rat pituitaries *in vitro* is of considerable interest.

B. GROWTH HORMONE-RELEASE INHIBITING HORMONE (SOMATOSTATIN)

Although there is good experimental evidence for the existence of a GH-releasing hormone, such a substance still awaits characterisation. While searching for it in hypothalamic extracts it has been noticed that some side fractions of these inhibited the secretion of RIA-GH. Following this observation, Brazeau *et al.* (1973) isolated the major active principle, named it somatotropin-release-inhibiting factor or somatostatin and characterised it as the cyclic tetradecapeptide H-Ala-Gly-Cys-Lys-Asn-Phe-Phe-Trp-Lys-Thr-Phe-Thr-Ser-Cys-OH and synthetised it (see Chapter 5). The synthesis of the peptide has also been described by Schally and his colleagues (Coy *et al.*, 1973). Somatostatin either in linear or cyclised form was shown to inhibit the resting or stimulated secretion of RIA-GH in rats, and to elevate

RIA-GH secretion *in vitro* by fragments of the pituitary from a patient with acromegaly obtained at time of hypophysectomy.

Synthetic somatostatin has been repeatedly shown to reduce or abolish the secretion of GH (dog, sheep, rat, cat and baboon), induced by sleep, exercise, electrical stimulation of the amygdala or ventromedial hypothalamus, insulin-induced hypoglycaemia or administration of L-dopa, arginine, morphine, pentobarbitone and adrenaline injected in the ventromedial nucleus (Brazeau and Guillemin, 1974). The biological half-life of somatostatin is short (about 4 min.) and it must be given by i.v. infusion, to demonstrate a measurable effect. Slow absorption preparations of somatostatin can be made by complexing the linear peptide with protamine zinc or mixing it with gelatine or arachis oil (Brazeau and Guillemin, 1974).

Studies on the mechanisms of action of somatostatin show it to act in inhibiting the secretion of GH directly at the level of adenohypophysial somatotroph cells beyond the adenylcyclase mechanism (Brazeau and Guillemin, 1974; Martin, 1974b). Its action is also unaffected by inhibitors of RNA-directed DNA synthesis or of protein synthesis. Somatostatin administration for limited periods of time inhibits the release but not the synthesis and storage of GH.

The actions of somatostatin have been studied in normal human subjects as well as in acromegalics. Circulating GH levels fall sharply in both healthy and acromegalic subjects, in a dose-dependent manner. After discontinuing the intravenous administration, it has been found that there is an augmented rise in GH in acromegalic patients. Such a rebound phenomenon may reflect the accumulation of a secretory stimulant (e.g. GHRH) during the period of somatostatin administration, which is then free to act when the somatostatin is removed (Carlson *et al.*, 1974).

Somatostatin also reduces TSH secretion and the TSH response to TRH. It does not affect the TRH induced rise in prolactin, but has been reported to lower the prolactin concentration in adenohypophysial cells when its secretion has been increased by hypothyroidism or by chronic administration of oestrogens; it has no effect on gonadotrophin secretion.

It is of considerable interest that somatostatin inhibits the release of three separate hormones controlling carbohydrate metabolism: it reduces the secretion of insulin and glucagon (the only substance known to inhibit both) by direct action on the pancreas, and also their response to glucose, tolbutamide and arginine administration. Whether this action has any physiological importance is uncertain, since relatively large doses of somatostatin are required to produce this effect *in vivo*. The demonstration by immunofluorescence technique that somatostatin may be present not only in the central nervous system (namely in nerve terminals of the (rat) hypothalamus see Fig. 3 in Chapter 2), but also in several peripheral tissues (D cells of the pancreas and other areas of the gut) suggests a possible regulatory role of the peptide at different levels of the organism, in inhibitory action on growth hormone secretion (Polak *et a* *et al.*, 1975a). It is interesting that somatostatin has been r the secretion of gastrin in response to feeding (Bloom *et al* acid (Barros D'Sa *et al.*, 1975) and of pepsin (Gomez-Pan *et al.*,

Although at present its action seems too diverse, somatostatin and its analogues may prove to be of therapeutic value. It seems probable that the therapeutic use of somatostatin will involve its long-term administration to patients. In view of these treatments over extended periods, data on toxicity are extremely important. Recently, caution has been suggested by reports of thrombocytopoenia and impaired platelet aggregation in the baboon and man after infusions of somatostatin (Koerker *et al.*, 1975; Besser *et al.*, 1975). This preliminary evidence has been questioned by Mielke *et al.* (1975) who present data indicating that when conventional concentrations are used, somatostatin does not affect coagulation and platelet function in man, and that its prolonged administration lacks ostensible toxicity.

IV. Influence of Metabolic Factors and Various Conditions (Exercise, Sleep, Stresses) on Growth Hormone Secretion Regulatory Control

A. BLOOD GLUCOSE CONCENTRATIONS

Wide variations in the rate of GH secretion have been demonstrated. An important metabolic signal inducing GH secretion in primates is hypoglycaemia and in man it has been shown that a falling plasma glucose level, though it remains within normal limits, will release GH. However, a rising glucose level or sustained hyperglycaemia over several hours will inhibit spontaneous GH secretion and some, but not all, types of induced GH discharges.

There are observations which point to the possibility that within the central nervous system, systemic hypoglycaemia or glucoprivation (e.g. that experimentally induced by a glucose derivative which inhibits the utilization of glucose in the tissues) are monitored by sensitive neurones which act as receptors for the regulation of GH secretion by the brain. The main glucoreceptor areas identified thus far, correspond to the ventromedial nucleus area and the lateral hypothalamus. Minor variations in plasma glucose that occur under normal basal conditions or in fasting, seem not to be an effective homeostatic regulator of GH secretion, and the release of GH in hypoglycaemia appears to be an "emergency" measure in the adult, representing an important defence mechanism against hypoglycaemia. Starvation in man consistently induces a rise of GH and this response of GH secretion, although probably not essential within narrow limits of energy demand, may become crucial in more severe situations (Merimee and Fineberg, 1974).

Foetal GH is altered by glucose administration in a similar way to the adult, but during early postnatal life the rate of GH secretion responds "paradoxically" to glucose loading (Cornblath *et al.*, 1965).

In non-primates the influence of glucose on GH secretion remains uncertain and confusing. There is, in the rat, for example, a fall in bioassayable GH in the pituitary as a consequence of insulin hypoglycaemia, but contradictorily, and in both the rat and the rabbit, hypoglycaemia induces a fall in plasma RIA-GH. Measurements made by radioimmunoassay, in the rabbit, also showed that insulin-induced hypoglycaemia after fasting caused a decline in plasma GH (McIntyre and Odell, 1974). The responses in the pig, as in the obese human, to altered glucose levels are sluggish. GH secretion in the mouse is apparently relatively independent of glucose concentrations. The dog behaves like the human. In severe metabolic situations which correspond to starvation, changes of GH secretion in animals other than man are highly variable, some showing a rise, others a fall and still others no change.

B. PROTEINS AND AMINOACIDS

Both the ingestion of a high protein meal and the oral or intravenous administration of relatively large amounts of aminoacids, stimulate GH secretion. Arginine is one of the most potent aminoacids causing GH release. Aminoacids also stimulate the release of insulin which precedes the release of GH, but the release of the two hormones is not causally related, since GH secretion occurs in response to arginine in juvenile diabetics in whom the secretion of insulin is low. A direct effect of aminoacids is also shown by experiments in the rat (Pecile *et al.*, 1972), in which injections of minute amounts of aminoacids (arginine, lysine) into the lateral ventricles of the brain elicited GH release without any concomitant decrease of blood glucose level.

The release of GH after a high protein meal is lower than that following the intravenous injection of a single aminoacid like arginine. Arginine brings about a change in the blood aminoacid pattern and not a balanced increase of aminoacidaemia; thus the secretion of GH induced by aminoacids may be more a mechanism for protecting against precipitous protein degradation, (with abnormal blood aminoacid patterns) than a mechanism for stimulating protein synthesis when precursors are available. The protection could derive from the provision of free fatty acids for the energy demands, through the fat-mobilizing effect of GH. This suggestion could help in the interpretation of the apparently paradoxical sustained high plasma GH levels in severe protein malnutrition.

C. PLASMA FREE FATTY ACIDS

In contrast to earlier studies, recent experiments in man have provided substantial evidence that the plasma free fatty acids (FFA) concentration and GH release are inter-related. Decrease of basal FFA concentrations induced by nicotinic acid administration provokes GH release. By contrast, elevated FFA concentrations completely suppress the stimulation of GH release by protein ingestion and inhibit the maximal GH released by arginine (Fineberg *et al.*, 1972). It has been also reported that infusion of fatty acids inhibits nocturnal GH secretion (Blackard *et al.*, 1971).

D. EFFECT OF DIET

The accumulated observations on the influence of metabolic signals like glucose, aminoacids and FFA on GH secretion have promoted research to determine the effect of diets on the secretion of GH in man. Merimee and Fineberg (1973) have reported that relatively short term changes in diet composition can alter GH secretion in normal subjects. A high carbohydrate diet appears to suppress GH secretion. This effect may be less after decreasing the protein content of the diet, but more detailed studies will be required to clarify this point.

E. EFFECT OF EXERCISE

Closely connected with the influence of these metabolic variations on GH release, is the GH secretion associated with exercise. In normal subjects it has been shown that exercise of sufficient intensity will invariably result in a release of GH after a latent period (Buckler, 1972): this has been used as the basis of a stimulatory test for GH release. Schalch and Reichlin (1968) for instance, have recommended vigorous stair climbing as a test of GH secretory reserve.

The importance of initiating and sustaining lipolysis in adipose tissue during exercise in order to spare the limited supplies of glucose, seems a reasonable interpretation of the promotion of GH secretion during exercise.

F. EFFECT OF SLEEP

In several species including man, unexplained bursts of GH secretion have been documented. These bursts occur in subjects maintained at rest and are not associated with any measurable change in the plasma metabolites (glucose, amino acids or FFA) and are not suppressed by hyperglycaemia. It appears likely, therefore, that such GH secretory bursts represent pulsatile hormone secretion which may be the result of primary neural activation (Martin, 1973). A particularly prominent burst of GH secretion occurs during deep sleep, 45 to 90 minutes after sleep onset, which reaches the peak before the first cycle of rapid-eye-movement (REM) sleep.

That this release is triggered by the condition of sleep itself and is not simply related to a free running circadian rhythm of GH secretion, is indicated by the fact that a delay in sleep until a later period of the night results in a delay in GH release. Moreover, if the subject is awakened after the first GH burst and then permitted to return to slow wave sleep (SWS), a second GH burst ensues. However, SWS deprivation prevents the appearance of GH release (Martin, 1973).

The mechanism and the significance of such GH release has not been elucidated. The possibility that SWS-related GH release is the result of non-specific cortical depression and consequent suppression of an inhibitory mechanism for GH control, has been considered. The possibility that activation of certain limbic brain areas can trigger SWS-induced GH release, may be related to the finding that during the induction of pentobarbital anaesthesia activation of several subcortical brain structures, including the ventromedial hypothalamic nucleus, the hippocampus and the amygdala, takes place. The anatomical areas related to the limbic system presumably provide the neural substrate of emotional and sleep-induced elevation in GH secretion.

G. EFFECT OF STRESS

The GH secretion which is related to metabolic homeostasis is unexpectedly responsive to even minor changes in the environment. A wide variety of stressful stimuli cause GH discharge in both man and monkey, in a manner reminiscent of the changes in ACTH secretion under similar circumstances (see Chapter 11). Surgery, acute trauma and arterial puncture are among the physical stimuli which raise plasma GH in man and the monkey. In the rhesus monkey it has been demonstrated that capture from the cage, pinching of the abdomen, a loud noise and confrontation with unfamiliar persons, all produce rapid elevations in plasma GH. Surprisingly, in the rat, plasma RIA-GH falls in response to stress and the pituitary RIA-GH content remains unchanged. But it may be noteworthy to add that, in contrast to these results, the pituitary GH content of the rat determined by bioassay (BA) decreases in various stress situations (Pecile *et al.*, 1972). Such findings would indicate that after secretion *in vivo* the rat's GH would rapidly lose its detectability by the RIA. The existence of electrophoretically separable forms of rat GH with disparate BA : RIA activity ratios and the RIA lability of GH, can explain the discrepancies between BA and RIA estimates of the hormone (Vodian and Nicoll, 1975).

Psychological events inducing apprehension or anxiety may alter GH release in man as they do ACTH release, the adrenal response occurring in a wide variety of psychiatric conditions (Brown and Reichlin, 1972; Chapter 11). The child exposed to a psychologically inadequate environment may develop prolonged, though potentially reversible, inhibition of GH secretion and GH secretory responses. Related to the above is the finding that rats handled for two weeks have a higher plasma GH than control rats, supporting the theory that handling promotes growth by increasing GH secretion.

The GH response to stressful stimuli seems to be controlled independently of the metabolic responses. Whereas the GH rise in response to exercise or to arginine is inhibited by glucose, hyperglycaemia does not prevent the GH elevation associated with apprehension or the stress of surgery. The hypothalamus appears to mediate this GH response to stressful stimuli (Pecile *et al.*, 1972). The mechanism for release of GH and of ACTH are independent, since a dissociation of the two responses is found after a variety of stimuli in man and the rhesus monkey (Brown and Reichlin, 1972). A possible teleologic significance of GH elevation after stress is that it may produce metabolic effects related to caloric requirements and to exert protective effects on tissues (Brown and Reichlin, 1972).

V. Effects of Growth Hormone and Other Hormones on Growth Hormone Secretion

A. Short-Loop Negative Feedback

It has been suggested that "short-loop" negative feedback mechanism (Chapter 3) exists for GH as for other pituitary trophic hormones. Bioassayable GH content of rat pituitary for instance, has been reported to be altered by either a single dose treatment (Müller and Pecile, 1966) or a repeated administration of the hormone.

In rats in which either pellets containing GH were implanted into the hypothalamus or GH-secreting tumours were implanted, pituitary GH content was reduced. Insulin hypoglycaemia has been shown, in animals pretreated with GH, to result in a smaller decrease in pituitary GH content. Exogenous administration of GH in both man and rhesus monkey reduces the GH responses to the provocative stimuli of exercise, arginine administration or insulin-induced hypoglycaemia (Hagen *et al.*, 1972a.) The hypothalamus is the anatomical site of this GH autoregulation; this is indicated by the decrease in hypothalamic GH-releasing activity and pituitary GH seen after implantation of GH in the rat ME, and by the appearance of GH-releasing activity in the peripheral plasma of long-term hypophysectomised rats. However, a direct feedback at the pituitary level cannot be excluded (Müller, 1973). Whether the GH itself is active in the negative feedback mechanism, or a component or metabolite or the GH-dependent substance *somatomedin*, is at present unknown.

B. Steroid Hormones

Secretion of GH is influenced by several steroid hormones. Plasma GH is higher after the usual provocative stimuli in women than in men.

Oestrogens are responsible for this sex difference and relatively brief administration of oestrogens will increase the GH response induced in men by arginine administration. While oestrogens are undoubtedly more potent, androgens also may influence GH release. Thus in prepubertal children with growth retardation, androgens may increase a blunted GH response to hypoglycaemia and in children who have normal hypoglycaemic responses, may elevate the fasting GH level (Deller *et al.*, 1970). Insulin or arginine-induced GH release in the adult man may be suppressed or blocked by progestins, but there are no effects on nocturnal GH discharge.

Corticosteroid excess occurring in the spontaneous hyperadrenocorticoidism or Cushing's syndrome, or after high-dose frequent corticosteroid administration, impairs the plasma GH response to hypoglycaemia and inhibits the sleep-related peak of secretion (Merimee and Rabin, 1973; Krieger and Glick, 1974).

C. THYROID HORMONES

Excess thyroxine suppresses both basal GH secretion and the GH response to hypoglycaemia, whereas thyroid deficiency only inhibits the latter (Young, 1973). TRH stimulates GH secretion from the perfused rat pituitary; this suggests that TRH and hypothalamic GH-releasing substances may share certain common structural features. The adenohypophysial somatotroph may thus be able to respond to either hormone although with varying affinity (Carlson *et al.*, 1974).

D. GLUCAGON

Glucagon has repeatedly been found to stimulate GH release. Some maintain that this is a direct action of glucagon on the hypothalamus or pituitary, because blood glucose enhancement by glucagon should block the rise in GH; but Podolsky and Sivaprasad (1972) suggest that a fall in raised blood glucose level will cause secretion of GH, even though the level it sinks to is not low enough itself to produce hypoglycaemic symptoms. They favour the view that release of GH is indeed related to the fall in blood glucose, for, when the blood glucose changes after glucagon were reduced by fasting for 48 hours, there was no GH response to glucagon, though there was still a GH response to arginine infusion.

VI. Role of Brain Monoamines in the Neurohumoral Control of Growth Hormone Secretion

Pharmacological studies have contributed to show that dopamine (DA) and noradrenaline (NA), and 5-hydroxytryptamine (5-HT) might participate in the chain of events which control the secretion of GH. Attempts have been made to alter neurotransmitter function by administration of the monoamines into the lateral ventricles of the brain; precursors of monoamines;

inhibitors of enzymes involved in monoamines biosynthesis; drugs which deplete monoamine storage or block their receptors, affect the mechanism of their re-uptake or influence their turnover rate, and monoamine agonists (see Chapter 3).

Studies in the rat and man have suggested a central adrenergic mechanism for GH regulation that has both stimulatory and inhibitory effects. Blockade of α-adrenergic receptors by phentolamine, prevented the increase in plasma GH that followed insulin-induced hypoglycaemia in normal subjects. In contrast, blockade of β-adrenergic receptors by propranolol augmented the levels of GH in hypoglycaemic subjects (Blackard and Heidingsfelder, 1968).

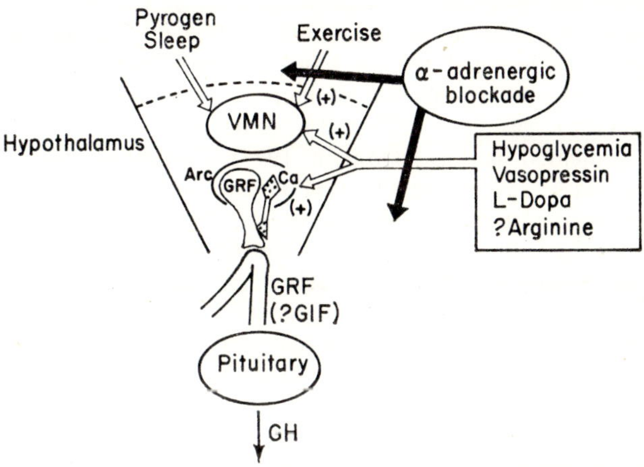

Fig. 1. Diagrammatic representation of adrenergic control mechanisms for GH secretion. VMN, ventromedial nucleus; Arc, arcuate nucleus; CA, catecholamines; GRF, GH-releasing factor; GIF, GH-release inhibiting factor (= somatostatin). (From: Martin, 1973.)

A lowering of bioassayable pituitary GH and of GH-releasing activity in the hypothalamus was reported after the injection of NA and DA into the lateral ventricle of the rat brain. Consistent rises in serum GH were observed after NA microinjection into the ventromedial nucleus of the hypothalamus of the baboon. The administration in man of L-dopa, the immediate precursor of DA, a drug that crosses the blood-brain barrier to increase brain levels of DA and NA, caused a significant rise in plasma GH levels (Boyd *et al.*, 1970). This stimulatory effect on GH secretion by L-dopa was not blocked by either oral or intravenous glucose, but was considerably reduced by concomitant infusion of the α-adrenergic receptor blocker, phentolamine (Kansal *et al.*, 1971). The effect of L-dopa on GH secretion has been proposed as a stimulation test for the diagnosis of GH deficiency (Gupta *et al.*, 1972). L-dopa has the advantage of few, if any, side effects, hence this simple provocative test procedure is becoming routinely utilised as the test of choice for GH release (Eddy *et al.*, 1974). Chlorpromazine blocks GH release possibly by a competitive inhibitory blockade of dopaminergic receptor sites in the hypothalamus. Furthermore, apomorphine, a drug which stimulates dopaminergic receptors, causes GH releases (Lal *et al.*, 1972; Brown *et al.*, 1973).

In a diagrammatic representation (Fig. 1) Martin (1973) has constructed a hypothetical scheme of the hypothalamic sites of action of several agents, including the adrenergic control mechanisms known to affect GH secretion. Stimuli such as exercise, hypoglycaemia, L-dopa, all probably release GH through an adrenergic mediation. Indeed, intravenous injection of phentolamine prevents the effect of all the above stimuli on GH release. The effects of α-adrenergic blockade on arginine-induced GH release are not clear and those on glucagon-induced release have not been reported, but the GH response to glucagon appears to be increased by β-adrenergic blockade. Sleep-related GH release is not affected by either α-adrenergic or β-adrenergic blockade as indicated in Fig. 1 (Martin, 1973). Catecholamines are involved in the neural pathways leading to GH release induced by stimulation of extrahypothalamic structures (Martin *et al.*, 1973). Depletion of central catecholamines by administration of α-methyl-p-tyrosine, an inhibitor of catecholamine synthesis, prevents the GH release that occurs after stimulation of the hippocampus or basolateral amygdala.

Further indication that monoamines may participate in the release of GH comes from evidence that 5-HT pathways may have an even larger direct role in GH secretory mechanisms than catecholaminergic pathways. GH secretion is stimulated in man by relatively small doses of the 5-HT precursor, 5-hydroxy-tryptophan (Imura *et al.*, 1973) and intraventricular administration of 5-HT causes GH release in the rat. The intraperitoneal administration of 5-hydroxy-tryptophan which stimulates GH secretion in the rat, is not effective after 5-HT blockade nor after melatonin, which being an *o*-methylated derivative of 5-HT could act as 5-HT receptor-site competitive antagonist in the hypothalamus. Cyproheptadine, methysergide or melatonin-induced 5-HT antagonism in man produces inhibitory effects on GH secretion similar to that observed in the rat, and provides further evidence for 5-HT and possibly pineal involvement in the control of GH secretion in man (Smythe and Lazarus, 1974). The reported 5-HT pathways for the control of GH secretion suggest that the GH release, observed in humans after the onset of slow-wave sleep, originates via 5-HT pathways.

VII. Effect of Drugs on Growth Hormone Secretion

The possibility that the release of GH may be altered by drugs affecting the hypothalamus has been envisaged in the preceding section concerning brain monamines. Pharmacological agents have been studied in addition to the drugs that clearly influence the α and β-adrenergic, dopaminergic or 5-HT systems involved in the hypothalamus or its connections.

A. PROSTAGLANDINS (PGS) AND CYCLIC NUCLEOTIDES

PG E_1 has been found to act directly upon the pituitary to elevate cyclic AMP levels by activation of adenyl cyclase; it does this independently of adrenergic receptors. Intravenous infusion of dibutyryl cyclic AMP or PG E_1 has been found to significantly increase plasma GH levels (Kato *et al.*, 1973) in man. Cyclic AMP, may also be equally or more effective in this respect. Taken together, the results are compatible with the concept that cyclic nucleotides are target cell mediators in the pituitary of the hypothalamic control of GH secretion.

B. FENFLURAMINE

Fenfluramine, a derivative of amphetamine, is used in the management of obesity. Its mechanism of action is uncertain (it has been attributed to either a direct metabolic effect on fat and carbohydrate metabolism or a depressant action on the centres in the central nervous system regulating appetite and food intake), but was found to abolish the rise in plasma GH following exercise. Fenfluramine decreases the concentration of cerebral amines particularly in the hypothalamus, and it may, therefore, interfere with GH release by affecting the adrenergic activity of the hypothalamus.

C. BROMOCRIPTINE

Bromocriptine has been studied as a new approach to the management of acromegaly (Liuzzi *et al.*, 1974; Chiodini *et al.*, 1975; Thorner *et al.*, 1975b). A single oral dose (2.5 mg) of the drug produced a marked fall in plasma GH levels in acromegalic patients, which is apparently sufficient to reduce greatly the damaging metabolic effects of excessive exposure to the tissues of GH. The therapeutic effect is maintained during continuous therapy, thus bromocriptine promises to be a safe, effective method of controlling GH levels in acromegaly.

The mechanism of action of bromocriptine in inhibiting GH release, suggests that it is a dopamine agonist, this may explain its pharmacological effects in the acromegalic patient.

D. INDOMETHACIN

Unsuspected drug influences on growth hormone release have also been documented. Indomethacin, a commonly used anti-rheumatic agent, was shown to increase serum GH levels as early as 30 minutes after the usual therapeutic dose (100 mg). An effect of indomethacine at the hypothalamo-hypophysial level seems possible, since another action of anti-inflammatory agents, their antipyretic effect, appears to follow an action in the hypothalamus (Syvälahti, 1974).

VIII. Variations of Growth Hormone Secretion During Development and at Different Ages During Life Span

Immunoreactive GH has been detected in the human pituitary gland as early as 68 days of gestation. The pituitary concentration and total content rise progressively from 10 to 14 weeks of gestation (pituitary GH content

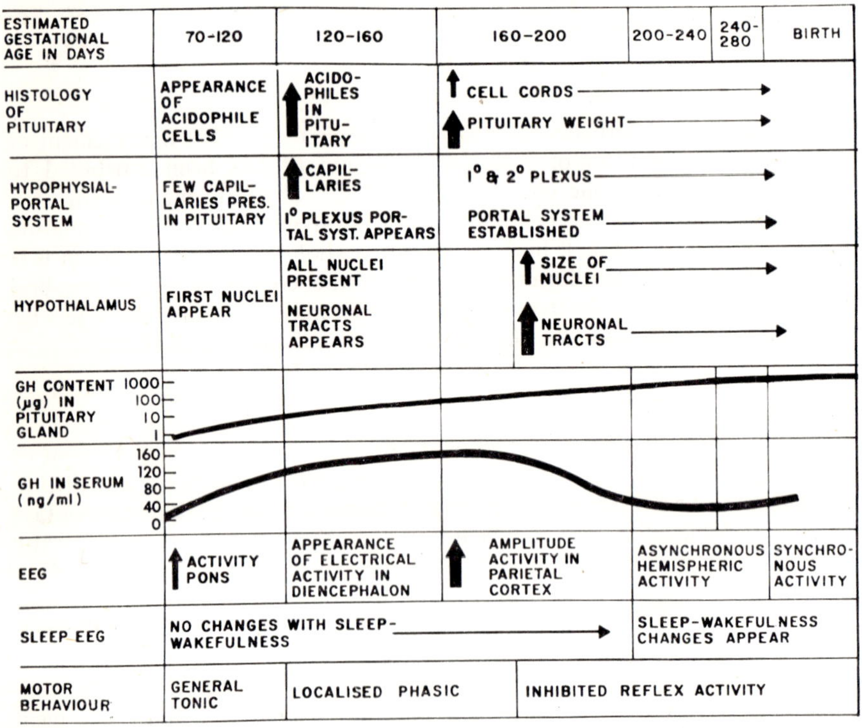

Fig. 2. Variations in growth hormone and related phenomena during foetal life (from Grumbach and Kaplan, 1972).

0·44 μg) to 30 to 40 weeks (pituitary GH content 577 to 675 μg). Pituitary glands of 1- to 12-month-old infants were similar in GH content to 25- to 29-week foetuses, and in concentration to 30- to 40-week foetuses. The concentration of GH in serum increases during gestational age and levels similar to those of adult acromegalic patients were found in all but a few of the foetuses studied from 70th day to term (Grumbach and Kaplan, 1973). After a rise between 70 and 100 days' gestation, there are strikingly elevated values between 100 and 200 days and then a decrease in the last few weeks. Grumbach and Kaplan show, in a scheme (Fig. 2), the interrelationship between the pattern of change in the serum GH concentration and the differentiation and development of the pituitary gland, of the hypothalamic monoaminergic network, of the hypophysial portal system and of the neurophysiological maturation of the central nervous system. The sequence of development of the neuroendocrine regulation of GH secretion by the adenohypophysis is interpreted by them in the following manner:

(1) The capacity of the foetal pituitary gland to synthetise and secrete GH occurs autonomously or is, at least in part, regulated by a hypothalamic growth hormone releasing factor (GRF).

(2) By mid-gestation a relatively unrestrained tonic release of GRF results in intense stimulation of GH secretion by the pituitary.

(3) In late gestation, especially during the last 10 weeks, inhibitory neural influences become operative.

(4) Further maturation of regulatory mechanisms which affect GH secretion occurs during the first few post-natal weeks. The acquisition of sleep-mediated growth hormone release, which may be used as an index of the attainment of a fully functional mechanism, occurs in infants over 3 months of age.

It is considered likely that the maturation of the hypothalamic neurotransmitter mechanisms, which affect GRF and somatostatin release post-natally, have a significant part in the scheme of foetal GH secretion outlined by Grumbach and Kaplan.

The high plasma GH levels in foetuses and newborn, probably reflect a generalised phenomenon, since observations similar to those reported for man have been made in mouse, rat, rabbit, dog, sheep, pig and cow. Furthermore, from the experiments on these other animals it appears almost certain that GH secretion in the foetus is controlled by the hypothalamus and the raised levels of plasma GH reflect a high degree of stimulation by its hypothalamic releasing factor. The marked fall in GH secretion rate after birth suggests that inhibitory mechanisms begin to exert their effect at this time (Rieutort, 1974; Wallace *et al.*, 1973).

While foetal GH is altered by glucose administration, as in the adult, during early post-natal life, GH levels respond "paradoxically" to glucose loading (there is a rise, rather than the expected fall in plasma GH levels after glucose administration). Within a few weeks after birth, plasma GH levels become lower and the sensitivity to glucose suppression, which probably represents a sign of hypothalamic maturity, develops. It is worth noting that the maturation of the regulatory mechanisms on GH secretion is influenced by thyroxine.

Thus thyroxine deficiency in early life is associated with a persistent alteration of the pituitary secretion of GH (Kikuyama *et al.*, 1974).

The function and the physiological significance of the high GH levels and the possible regulation of GH secretion during foetal development, remain unexplained. The absence of the pituitary in the human foetus or spontaneous or experimental apituitarism in vertebrates, is, however compatible with continued near-normal somatic growth up to the time of birth or hatching (Pecile and Müller, 1966). Anencephalic infants do not have elevated or reduced GH levels and are of normal size at birth, as are ateliotic dwarfs with isolated GH deficiency. These findings suggest that GH concentration probably does not control the rate of growth, although the presence of growth hormone is essential for normal growth after birth and excess GH results in increased growth.

In pubertal children the basal plasma GH concentration is not greatly different from that reported for adults, but more secretory episodes may occur during the day, associated with physical activity. GH continues to be secreted after the period of skeletal growth has been completed. No significant changes in pituitary GH content with age are evident. Octogenarians, for instance, have nearly as much GH as the rapidly growing child and GH secretory responses to provocative stimuli are unimpaired, even in elderly individuals.

Tissue responsiveness to GH and to other hormones seems to be more important in determining the pattern of childhood growth than variations in GH secretion. However, subnormal GH secretory patterns are related to deficient growth rates. The few instances in which apparently normal growth occurred without evidence of GH secretion, are at present unexplained.

IX. Abnormalities of Growth Hormone Secretion

Hypo-or hypersecretion of GH may occur in children and in the adult, and may be due to abnormalities either of basal secretion or of phasic GH release.

A. HYPOSECRETION OF GH

Growth failure with dwarfism may occur as the consequence of primary pituitary disturbances, but in some cases, hypothalamic disorders are responsible for the GH deficiency. Hypothalamic tumours or lesions due to inflammatory conditions within the hypothalamus, may be followed by reduced or undetectable basal GH levels and failure of GH release after insulin-induced hypoglycaemia or arginine infusion. In patients with anatomical malformations or cerebral structures, GH deficiency has been recognised. Although awaiting the availability of purified or synthetic GH-releasing hormone before it can be confirmed, primary disturbances of the hypothalamic control in several forms of hyposomatotrophic dwarfism is indicated.

The interesting cases reported of reduced GH secretion due to emotional factors (maternal deprivation syndrome) and the demonstration by Imura *et al.* (1971) that normal GH secretion has been restored in a patient by treatment with a β-adrenergic blocker (propranolol), reinforces the concept of the nervous control of GH secretion and of the importance of adrenergic system in the regulatory mechanisms of GH release.

B. HYPERSECRETION OF GH

Gigantism and acromegaly are both pathological conditions due to hypersecretion of GH associated with adenomas of the anterior pituitary. Recently evidence has been accumulating, pointing to the possibility that in at least some cases, the primary disturbance is at the hypothalamic level. An increased or inappropriate secretion of GH-releasing hormone or perhaps more likely, a deficiency of somatostatin, could be responsible for excessive GH secretion. A material with GH releasing activity *in vitro* was shown to be present in the peripheral plasma of acromegalic patients (while it is constantly absent in normal subjects). An inefficiency in the short-loop feed-back mechanism through which GH plasma levels may reduce GH secretion, has been suggested. Some acromegalic patients exhibit little fluctuation of plasma GH throughout a 24-hour period; in some others, GH levels are strikingly variable with abrupt rises occurring at short intervals. The sleep-related peak of GH secretion is absent in almost all cases. The response to the usual stimuli for GH release may persist in acromegaly although it is highly variable. Further elevation of plasma GH concentrations can be provoked by hypoglycaemia and arginine infusion. Glucose administration may depress GH secretion but in some cases it may provoke a paradoxical rise in plasma GH. Thus, pituitary GH secretion appears to be not entirely autonomous in many, if not most patients with acromegaly, but disorders of regulation by metabolic stimuli, which require hypothalamic participation, undoubtedly exist. Some changes in GH secretion of acromegalic patients compared with normal subjects have been noticed, using the recently proposed provocation test with L-dopa. An acute suppression of plasma GH levels after L-dopa administration has been reported (Martin, 1973).

C. IN THYROID DISEASES

In hyperthyroidism a decrease of GH secretion has been observed which returns to normal when the patient becomes euthyroid. The reason for this decreased GH secretion in the untreated state, probably lies in a disturbance of brain-hypothalamic-pituitary function, which is either induced by excessive thyroid hormone secretion, or which may be an integral part of the disease (Finkelstein *et al.*, 1974).

In prolonged hypothyroidism plasma GH falls to very low or undetectable levels. GH secretory responses are suppressed. It appears that hypothyroidism directly affects the ability of the somatotroph to produce GH and this decreased storage and secretion of the hormone is not necessarily mediated via a lowered hypothalamic releasing factor activity (Peake *et al.*, 1973a). However, the precise mechanism of action of thyroxine deficiency in affecting GH secretory responses remains to be established.

D. IN NUTRITIONAL STATES

GH secretion in protein-calorie malnutrition has been studied. Basal plasma GH levels were elevated in kwashiorkor and returned to normal levels after treatment. The basal plasma GH in marasmic children was normal and there was no change after nutritional rehabilitation. The increase in plasma GH during sleep was inadequate in marasmic children on admission and this did not improve even after three to four weeks of nutritional rehabilitation. Thus disturbances in nervous control of GH secretion in protein-calorie malnutrition are suggested (Raghuramulu and Jaya Rao, 1974).

Obese subjects have decreased basal GH levels and decreased responsiveness to a variety of stimuli that cause GH release in normal subjects, such as hypoglycaemia, arginine infusion, exercise, oral glucose and sleep. A lack of responsiveness to L-dopa was also found in obese subjects, which would imply that the decreased sensitivity to other stimuli capable of provoking GH release, is not secondary to decreased levels of the hypothalamic neurotransmitter agents believed to be involved in GH regulation (Fingerhut and Krieger, 1974).

E. DIABETES

Plasma GH concentrations have been extensively studied in diabetic patients. Contradictory results have been reported, both in the juvenile-onset and maturity-onset types of the disease. Diurnal plasma GH concentration is undoubtedly increased in juvenile diabetes. In maturity-onset diabetes a possible hypersecretion, mainly in poorly controlled patients, may be obscured by the often co-existent obesity which blunts the growth hormone response. An excessive GH response to muscular exercise is displayed by both juvenile-onset and maturity-onset diabetics (Hansen, 1973a; Passa *et al.*, 1974) and this suggests the existence of a disturbance in the control of GH secretion.

The factor(s) responsible for the elevated GH levels of prediabetics are unknown, but an alteration of the mechanism(s) of nervous secretory regulation cannot be excluded.

A role of GH in diabetic ketoacidosis has been recently reported (Gerich *et al.*, 1975a, b). Whereas increased glucogenesis during insulin lack is mainly attributable to glucagon, lipolysis and ketogenesis may result from either glucagon or GH. Insulin lack is undoubtedly the prerequisite for development of diabetic ketoacidosis; nevertheless, the presence of glucagon or growth hormone is proved necessary in initiating, both at the periphery and in the liver, the enzymatic changes that lead to the metabolic abnormalities observed in ketoacidosis.

It is of interest that somatostatin infusion in ketoacidotic insulin-deprived patients, in addition to the control of hyperglycaemia, diminishes lipolysis and ketogenesis. Such effects are due to the suppression by the polypeptide both of glucagon and GH secretion. Although the actions of somatostatin are too widespread to allow for the use of the peptide in diabetes, new synthetic analogues may provide a long-acting preparation with selective suppressive effects on glucagon and GH which might be a big advance in the management of diabetes.

F. NERVOUS DISEASES

In diseases of the nervous system with neural lesions and also in schizophrenia, abnormal GH responses to the usual stimuli have been described and/or suppression of nocturnal GH peaks. This suggests that brain biogenic amine metabolism alterations are the main factors responsible for both sleep alteration and impairment of nocturnal GH release (Vigneri *et al.*, 1974). Elevation in plasma GH levels are found in the diencephalic syndrome of early infancy, a condition which is associated with gliomas localised to the anterior hypothalamus which may alter normal inhibitory mechanism(s) for GH control (Salmon, 1972).

X. Summary and Conclusions

The control of growth hormone secretion by the hypothalamus is clearly emerging from a number of experimental and clinical data.

Although the precise sites in the hypothalamus concerned with GH control are not yet defined, considerable information has been accumulated from different experimental approaches that hypothalamic neural mechanisms are implicated in the regulation of GH secretion.

Hypothalamic hypophysiotrophic hormones, synthesised by neurosecretory neurones and released from nerve endings in the median eminence into the hypophysial portal circulation, are among the instruments of the GH control. While GH-releasing hormone still awaits chemical characterisation, the GH-release inhibiting hormone (somatostatin) has been isolated and synthesised. In addition to its inhibitory effect on GH secretion, somatostatin has many interesting extra-hypophysiotrophic effects: it reduces the secretion of insulin and glucagon, prolactin, gastrin, gastric acid and pepsin and thus may prove to be of fundamental physiological importance also in the control of endocrine secretion in the pancreas and the gut, and perhaps in the pathogenesis of diabetes mellitus.

Metabolic factors have been demonstrated to influence GH secretion regulatory control. Falling plasma glucose levels stimulate GH secretion, rising glucose levels or sustained hyperglycaemia inhibit spontaneous GH secretion and some, but not all, types of induced GH discharges. An increase of aminoacid blood concentration provokes GH release. By contrast, elevated FFA concentrations may reduce or suppress GH release.

Exercise of sufficient intensity induces GH release. Bursts of GH secretion occur during deep sleep. They are not associated with any measureable change in the plasma metabolites (glucose, amino acids or FFA) and may represent a primary neural activation. The anatomical areas related to the limbic system presumably provide the neural substrate of sleep-induced elevation in GH secretion.

Unexpectedly, a wide variety of stressful stimuli cause GH discharge in both man and monkey in a manner reminiscent of the changes in ACTH secretion under similar circumstances. Psychological events inducing apprehension or anxiety may also alter GH release in man. The GH response to stressful stimuli which seems to involve the hypothalamus, appears to be controlled independently of the metabolic responses.

GH itself, or a component or metabolite, or the GH-dependent substance somatomedin, is active in the negative feed-back mechanism controlling GH secretion at the hypothalamic level and perhaps also at pituitary level.

Secretion of GH is influenced by excess or defect of various hormones such as oestrogens, progestins, androgens, corticosteroids, thyroid hormones and glucagon.

In the hypothalamic neurohumoral control of growth hormone secretion, brain monoamines such as dopamine, noradrenaline and 5-hydroxytryptamine have an important role. A central adrenergic mechanism for GH regulation has both stimulatory and inhibitory effects. Stimulation of dopaminergic receptors causes GH release. Catecholamines are involved in the neural pathways leading to GH release induced by stimulation of extrahypothalamic structures. Tryptaminergic pathways may have an even larger direct role in GH secretory mechanisms than catecholaminergic pathways.

Pharmacological agents that clearly influence the α and β-adrenergic, dopaminergic or 5-HT systems involved in the hypothalamus or its connections, may have an effect on growth hormone secretion.

Prostaglandins, cyclic nucleotides and drugs such as fenfluramine, indomethacin, bromocriptine have been found to alter GH secretion. It is worth noting that bromocriptine administration in acromegalic patients promises to be a safe, effective method, in controlling GH levels so that the damaging metabolic effects of excessive exposure to the tissues of GH are greatly reduced.

GH secretion during foetal and early post-natal development is different from that of the adult; and GH levels paradoxically rise during early post-natal life. These variations are interpreted as due to the maturation of the hypothalamic regulatory mechanisms which affect GH secretion control.

Hypo- or hypersecretion of GH may occur in children and in the adult and may be due to abnormalities either of basal secretion or of phasic GH release. Alterations of GH secretion have been reported in thyroid diseases, in nutritional states, in diabetes, in diseases of nervous system with neural lesions and also in psychiatric disorders. The possible involvement of the neuroendocrine control in the majority of the abnormalities of GH secretion is suggested on the basis of growing evidence. Pharmacological studies, both in laboratory animals and in man, will continue to be of valuable help in clarifying the multivarious aspects of GH regulatory control in health and disease.

CHAPTER 11

Control of Corticotrophin (ACTH) Secretion

M. T. JONES

I. Introduction

The dependence of corticotrophin (ACTH) secretion on hypothalamic control was demonstrated by Green and Harris (1949), and Harris and Johnson (1950). It was postulated that the release of specific neurohormones into the portal vessels controlled the release of ACTH from the adenohypophysis (Harris, 1955; Ganong, 1963). In this chapter, current knowledge on the nature and control of this neurohormonal mechanism will be reviewed.

II. Corticotrophe Cells; ACTH and Related Peptides; Assay of ACTH

A. Corticotrophe Cells

The cells that synthesise and secrete ACTH are known as corticotrophes; they form some 1% of the adenohypophysis, are chromophobic with an irregular outline with few granules which are greatly increased in number after adrenalectomy (Siperstein and Allison, 1965; Siperstein and Miller, 1973). Siperstein (1963) using tritiated glycine has shown that protein turnover in these "adrenalectomy cells" is increased. In the pars distalis of the rat, the corticotrophes are localised towards the periphery of the gland (Baker *et al.*, 1970). Whilst most studies have used the rat, the corticotrophes in man have been studied and their distribution described (Phifer *et al.*, 1970; Hachmeister *et al.*, 1971). ACTH has also been described in the pars intermedia of the rat (Gosbee *et al.*, 1970) but it is now clear that this corticotrophin-like intermediate peptide (CLIP) is the 18–39 amino acid sequence of ACTH and it has been proposed that ACTH is the precursor for both α-MSH and CLIP (Scott *et al.*, 1974a).

Pituitary corticotrophes also synthesise and release additional MSH-like peptides of which β-lipotrophin (LPH) is most important in man. It contains 91 amino acids of which β-MSH is part of the sequence. Secretion parallels that of ACTH (see also Chapter 12).

B. Adrenocorticotrophin (ACTH, Corticotrophin)

The existence of a pituitary adrenocorticotrophic hormone became evident when Smith (1927), showed that removal of the pituitary gland in the rat resulted in atrophy of adrenal cortex. This could be prevented by administration of alkaline extracts of the pituitary gland (Evans, 1933), or more purified protein preparations from pituitary tissue (Collip *et al.*, 1933). In the 1940s ACTH was further purified and the final determination of the 39 amino acid sequence of ACTH was achieved by Bell (1954) and synthesis of fragments was reported by Hofmann *et al.* (1961) and Li *et al.* (1962). The validity of the structural analysis of ACTH was shown by the complete synthesis of 1–39 ACTH by Schwyzer and Sieber (1963). The work of Bell *et al.* (1956) on peptic digestion of purified porcine ACTH, suggested that the terminal carboxyl portion of the peptide was not necessary for full biological activity; this resides in the first 18 amino acids of the molecule (Ramachandran *et al.*, 1965). The first 14 residues form the active site of the peptide and positions 15–18 increase the affinity for ACTH of the adrenal cells (Nakamura, 1972).

C. THE ASSAY OF ACTH

Givens and Ney (1968) have reviewed the various methods previously available for the bioassay of ACTH. Sayers and his associates have developed a technique for separation of adrenocortical cells and used them for the bioassay of ACTH (Sayers *et al.*, 1971); the method is highly sensitive but its specificity has to be ascertained. Radio-immunoassay (Berson and Yalow, 1968; Landon and Greenwood, 1968) has been used widely for the measurement of ACTH in man and the rat. The results obtained with radio-immunoassay and bioassay have been compared and there are some discrepancies (Besser *et al.*, 1971); it is likely that immunoreactive fragments of ACTH not active biologically, may be measured by immunoassay.

Of the newer techniques available, the cytochemical assay appears to be the most sensitive, specific and precise (Alaghband-Zadeh *et al.*, 1974); this method enables the measurement of normal resting values of ACTH in 100 μl of plasma in both man and the rat (Daly *et al.*, 1974a; Buckingham and Hodges, 1974b). The original method depended on the use of adrenal wedges which greatly limited the number of samples that could be processed, but the use of adrenal sections (Alaghband-Zadeh *et al.*, 1974) has overcome this handicap. The cytochemical assay has been compared with other assay systems (Rees *et al.*, 1973).

In this chapter the term "ACTH" is used when ACTH release has been measured directly and "corticotrophic response" when adrenal corticosteroids have been measured as a parameter of ACTH secretion.

III. CORTICOTROPHIN RELEASING FACTOR (CRF)

A. THE NATURE OF CRF

Evidence accumulated by G. W. Harris and others, over 25 years, provided convincing arguments for the neuro-humoral theory for the regulation of the anterior pituitary gland (Harris, 1955; Chapter 1). The first direct evidence for the neuro-humoral regulation of ACTH release came from the work of Saffran *et al.* (1955) who showed that both the posterior pituitary gland and hypothalamic extracts contained material which could stimulate ACTH release *in vitro*. The nature of CRF is still not known. Problems associated with its purification include instability, the difficulty of its assay, contamination with ACTH and the fact that substances other than CRF can release ACTH (Schally *et al.*, 1958). The satisfaction of some or all of the following criteria will result in increasing certainty that a particular material is "the" CRF (see also Chapter 5).

(a) The material releases ACTH preferentially from the anterior pituitary both *in vivo* and *in vitro*.

(b) It is present in portal blood in concentrations greater than in peripheral blood.

(c) Its concentrations vary in accordance with what is known or assumed about influences on ACTH secretion, a rise under stress and a circadian rhythm. (Direct estimates of such changes in portal blood will not be possible since the procedure for collection of the blood is itself stressful and stimulates CRF release).

(d) Its ability to release ACTH from the anterior pituitary is reduced by corticosteroids.

(e) The material exists in specific cells in the hypothalamus, whose activity is modified by application of corticoids, ACTH and neurotransmitters which are known to stimulate and inhibit CRF synthesis and secretion.

(f) The hypothalamus contains ribosomal mechanisms or enzymes able to catalyse the synthesis of the hormone and peptidases controlling its degradation.

B. NEUROHYPOPHYSIAL CRFs

Schally *et al.* (1958) showed that porcine neurohypophysial extracts contain CRF material (other than vasopressin), as shown by its ability to release ACTH both *in vivo* and *in vitro*. It contained the same amino acid residues as lysine vasopressin with the addition of histidine and serine. The material had pressor activity of 14–40 units/mg and released ACTH from the anterior pituitary *in vitro* in doses as low as 1 ng/ml. Similar but less pure material was obtained from the same source by Guillemin *et al.* (1957). Subsequently, Guillemin and Schally (1963) obtained two fractions which they designated α-CRF and β-CRF. β-CRF contained serine and histidine in addition to the amino acids of lysine vasopressin. β-CRF is a more potent releaser of ACTH in the rat and is active *in vivo* in amounts of 10 ng. Fractionation has shown that it is separable from vasopressin (Schally *et al.*, 1958).

CRF stimulates corticosteroid release in chlorpromazine-morphine-nembutal treated, but not in dexamethasone-morphine-nembutal treated or hypophysectomised rats. This shows that it has no direct ACTH-like action on the adrenals and that its action on the adenohypophysis is suppressible by corticosteroids (Schally *et al.*, 1958). α-CRF, on the other hand, can release corticosteroid even in hypophysectomised animals; thus it has ACTH-like activity in addition to its CRF activity (Schally *et al.*, 1958).

C. RELATIONSHIP BETWEEN VASOPRESSIN AND CRF

Vasopressin may influence hypothalamo-pituitary-adrenocortical function but despite earlier claims there is no doubt that this neurohormone is not CRF. Thus, McDonald *et al.* (1957) failed to demonstrate a correlation between endogenous ADH and ACTH secretion in man in response to nicotine, water loading and water deprivation. Recently it has been shown that material of hypothalamic origin with considerable CRF-like activity (Guillemin, 1971; Gillham *et al.*, 1975) has little ADH activity.

Although CRF and vasopressin are separate entities, there is evidence of some relationship between them, and it has been suggested that vasopressin may play a permissive role in the regulation of ACTH secretion (de Wied, 1961). Following the observation that infusion of vasopressin into the median eminence will release ACTH in doses that are ineffective when infused into the anterior pituitary, it was suggested that it may release CRF (Hedge *et al.*, 1966). The weakness of this evidence, however, is that infusion into the median eminence may be the best method of perfusing the anterior pituitary (see Chapter 3). The fact that vasopressin fails to release CRF activity when added to the rat hypothalamus *in vitro* (Burden *et al.*, 1975) is also against this proposition. What appeared to be more likely was that vasopressin potentiates ACTH release by CRF (Yates *et al.*, 1971). This has recently been confirmed (Buckingham and Hodges, 1977) using the anterior pituitary *in vitro*. This potentiation is not unique to vasopressin since CRF will also increase its own ACTH-releasing activity (M.T. Jones, J.L. Burden and F.W. Hillhouse, unpublished observation). It has been shown that some fibres from the supraoptic nucleus terminate in the median eminence (Parry and Levitt, 1973), and cannulated portal vessels in the monkey contain high concentrations of vasopressin (Zimmerman *et al.*, 1973). It is possible, therefore, that in stresses where both CRF and vasopressin are secreted, the latter may play a role in the release of ACTH. It is likely that the potentiation by vasopressin may help to over-ride corticoid feedback at the adenohypophysis.

Animals genetically deficient in vasopressin are able to respond to stress by secreting ACTH and the reduced plasma corticosterone response in such animals is, at least in part, due to reduced adrenocortical sensitivity to ACTH. The amount of vasopressin required to release ACTH is very high and most of the ring and tail fragments of oxytocin and vasopressin that have been synthesised (see Chapter 5) do not stimulate ACTH release. Deaminopressinamide and pressinoic acid are inconsistent in their CRF-like activity (Pearlmutter *et al.*, 1974).

Vasopressin is associated with a neurophysin in dense core vesicles. Wittkowski and Bock (1972) have shown that adrenalectomy and corticoid therapy influence the appearance of such vesicles in the median eminence. If these represent CRF vesicles then it would appear that CRF is associated with a carrier protein.

Crude hypothalamic extracts containing CRF will release ACTH from the anterior pituitary *in vitro* without a pre-incubation period, but vasopressin fails to stimulate ACTH under these conditions (Sadow, 1973). The addition of prostaglandin E_1 to the medium, however, potentiates the vasopressin to stimulate ACTH secretion; whereas prostaglandin E_1 has no effect on the CRF-induced release of ACTH (Sadow and Penn, 1973). The two neurohormones must therefore act via different mechanisms on the release of ACTH. Yates and Maran (1975) suggest that vasopressin acts early in the

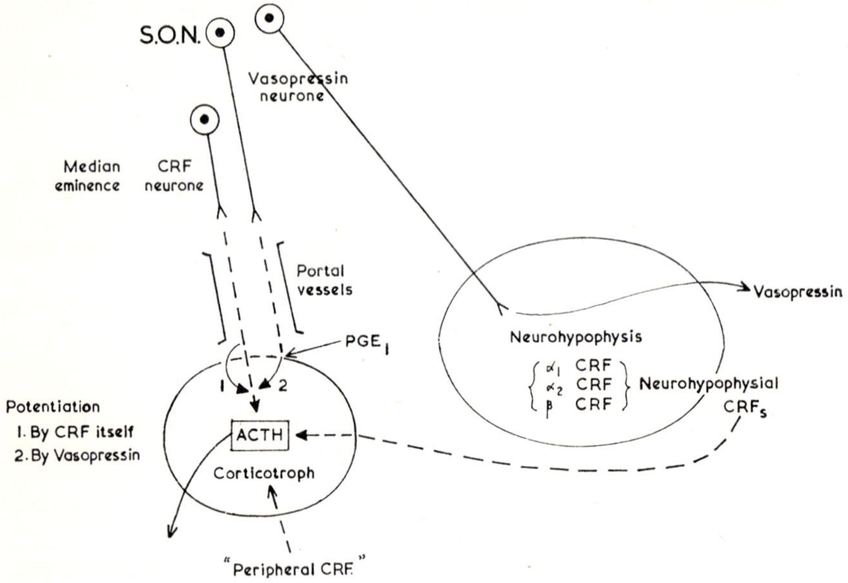

Fig. 1. Relationship between neurohypophysial peptides, CRF and ACTH.

events leading to ACTH secretion, and in this way it can potentiate the ACTH releasing activity of CRF. This suggestion also explains why basal hypothalamic lesions reduce the sensitivity of the anterior pituitary to vasopressin.

The data discussed in this section are summarised in Fig. 1.

D. NON-NEURAL INPUTS TO THE HYPOTHALAMUS AND ANTERIOR PITUITARY

Humoral or blood-borne factors other than CRF have also been implicated in the secretion of ACTH. Angiotensin II has been implicated in the release of CRF from the hypothalamus in response to haemorrhage (Gann and

Cryer, 1973; Redgate *et al.*, 1973), but it fails to release CRF from the hypothalamus *in vitro* and similar negative results have been obtained when attempts were made to release CRF with vasopressin (Burden *et al.*, 1975). However, vasopressin does stimulate ACTH release from the anterior pituitary and will potentiate the response to CRF (Yates *et al.*, 1971), an observation that has recently been confirmed using the anterior pituitary *in vitro* (Buckingham and Hodges 1977). It appears, therefore, that of the two humoral factors claimed to activate the corticotrophic response to stress, only the action of vasopressin has been substantiated. Vasopressin normally plays only a minor part in the regulation of ACTH secretion; both CRF and vasopressin are released in some forms of stress, such as haemorrhage, and under these circumstances the vasopressin may potentiate the activity of CRF. This potentiation may be one of the circumstances under which corticoids fail to modulate the corticotrophic response, giving rise to the concept of stresses that by-pass corticoid feedback. Corticoids do feedback at the anterior pituitary level, as well as at the hypothalamic level, but under the conditions being discussed the feedback at the anterior pituitary is swamped by the potentiating action of vasopressin.

E. "PERIPHERAL CRF"

Witorsch and Brodish (1972) have claimed that ACTH can be released by a peripheral or tissue CRF. However, it is difficult to understand the mechanism of release of this peripheral CRF, or from which tissue it is being released. These authors have shown that it is released even when the hypothalamus has been extensively destroyed. An alternative explanation may be that secretomotor or vasomotor actions of the sympathetic nervous system to the anterior pituitary are involved. Lastly, circulating kinins or the effects of hypoxia on the pituitary gland may play a role, or vasopressin may be released from the posterior pituitary gland. Many of these possibilities have been excluded and the release of peripheral CRF is now a fitting explanation.

F. HYPOTHALAMIC CRF PREPARATIONS

Although the physiological significance of the neurohypophysial CRF-like peptide is unknown, it seems likely that CRF obtained from the hypothalamus will be the true releasing hormone (see Chapter 5). The first effort to purify hypothalamus CRF was made by Guillemin and Schally (1963), using extracts of sheep hypothalamus following gel filtration on Sephadex G-25 and countercurrent distribution. Two zones of CRF activity were obtained. The claim of Dhariwal *et al.* (1966a) that there is only one CRF in sheep hypothalamus has been discounted by Guillemin (1967), since countercurrent distribution which separates the α- and β-CRF was not carried out. It is unlikely that the two hypothalamic CRFs are identical with those from the posterior lobe, as hypothalamic α-CRF has no intrinsic ACTH-like activity, whereas (as mentioned above) posterior lobe α-CRF does. The peptides show the opposite; that of posterior lobe origin having no ACTH-like activity, but hypothalamic β-CRF being able to stimulate the adrenal directly. Both Schally and Guillemin appear to favour hypothalamic α-CRF as the likely candidate for the true releasing hormone (see discussion following Chan *et al.*, 1970).

Rat median eminence has also been used as a source of CRF (Chan *et al.*, 1969) and fractionation on Sephadex G-25 showed some activity in the

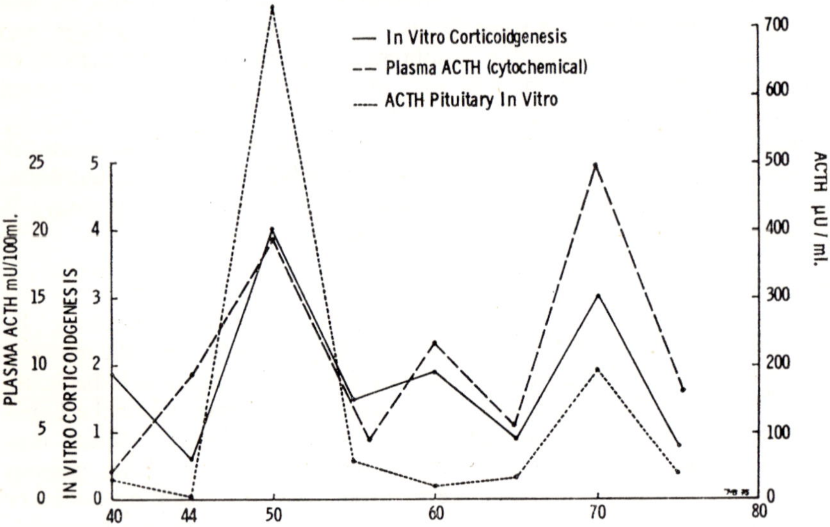

Fig. 2. Elution pattern of CRF on Sephadex G-25.

void volume, but most appeared in a low molecular weight fraction. It is not possible to see any identity between these two peaks and Schally's α- and β-CRF.

A different approach to the isolation of CRF has been adopted by Gillham *et al.*, (1975), which is based on the ability of 5-HT to release CRF selectively from the rat hypothalamus *in vitro* (Jones *et al.*, 1976a). This reduces problems of contamination with other releasing hormones and hence eases purification procedures. Two peaks of CRF activity have been demonstrated on Sephadex G-25 and two bands on high voltage electrophoresis. Fraction A is about 2,500 MW and fraction B 1,200 MW as judged by gel filtration (Fig. 2). Neither fraction has a significant effect on TSH, LH, FSH or prolactin secretion. A preliminary study of the amino acid residues suggests that fraction A is similar to but quite distinct from α-CRF.

G. MODE OF ACTION OF CRF

The mechanism whereby CRF causes ACTH synthesis and release is not known and is unlikely to be understood until CRF has been isolated. It has been suggested that as with other hypothalamic hormones, (see Chapters 4 and 5) cyclic AMP is involved as a messenger (Fleischer *et al.*, 1969; Hedge, 1971), and that protein synthesis is not involved in its rapid action (Arimura *et al.*, 1969). An elevated potassium concentration depolarises the cell membrane and releases ACTH which is Ca^{++} dependent (Kraicer *et al.*, 1969). Whatever the mechanism of ACTH release may be, it is usually associated with increased synthesis (Vernikos-Danellis, 1964), but whether this is due to the direct action of CRF or the result of stimulus-secretion coupling remains to be determined.

H. THE ASSAY OF CRF

The profusion of assay systems for CRF is a monument to the limitations of existing methods for its measurement. The methods currently in use fail to satisfy one or more of the requirements of specificity, sensitivity, accuracy and convenience. No single method is entirely satisfactory and it is probably wise to use more than one assay system to screen substances for CRF activity. Reviews have appeared covering the principal *in vitro* (Guillemin and Vale, 1970) and *in vivo* assay systems (Munson and Tai Chan Peng, 1970; Vernikos-Danellis and Marks, 1970). A problem associated with *in vivo* assay is to exclude interference from endogenous CRF. A variety of methods has been used to achieve this, including pretreatment with corticosteroids (Vernikos-Danellis and Marks, 1970), basal hypothalamic lesions (Witorsch and Brodish, 1972) and pituitary transplants or grafts (Sirret and Purves, 1973; Fortier, 1958). As an alternative, a mixture of some or all of the following drugs has been used: morphine, nembutal, chlorpromazine and dexamethasone (Munson and Briggs, 1955; Dhariwal *et al.*, 1969). The use of corticoid pre-treated animals is based on the assumption that corticoids act on the hypothalamus and not on the anterior pituitary. This assumption is not valid, as recent reports show that corticoids feedback at both the pituitary and hypothalamic levels (see later; Bradbury *et al.*, 1974; Sayers and Portanova, 1974; de Kloet *et al.*, 1974).

The alternative is to use preparations of the anterior pituitary *in vitro* (Saffran *et al.*, 1955) or in tissue culture (Guillemin and Vale, 1970). Superfusion of the anterior pituitary has been used as a sophistication of the original method and has increased the sensitivity of the *in vitro* assay (Saffran, 1974). Another method of increasing the sensitivity of the pituitary *in vitro* is to prime the tissue with ADH prior to the assay, which enhances CRF-induced ACTH release (Buckingham and Hodges, 1977). The most sensitive assay utilises an isolated anterior pituitary cell system, prepared from adrenalectomised rats (Sayers and Portanova, 1974).

A major problem still outstanding is the lack of a satisfactory standard. CRF that is currently available is of low purity. The various problems associated with the satisfactory assay of CRF will not be resolved till the neurohormone has been isolated, characterised and is available as a standard. Meanwhile, CRF activity is best expressed in terms of ACTH-releasing ability in a given assay system.

IV. Neurotransmitters Involved in CRF Regulation

The activity of the CRF neurone is determined by the summation of excitatory and inhibitory neural inputs, which arise in various parts of the central system and are responsible for the stress induced and episodic ACTH secretion. Many neurotransmitters may be involved in CRF regulation, since there are several polysynaptic pathways involved in the stress mechanism. These include somatic and visceral reflex pathways as well as those influenced by psychological factors such as anxiety and fear. Discussion will be limited here to the hypothalamus. The techniques used to study the role of neurotransmitters in the endocrine hypothalamus have been discussed in Chapters 3 and 4.

The location of cell bodies of the CRF neurones within the hypothalamus is as yet unknown and is likely to remain so until techniques such as immunofluorescence (see Chapter 4) can be used to identify the cells. It will then be possible to study the response of these cells to iontophoretically applied neurotransmitters (see Chapter 2) and the results correlated with morphological identification of nerve terminals around the dendrites and cell bodies. In the meantime, data based on the indirect *in vivo* and *in vitro* correlates must be used to implicate the neurotransmitters regulating CRF activity.

A. Excitatory Neurotransmitters

The release of CRF is an integral part of the stress response and it is therefore difficult to distinguish between the effects of the experimental procedure and the excitatory effect of a putative neurotransmitter; the problem is compounded by the probable episodic nature of CRF release.

(1) *Acetylcholine*

There is evidence that acetylcholine, when injected into the third ventricle, releases CRF (Krieger and Krieger, 1970) and implantation of atropine into the anterior hypothalamus inhibits CRF and ACTH release, without inhibiting vasopressin secretion. Acetylcholine added to the hypothalamus *in vitro*, in physiological concentrations (pg/ml), causes synthesis and release of CRF-activity (Jones *et al.*,1976a). In slightly higher doses acetylcholine will also release CRF activity from isolated hypothalamic nerve terminals synaptosomes (Bennett and Edwardson, 1975). The action of acetylcholine on the rat hypothalamus *in vitro* is blocked paratially by atropine and completely by hexamethonium, suggesting that CRF is released by a mixed, cholinergic receptor mechanism which is predominantly nicotinic; this view is supported by the observation that bethanacol, a pure muscarinic agent, does not alter the release of CRF activity even in doses of 100 ng/ml (Jones *et al.*, 1976a).

The observations by Hedge and Smelik (1968) and Hedge and de Wied (1971), that atropine implants in the anterior hypothalamus will abolish the corticotrophic response to stress, could therefore be explained by inhibition of muscarinic receptors and the membrane stabilising effect of atropine in high concentrations.

(2) 5-*hydroxytryptamine* (5-HT)

The implantation of 5-HT into various regions of the CNS has been shown to result in the activation of the pituitary-adrenal system in both the cat (Kreiger and Kreiger, 1970) and the guinea-pig (Naumenko, 1968). The administration of 5-HT into the lateral ventricle has been shown to block the response to ether and surgical stress in both the dog (van Loon *et al.*, 1971) and the rat (Vermes *et al.*, 1973b). In experiments on conscious rats with chronically implanted intraventricular cannulae, the administration of 5-HT into the lateral ventricle activates the adrenocortical system: this effect is not affected by prior treatment with reserpine, but is abolished by nembutal anaesthesia (Abe and Hiroshige, 1974). As the experiments of van Loon *et al.* (1971) and Vermes *et al.* (1973b) were performed on anaesthetised animals, this may account for the discrepancy in the results obtained. In guinea-pigs, the injection of 5-HT into the lateral ventricles causes an increase in plasma corticosteroids, both in intact and midbrain-sectioned animals (Naumenko, 1968). This suggests that the stimulatory action of 5-HT on the pituitary-adrenal system is due to a central action and not a peripheral one. However, in all these experiments the doses of 5-HT used were extremely high and, as with all intraventricular injection studies, the precise site of action is undetermined. Addition of 5-HT to hypothalamus incubated *in vitro* causes an increase in the secretion of CRF activity (Burden *et al.*, 1974b; Jones *et al.*, 1976a, b), and the amounts required to do this are within the physiological range (Stoner and Elson, 1971; Vermes *et al.*, 1973a). The effect of 5-HT on CRF secretion *in vitro*, appears to be mediated by specific serotoninergic receptors, since methysergide antagonises the action of 5-HT (Jones *et al.*, 1976a). The effects of cholinergic blockade on the secretion of CRF activity in response to 5-HT have also been investigated (Burden *et al.*, 1974b; Hillhouse *et al.*, 1975; Jones *et al.*, 1976a). Both hexamethonium and atropine reduced the response to 5-HT to basal levels; this suggests that the 5-HT acts via a cholinergic interneurone which contains specific serotoninergic receptors, since the effect is blocked by methysergide. In the rat, cell bodies of the acetylcholine containing neurones are thus situated within the hypothalamus. This partly agrees with the conclusions reached by Naumenko (1968) and Papova *et al.* (1972), who suggested that 5-HT acted directly on the hypothalamus, although it was thought that other neurotransmitters might be involved in addition.

A number of reports have appeared implicating ACTH and 5-HT in the control of the circadian periodicity of the hypothalamo-pituitary-adrenal axis (Krieger and Rizzo, 1969; Scapagnini *et al.*, 1970). The administration of anticholinergic agents blocks the afternoon peak of plasma corticosteroid in the cat (Krieger *et al.*, 1968), suggesting the participation of central cholinergic structures in the circadian rhythm of secretion. Further studies have revealed that drugs, which are able to affect the metabolism of 5-HT, or its action, were also capable of preventing the daily rise in plasma corticosteroid levels (Krieger and Rizzo, 1969). Drugs which depleted brain noradrenaline levels had no effect on the circadian periodicity of ACTH secretion, and none of the drugs used had any effect upon the stress response to insulin hypoglycaemia or to Pseudomonas polysaccharide. This suggests that the central cholinergic and serotoninergic pathways involved in the circadian rhythm of ACTH release, are different from those which mediate certain types of stress responses.

B. INHIBITORY NEUROTRANSMITTERS

(1) *Noradrenaline*

Pharmacological experiments, using drugs which are known to affect catecholamine synthesis and metabolism, have suggested a role for hypothalamic catecholamines in the control of ACTH secretion (Ganong, 1972; Scapagnini and Preziosi, 1972; van Loon, 1973). Ganong (1972) has shown an inverse relationship between plasma corticosterone concentrations and hypothalamic noradrenaline levels in the rat. However, not all reports have supported this finding (Kaplanski *et al.*, 1972; Lippa *et al.*, 1973; Ulrich and Yuwiler, 1973). The interpretation of some of these experiments using catecholamine inhibitors may be complicated by the possible developments of "denervation hypersensitivity", which is a well-documented phenomenon in the peripheral nervous system. Thus, small amounts of catecholamines might be able to maintain normal function because of an increased sensitivity of the hypothalamic neurones. Another possibility is that other neuroinhibitory pathways, which are not catecholaminergic, might be able to take over the role of noradrenergic pathways and help to maintain the functional integrity of the system. This would seem possible as both GABA and melatonin are able to inhibit the release of CRF from the rat hypothalamus *in vitro* (see following section).

Probably the most convincing evidence in favour of a central noradrenergic inhibitory control of ACTH secretion, is that of Fuxe and his coworkers (Fuxe *et al.*, 1973). The 6-hydroxydopamine induced degeneration of the dorsal noradrenergic pathways of the brain, had no effect on either the hypothalamic catecholamine fluorescence or on the normal activity of the pituitary-adrenal axis. However, when the ventral noradrenergic pathway was destroyed there was a considerable loss of hypothalamic noradrenergic terminals, which resulted in an enhanced ether stress response at one day after the lesion but not at one week. Again this suggests that either denervation hypersensitivity takes place, or that other neuro-inhibitory pathways are able to compensate for the loss of the noradrenergic pathways.

Noradrenaline in doses of 3–10 ng/ml causes a reduction in the release of CRF in response to acteylcholine from the hypothalamus *in vitro*. Dopamine and histamine have no effect on the release of CRF activity, but the action of noradrenaline is antagonised by phentolamine (Hillhouse *et al.*, 1975; Jones *et al.*, 1976a). This is in agreement with the results of Scapagnini and Preziosi (1973), who investigated the effect of α and β blocking agents on the noradrenergic inhibition of ACTH secretion in the rat and concluded that an α-receptor mechanism was involved. The systematic administration of phentolamine produced an increase in plasma corticosteroid levels and the intraventricular administration of a systemically ineffective dose had a similar effect. Propranalol had no effect.

There is some evidence that dopamine is also involved in the inhibition of CRF release (Edwardson and Bennett, 1974), but the weight of evidence is in accordance with the hypothesis of Ganong (1972), that there is a central noradrenergic neuro-inhibitory pathway to CRF release mediated via an α-adrenergic receptor mechanism on ACTH secretion (Scapagnini *et al.*, 1971).

(2) *The Putative Amino Acid Neurotransmitters*

Although the role of certain amino acids as neurotransmitters in the central nervous system is rapidly becoming established (see Chapter 3), their role as potential modulators of CRF secretion has received little attention. The effect of several of these amino acids on the secretion of CRF from the hypothalamus *in vitro* has, however, been reported (Burden *et al.*, 1974a; Jones *et al.*, 1976a). The basal release of CRF was not altered by the addition to the incubating medium of aspartate, glutamate, glycine or GABA, but GABA reduced the stimulatory action of acetylcholine on release of CRF activity. It is noteworthy that GABA is a more potent inhibitory neurotransmitter than noradrenaline in the hypothalamus *in vitro* test system. Since the response to both 5-HT and to acetylcholine is inhibited by GABA, it is suggested that GABA acts directly on the CRF neurones, which must contain specific GABA receptors, as the effect is abolished by picrotoxin (Jones *et al.*,

1976a, b). This has been supported *in vivo* by the use of the intraventricular administration of both GABA and picrotoxin (Makara and Stark, 1974). Yagi and Sawaki (1975) have shown recurrent inhibition via GABA interneurones in the tubero-infundibular system; it is possible that these may form the anatomical basis for the inhibition of the CRF neurone by the ultra-short loop feedback.

(3) *Melatonin*

Studies in which the pineal was removed or where melatonin was implanted or injected intraventricularly, support the conclusions that melatonin inhibits CRF release (Motta *et al.*, 1971b). Melatonin inhibits the release of CRF activity from the hypothalamus *in vitro* in response to both 5-HT and acetylcholine (Jones *et al.*, 1976a, b) and this may reflect a physiological role. It has been suggested that melatonin acts as an antagonist on the 5-HT receptor but since melatonin inhibits the CRF releasing action of acetylcholine as well as 5-HT (Jones *et al.*, 1976a), the effect must be distal to the reception. The finding that melatonin and 5-HT have opposing effects upon CRF release has a significance with respect to experiments in which *p*-chlorophenylalanine (*p*CPA) has been used to study the role of 5-HT in the central control of CRF secretion (Bhattacharya and Marks, 1970; Marks *et al.*, 1970; Vernikos-Danellis *et al.*, 1970; Scapagnini *et al.*, 1971; Van Delft *et al.*, 1973). The pretreatment of animals with *p*CPA would interfere with melatonin as well as 5-HT synthesis and the results of such experiments must therefore be interpreted with caution. It is possible that variations in the levels of these two indoleamines may be important in the control of the circadian rhythm of CRF release by the light-dark cycle (see Section VI).

C. INTERACTION BETWEEN THE INHIBITORY ACTIONS OF NORADRENALINE, GABA AND MELATONIN: A MODEL

Experiments based on the use of the hypothalamus *in vitro* show that, in the rat at least, these three inhibitory substances exert their effects through independent receptor mechanisms. Thus picrotoxin has no effect on the inhibitory activity of noradrenaline or melatonin, whilst phentolamine is without effect on the inhibition of CRF caused by GABA or melatonin (Jones *et al.*, 1976a, b). A model of the neurotransmitters involved in CRF release at the hypothalamus is shown in Fig. 3. The model is highly tentative and is based on data obtained using the rat hypothalamus (*vide supra*) *in vitro* and data selected from *in vivo* experiments by other workers using several species. The anatomical site of the CRF cells is unknown and their location in the model, therefore, is unspecified.

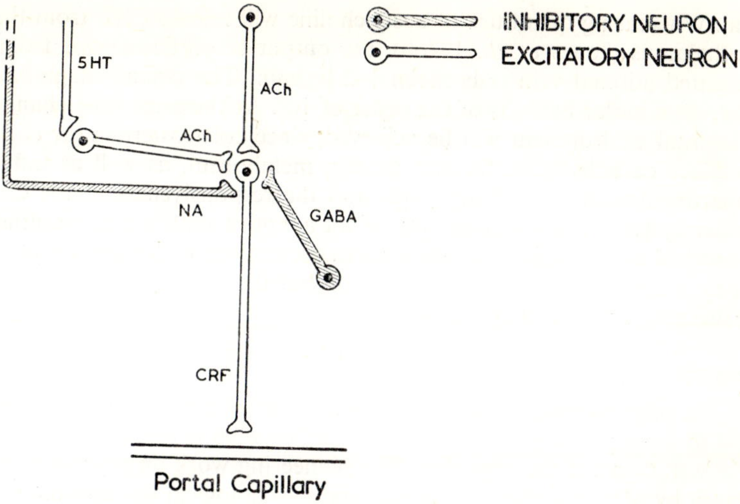

Fig. 3. A proposed model of the neurotransmitters involved in the release of CRF at the hypothalamus. 5 HT, 5-hydroxytryptamine; ACh, acetylcholine; NA, noradrenaline; GABA, γ-aminobutyric acid.

V. The Hypothalamo-Pituitary-Adrenocortical Axis

A. Components of the Axis

The axis consists of hypothalamus, pituitary and adrenal. The first step in the activation of the system involves the release of acetylcholine, since the final excitatory pathway to the CRF neurone is cholinergic (Jones *et al.*, 1976a). Such an activation may also be associated with a decrease in the neuro-inhibitory discharge in the noradrenergic pathway to the CRF neurone (Ganong, 1970). Acetylcholine causes the release of CRF activity and this is accompanied by an increased synthesis of the hormone (Bradbury *et al.*, 1974).

Vernikos-Danellis (1964; 1965a, b), in a series of experiments, has studied the dynamics of CRF and ACTH under conditions of stress and shown that for each hormone increased release is associated with accelerated synthesis. Thus, there is an increase of CRF activity in the median eminence 1–2 minutes after the application of ether stress; this is followed by an increase in anterior pituitary ACTH content within 2·5 minutes. The concentration of corticosterone in the adrenal gland is low under basal conditions and ACTH causes a rapid synthesis of the steroid. The increased release may be a simple spillover into the blood from the rising levels in the adrenal gland, or a secretory process is involved.

Each stage in the above processes involves amplification of the preceding signal. Whereas pg quantities of acetylcholine will release CRF from the rat hypothalamus (Jones *et al.*, 1976b), the output of corticosterone from the cannulated adrenal vein is between 1–2 μg/min. The overall magnification factor, on a molar basis, is of the order of 10^6. At the same time changes in the internal environment will be achieved, since corticosterone (or cortisol) will affect carbohydrate, fat and protein metabolism, as well as influence compartmentalisation of body fluids and the responsiveness of the cardiovascular system. These changes are of an adaptive nature and essential for the survival of the organism, since adrenalectomised or hypophysectomised animals are unable to withstand anything but the blandest of environments and succumb under conditions of stress.

B. STRESS

To the physiologist, stress is a stimulus to adrenaline or ACTH secretion, whilst to the psychologist it implies a threat, or what is perceived as a threat, to the well-being of the organism. The pioneering work on psychoendocrine research by Mason (1968) has shown that psychological or emotional stress is a major cause of adaptive changes in the endocrine system. It is likely that all forms of stress, applied to man or animal, have important psychological components and what is stressful to one individual may not be to another. It is usually accepted that individuals will vary in their perception and response to stress, but it is not usually realised that the same person may show widely differing responses to various physiological parameters (Jones *et al.*, 1970); some of this variation may be associated with differences in body build (Bridges and Jones, 1973).

(1) *Stress Intensities*

Most laboratories engaged in work on control of the hypothalamo-pituitary-adrenocortical axis have made little attempt to use graded or quantifiable stress, though there are some exceptions (e.g. Gann and Cryer, 1973).

The intensity of ACTH secreting stimuli influences the ability of corticosteroids to suppress the corticotrophic response to stress. High intensity, but not low intensity stresses, will saturate the corticosteroid delayed feedback mechanism (Yates and Maran, 1975; see Section C (2) later).

(2) *Neural and Systemic Stresses*

Attempts have been made to classify stimuli that activate the hypothalamo-pituitary-adrenocortical axis into "neural" and "systemic" stresses. The term "neural" stress implies that the stimulus excites ACTH release by means of nervous reflex pathways acting on the CRF neurones. "Systemic" stresses resulting from metabolic changes or tissue damage, probably induce ACTH discharge both by neural reflex (via CRF) and direct effect on the anterior pituitary itself.

The neural pathways involved in the stress mechanism depend upon the nature of the stimulus, the number and nature of sensory and visceral reflexes involved and the part played by psychological factors. It is doubtful whether any stress applied in man, or in animals, is devoid of psychological influences. It is not possible, therefore, to deal with all the pathways involved in the total stress response but it is possible to determine some that terminate in the hypothalamus and which carry fibres influencing CRF secretion (Dallman and Yates, 1968). It is not possible at the present time, to describe in a definite way the relationship between the neural pathways and neurotransmitters implicated in CRF secretion.

Some stresses activate neural pathways which then reach the hypothalamus from the mid-brain and the pons and have traversed the mesencephalic and hind-brain reticular formation from the spinal cord (Anderson *et al.*, 1957; Gibbs 1969; Taylor, 1969; Allen *et al.*, 1973). The afferent fibres responsible for the increase in ACTH release produced by leg fracture, probably travel in the lateral spinothalamic tract and the increase is prevented by cord section or by hemisection of the cord on the side opposite to the leg fracture (Gibbs, 1969). Gibbs has successfully followed the pathways mediating the corticotrophic response to leg fracture into the midbrain, where it is known that collaterals from the spinothalamic tracts enter the reticular formation. Stimulation of the reticular formation releases ACTH (Mason, 1968), as does stimulation of the dorsal bundle of Schutz and the mammillary peduncle (Ganong, 1970). Thus, the fibres enter the reticular formation and traverse to the ventral hypothalamus and activate CRF release.

The visceral reflexes involved in stresses such as haemorrhage, originate in baroreceptors in the carotid and aortic arch, as well as volume receptors in the atria and large veins (Gann and Cryer, 1973). It has been suggested that both the carotid and vagal afferents have a tonic inhibitory effect on ACTH release and interact centrally to influence the corticotrophic response to stress (Gann and Cryer, 1973).

The amygdaloid region is said to be concerned with genesis of emotions and ablation of the amygdaloid area causes the release of ACTH, but leaves the response to surgical stress (under anaesthesia) unaffected (Ganong, 1970). Stimulation of the amygdaloid area causes the release of ACTH but stimulation of the hippocampus causes inhibition (Knigge, 1961; Knigge and Hayes, 1963). The various connections between these and other related areas and the hypothalamus have been reviewed (Raisman, 1970b; see Chapters 1 and 3).

C. NEGATIVE FEEDBACK CONTROL OF CRF AND ACTH

The earliest work on the influence of corticosteroids on pituitary-adrenocortical activity was by Ingle and associates, who showed that administration of adrenocortical extracts to the intact rat resulted in adrenocortical atrophy, whilst iniection of anterior pituitary with the adrenocortical extracts did not (Ingle and Kendall, 1937; Ingle *et al.*, 1938). These results suggested that corticosteroids inhibited ACTH secretion and that ACTH secretion could be regulated by adrenocortical secretion. Evidence for the existence of such a negative feedback mechanism was subsequently provided by the following observations.

(a) Adrenalectomy results in an elevated blood level of ACTH (Gemzell *et al.*, 1951; Sydnor and Sayers, 1954; Dallman *et al.*, 1972; Buckingham and Hodges, 1974a, b). Patients suffering from Addison's disease, due to adrenocortical atrophy, have similarly elevated levels (Bethune *et al.*, 1956; Liddle *et al.*, 1961).

(b) The high circulating level of ACTH produced by adrenalectomy can be prevented by replacement therapy with physiological doses of corticosteroids (Dallman *et al.*, 1972; Dallman and Jones, 1973). Glucocorticoid replacement therapy also reduces blood ACTH levels in Addisonian patients.

(c) The release of ACTH in response to stress is greatly exaggerated in the adrenalectomised animal or patient.

(d) Pre-treatment of animals with physiological doses of corticoids will reduce the ACTH release in response to moderate stresses (Dallman and Yates, 1969; Jones *et al.*, 1972).

(e) Implantation of corticosteroids into hypothalamus or infusion *in vivo* in the rat into the anterior pituitary gland, causes suppression of ACTH secretion (Rose and Nelson, 1956; Smelik and Sawyer, 1962).

(1) *Early Theories on the Nature of Corticosteroid Feedback Action*

(a) *The Sayers' Hypothesis.* The first attempt to explain the mechanism whereby adrenocortical secretions may control ACTH release was made by Sayers and Sayers (1947), who proposed that when animals are subjected to stress there is an increased utilisation of circulating corticosteroids and that the resulting drop in plasma corticoids serves as a stimulant to ACTH secretion. It is now known that this hypothesis is untrue, since stress causes a rise not a fall, in plasma corticosteroid concentration.

(b) *The Variable Set-Point Hypothesis.* Yates *et al.*, (1961) found that the i.v. injection of small doses of corticosterone in the rat partially or completely inhibited the corticotrophic response to a stress applied 15 seconds after steroid administration. Yates *et al.* (1961) proposed that stress results in a "re-set of the negative feedback control of plasma corticosteroid concentration to regulate at higher levels". This theory resembled the variable setpoint hypothesis for the regulation of body temperature. The theory proposed that, in response to stress, ACTH secretion continues until plasma corticosteroids concentration reaches the new threshold. This attractive hypothesis overcame many of the objections to the original theory of Sayers. It accounted for the rapidity of the corticotrophic response and the rise in plasma corticosteroid concentration. Since the hypothesis assumes that the rate of adrenocortical activity is regulated by the difference between the concentration of the corticosteroids in the plasma and the controller set-point, the fact that several workers required large doses of corticoids to inhibit ACTH release could be explained because of a low steroid level at the time of stress. However this re-set hypothesis became untenable when it was shown that corticosterone, in doses sufficient to raise the blood concentration up to four times the maximal level induced by stress, did not diminish the corticotrophic response (Hodges and Jones, 1963; Smelik, 1963a). It was shown later that inhibition of the corticotrophic response to stress occurred when the steroid level had fallen (Smelik, 1963b; Hodges and Sadow, 1967).

(2) *Fast* (*Rate-sensitive*) *and Delayed* (*Proportional*) *Feedback Control*

The next advancement in our understanding of corticosteroid feedback control of ACTH, was made by Dallman and Yates (1969), who, using a continuous infusion of corticosterone, showed that there exists a short delay, rate-sensitive, feedback control element associated with the rate of rise of the plasma corticosteroid concentration. During the phase when the corticosterone concentration was rising at its fastest rate (first 5 minutes), there was an 80% inhibition of the stress response; during the slower phase (5–20 minutes) this inhibition disappeared. When the corticosterone concentration had reached a plateau (20 minutes onwards), there was again no inhibition of the stress response until 120 minutes from the start, when 85% inhibition became evident (Fig. 4). Dallman and Yates suggested that the early inhibition was due to the proposed rate-sensitive or fast feedback and the later inhibition was a proportional or a delayed feedback mechanism. The later inhibition is probably similar to the inhibition which occurs some hours after corticosteroid administration (Hodges, 1953; Smelik, 1963b). The importance of this proposal by Dallman and Yates is two-fold. First, it suggests that ACTH secretion, like many other physiological variables, is regulated by both rate-sensitive and proportional feedback mechanism. Secondly, it implies, that the failure of others to demonstrate feedback was probably due to the application of stress during the "silent" period between fast and delayed feedback. It would appear, therefore, that ACTH secretion is regulated by two temporally distinct feedback mechanisms: fast feedback, which occurs within minutes of administration at a time of rising plasma corticosteroids; and delayed feedback, which occurs hours later when the plasma corticosteroids may be high, declining or low.

The rate of increase required to reduce the corticotrophic response to surgical stress has been defined (Jones *et al.*, 1972). When the stress increments in adrenal corticosterone (reflecting endogenous ACTH levels) are plotted (Fig. 5) against the rate of rise in plasma corticosterone concentration (exogenously induced) at the time of stress, a suppression of the stress response to levels, not significantly different from control values, occurs when the rate of rise equals or exceeds 13 ug/l/min.

The two principle glucocorticoids secreted by the adrenal gland, cortisol and corticosterone, are agonists on fast feedback and inhibit ACTH secretion during the rising phase in plasma concentration (Jones *et al.*, 1974). The other

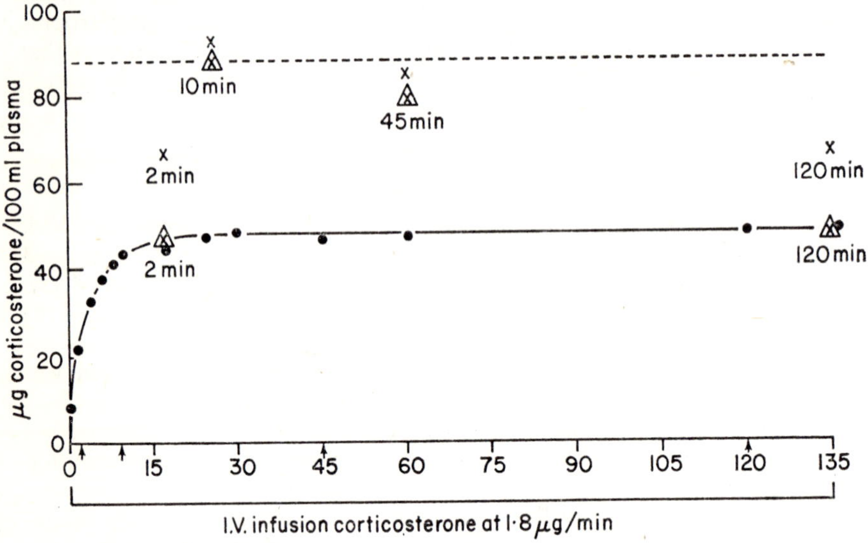

Fig. 4. Rate sensitive and delayed negative feedback. (From: Dallman and Yates, 1969.)

steroids secreted by the adrenal gland do not induce fast feedback and neither do many of the other synthetic glucocorticoids (Tiptaft and Jones, 1975; Hillhouse *et al.*, 1974).

(3) *Receptor Mechanisms*

Fast and delayed feedback have different dynamics and the question arises whether the two mechanisms involve the same or different receptors; this has been investigated by testing the effect of different steroids on stress-induced release of ACTH, minutes or hours after their administration (Jones *et al.*, 1972, 1974; Tiptaft and Jones, 1975). The 11-deoxycorticosterone (DOC) and 11-deoxycortisol (S), have no fast feedback action and indeed antagonise the fast feedback action of endogenously secreted or exogenously administered corticosterone or cortisol. Both DOC and S suppress the corticotrophic

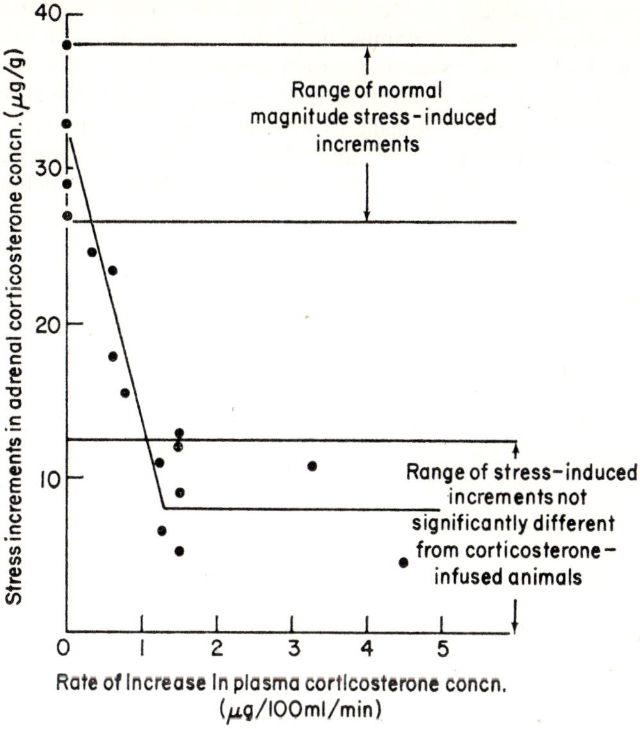

Fig. 5. Relationship between magnitude of stress response and rate of increase of plasma corticosterone. (From: Jones *et al.*, 1972.)

response to stress if they are given some 2–4 hours prior to the stimulus (Jones *et al.*, 1974). There are many other examples of steroids that have no effect on fast feedback but produce inhibition some hours after administration (Tiptaft and Jones, 1975). It is concluded, therefore, that the receptors for fast and delayed feedback must be different.

(4) *The Hypothalamus* in vitro *as a Test System for Corticosteroid Feedback*

The rat hypothalamus remains viable for at least 3 hours *in vitro* (Bradbury *et al.*, 1974; Jones *et al.*, 1976b) and has been used to test the role of corticoids and neurotransmitters in the fast and delayed control of CRF secretion (Jones *et al.*, 1976b).

Fast feedback can be studied by examining the effect of adding corticosteroids simultaneously with the neurotransmitter on the secretion of CRF activity. Delayed feedback can be studied by either pretreatment 2–24 hours before removal of the hypothalamus, or by the addition of the steroid *in vitro* 2 hours before stimulation with the neurotransmitter. From such work it has been possible to determine structure-activity relationships for the two receptor mechanisms (Fig. 6). Antagonism to delayed feedback has been demonstrated by 17 α-OH progesterone and 11 α-OH epimer of cortisol, but the structure-activity relationship has not yet been completely defined. It is

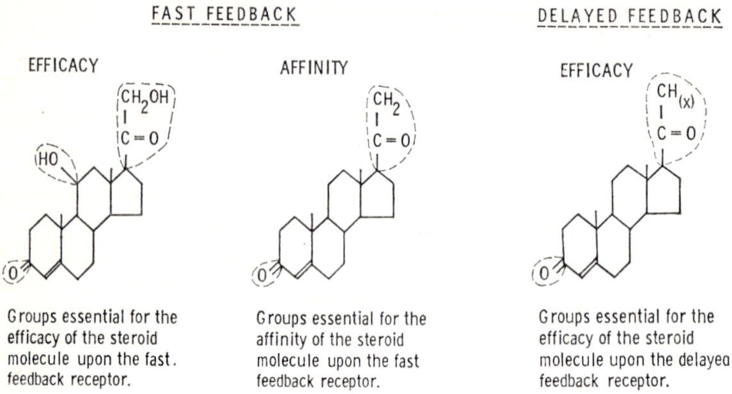

Fig. 6. Steroid-receptor structure-activity relationships: (a) for fast feedback; (b) for delayed feedback.

possible that steroids active on the delayed feedback mechanism may have been metabolised to a highly effective feedback steroid. If this occurs, then it presumably takes place in the hypothalamus since these steroids are effective *in vitro* (Jones *et al.*, 1976c).

(5) *Sites of Fast and Delayed Feedback*

Sayers and Portanova (1974) have shown that corticosterone reduces CRF-induced release of ACTH from anterior pituitary cell within 15 minutes, which suggests that fast feedback may occur at the pituitary level. Fast feedback at this level has been confirmed using the basal hypothalamic-lesioned rat (Jones *et al.*, 1976c). Fast feedback inhibition of CRF secretion has been demonstrated using the hypothalamus *in vitro* and hypothalamic synaptosomes *in vitro* (Edwardson and Bennett, 1974). It may be concluded that fast feedback is exerted at both hypothalamic and anterior pituitary levels, reducing both CRF secretion and pituitary response to CRF.

Delayed corticosteroid feedback has been extensively investigated and again there is evidence for both anterior pituitary and hypothalamic feedback. Rose and Nelson (1956) found that infusion of cortisol into pituitary fossa caused adrenal atrophy, whereas infusions into the subarachnoid space was ineffective. Some investigators have reported that cortisol reduces the response of the pituitary to CRF (Royce and Sayers, 1958; Rochefort *et al.*, 1959). Dexamethasone had been shown to suppress ACTH release from the anterior pituitary (de Kloet *et al.*, 1974). Implantation of cortisol (Smelik and Sawyer, 1962) in rabbits is ineffective when the implant is placed in the anterior pituitary, but is extremely effective when placed in the hypothalamus; this argues for a hypothalamic site of delayed feedback. Cortisol injections (Vernikos-Danellis, 1965a) and dexamethasone implants in the hypothalamus

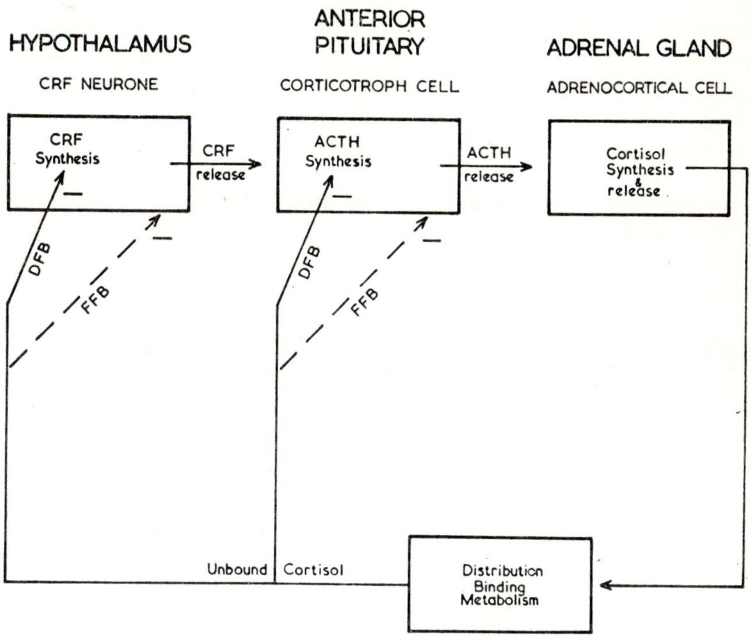

Fig. 7. Model of fast (FFB) and delayed (DFB) feedback.

(Chowers *et al.*, 1967) decrease hypothalamic CRF as well as pituitary and plasma ACTH concentrations. More direct evidence for delayed corticoid feedback at the hypothalamus comes from experiments carried out using the hypothalamus *in vitro*, following *in vivo* pretreatment or *in vitro* preincubation with corticoids prior to stimulation of release with an excitatory neurotransmitter (Jones *et al.*, 1976c). In summary, there is therefore, evidence for fast and delayed feedback at both the hypothalamus and the anterior pituitary (Fig. 7).

Labelled corticosteriods are taken up by various parts of the brain, particularly the hippocampus and septum (McEwen *et al.*, 1969, 1970, 1972). In view of the association of these areas with behavioural responses, it is likely that the effect of corticosteroids at these sites is related to behavioural rather than feedback regulation. It is unnecessary to postulate that corticoids acting on limbic structures cause organisational changes in the hypothalamus which mediate feedback activity, since corticosteroids added to the medium of the hypothalamus *in vitro* produce both fast and delayed feedback action (Jones *et al.*, 1976c).

(6) "*De-sensitisation*" *and Corticoid Fast Feedback*

Jones *et al.* (1972) and Burden *et al.* (1975), have reported a curious property of fast feedback which they have called "desensitisation". On increasing either the dose of steroid or the rate of infusion, the fast feedback action is curtailed. The suggested explanation for this phenomenon is that the number of receptors is limited and that once they are all occupied any further increase in the plasma corticoid level will not affect the system in any way, since the number of associations per unit time cannot be increased and is dependent solely upon the agonist-receptor dissociation constant. Thus, the feedback signal generated by the receptors cannot be increased by an increase in plasma corticosteroid concentration. This explanation is based on Paton's principle of the "rate theory" of drug action (Paton, 1961), which proposes that it is the dynamic act of association which produces the pharmacological effect, not the "status quo" of a molecule or drug united with its receptor. When a drug-receptor complex is formed, Rang (1971) has proposed that the receptor undergoes a conformational change from the "R" to the "R^1" form and this has to change back to the "R" form before further association with the drug can take place. On the Rang hypothesis the rate of change back to the "R" form of the fast feedback receptor is slow and limits the number of receptors able to generate the fast feedback signal. Whatever explanation is accepted, de-sensitisation of fast feedback has clear teleological implications in terms of CRF-ACTH regulation. Stresses of great intensity with high neural drive will swamp corticosteroid fast feedback and ensure persistent levels of ACTH and cortisol in the plasma, which may aid survival of the organism in a life-threatening situation.

(7) *Mechanism of Action of Fast and Delayed Feedback*

(a) *Fast Feedback*. Previous studies had shown that secretion of CRF is inhibited by two neurotransmitters – noradrenaline and GABA. Corticosterone might act by excitation of one of these inhibitory systems. This possibility has been examined using the hypothalamus *in vitro* and stimulating output of CRF activity with acetylcholine in the presence of phentolamine and picrotoxin. The action of corticosterone was unaffected and it it was thus concluded that the steroid does not act via a neural inhibitory pathway (Jones *et al.*, 1976c). Corticosterone prevents both acetylcholine and 5-HT induced release of CRF from the rat hypothalamus *in vitro*, but has no influence on release induced by K^+. This is taken to mean that the steroid stabilises the CRF cell membranes (Jones *et al.*, 1976c) and inhibits CRF release.

(b) *Delayed Feedback.* Delayed feedback at the hypothalamus is probably due to corticosterone influence on the rate of CRF synthesis, since the output of CRF in response to both acetylcholine and electrical stimulation is greater than the original content of the tissue, and the output is reduced to that seen in intact animals by pretreatment with corticosterone (Bradbury *et al.*, 1974). At the pituitary level corticosteroids are thought to exert their delayed feedback mainly on a step in protein synthesis (Arimura *et al.*, 1969) and hence on the synthesis of ACTH (Hodges and Vernikos, 1960); this presumably explains why there is a delay.

(8) *Integration of Fast and Delayed Feedback Mechanisms*

An unanswered question is the relative roles and integration of the fast and delayed feedback mechanisms in the control of ACTH secretion. A suppression of ACTH release occurring some hours after glucocorticoid administration, has been repeatedly seen in all species including man. Because of its long latency the delayed feedback mechanism cannot be the mechanism which regulates rapid responses such as the stress-induced release of ACTH. Furthermore, this delayed suppression occurs only with doses in the upper limit of the physiological range, and this is the form of feedback which results in the therapeutically unwanted suppression of the stress response.

Fast feedback can be demonstrated with small doses of corticosteroids (Dallman and Yates, 1969) and lasts up to 20 minutes, which approximates to the duration of corticotrophic response (Jones *et al.*, 1972). It would appear, at first sight, that fast feedback alone could provide the necessary regulation of the stress response. More experiments are, however, needed to define the relative amounts of steroid required for the regulation of the stress response.

Espiner *et al.* (1972) have shown that in conscious sleep the corticotrophic response to immobilisation stress is biphasic. There is a rapid increase followed by an immediate decline to a basal level within 20–30 minutes. This is followed by a rise from 1 to 15 hours before a final decline again to the original level. Yates and Maran (1975) have suggested that the best explanation for this pattern of secretion is the stress stimulation being modulated by first, the fast, and then the delayed, feedback mechanisms.

A clearer example of how the two feedback mechanisms may integrate to modulate the stress response, is illustrated by experiments in which the effects of ACTH secretion of sham operation, adrenalectomy and adrenalectomy followed by replacement therapy are compared (Dallman *et al.*, 1972; Dallman and Jones, 1973). Adrenalectomy results in elevations of blood ACTH levels for up to 2 hours, but this response is converted to one similar to sham operation if adrenalectomy is followed by a single injection of a physiological dose of corticosterone (Fig. 8). In this experiment the delayed feedback was presumably provided by endogenous corticosterone secretion in the hours preceding adrenalectomy. The dose of corticosterone used for replacement was probably within endogenous secretion rate, since adrenal vein cannulation experiments have shown that one adrenal can secrete 1–2 μg/min of corticosterone for up to one hour under surgical stress (M. T. Jones and R. B. Neame, unpublished observations). An important point is that the steroid

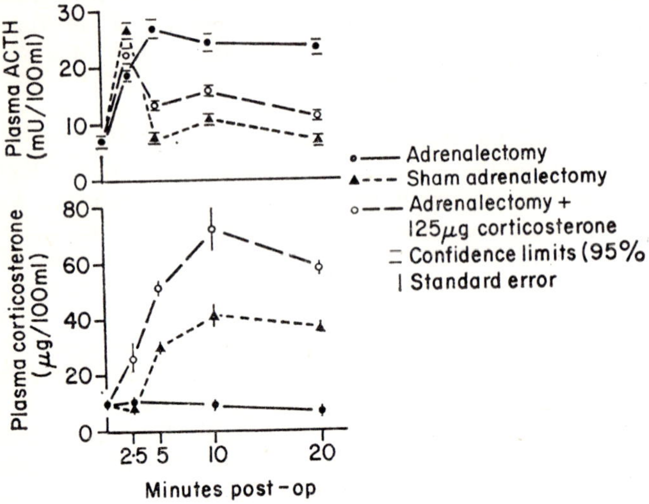

Fig. 8. Immediate effects of adrenalectomy on ACTH levels. (From: Dallman *et al.*, 1972.)

was administered after removal of the adrenal glands: this is a consideration that has been ignored by others who have administered the corticoid before the stimulus to ACTH secretion.

The physiological basis for the inter-relationship between the two mechanisms in the hypothalamus might be that the fast feedback reduces the number of neural impulses releasing CRF, whilst the delayed feedback controls the synthesis of CRF and thus the amount released per impulse.

(9) *Negative Feedback Control and the Episodic Secretion of ACTH*

ACTH is secreted in an episodic fashion which shows ultradian as well as circadian periodicity (Section VI). The question arises whether this is in any way influenced by corticosteroid feedback. The basic pattern, of elevated blood ACTH levels in the period before waking and low values in the period leading up to sleep, is unaffected by the presence or absence of the adrenal glands. However, in the absence of the adrenals the levels are higher at all points and each episode of ACTH secretion is prolonged. It is also possible to modify or abolish the circadian rhythm with both natural and synthetic glucocorticoids (Ceresa *et al.*, 1970). These facts suggest that episodic ACTH secretion in the unstressed state may be modified by corticosteroid negative feedback mechanisms, the initial pattern being engendered by rhythms that have their origins within the central nervous system.

(10) *The Influence of Adrenalectomy on ACTH Secretion*

Bilateral adrenalectomy is associated with an immediate and prolonged secretion of ACTH, associated with a fall in its pituitary stores beyond that seen in the sham operated animal (Redgate, 1967; Dallman *et al.*, 1972;

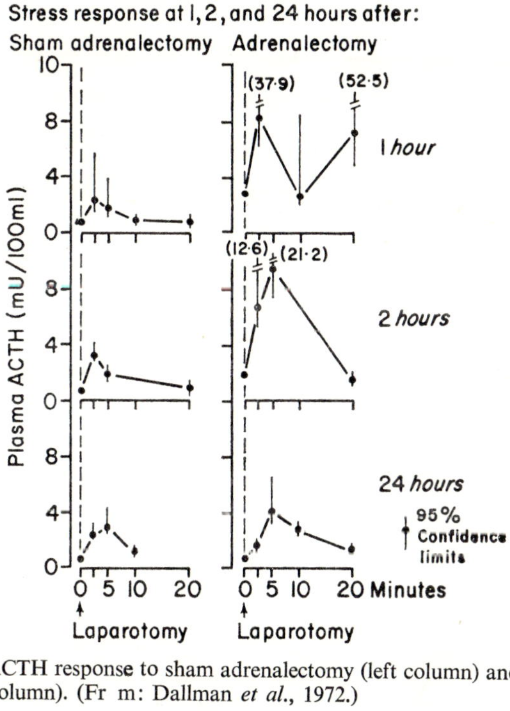

Fig. 9. Plasma ACTH response to sham adrenalectomy (left column) and bilateral adrenalectomy (right column). (Fr m: Dallman *et al.*, 1972.)

Dallman and Jones 1973; Buckingham and Hodges, 1974a, b). Another characteristic effect of bilateral adrenalectomy is hypersecretion of ACTH in response to stress, an effect that would be predicted if the system was converted from a closed to an open loop (Fig. 9; and see Chapter 3).

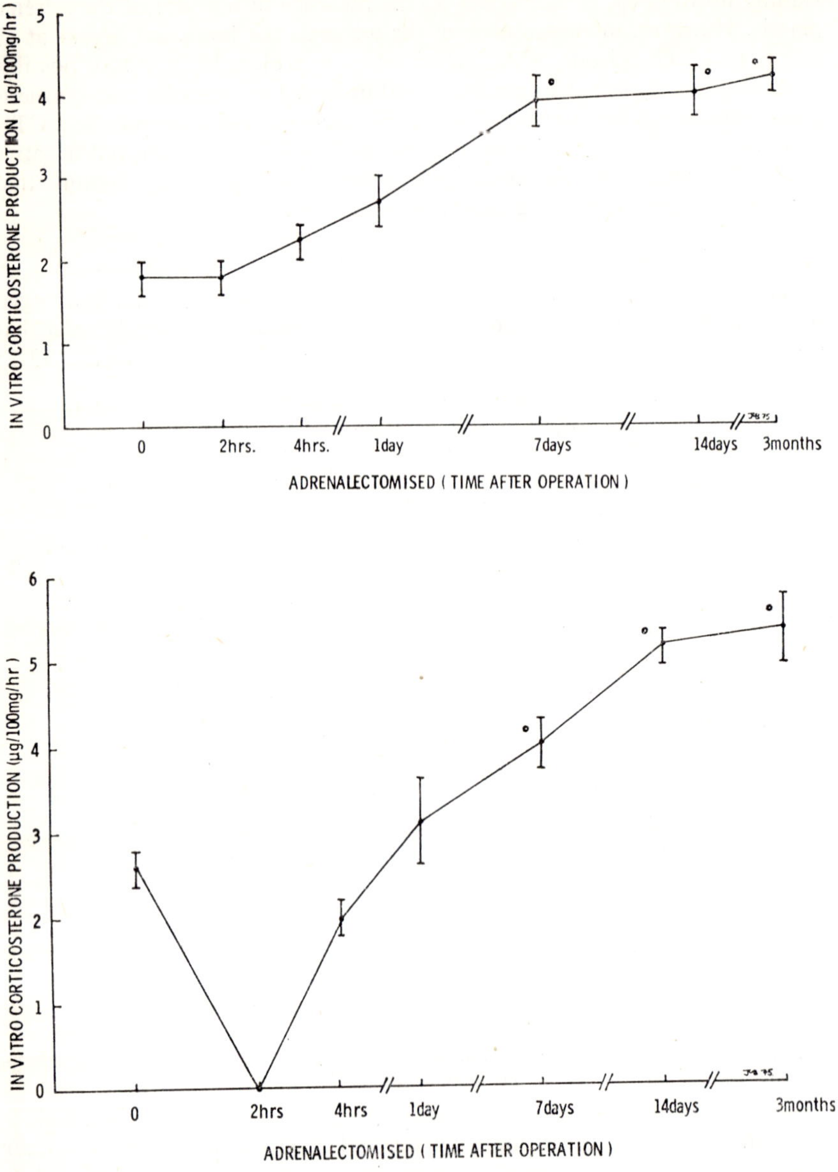

Fig. 10. Effect of acetylcholine on hypothalamic CRF release *in vitro* following adrenalectomy. (From: Jones *et al.*, 1976.) Upper panel: Basal secretion. Low panel: Effect of acetylcholine. Asterisks indicate values significantly higher than controls ($P<0{\cdot}01$).

Hypothalamus from adrenalectomised rats incubated *in vitro*, show an increased secretion of CRF activity in response to an excitatory neurotransmitter (Fig. 10). This shows that the hyper-secretion of ACTH that occurs in adrenocortical insufficiency has its origins in the hypothalamus. Adrenalectomy also results in changes in several sensory modalities, but whatever changes are induced in these, the net change is unlikely to play as important a role as the loss of feedback action of corticosteroids. Further evidence that the site of action of corticosteroid feedback is on the CRF cell, is that the response to acetylcholine is enhanced by adrenalectomy, the final neural pathway to CRF release being cholinergic (see Section IV).

(11) *Fast Feedback in Man*

It has recently become possible to measure circulating ACTH in man under basal as well as stress conditions, and though a number of reports have appeared in which the effect of corticosteroids on circulating ACTH have been studied, few have investigated the fast feedback mechanism. However, Daly *et al.* (1974a) have determined plasma ACTH level in two normal subjects before and after i.v. dexamethasone 0·5 mg or cortisol 5 mg. There was an immediate "switch off" of ACTH secretion and the plasma concentration continued to decline for up to one hour. These results indicate that the basal secretion mechanism for ACTH, in man, is responsive to the fast feedback action of corticosteroids. The same authors (Daly *et al.*, 1974a) report a negative correlation between the plasma levels of ACTH and corticosteroids: this finding is compatible with the view that a corticosteroid negative feedback mechanism is one of the regulatory systems controlling ACTH secretion. Rees *et al.* (1973) have also reported data which show a rapid fall in plasma ACTH concentration following the administration of cortisol, but the dose used (100 mg) was high and unphysiological.

(12) *The Effects of Normal Human Pregnancy*

The free, unbound fraction of plasma cortisol is raised during pregnancy, as is the concentration of circulating ACTH. The levels lie mostly at the upper range of normal, with a few at the lower end of the range seen in Cushing's syndrome (Burke and Roulet, 1970; Rees *et al.*, 1975). Although the concentration of cortisol binding globulin (CBG) in plasma is elevated, the free cortisol is still higher, suggesting that the normal feedback regulation of ACTH has been altered. Since only the free moiety is biologically active, the tissues are subjected to higher than normal glucocorticoid stimulation during normal pregnancy. The possible biological advantage of this is not known, but it may benefit the developing foetus. The high free levels of cortisol may, however, account for the mild Cushingoid features (such as striae) that are encountered in pregnant women.

Two explanations for the co-existence of high ACTH and high plasma cortisol are possible. The first is that much of the circulating ACTH is of placental not pituitary origin and is therefore not subject to normal feedback regulation; the placenta does contain a great deal of the hormone but the ability of the tissue to synthesise ACTH has yet to be demonstrated (Rees *et al.*, 1975). The second hypothesis suggests a reduced sensitivity of the hypo thalamo-pituitary system to the feedback action of cortisol; this could result from interference of progesterone on fast feedback and 17-OH-progesterone on the delayed feedback action of cortisol, thus effectively re-setting the feedback controller to function at a higher cortisol level (Burden *et al.*, 1975).

(13) *Short Loop* (*ACTH*) *Feedback*

The suggestion has been made by several authors (Hodges and Vernikos, 1958; Kitay *et al.*, 1959; Brodish and Long, 1962; Motta, 1964; Martini *et al.*, 1968), that further secretion of ACTH may be inhibited by a direct feedback mechanism involving the peptide hormone itself (see Chapter 3). Anterior pituitary homografts placed in the third ventricle cause a depression in ACTH secretion, as shown by nuclear shrinkage of the zona fasciculata (Szentágothai *et al.*, 1972). Brodish and Long (1962), using a cross-circulation technique, showed that after hypophysectomy CRF appeared in the general circulation within one hour, and this was diminished by a lesion in the hypothalamus. Jones *et al.* (1976b) have shown that ACTH, added to the rat hypothalamus *in vitro*, prevents the CRF release induced by 5-HT. All these data have been interpreted as evidence for a hypothalamic short loop feedback. The doses found to be effective in the experiments of Jones *et al.*, suggest that ACTH in the concentrations achieved under stress may modulate CRF secretion.

D. THE ROLE OF THE FOETAL HYPOTHALAMO-PITUITARY-ADRENOCORTICAL AXIS IN PARTURITION

The foetal adrenocortical system is believed to be partly responsible for the initiation of parturition in the sheep, since this is delayed by adrenalectomy and hypophysectomy and hastened by the administration of ACTH and glucocorticoids (Dorst and Holm, 1968; Liggins, 1968). At the time of delivery, plasma cortisol and ACTH are elevated in the sheep foetus (Nathanielsz *et al.*, 1972; Alexander *et al.*, 1971, 1973). However, observations in other species and in man, make the role of the foetal adrenocortical system questionable, for adrenocortical activity decreases before birth in the rat foetus (Jost, 1966, 1973). There is a rise in plasma cortisol in spontaneously born babies (Murphy, 1973), but the length of gestation is not altered in congenital adrenal hyperplasia (Price *et al.*, 1971), congenital aplasia of the pituitary (Swaab and Honnebier, 1974) or in human anencephalics (lacking a functional hypothalamus) without hydramnios (Honnebier and Swaab, 1973). Thus, the sheep may be the exception, occupying one end of the spectrum, where adrenocortical activity is important in controlling the length of gestation, with man occupying the other end, where the system plays only a role in enabling the foetus to withstand the stresses associated with birth.

VI. Adrenocortical Rhythms

ACTH (and presumably CRF) is secreted episodically, with a periodicity of approximately 3 hours, which increases during the circadian peak of pituitary-adrenocortical activity. These ultradian bursts are so wide apart that for about 25% of the time the secretion rate of adrenocortical steroids is negligible (Hellman *et al.*, 1970; Krieger *et al.*, 1971; Weitzmann *et al.*, 1971). The circadian ryhthm is such that plasma ACTH and cortisol show peak levels during the sleep period shortly before waking and low levels shortly before the sleep period (Conroy and Mills, 1970). The individual episodes of ACTH secretion can be detected even at the peak of the circadian rhythm, but because of the longer half life of cortisol, and its binding to cortisol binding globulin in plasma which "damp" down the peaks, there is fusion of the separate bursts in plasma cortisol levels at the height of the circadian rhythm.

The influences that entrain the circadian pituitary-adrenocortical rhythm include the sleep/wakefulness cycle and the dark/light cycle which are usually reinforced by each other in normal life. When these two cycles are out of phase, for instance in night-workers, the actions of both may be seen; a peak occurs before awaking and another at the time of increased lighting (Orth and Island, 1969). Few observations have been made on individuals in complete isolation from light/dark for long periods (Conroy and Mills, 1970). One such subject, isolated for three months underground, showed no circadian rhythm in plasma cortisol (Mills, 1964). The disturbance in the normal cortisol rhythm that occurs by crossing time zones, may take a week to recover (Conroy and Hall, 1969).

There is a circadian variation in the ability of both natural and synthetic corticoids to suppress ACTH secretion. In both man and rat, their maximum inhibitory effect is exerted when they are administered 4–8 hours before the circadian peak (Grant *et al.*, 1965; Nichols *et al.*, 1965; Retiene *et al.*, 1967). Corticoids administered for therapeutic reasons when given only in the morning, give relief of symptoms equal to the same dose divided between morning and evening. There is, therefore, a sound argument for restricting administration to the morning only, thus limiting the dangers of pituitary suppression. There is also a circadian variation in the magnitude of the corticotrohpic response to stress in man, the response being at its highest when the plasma cortisol is at its nadir (Clayton *et al.*, 1963; Takene *et al.*, 1966).

It has been suggested that a variation in the sensitivity of the feedback system to corticoids is responsible for generating the circadian rhythm of ACTH. This suggestion is difficult to reconcile with what is known of the influences that entrain the pituitary-adrenocortical system. A simpler and better model is one in which a periodic drive to CRF is considered to cause periodic changes in ACTH secretion. the duration of each burst of activity being modulated by corticoid fast feedback. The periodicity of CRF secretion is considered to be the result of a summation of impulses from higher centres on to the CRF neurone.

It is established that the circadian rhythm of the hypothalamo-pituitary-adrenocortical axis is regulated by pathways that enter the anterior region of the hypothalamus. Lesions of this region abolish the rhythm without affecting most of the stress responses (Slusher, 1964; Halász *et al.*, 1967a,b; Palka *et al.*, 1969). However, the pacemaker which regulates the circadian rhythm has not been located.

Burns (1975) has suggested that changes in plasma ACTH occurs during the menstrual cycle, the levels in mid-cycle being significantly higher than those at menstruation. This was interpreted to mean that oestrogens affect ACTH production, but an alternative explanation would be the high levels of 17-OH progesterone, as this steroid is known to interfere with the delayed feedback action of cortisol in experimental animals (Burden *et al.*, 1975).

VII. Clinical Aspects

A. GLUCOCORTICOID AND ACTH THERAPY

ACTH and synthetic glucocorticoids are used for the control of chronic inflammatory diseases. Both forms of therapy are efficacious but not without danger of being associated with suppression of the hypothalamo-pituitary axis. However, the deficiency that results from ACTH therapy is less severe than that accompanying steroid treatment. Thus Daly *et al.* (1974b) found that patients undergoing long-term ACTH treatment, showed a normal cortisol response to insulin-induced hypoglycaemia. Although the increase in blood ACTH in response to insulin was less than in the controls, the integrated amount over 9 minutes was not different from normal. Patients on therapeutically equivalent doses of prednisolone show reduction in cortisol, and ACTH secretion. The explanation for this difference between the two forms of treatment is not obvious, since ACTH releases cortisol and this should suppress ACTH in much the same way as prednisolone does. Daly *et al.* (1974b) suggest that as prednisolone is given in two divided doses (morning and evening), this doubles the feedback signal and suppresses hormone secretion. ACTH therapy, on the other hand, is given as a single injection in the morning, and the suppression has worn off by the following day when the next dose is administered. Thus, the inhibition of endogenous ACTH is intermittent and not continuous as is the case with prednisolone. Myles *et al.*, 1971 had previously shown that the same total dose of prednisolone, given in two divided doses, caused greater suppression than a single dose in the morning. Almost equal benefit was obtained from a therapeutic point of view with the single dose regime in the treatment of rheumatoid arthritis.

Long-term treatment with corticoids may induce suppression of the hypothalamo-pituitary adrenal system which, on withdrawal of therapy, results in adrenocortical insufficiency (Salassa *et al.*, 1953). The severity of adenohypophysial basophil degranulation, associated with this insufficiency, correlated with the duration and the quantity of steroid administered. A few patients with functional impairment may develop severe (sometimes irreversible) post-operative shock unless they are given glucocortoid cover. It is essential, therefore, to have some reliable tests of the integrity of the stress-responsiveness of the system (e.g. insulin hypoglycaemia). Glucocorticoid therapy may also cause a suppression of the normal circadian rhythm of plasma ACTH and cortisol. If there is adrenocortical atrophy, the adrenocortical response to ACTH (or α 1–24 ACTH) will also be depressed. Even a good response to insulin-induced hypoglycaemia, however, does not of necessity guarantee that the patient will respond normally to surgical stress, but it does suggest that daily stresses of moderate intensity can provoke adequate ACTH and cortisol secretion.

Normal adrenal responsiveness to ACTH is maintained by the episodic and stress-induced secretion of endogenous ACTH. A failure of ACTH secretion precedes the reduction in adrenal responsiveness and pituitary (and hypothalamic) recovery must occur prior to a restoration of adrenal function. Support for this logical sequence is provided by Graber *et al.* (1965) who measured ACTH in corticoid treated patients and found that the pituitary recovered before the adrenal, after steroid withdrawal.

B. TESTING THE HYPOTHALAMO-PITUITARY-ADRENAL AXIS

The function of the whole axis can be assessed by using insulin-induced hypoglycaemia, for which a standard procedure has been described by Landon *et al.*, (1963). Other tests, such as the response to metyrapone (the 11-β hydroxylase inhibitor) or vasopressin, have certain disadvantages. Metyrapone tests only the integrity of the feedback elements and tells nothing about the stress mechanism, whilst the relevance of a response to vasopressin is doubtful, since it is not the true CRF and probably tests only the ability of the anterior pituitary to release ACTH. It is not surprising, therefore, that several reports show patients giving different responses in the various test (Kaplan, 1963; Brinck-Johnson *et al.*, 1963; Daly *et al.*, 1967).

C. DISEASES OF THE HYPOTHALAMO-PITUITARY-ADRENAL AXIS

Cushing's syndrome is due to the excessive secretion of corticosteroids. It can be caused by an autonomous neoplasm of the adrenal gland or an inappropriately secreting ACTH tumour of the anterior pituitary gland or an autonomous tumour elsewhere in the body (e.g. oat-cell carcinoma of the lung). These three causes can be differentiated by measuring ACTH before and after a large dose of dexamethasone and by evidence of a tumour at a specific site (Liddle, 1967).

Autonomous adrenal tumours (benign or cancerous) function independently of ACTH and a large dose of dexamethasone has no influence on their function. A typical patient has a high plasma cortisol concentration associated with subnormal plasma ACTH levels and the plasma cortisol level is unaffected by either dexamethasone or ACTH administration.

A patient with an ACTH-secreting pituitary tumour has high plasma concentrations of ACTH and cortisol, but both are suppressed by large doses of dexamethasone. The patient shows a marked rise in plasma ACTH and cortisol concentrations in response to metyrapone. The recent successful treatment of patients with cyproheptadine (a 5-hydroxtryptamine receptor blocker), suggests that inappropriate CRF secretion may be the primary dysfunction in this form of Cushing's syndrome (Krieger *et al.*, 1975).

Ectopic ACTH-secreting tumours can be diagnosed by demonstrating high circulating concentrations of ACTH and cortisol that are uninfluenced by large doses of dexamethasone and the finding of a non-pituitary tumour that contains a high concentration of ACTH.

VIII. Summary

Corticotrophin releasing factor is responsible for the release and synthesis of ACTH from the anterior pituitary. CRF is released in an episodic fashion with a variable frequency which results in a circadian rhythm in plasma ACTH and cortisol levels.

Stress increases neural input to the hypothalamus and the release of CRF and ACTH is then modulated by the corticosteroid negative feedback mechanism. The neural pathways which control CRF secretion are in the process of being defined, as well as the nature of the neurotransmitters involved. An understanding of the neurotransmitters regulating CRF should lead to new methods of treating Cushing's syndrome due to inappropriate secretion of ACTH from a pituitary tumour. It may also help in the treatment of Addison's disease of anterior pituitary origin.

The role of ADH in the control of ACTH is debatable, but there can be no doubt that it is not CRF. There are fibres from the magno-cellular nuclei that terminate at the portal vessels and it is likely that ADH from these may be released in certain stresses and potentiate the action of CRF.

The fact that there is no correlation in structure-activity relationship between the anti-inflammatory and feedback effects of corticosteroids, may lead to the future development of anti-inflammatory steroids which do not suppress ACTH and GH secretion.

However, the greatest advancement will come when CRF has been isolated and synthesised and the therapeutic benefits both from testing and correcting the functional ability of the corticotroph cells of the anterior pituitary can be realised. The release of CRF is one of the most rapid of endocrine responses and is involved in the stress response. It will be interesting to find whether this peptide has concomitant effects on behaviour, since stress has profound psychological as well as biochemical effects in man and other species.

CHAPTER 12

Control of Melanocyte-Stimulating Hormone (MSH) Secretion

S. TALEISNIK

I. Introduction

Most early studies of the regulation of the function of pars intermedia were on lower vertebrates (see Etkin, 1967; Kastin *et al.*, 1974b; and Hadley and Bagnara, 1975, for review), probably because the functional state of the gland can be evaluated by the changes in skin pigmentation. In the last decade, however, the pars intermedia function in mammals has received increased attention in regard to both mechanisms of control and physiological significance and an excellent review has recently been published (Howe, 1973). However, many of the control mechanisms are still poorly understood and the information on the concentration of MSHs in blood is scanty owing to a lack of adequate techniques for measuring circulating levels of the hormones.

The objective of this chapter is to present the current state of information on the regulatory mechanisms of MSH secretion operating in mammals. Studies on lower vertebrates will also be considered when they are pertinent to the enunciation of general concepts.

II. Morphological Aspects

The pars intermedia of the hypophysis is a structure composed of several layers of epithelial cells in some mammals (rat, dog and rabbit) but is almost absent in others (whale, elephant). In man only a vestigial pars intermedia exists; the intermediate cells are dispersed in the pars distalis.

A. Cytology

Light-microscope studies have shown more than one type of cell (Wingstrand, 1966; Howe, 1973). Recent electron microscope descriptions of the pars intermedia of amphibia (Saland, 1968) and mammals (Howe and Maxwell, 1968; Vincent and Anand-Kumar, 1969; Cameron and Foster, 1971; Naik, 1972) have indicated at least two types of cells.

(1) *Type I or Glandular Cells*

The characteristic of these predominant cells is the presence of granules in their cytoplasm. Two types of granules can be recognised; an electron-dense membrane-bound granule and a relatively clear vesicle. The electron-dense granules, about 100–250 nm diameter are mainly confined to the Golgi apparatus, whereas the cytoplasmic vesicles of similar or larger size are usually very numerous and almost fill the cell.

The glandular cells show a positive reaction with anti-MSH antibodies and probably synthesise and secrete MSH. Immunocytochemical staining indicates that MSH is localised in the dense granules and cytoplasmic vesicles. Stain is never observed in the core granules inside the Golgi complex, which may indicate that MSH is transferred to the granules once they have been liberated from the Golgi cisternae (Naik, 1973).

(2) *Type II, Interstitial or Stellate cells*

These smaller, and usually less common, cells have relatively dense irregular shaped nuclei and thin rims of cytoplasm with long processes extending between neighbouring Type I cells. They appear to be linked to each other and to the glandular cells by junctional complexes. At the border of the hypophysial cleft where they are very common, the microvilli extend into the colloid of the cleft. The cytoplasm of these cells never contains secretory granules. The functional significance of the interstitial cells is uncertain.

(3) *Corticotrophic Cells*

Another type of cell, smaller than the glandular cells and morphologically similar to the corticotrophic cells of the pars distalis, is found in the pars intermedia of the rat and the mouse (Naik, 1972; Stoeckel *et al.*, 1973). These cells are concentrated in the rostral zone of the pars intermedia, where it joins the pars distalis, but are also found scattered throughout the gland. Immunohistochemical staining for ACTH is positive, not only in these cells, but also in almost all the parenchymal cells of the pars intermedia (Moriarty and Halmi, 1972; Naik, 1973).

B. INNERVATION

The pars intermedia is innervated by fibres which arise from perikarya located in the hypothalamus and reach the gland via the pars nervosa. Electron microscopic preparations show at least two types of neuronal endings in the pars intermedia of both lower vertebrates (Saland, 1968; Nakai and Gordman, 1969) and mammals, (Howe and Maxwell, 1968; Vincent and Anand-Kumar, 1969; Belenky *et al.*, 1970; Cameron and Foster, 1971) which can be recognised by their characteristic granules. One type contains uniformly electron-dense elementary granules of 110–250 nm diameter and has been referred to as neurosecretory. The (Gomori positive) fibres have also been described using the light microscope. The endings of these fibres may be found in the connective tissue layer between the pars nervosa and pars intermedia, or in the septa from this layer which penetrate the gland, where they are separated from the glandular cells by a thick basement membrane. Direct contacts between the neurosecretory endings and the glandular cells of the pars intermedia have also been observed.

A second type of neuronal ending contains synaptic vesicles of 50–90 nm diameter with electron-dense granules that are separated from the limiting membrane by a clear space. Morphologically these granules are similar to those described carrying monoamines and thereby the neurones are considered as monoaminergic. Electron microscopic autoradiographic studies after systemic injection of (^{3}H)-adrenaline indicate that these endings take up the labelled amine selectively (Doerr-Schott and Follenius, 1970; Hopkins, 1971) thus indicating their adrenergic nature. Histochemical fluorescence methods have provided further evidence of an adrenergic innervation of the pars intermedia. Both in amphibia and mammals most of the cells of pars intermedia are surrounded by a rich plexus of delicate fluorescent varicose fibres (Enemar *et al.*, 1967; Björklund and Falck, 1969). The fluorescent fibrous structures cannot be demonstrated after reserpine treatment, suggesting that dopamine and noradrenaline are stored in them. Microspectro-fluorometrically these nerve fibres exhibit the characteristics of noradrenaline and dopamine (Weman and Nobin, 1973).

The endings of neurosecretory and monoaminergic fibres also contain agranular synaptic vesicles of 30–40 nm diameter but endings can be found that contain only these latter vesicles. Such terminals have been considered to belong to a cholinergic type of neurone innervating the pars intermedia (Bargmann *et al.*, 1967; Vincent and Anand-Kumar, 1969). However, electron histochemical studies failed to detect acetylcholinesterase activity in the neurones, a fact which argues against the presence of a cholinergic system in the pars intermedia (Whitaker and La Bella, 1973). Since the dark-cored granules in the monoaminergic type of endings are generally grouped together it is possible to find sectors of these fibres containing only clear synaptic vesicles.

Both neurosecretory and monoaminergic fibres show synaptic contacts with the secretory cells. Single secretory cells may have synaptic contacts with two different kinds of nerve endings, neurosecretory and monoaminergic. In general, the neurosecretory endings are more numerous near the pars nervosa than in the rest of the pars intermedia, whereas, the monoaminergic endings are found throughout the gland. The degree of innervation of the pars intermedia varies from one species of animals to another, from glands receiving a profuse innervation, the cat (Bargman *et al.*, 1967), to those very poorly innervated, the lizard, *Klauberina riversiana* (Rodriguez and La Pointe, 1970). In the latter situation the fibres end at the vascular septum, interposed between the pars nervosa and the pars intermedia. In such cases, neurosecretory substances may reach the cells, lacking direct innervation, by diffusion or via short blood vessels.

C. BLOOD SUPPLY

The pars intermedia is poorly vascularised in most vertebrates (Wingstrand, 1966; Howe, 1973). A vascular plexus situated in the conjuctive layer between the pars nervosa and pars intermedia supplies vessels which enter into the intermedia. Capillaries are rare in many mammals, including the rat; however, in some like the rabbit and sheep, the lobulated pars intermedia is supplied by capillaries which follow the septa of conjunctive tissue forming a plexus around the lobules. In general, the blood reaching the pars intermedia is related to the pars nervosa and independent of the hypothalamo-hypophysial portal system, although some capillaries originating in the caudal extension of this system supply the rostral zone. How substances reach pars intermedia cells not in direct contact with the vessels, or the secretions of these cells reach the periphery, is not understood. The extracellular channels between parenchymal cells described in Maxwell, 1968) may be important.

III. MELANOCYTE-STIMULATING HORMON

Two types of MSH have been isolated from pituitary gla α-and β-MSH. Detailed reviews on this subject have been written (Harris, 1966; Lande and Lerner, 1967). α-MSH is a straight chain of 13 amino acids and β- MSH consists of 18 amino acids. They have in common a sequence of 7 amino acids (the hepatapeptide core) which is also present in ACTH

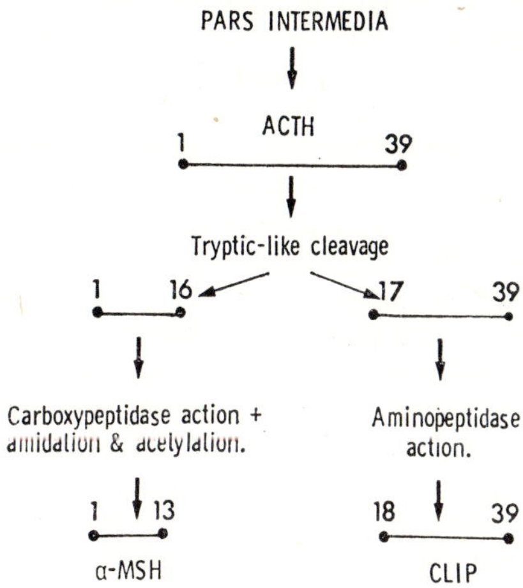

Fig. 1. Proposed mechanism for the formation of α-MSH. (From: Scott *et al.*, 1973.)

and β- and γ-lipotrophin and is responsible for the MSH-like activity of these peptides.

It has been proposed that in the pars intermedia ACTH might act as a precursor in the formation of α-MSH (Scott *et al.*, 1973, 1974b). According to this view ACTH is split intracellularly into two fragments one of which (the N-terminal) is α-MSH. The C-terminal fragment is converted to the 18–39 portion of ACTH by the action of an aminpeptidase: this peptide has been termed corticotrophin-like intermediate lobe peptide (CLIP) (Fig.1). It is further proposed that β- and γ-lipotrophins are precursors of β-MSH. In the human β-MSH is residues 41–58 of these peptides and it it is now thought that the biological and immunological activity formerly attributed to a β-MSH in man is, in fact, associated with β-lipotrophin.

IV. Neural Mechanisms of Control of MSH Secretion

A. BRAIN INFLUENCES

Several procedures have been used to disconnect the pars intermedia from the brain, in order to assess the influence exerted by the central nervous system on its functional activity. These include transection of the pituitary stalk, destruction of the hypothalamic median eminence and transplantation to a site remote from the hypothalamus (see Chapters 1 and 2). Transection of the hypothalamo-hypophysial tract or injury of the hypothalamus, in both immature and adult amphibia, leads to hyperactivity of the pars intermedia cells, characterised by long-lasting skin darkening and loss of their capacity to respond to changes in light environment (Etkin, 1967). It is believed that these effects are due to an increased release of MSH, resulting from the removal of a sustained inhibition exerted by the central nervous system on the pars intermedia. In animals with cut hypothalamic connections, the MSH content in the posterior lobe markedly decreases (Ito, 1968; Kastin *et al.*, 1974b), while electron-microscopic studies show a well-developed, rough-surface endoplasmic reticulum and a reduction in cytoplasmic granules in pars intermedia glandular cells, suggesting that both the release and synthesis of MSH increases (Ito, 1968). Similar hyperactivity has been observed in mammals in the pars intermedia after removal of the pituitary gland from the direct influence of the central nervous system. Several species show hypertrophy and often hyperplasia of the pars intermedia after stalk section (see Kastin *et al.*, 1974b). The ultrastructure of the secretory cells in mammals with sectioned stalks or hypothalamic lesions are similar to those described in lower vertebrates (Vincent, 1969; Ooki *et al.*, 1973). The MSH concentration in serum rises in rats after lesioning the hypothalamus in the median eminence, whereas the pituitary content of the hormone initially falls but is later restored to normal values, in spite of the sustained secretion of MSH (Taleisnik *et al.*, 1967; Carrillo *et al.*, 1973). Transplantation of the pituitary gland to a site remote from the brain results in similar functional and structural signs of hyperactivity of the pars intermedia (Rust and Meyer, 1969; Perryman, 1974). These findings indicate that when the pituitary gland is disconnected from the brain, the intermediate cells are freed from inhibition and MSH is released in greater amounts into the circulation. The content of the hormone in the gland falls as a consequence of the increased release, but afterward it is restored due to an increase in synthesis.

Another procedure to disrupt the brain influence on the pars intermedia is by pharmacological suppression of hypothalamic activity. Reserpine treatment, for example, enhances the secretion of MSH in lower vertebrates and rats (Tomatis and Taleisnik, 1968b).

The site in the hypothalamus from which the inhibitory neurones originate has not been completely established. Lesions or transections at various levels in the hypothalamus of lower vertebrates, suggest that the inhibitory neurones originate from the preoptic area (Voitkevich, 1962; Enemar *et al.*, 1967). Extirpation of the pars magnocelullaris of the preoptic nucleus from which neurosecretory fibres originate does not however, result in a loss of control of MSH release (Dierickx, 1967; Compher and Dent, 1973), whereas extirpation of a monoaminergic nucleus causes darkening of the skin (Goos, 1969).

In the rat the neurones of the paraventricular nuclei are responsible for an inhibitory action exerted by the hypothalamus on the pars intermedia. After lesioning these nuclei, MSH levels rise in serum and decrease in the pituitary (Taleisnik *et al.*, 1967; Taleisnik and Celis, 1973) and there are concomitant morphological signs of hyperactivity in the cells of the pars intermedia (Howe and Thody, 1969). The release of MSH induced by the injection of hypertonic saline is also prevented, which indicates that the paraventricular neurones are part of a pathway activated by MSH releasing stimuli.

The paraventricular nuclei in mammals are considered to be homologous with the preoptic nuclei of amphibia, which may suggest a common site of location of inhibitory neurones in the different phyla.

B. MONOAMINES

There is some evidence to indicate that the release of MSH is affected by catecholamines through a mechanism acting directly on the pars intermedia cells. Thus, in toads, *Bufo arenarum*, bearing a posterior lobe implanted into the eye, the melanocyte index reading of the skin increases after the sympathetic fibres innervating the eye are cut, but it decreases after local application of dopamine or noradrenaline (Iturriza, 1967). Also, the paling effect produced by the injection of adrenaline, noradrenaline or dopamine into the cranial fluid in fish (*Scyliorhinus canicula*), adapted to a black background, is believed to be due to an action on the pituitary gland, as the agents are effective even when the hypothalamo-hypophysial tract has been transected or the animals are treated with reserpine (Wilson and Dodd, 1973). Consistent with these findings is that the secretion of MSH from frog (*R. pipiens*) posterior lobes or rat pituitaries *in vitro* is inhibited by the presence of adrenaline, noradrenaline or dopamine in the incubation medium. It appears that α-adrenergic receptors are involved in this inhibitory action, whereas β-adrenergic receptors are implicated in the stimulation of MSH secretion by catecholamines (Hadley *et al.*, 1975). Beside these effects on the pars intermedia, catecholamines inhibit MSH release in the rat by an action at the hypothalamic level via the release of MIF (Taleisnik *et al.*, 1973).

C. HYPOTHALAMIC PEPTIDES

Extracts of hypothalamic tissue were found either to promote or to block the release of MSH, depending on the condition of the animals from which the hypothalami were obtained; accordingly a MSH-releasing factor and a MSH-release inhibiting factor have been proposed.

(1) *MSH-release Inhibiting Factor* (*MIF*)

The hypothalamic extracts containing MIF block the release of MSH from pituitary glands *in vivo* and *in vitro*. MIF prevents the depletion of pituitary MSH and the rise in plasma MSH induced by the i.v. injection of hypothalamic extracts containing MRF (Taleisnik and Tomatis, 1967a, b) or by nembutal-morphine treatment (Kastin *et al.*, 1969b); it also prevents the fall in pituitary MSH found in nursing rats (Taleisnik and Tomatis, 1967b). Added into the medium of incubated rat pituitaries, it depresses MSH release (Bower *et al.*, 1971).

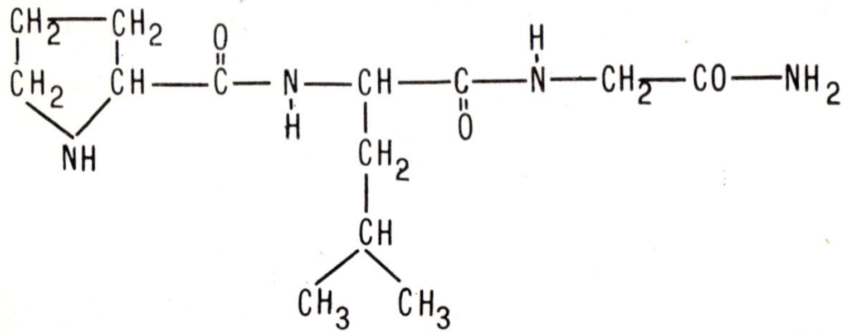

Fig. 2. Prolyl-leucyl-glycinamide (MSH-Release Inhibiting Factor, MIF).

The tripeptide Pro-Leu-Gly-NH_2 (Fig. 2) was shown to inhibit MSH release (Celis *et al.*, 1971b) and later its isolation from stalk-median eminence extracts was reported (Nair *et al.*, 1971a). A pentapeptide Pro-His-Phe-Arg-Gly-NH_2 possessing MIF activity was also isolated, but it is at least 20 times less active than the tripeptide in inhibiting MSH release (Nair *et al.*, 1972; Chapter 5).

Pro-Leu-Gly-NH_2 in ng doses inhibits the depletion of pituitary MSH induced by the i.v. injection of MRF or the i.v. injection of hypertonic saline in rats; the effect is dose dependent (Celis *et al.*, 1971b). When injected in rats with lesions of the median eminence, there is a lowering of the elevated plasma MSH level and an increase in pituitary MSH concentration, indicating a direct action on the pituitary gland *in vivo* (Celis *et al.*, 1973). Pro-Leu-Gly-NH_2 also blocks the secretion of MSH from pituitary glands incubated *in vitro* (Celis *et al.*, 1971b). Applied directly to the surface of the pituitary it produces lightening of frogs (*Rana pipiens*) darkened by destruction of the hypothalamus (Kastin *et al.*, 1971b). However, several authors have failed to find any effect of this tripeptide on MSH secretion *in vitro* or *in vivo*.

The MIF content in rat hypothalamus varies under different experimental conditions. It rises after suckling (Taleisnik and Tomatis, 1968), vaginal stimulation (Taleisnik and Tomatis, 1970), or reserpine injection (Tomatis and Taleisnik, 1968b), but falls after the injection of oestrogen or progesterone (Taleisnik and Tomatis, 1969, 1970) or the injection of noradrenaline into the 3rd ventricle through chronic implanted cannulas into the brain in spayed rats (Taleisnik *et al.*, 1972). During the oestrous cycle of the rat there is maximal value on the morning of pro-oestrus (Tomatis and Taleisnik, 1968a). While pituitary MSH depletion induced by suckling or vaginal stimulation is associated with high hypothalamic concentrations of MIF, that induced by oestrogen or progesterone is associated with low MIF concentrations. In the former conditions, the release of the inhibiting factor may be blocked and thereby the store in the neurones be increased; in the latter, the synthesis of the factor would be decreased. Apparently the stimuli which promote MSH secretion may also affect either the release or the synthesis of MIF. This double mechanism could imply that neurogenic stimuli would permit a rapid adjustment of MSH secretion, whereas the steroids would determine long-lasting effects.

(2) *MSH-releasing Factor* (*MRF*)

A factor stimulating the release of MSH is also found in the hypothalamus. Acidic extracts of stalk-median eminence from male rats, injected intravenously, induces a depletion of the MSH content of the pituitary and an increase in plasma melanophoretic activity. The presence of a response in rats with denervated pituitaries indicates a direct action of the factor at the pituitary level (Taleisnik and Orias, 1965; Taleisnik and Tomatis, 1967a). Studies of its localisation in the hypothalamus show highest concentration in the area of the paraventricular nuclei (Taleisnik *et al.*, 1966).

Part of the oxytocin molecule, H-Cys-Tyr-Ile-Gly-Asn-OH, has been reported to induce MSH release (Celis *et al.*, 1971a). Injected i.v. into rats, it produces a rise in serum MSH concentration and a depletion of the MSH pituitary content; it is active at ng doses and the effect is dose dependent. This pentapeptide is also active in animals with denervated pituitaries. No indication exists, however, about its similarity with the natural substance present in the hypothalamus. (see also chapter 5.)

(3) *Formation of MRF and MIF*

Enzymes present in the hypothalamus are involved in the formation of the hypothalamic factors that regulate MSH secretion. Incubation of fresh hypothalamic extracts at 37°C, pH 7·0, results in a product which inhibits the release of MSH. This effect depends on the interaction of two substances, one of which is heat-labile and non-diffusible and the other heat-stable and diffusible. Both substances, though present in hypothalamic tissue, are not found in cerebral cortical tissue. Oxytocin can act as the heat-stable substance, whereas the non-diffusible substance may be a peptidase. This enzyme which is referred to as E_I, is attached to the microsomal and supernatant fractions (Celis and Taleisnik, 1971a). A second enzyme referred to as E_R, is attached to the mitochondrial fraction and yields, upon incubation with oxytocin, a factor that induces MSH release (Celis and Taleisnik, 1971b). Oxytocin can thus serve as substrate for both enzymes. The hydrolysis by E_I involves the release of the C-terminal tripeptide, Pro-Leu-Gly-NH_2, whereas the appearance of releasing activity after incubation with E_R may be related to the oxytocin five-membered ring production. Certain oxytocin analogues and intermediates can also serve as substrate for E_I. All compounds having a Pro-Leu-Gly-NH_2 terminal yield MIF, whereas those lacking this terminal do not. Vasopressin can also serve as a substrate of E_I but the factor formed is less potent in inhibiting MSH release than that formed from oxytocin (Celis *et al.*, 1971b). The activity of both enzymes, is oestrogen-dependent and varies during the oestrous cycle in the rat.

In the light of these findings it has been proposed (Celis and Taleisnik, 1971a) that the neurosecretory substances which regulate MSH secretion are formed by enzymatic degradation of a precursor molecule or prohormone, oxytocin. The available evidence indicates that both enzymes compete for the same substrate. When these enzymes are together in the presence of a limited oxytocin supply, either MRF or MIF will be formed, depending on the prevailing enzyme in the system.

D. THE NERVOUS SYSTEM AND THE PARS INTERMEDIA

Evidence from studies in anurans (Etkin, 1967) seems to discount the possibility of a hypothalamic control over the pars intermedia via the portal vessels. Section of the infundibular stalk produces only transitory impairment of adenohypophysial function due to the re-establishment of the vascular connections (see Chapter 1), whereas similar lesions result in long-lasting darkening of amphibian skin, suggesting that the portal system is unlikely to mediate the hypothalamic influences on the pars intermedia as it mediates those on the pars distalis. The nerve fibres penetrating the gland can therefore be regarded as essential for the transfer of information from the hypothalamus to the pars intermedia. Although these nerve fibres are usually in contact with the pars intermedia secretory cells this is not always so, suggesting that the neurosecretory products may reach the cells not only by direct contacts but also by diffusion from neighbouring regions where these fibres end. Nevertheless, the distribution seems to be limited to certain areas of the pars intermedia as shown by the fact that unilateral lesions of the paraventricular nuclei in the rat, produce changes in MSH content of the homolateral hemipituitary gland without affecting the contralateral half (Taleisnik and Celis, 1973).

V. Feedback Control of MSH Secretion

Several pieces of evidence suggest that MSH can exert a direct feedback inhibition on its own secretion. Pituitary MSH content in rats was found to increase after a single i.v. injection of MSH (Kastin and Schally, 1967) or to decrease after long-term treatment (Frieden, 1951). The acute effect of MSH may be to inhibit MSH secretion, whereas the chronic administration may block synthesis. Blockade of MSH release from cultured rat pituitaries (Kastin *et al.*, 1971) or toad, *Bufo arenarum*, neural lobes (Iturriza, 1973) also occurs after increasing MSH concentration in the incubation medium, indicating a direct action on the gland. This auto-feedback inhibition *in vitro* was not, however, confirmed (Hadley *et al.*, 1975). Further support to the concept of an auto-inhibition comes from electron-microscopic studies. The ultrastructural changes in the pars intermedia of the frog, *Rana pipiens*, maintained on a black background, suggestive of increased synthesis and release of MSH, are prevented by the increase in circulating MSH (Perryman, 1974). The importance of this mechanism in setting the circulating levels of MSH in physiological conditions remains to be determined.

VI. Stimuli Affecting MSH Secretion

A. STRESS

Stressful stimuli cause pituitary MSH release. High plasma β-MSH levels were found in patients subjected to abdominal surgery (Abe *et al.*, 1969). A rise in plasma MSH concentration occurs in rats exposed to ether anæsthesia (Kastin *et al.*, 1969b) while the hormone content in the pituitary gland decreases or is unchanged. During a long lasting stress, as in mice kept in an adverse environment, the MSH content in the pituitary gland increases (Francis and Peaslee, 1974), perhaps due to enhanced hormone synthesis.

B. LIGHT

Experiments on amphibia and fish have provided good evidence for the inhibitory influence of light on the release of MSH. Animals of these species kept on a black background, become dark and show a high melanophore index indicating pituitary MSH release. After they are returned to the white background the release of MSH ceases and they become pale. During the increased release phase the pituitary hormone content decreases but returns to normal values when the animals are transferred to the white background (Burgers *et al.*, 1963; Thornton, 1971). The ultrastructure of the secretory cells of the pars intermedia of black adapted anurans also suggest increased secretion and synthesis of MSH (Saland, 1968; Perryman, 1974).

Light may also affect the secretory activity of the pars intermedia in mammals. The pituitary MSH content rises in rats exposed to constant illumination and decreases in those kept in the dark (Kastin *et al.*, 1967). Although no changes in plasma MSH concentration were found under these experimental conditions, the pituitary hormone changes may indicate that in the rat, as in amphibia, light deprivation leads to an enhanced secretion of MSH. However, opposite conclusions can be drawn from studies in the short-tailed weasel. Decrease in day length causes a white pelage to develop, while increase in day-length either naturally or artificially initiates growth of brown hair (Rust, 1965). As it is suggested that brown hair growth is associated with an increased release of MSH, it is possible that light favours the secretion of MSH in these animals. The effect of light may be mediated by pineal gland (Section VII, D).

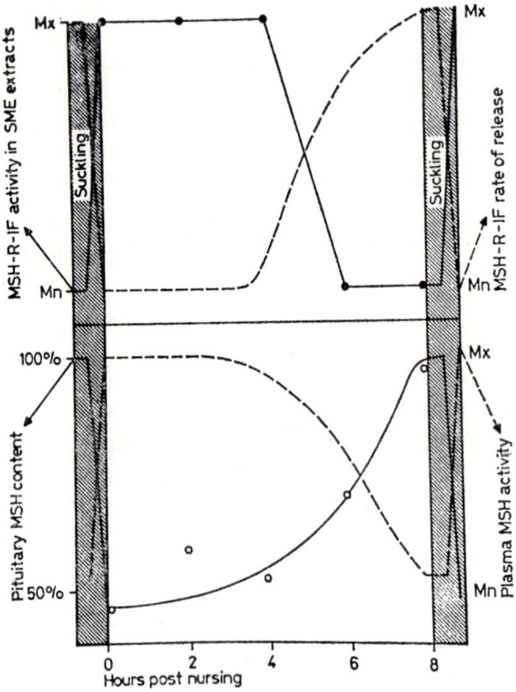

Fig. 3. Variation of pituitary MSH and hypothalamic MIF content of lactating rats after suckling. In the lower part of the figure each circle represents pituitary MSH content expressed as percentage values of non-nursing controls. In the upper part of the figure each point represents the MIF content in stalk median eminence (SME) extracts. The interrupted lines shows presumptive variations of MSH concentration in plasma and MIF release. The shaded are corresponds to the nursing period. (From: Taleisnik and Tomatis, 1968.)

C. SUCKLING

Suckling stimuli in lactating rats causes a decrease of the pituitary MSH content which remains low four hours after suckling. Simultaneously detectable levels of MIF are found in the hypothalamus, whereas in pre-nursing animals the level of MIF is undetectable (Fig. 3). These findings suggest that the suckling stimulus blocks the release of MIF, which accumulates in the hypothalamic neurones and MSH is released from the pituitary (Taleisnik and Orías, 1966; Taleisnik and Tomatis, 1968).

D. SOUND

A suckling mother and her litter induce a decrease in pituitary MSH content in lactating rats kept in nearby cages (Deis and Orías, 1968). Deaf rats show no such changes underlining the auditory nature of the stimulus.

E. TOUCH

Stimulation of the rat vagina at oestrus with a glass rod results in a decrease in pituitary MSH content one hour later. This stimulus also provokes an increase in MIF in the hypothalamus perhaps by blocking its release (Taleisnik and Tomatis, 1970). The effect appears to be mediated by 5-HT containing neurones, since the response failed to occur in animals in which parachlorophenylalanine was intraperitoneally injected to deplete brain 5-HT or in those in which methysergide, a 5-HT antagonist, was injected into the 3rd ventricle through a chronic implanted cannulae into the brain (Taleisnik *et al.*, 1973).

Copulation is also accompanied by a drop in MSH concentration of the pituitary of both male and female rats (Taleisnik and Tomatis, 1970).

F. HYPERTONIC SALINE

The degree of development of the pars intermedia is inversely related to the amount of water available to the animal. A well-developed pars intermedia is found in all desert rodents, whereas those which spend much of their life in water have very small glands (Legait and Legait, 1962).

Cellular changes in the pars intermedia occur in the rat after dehydration or the ingestion of hypertonic saline. They are characterised by the increase in the number of Type 2 cells and hypertrophy and degranulation of Type 1 cells, followed in a later stage by reduction in size and an increased intensity of stain by periodic acid-Schiff (Duchen, 1962; Howe and Thody, 1970). Electron microscopy shows that cytoplasmic vesicles decrease in number and that high electron dense granules are distributed throughout the cytoplasm (Ziegler, 1963; Kobayashi, 1964). The functional significance of these changes is not understood, although they seem to be related to a greater activity of the cells. Hypertonic saline administration also induces release of MSH. Following i.v. injection of 2% NaCl in rats the serum MSH concentration rises and the content of the hormone in the pituitary gland falls (Taleisnik *et al.*, 1973). This effect, however, does not occur in animals in which the paraventricular nuclei have been lesioned, indicating that paraventricular neurones are involved in the neuronal pathway activated by hypertonic saline (Taleisnik and Celis, 1973). This effect of hypertonic saline is mediated by monoaminergic neurones. When catecholamines are depleted by reserpine treatment or adrenergic α-receptors are blocked by the injection of dibenamine, hypertonic saline fails to induce MSH secretion. A 5-HT containing neurone has also been implicated (Taleisnik *et al.*, 1972, 1973).

VII. Other Endocrine Influences

A. OVARY

The pars intermedia and MSH secretion are influenced by the oestrogens. Several studies have shown a pronounced hypertrophy of the pars intermedia in the hamster after long-term administration of stilboestrol. The concentration of MSH in the pituitary gland of ovariectomised rats is twice that of animals at oestrus, and oestrogen administered to spayed rats significantly decreases pituitary MSH content. Such an effect appears a few hours after a single injection of the steroid but after several days of treatment the MSH content is increased (Karkun and Sen, 1965; Taleisnik and Tomatis, 1969). It would appear that the initial action of oestrogen is to deplete the pituitary content of the hormone by stimulating its release, but later the levels are restored by increased synthesis.

Oestrogen may affect the MSH content of the pituitary by acting either directly on the gland or through the hypothalamus; not only is this effect present when the pituitary gland has been disconnected from the central nervous system, by a lesion in the media eminence, but also, there is a depletion of hypothalamic MIF after oestrogen administration (Taleisnik and Tomatis, 1969).

Associated with this action are the changes in pituitary MSH content observed during the oestrous cycle of the rat, when the maximum is found on the morning of proestrus. The level falls by the afternoon of this day reaching a minimum at oestrus (Tomatis and Taleisnik, 1968a).

Progesterone can also affect MSH secretion. In rats made pseudopregnant by vaginal stimulation, the pituitary MSH content falls for at least six days but returns to the pre-stimulation levels by the tenth day of pseudopregnancy. However, if ovariectomy is performed soon after the vaginal stimulation, pituitary MSH returns to normal at least two days later. The low level is maintained if progesterone is given (Taleisnik and Tomatis, 1970). These findings imply that the effect of progesterone is either due to enhanced MSH release or to a block in MSH synthesis. The fact that progesterone depletes MIF content of the hypothalamus, favours the view of an action on the release. An increased blood MSH concentration and urinary excretion in pregnant women (Shizume and Lerner, 1954; Dahlberg, 1963) and a decrease in pituitary MSH during pregnancy in rats (Karkun and Sen, 1965), may be due to a progesterone effect.

B. ADRENAL CORTEX

When the plasma concentration of ACTH and MSH in patients with a variety of disorders of the pituitary-adrenal axis is compared, the conclusion can be drawn that the the secretion of "β-MSH" and ACTH in man is regulated by the same mechanism. In conditions with decreased or absent adreno-cortical function such as Addison's disease, or after adrenalectomy, high plasma levels of "MSH" are observed, while low plasma levels occur in patients with increased secretion of cortisol or following administration of corticosteroids to normal subjects or to patients with Addison's disease. Plasma "β-MSH" is also elevated in Cushing's syndrome, due to pituitary ACTH excess and low in patients with adrenal tumours (Shizume and Lerner, 1954; Abe *et al.*, 1969). It would appear that a good correlation between plasma ACTH and "β-MSH" exists in conditions associated with chronic alterations of adrenal steroid secretion. However, the acute changes in cortisol secretion are not consistently accompanied by changes in "MSH" release. Thus, blocking of cortisol secre- by injecting the 11-β-hydrozylase blocking substance, metyrapone, was not always followed by an increase in urinary excretion of "MSH" (Akinci *et al.*, 1962) and the administration of high doses of ACTH was not associated with changes in plasma "β-MSH" concentration over a period of six hours (Abe *et al.*, 1969). As noted earlier (p. 425) it is now thought that "β-MSH" activity in man is associated with a much larger peptide, human β lipotrophin.

In the rat changes that influence the secretion of corticotrophin (Chapter 11) may also affect the pars intermedia. Following adrenalectomy there is an increase in the volume of the pars intermedia and in the proliferative activity of the intermediate cells. The opposite effects result from cortisol admini-stration. However, neither adrenalectomy nor cortisol treatment influence pituitary MSH content and plasma MSH concentration (Gosbee *et al.*, 1970). Furthermore, the administration of corticosteroids to rats fails to affect the release of MSH induced by morphine-pentobarbitone treatment or ether anæsthesia (Kastin *et al.*, 1969b), nor to influence the drop in pituitary MSH associated with suckling in lactating rats (Taleisnik and Orías, 1966).

C. THYROID

There is some evidence which suggests that the thyroid gland may influence the function of the pars intermedia. Thyroidectomy in mice produces an increase in the volume of pars intermedia cells with cyctological signs of enhanced secretory activity (Roux, 1972). In the rat thyroidectomy or injection of thiouracil are reported to increase pituitary MSH content, (Giroud *et al.*,

1944), although the opposite result has been reported by others (Frieden, 1951). A specific uptake of thyroxine has been demonstrated for the pars intermedia of the frog, *Rana esculenta* (Céhovic, 1967).

D. PINEAL GLAND

Reports on the influence of the pineal on MSH secretion are contradictory. In the rat, pinealectomy increases pituitary MSH content and the injection of melatonin induces a depletion of MSH, suggesting that melatonin promotes the release of MSH. Furthermore, pinealectomy prevents the fall in pituitary MSH content caused by continuous darkness. Since the pineal is activated by light deprivation, it might be that an increased melatonin secretion is responsible for the release of MSH in rats kept in the dark (Kastin *et al.*, 1974b). However, Tilders and Smelik (1975) failed to show a role of the pineal gland in the control of pituitary MSH content and release. In the short-tailed weasel, however, the growth of brown hair which is induced by long-day photo-periods and depends on MSH secretion, is inhibited by the injection of melatonin and a white pelage develops, as after the exposure to short-day photo-periods. This inhibitory effect of melatonin on MSH secretion seems to be exerted at the hypothalamic level, since melatonin does not prevent the growth of brown hair in animals bearing a transplanted pituitary (Rust and Meyer, 1969). The difference in the action of melatonin on MSH secretion in the rat and short-tailed weasel has not been explained.

VIII. Integration of Neural Influences on MSH Secretion

There is much evidence supporting the concept that MSH secretion in both lower vertebrates and mammals is subjected to inhibition by the hypothalamus. After this inhibition is removed (see Section IV A), increased amounts of MSH are released and morphological signs of hyperactivity in the pars intermedia cells appear. Stimulatory mechanisms may also be operating. For example, it was observed in the toad, *Bufo bufo*, that transections of the brain in front of the optic chiasma, produce loss of the capacity of these animals to respond with darkening of the skin when placed in a black background (Jørgensen and Vijayakumar, 1971), it being suggested that such lesions have interrupted neurones which stimulate MSH secretion. Furthermore, if the secretion of MSH is controlled only by an inhibitory mechanism, then the suppression of this mechanism should result in a maximal rate of MSH secretion. However, in rats with lesions in the median eminence, the high levels of MSH in blood can be further increased by injecting MRF. Therefore, even if the suppression of the hypothalamic inhibition is sufficient to initiate and maintain a high level of MSH secretion, a superimposed facilitatory mechanism could enable faster and more intense responses to occur. Support for the concept of a dual control mechanism for MSH secretion has come from studies on the spontaneous single unit firing activity in the pars intermedia of the frog, *Rana pipiens* (Oshima and Gorbman, 1969; Dawson and Ralph, 1974). Two populations of neurones have been described, one of which is light sensitive and the other light independent; it is believed that they produce opposite effects on MSH secretion. The facilitatory mechanisms seems to be mediated by a hypothalamic releasing factor and/or by catecholamines through β-adrenergic receptors, but the circumstance and mode of operation remain unclear. The inhibitory mechanism appears to

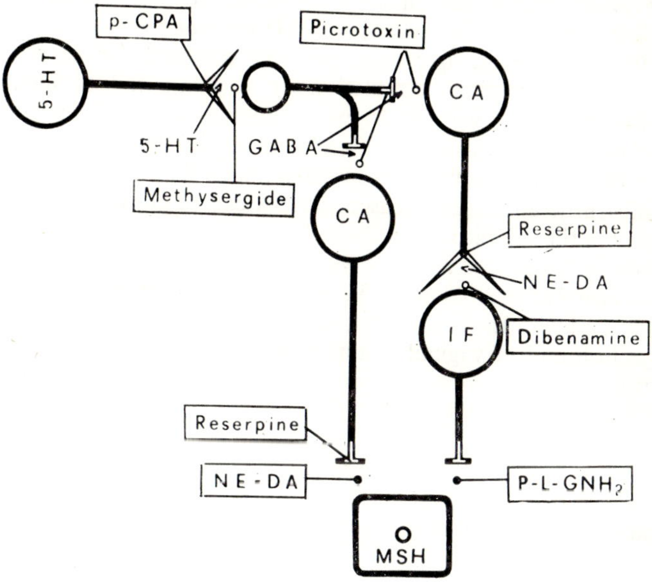

Fig. 4. Neuronal circuit involved in the inhibitory mechanism of MSH secretion. CA: catecholamine; NE: noradrenaline; DA: dopamine; 5HT: 5-hydroxytryptamine; p-CPA: p-chlorophenylalanine; GABA: γ-aminobutyric acid; P.L.GNH$_2$: prolyl-leucyl-glycin-amidc (MIF). Arrows indicate agents activating or inhibiting the next neurone; circles: agents which block afferent impulses to the next neurone.

depend on the action of catecholaminergic neurones involving α-adrenergic receptors. Interruption of the influences of these neurones such as treatment with reserpine, results in an enhanced secretion of MSH and causes reduced responses to MSH-releasing stimuli. Contrariwise noradrenaline or dopamine injected into the third ventricle prevent the release of MSH (Taleisnik *et al.*, 1972). Catecholaminergic neurones can affect MSH secretion by acting directly on the pars intermedia cells as well as at the hypothalamic level through the neurones secreting MIF. The demonstration of monoaminergic fibres ending in contact with the cells of the pars intermedia and experiments showing an inhibition of MSH secretion from isolated glands by local application of noradrenaline or dopamine (see Section IV, B), favour the view of a direct action on the pars intermedia cells. Increase in the hypothalamic content of MIF after reserpine treatment as well as depletion after noradrenaline injection into the third ventricle, substantiate a hypothalamic site of action. These latter effects have led to the suggestion that catecholaminergic neurones activate the release of MIF (Taleisnik *et al.*, 1972). Although both MIF and catecholamines act on pars intermedia cells, the relative importance and possible synergy in the control of MSH are unknown.

The catecholaminergic neurones are also essential for certain stimuli. Hypertonic saline, vaginal stimulation or suckling, fail to induce MSH release after treatment with reserpine or after α-adrenergic receptors have been blocked. The stimuli appear to activate an inhibitory pathway, which, curtailing the afferent inflow of catecholaminergic neurones, reduces the amount of neurotransmitter released. GABA can be considered as the transmitter agent in the inhibitory synapses. This mechanism, like other inhibitory mechanisms in the central nervous system, can be blocked by the injection of picrotoxin (see Table 2, Chapter 3). A serotonin-containing neurone seems to be associated with the inhibitory chain, in view of the failure of hypertonic saline injection and of vaginal stimulation to induce the release of MSH in animals treated with para-chlorophenylalanine or with the serotonin antagonist, methysergide (Taleisnik *et al.*, 1973).

The results reported in the foregoing account can be arranged in a neuronal chain as illustrated in Fig. 4. The activity of this circuit will depend not only on the incoming neuronal stimuli, but also on the hormonal background in the animal.

IX. Summary and Conclusions

The secretion of MSH from the pars intermedia of the pituitary gland, both in mammals and lower vertebrates in resting conditions, is blocked by a hypothalamic inhibitory mechanism. Suppression of the hypothalamic influences enables the pars intermedia cells to exercise their intrinsic capacity to secrete high levels of MSH. The hypothalamic control of the pars intermedia, unlike that of the pars distalis, is mediated by nerve fibres which end in contact with the secretory cells or in their vicinity. The ultrastructure of the endings of these nerves provide evidence of a double innervation of the gland by neurosecretory and catecholaminergic neurones. Pharmacological studies indicate that inhibition of MSH secretion can be induced both by catecholamines and by substances of peptidergic nature present in the hypothalamus (MIF). It appears that catecholamines inhibit MSH release through α-adrenergic receptors, acting both at pars intermedia secretory cells and at hypothalamic level. At this latter level they activate neurones which secrete

MIF. The relative importance of these two mechanisms in the control of MSH secretion remains to be established.

Evidence from pharmacological and physiological experiments would suggest that MSH secretion may also be affected by a stimulatory mechanism.

Interoceptive and exteroceptive stimuli of different sensory modalities influence the secretion of MSH. Stimuli causing MSH release probably operate by an inhibitory mechanism which curtails the volley of inputs of the catecholaminergic neurones, thus removing hypothalamic inhibition. A 5-HT neurone appears to be implicated in this mechanism.

REFERENCES

A

Abe, K. and Hiroshige, T. (1974). Changes in plasma corticosterone and hypothalamic CRF levels following intraventricular injection or drug induced changes of brain biogenic amines in the rat. *Neuroendocrinology*, **14**, 195–211.

Abe, K., Nicholson, W. E., Liddle, G. W., Orth, D. N. and Island, D. P. (1969). Normal and abnormal regulation of β-MSH in man. *J. Clin. Invest.* **48**, 1580–1585.

Acher, R. (1966). Chemistry of neurohypophysial hormones. *In* "The Pituitary Gland" (G. W. Harris and B. T. Donovan, eds) Vol. 3, pp. 269–287. Butterworths, London.

Acher, R. (1974a). L'évolution moléculaire au niveau des protéines. *Biochemie*, **56**, 1–19.

Acher, R. (1974b). Chemistry of the neurohypophysial hormones: an example of molecular evolution. *In* "Handbook of Physiology" Section 7, Vol. IV (E. Knobil and W. H. Sawyer, eds) Part 1, pp. 119–130. American Physiological Society, Washington, D.C.

Acher, R., Chauvet, J. and Lenci, M. T. (1960). Isolément de l'ocytocine du poulet. *Biochim. biophys. Acta*, **38**, 344–345.

Acher, R., Chauvet, J. and Chauvet, M. T. (1970). Phylogeny of the neurohypophysial hormones: the avian active peptides. *Eur. J. Biochem.* **17**, 509–513.

Acher, R., Chauvet, J. and Chauvet, M. T. (1972). Phylogeny of the neurohypophysial hormones. Two new active peptides isolated from a cartilaginous fish, *Squalus acanthias*. *Eur. J. Biochem.* **29**, 12–19.

Acher, R., Chauvet, J. and Chauvet, M. T. (1973). Phylogeny of the neurohypophysial hormones. The active peptides of a primitive fish, the Sturgeon (*Acipenser* sp.) *Eur. J. Biochem.* **40**, 585–589.

Adams, J. H., Daniel, P. M. and Prichard, M. M. L. (1963). The volumes of pars distalis, pars intermedia and infundibular process of the pituitary gland of the rat, with special reference to the effect of stalk section. *Quart. J. exp. Physiol.* **48**, 217–234.

Addison, T. (1855). "On the Constitutional and Local Effects of Disease of the Suprarenal Capsules". S. Highley, London.

Addison, W. H. F. (1917). The cell changes in the hypophysis of the albino rat after castration. *J. comp. Neurol.* **28**, 441–461.

Ågmo, A. (1974). Uptake of iodinated oxytocin by some tissues and organs in the male rat. *Acta Physiol. Scand.* **91**, 330–338.

Aiyer, M. S. and Fink, G. (1974). The role of sex steroid hormones in modulating the responsiveness of the anterior pituitary gland to luteinizing hormone-releasing factor in the female rat. *J. Endocr.* **62**, 553–572.

Aiyer, M. S., Chiappa, S. A. and Fink, G. (1974a). A priming effect of luteinizing hormone releasing factor on the anterior pituitary gland in the female rat. *J. Endocr.* **62**, 573–588.

Aiyer, M. S., Fink, G. and Greig, F. (1974b). Changes in the sensitivity of the pituitary gland to luteinizing hormone releasing factor during the oestrous cycle of the rat. *J. Endocr.* **60**, 47–64.

Ajika, K., Kalra, S. P., Fawcett, C. P., Krulich, L. and McCann, S. M. (1972). The effect of stress and nembutal on plasma levels of gonadotrophins and prolactin in ovariectomized rats. *Endocrinology*, **90**, 707–715.

Akinci, T., Apostolakis, M. and Tamm, J. (1962). The effect of the 11β-hydroxylase blocking substance, SU-4885, on the urinary excretion of 17-hydroxycorticosteroids and of melanocyte stimulating hormone-like activity. *Acta endocr., Copenh.* **39**, 253–261.

Alaghband-Zadeh, J., Daly, J. R., Bitensky, Lucille and Chayen, J. (1974). The cyto-chemical section assay for corticotrophin. *Clin. endocr.* **3,** 319–327.

Alberti, K. G. M. M. (1973). Inhibition of insulin secretion by somatostatin. *Lancet,* **ii,** 1299–1301.

Alexander, D. P., Britton, H. G., Forsling, M. L., Dixon, D. A. and Ratcliffe, J. G. (1971). The concentration of ACTH, vasopressin and oxytocin in the foetal and maternal plasma of the sheep in the latter half of gestation. *J. Endocr.* **49,** 179–180.

Alexander, D. P., Britton, H. G., Forsling, M. L., Dixon, D. A. and Ratcliffe, J. G. (1973). ACTH and vasopressin in foetal sheep and the response to stress. *In* "The Endocrinology of Pregnancy and Parturition: Experimental Studies in the Sheep" (C. G. Pierrepoint, ed.) pp. 112–125. Alpha Omega Alpha Publishing, Cardiff.

Allen, J. P., Cook, D. M., Kendall, J. W. and McGilvra, R. (1973). Maternal-fetal ACTH relationship in man. *J. Clin. Endocr. Metab.* **37,** 230–234.

Allen, J. P., Greer, M. A., McGilvra, R., Castro, A. and Fisher, D. A. (1974). Endocrine function in an anencephalic infant. *J. Clin. Endocr. Metab.* **38,** 94–98.

Amatruda, T. T. Jr. and Upton, G. V. (1974). Hyperadrenocorticism and ACTH-releasing factor. *Ann. N.Y. Acad. Sci.* **230,** 168–180.

Amoroso, E. C. and Marshall, F. H. A. (1960). External factors in sexual periodicity. *In* "Marshall's Physiology of Reproduction" (A. S. Parkes, ed.) 3rd edn Vol. II, Pt. 2, 707–831. Longmans, London.

Amoss, M., Burgus, R., Blackwell, R., Vale, W., Fellows, R. and Guillemin, R. (1971). Purification, amino acid composition and N-terminus of the hypothalamic luteinizing hormone-releasing factor (LRF) of ovine origin. *Biochem. Biophys. Res. Commun,* **44,** 205–210.

Anand Kumar, T. C. (1973). Cellular and humoral pathways in the neuroendocrine regulation of reproductive function. *J. Reprod. Fert. Suppl.* **20,** 11–20.

Anderson, E. and Haymaker, W. (1974). Breakthroughs in hypothalamic and pituitary research. *Progr. Brain Res.* **41,** 1–60.

Anderson, E., Bates, R. W., Hawthorne, E., Haymaker, W., Knowlton, R., Rioch, McK., Spence, W. T. and Wilson, H. (1957). The effects of midbrain and spinal cord transection on endocrine and metabolic functions with postulation of a midbrain hypothalamo-pituitary activating system. *Recent Prog. Horm. Res.* **13,** 21–66.

Anderson, M. S., Bowers, C. Y., Kastin, A. J., Schalch, D. S., Schally, A. V., Snyder, P. J., Utiger, R. D., Wilber, J. F. and Wise, A. J. (1971). Synthetic thyrotropin-releasing hormone. *N. Engl. J. Med.* **285,** 1279–1283.

Andersson, B., Ekman, L., Gale, C. C. and Sundsten, J. W. (1962). Activation of the thyroid gland by cooling of the pre-optic area in the goat. *Acta Physiol. Scand.* **54,** 191–192.

Anon. (1970). The effects of female company on beard growth in man. *Nature, Lond.* **226,** 869.

Anton-Tay, F. and Wurtman, R. J. (1971). Brain monoamines and endocrine function. *In* "Frontiers of Neuroendocrinology" (L. Martini and W. F. Ganong, eds) pp. 45–66. Oxford University Press, New York.

Aono, T., Kurachi, K., Mizutani, S., Hamanata, Y., Vozumi, T., Nakasima, A., Koshiyama, K. and Matsumoto, K. (1972). Influence of major surgical stress on plasma levels of testosterone, luteinizing hormone and follicle stimulating hormone in male patients. *J. Clin. Endocr. Metab.* **35,** 535–541.

Arai, Y. and Yamanouchi, K. (1975). Possible role of anterior inputs to the medial basal hypothalamus (MBH) in regulating prolactin release during pseudopregnancy in the rat. *Brain Res.* **83,** 51–58.

Arimura, A. and Schally, A. V. (1974). Hypothalamic LH- and FSH-releasing hormone: re-evaluation of the concept that one hypothalamic hormone controls the release of LH and FSH. *In* "Biological Rhythms in Neuroendocrine Activity" (M. Kawakami, ed.) pp. 73–90. Igaku Shoin, Tokyo.

Arimura, A. and Schally, A. V. (1975). Immunological studies on hypothalamic hormones with special reference to radioimmunoassays for TRH, LH-RH and GIF. *In* "Hypothalamic Hormones" (M. Motta, P. Crosignani and L. Martini, eds) pp. 27–42. Academic Press, New York and London.

Arimura, A. and Schally, A. V. (1976). Increase in basal and thyrotropin-releasing hormone (TRH) by passive immunization with antiserum to somatostatin in rats. *Endocrinology*, **98**, 1069–1072.

Arimura, A., Saito, T. and Schally, A. V. (1967). Assays for corticotrophin-releasing factor (CRF) using rats treated with morphine, chloropromazine, dexamethasone and nembutal. *Endocrinology*, **81**, 235–245.

Arimura, A., Bowers, C. V., Schally, A. V., Saito, M. and Miller, M. C. (1969). Effects of corticotrophin-releasing factor, dexamethasone and actinomycin D on the release of ACTH from rat pituitaries *in vivo* and *in vitro*. *Endocrinology*, **85**, 300–311.

Arimura, A., Debeljuk, L. and Schally, A. V. (1972). LH release by LH-releasing hormone in golden hamsters at various stages of the estrous cycle. *Proc. Soc. exp. Biol. Med.* **140**, 609–612.

Arimura, A., Sato, M., Kumasaka, T., Worobec, R. B., Debeljuk, L., Dunn, J., and Schally, A. V. (1973). Production of antiserum to LH-releasing hormone (LH-RH) associated with gonadal atrophy in rabbits: development of radioimmunoassay for LH-RH. *Endocrinology*, **93**, 1092–1103.

Arimura, A., Debeljuk, L. and Schally, A. V. (1974a). Blockade of the preovulatory surge of LH and FSH and of ovulation by anti-LH-RH serum in rats. *Endocrinology*, **95**, 323–325.

Arimura, A., Kastin, A. J., Gonzalez-Barcena, D., Siler, J., Weaver, R. E. and Schally, A. V. (1974b). Disappearance of LH-releasing hormone in man as determined by radioimmunoassay. *Clin. Endocr.* **3**, 421–425.

Arimura, A., Kastin, A. J., Schally, A. V., Sato, M., Kumasaka, T., Taoi, Y., Nishi, N. and Ohkura, K. (1974c). Immunoreactive LH-releasing hormone in plasma: midcycle elevation in women. *J. Clin. Endocr. Metab.* **38**, 510–513.

Arimura, A., Sato, H., Dupont, A., Nishi, N. and Schally, A. V. (1975a). Somatostatin: Abundance of immunoreactive hormone in rat stomach and pancreas. *Science, N.Y.* **189**, 1007–1009.

Arimura, A., Sato, H., Coy, D. H. and Schally, A. V. (1975b). Radioimmunoassay for GH-releasing inhibiting hormone. *Proc. Soc. exp. Biol. Med.* **148**, 784–789.

Arnold, W. H., White, W. and Flouret, G. (1973). An improved solid-phase synthesis of the luteinizing hormone-releasing hormone/follicle stimulating hormone-releasing hormone (LH-RH/FSH-RH). *J. Med. Chem.* **16**, 1044–1055.

Aschheim, P. (1976). Aging in the hypothalamic-hypophyseal-ovarian axis in the rat. *In* "Hypothalamus, Pituitary and Aging" (A. V. Everitt and J. A. Burgess, eds) pp. 376–418. C. C. Thomas, Springfield.

Aschner, B. (1909). Demonstration von Hunden nach Extirpation der Hypophyse, *Wien. klin. Wschr.* **22**, 1730–1732.

Aschoff, J. Ed. (1965). "Circadian Clocks". North-Holland, Amsterdam.

Atwell, W. J. (1926). The development of the hypophysis cerebri in man, with special reference to the pars tuberalis. *Am. J. Anat.* **37**, 159–193.

Averill, R. L. W. (1970). Disappearance of ^{131}I-labelled thyroxine (^{131}I-T4) from agar pellets implanted in the hypothalamus of rats. *Endocrinology*, **87**, 176–178.

Averill, R. L. W. and Salaman, D. F. (1967). Elevation of plasma thyrotropin (TSH) during electrical stimulation in the rabbit hypothalamus. *Endocrinology*, **81,** 173–178.

B

Baba, Y., Arimura, A. and Schally, A. V. (1971). On the tryptophan residue in porcine LH and FSH-releasing hormone. *Biochem. Biophys. Res. Commun.* **45,** 483–487.

Babiker, M. M. and Rankin, J. C. (1973). Effects of neurohypophysial hormones on renal function in the freshwater and seawater adapted eel (*Anguilla anguilla* L.). *J. Endocr.* **57,** xi-xii.

Bachelard, H. S. (1974). "Brain Biochemistry". Chapman and Hall, London.

Baird, D. T., Baker, T. G., McNatty, K. P. and Neal, P. (1975). Relationship between the secretion of the corpus luteum and the length of the follicular phase of the ovarian cycle. *J. Reprod. Fert.* **45,** 611–619.

Bajusz, S., Turau, A., Fauszt, I. and Juhasz, A. (1973). Side chain protection in the synthesis of peptide amides. *In* "Peptides 1972" (H. Hanson and H. D. Jakubke, eds) pp. 93–99, North-Holland, New York.

Baker, B. L., Pek, S., Midgley, A. R. and Gersten, B. E. (1970). Identification of the corticotrophin cell in the rat hypophysis with peroxidase-labelled antibody. *Anat. Rec.* **166,** 557–567.

Baker, B. L., Dermody, W. A. and Reel, J. R. (1974). Localization of luteinizing hormone-releasing hormone in the mammalian hypothalamus. *Am. J. Anat.* **139,** 129–134.

Baker, J. R. (1938). The evolution of breeding seasons. *In* "Evolution" (G. R. de Beer, ed.) pp. 161–177. Oxford University Press.

Bakke, I. L., Lawrence, N. and Robinson, S. (1972). Late effects of thyroxine injected into the hypothalamus of the neonatal rat. *Neuroendocrinology*, **10,** 183–195.

Balsam, A. and Leppo, L. E. (1974). Augmentation of the peripheral metabolism of L-triiodothyronine and L-thyronine after acclimatization to cold. *J. Clin. Invest.* **53,** 980–987.

Bandaranayake, R. C. (1971). Morphology of the accessory neurosecretory nuclei and of the retrochiasmatic part of the supraoptic nucleus of the rat. *Acta anat.* **80,** 14–22.

Barbato, T., Lawrence, A. M. and Kirstein, L. (1974). Cerebrospinal fluid stimulation of pituitary protein synthesis and growth hormone release *in vitro*. *Lancet*, **i,** 599–600.

Barbeau, A. (1975). Potentiation of levodopa effect by intravenous L-prolyl-L-leucyl-L-glycineamide in man. *Lancet*, **ii,** 683–684.

Barden, N. and Labrie, F. (1973a). Receptor for thyrotropin-releasing hormone in plasma membranes of bovine anterior pituitary glands. *J. Biol. Chem.* **248,** 7601–7606.

Barden, N. and Labrie, F. (1973b). Cyclic adenosine 3′5 monophosphate dependent phosphorylation of ribosomal proteins from bovine anterior pituitary gland. *Biochemistry, N.Y.* **12,** 3096–3012.

Bargmann, W. (1949). Über die neurosekretorische Verknüpfung von Hypothalamus und Neurohypophyse. *Z. Zellforsch.* **34,** 610–634.

Bargmann, W. and Scharrer, E. (1951). The site of origin of the hormones of the posterior pituitary. *Am. Scient.* **39,** 255–259.

Bargmann, W., Lindner, E. and Andres, K. H. (1967). Uber Synapsen an endokrinen Epithelzellen und die Definition Sekretorischer Neurone. Untersuchungen am Zwischenlappen der Katzenhypophyse. *Z. Zellforsch.* **77,** 282–298.

Barker, J. L., Crayton, J. W. and Nicoll, R. A. (1971). Noradrenaline and acetylcholine responses of supraoptic neurosecretory cells. *J. Physiol., Lond.* **218,** 19–32.

Barnea, A., Ben-Jonathon, N. and Porter, J. C. (1975). Synaptosomal localization of TSH releasing hormone (TRH) and LH releasing hormone (LRH) in the hypothalamus. *Fed. Proc. Fed. Am. Soc. Biol.* **34,** 239.

Barraclough, C. A. (1967). Modifications in reproductive function after exposure to hormone during the prenatal and early postnatal period. *In* "Neuroendocrinology" (L. Martini and W. F. Ganong, eds) Vol. II, pp. 61–99. Academic Press, New York and London.

Barraclough, C. A. and Cross, B. A. (1963). Unit activity in the hypothalamus of the cyclic female rat: effect of genital stimuli and progesterone. *J. Endocr.* **26,** 339–359.

Barraclough, C. A. and Gorski, R. A. (1961). Evidence that the hypothalamus is responsible for androgen-induced sensitivity in the female rat. *Endocrinology*, **68,** 68–79.

Barros D'Sa, A. Ay, Bloom, S. R. and Baron, J. H. (1975). Direct inhibition of gastric acid by growth hormone release-inhibiting hormone in dogs. *Lancet*, **i,** 886–887.

Barry, J., Dubois, M. P. and Poulain, P. (1973a). LRF producing cells of the mammalian hypothalamus: a fluorescent antibody study. *Z. Zellforsch mikrosk. Anat.* **146,** 351–366.

Barry, J., Dubois, M. P., Poulain, P. and Leonardelli, J. (1973b). Caracterisation et topographie des neurones hypothalamiques immunoréactifs avec des anticorps anti-LRF de synthèse. *C. R. hebd. Séanc. Acad. Sci., Paris*, **276,** 3191–3193.

Barry, J., Girod, Chr. and Dubois, M. P. (1975). Topographie des neurones élaborateurs de LRF chez les primates. *Bull. Ass. Anat., Paris*, **59,** 103–110.

Bassiri, R. M. and Utiger, R. D. (1972a). The preparation and specificity of antibody to thyrotropin releasing hormone. *Endocrinology*, **90,** 722–727.

Bassiri, R. and Utiger, R. D. (1972b). Serum inactivation of the immunological and biological activity of thyrotropin releasing hormone (TRH). *Endocrinology*, **91,** 657–664.

Bassiri, R. and Utiger, R. D. (1973). Metabolism and excretion of exogenous thyrotropin releasing hormone in humans. *J. Clin. Invest.* **52,** 1616–1619.

Bassiri, R. and Utiger, R. D. (1974). Thyrotropin-releasing hormone in the hypothalamus of the rat. *Endocrinology*, **94,** 188–197.

Bast, J. D. and Greenwald, G. S. (1974). Daily concentrations of gonadotrophins and prolactin in the serum of pregnant or lactating hamsters. *J. Endocr.* **63,** 527–532.

Baudoin, L. M. Y., Meyer, P. and Worcel, M. (1974). Influence of prostaglandins and oestrus cycle on the spasmodic action of angiotensin II and oxytocin on rat uterus. *Br. J. Pharmac.* **51,** 121P–122P.

Bauer, K. (1974). Degradation of thyrotropin releasing hormone (TRH). Its inhibition by Pyroglu-His-OCH_3 and the effect of one inhibitor in attempts to study the biosynthesis of TRH. *In* "Lipmann Symposium: Energy, Biosynthesis and Regulation in Molecular Biology" (W. de Gruyter, ed.) pp. 53–62. Springer-Verlag, Berlin, New York.

Bauer, K. and Kleinkauf, H. (1974). Degradation of thyrotropin-releasing hormone (TRH) by serum and hypothalamic tissue and its inhibition by analogues of TRH. *Hoppe-Seyler's Z. physiol. Chem.* **355,** 1173–1176.

Baumgarten, H. G., Björklund, A., Holstein, A. F. and Nobin, A. (1972). Organisation and ultrastructural identification of the catecholamine nerve terminals in the neural lobe and the pars intermedia of the rat pituitary. *Z. Zellforsch.* **126,** 483–517.

Bayliss, W. M. and Starling, E. H. (1902). The mechanism of pancreatic secretion. *J. Physiol., Lond.* **28,** 325–353.

Beck, C., Larkins, R. G., Martin, T. J. and Burger, H. G. (1973). Stimulation of growth hormone release from superfused rat pituitary by extracts of hypothalamus and of human lung tumours. *J. Endocr.* **59,** 325–333.

Belenky, M. A., Konstantinova, M. S. and Polenov, A. L. (1970). On the neurosecretory and adrenergic fibers in the intermediate lobe of the hypophysis in albino mice. *Gen. Comp. Endocr.* **15,** 185–197.

Bell, P. H. (1954). Purification and structure of β-corticotrophin. *J. Amer. Chem. Soc.* **76,** 5565–5566.

Bell, P. H., Howard, K. S., Shepherd, R. G., Finn, M. B. and Meisenhelder, J. H. (1956). Studies with corticotropin. II Pepsin degradation of β-corticotropin. *J. Am. Chem. Soc.* **78,** 5059–5066.

Ben-Jonathan, N., Mical, R. S. and Porter, J. C. (1973). Superfusion of hemipituitaries with portal blood. I. LRF secretion in castrated and diestrous rats. *Endocrinology,* **93,** 497–503.

Ben-Jonathan, N., Mical, R. S. and Porter, J. C. (1974). Transport of LRF from CSF to hypophysial portal and systemic blood and the release of LH. *Endocrinology,* **95,** 18–25.

Bennett, G. W. and Edwardson, J. A. (1975). Release of corticotrophin releasing factor and other hypophysiotrophic substances from isolated nerve endings (synaptosomes). *J. Endocr.* **65,** 33–44.

Bennett, G. W., Edwardson, J. A., Holland, D. T., Jeffcoate, S. L. and White, N. (1975). The release of immunoreactive luteinizing hormone-releasing hormone and thyrotrophin releasing hormone from hypothalamic synaptosomes. *Nature, Lond.* **257,** 323–325.

Benoit, J. (1934). Activation sexuelle obtenue chez le Canard par l'éclairment artificiel pendant la période de repos génital. *C. R. Acad. Sci.* **199,** 1671–1673.

Benoit, J. and Assenmacher, I. (1955). Le contrôle hypothalamique de l'activité pré-hypophysaire gonadotrope. *J. Physiol., Paris,* **47,** 427–567.

Bentley, P. J. (1974). Actions of neurohypophysial peptides in amphibians reptiles and birds. *In* "Handbook of Physiology", Section 7, Vol, IV (E. Knobil and W. H. Sawyer. eds) Part I, pp. 545–563. American Physiological Society, Washington, D.C.

Berg, G. R., Utiger, R. D., Schalch, D. S. and Reichlin, S. (1966). Effect of central cooling in man on pituitary-thyroid function and growth hormone secretion. *J. Appl. Physiol.* **21,** 1791–1794.

Berl, T., Anderson, R. J., McDonald, K. M. and Schrier, R. W. (1974). *In vivo* evidence for an antagonism between vasopressin (VP) and prostaglandin (OG) in the mammalian kidney. *J. Clin. Invest.* **53,** 6a.

Bern, H. (1972). Some questions on the nature and function of cranial and caudal neuro-secretory systems in non-mammalian vertebrates. *Progr. Brain Res.* **38,** 85–93.

Bernal, J. and Escobar del Rey, F. (1974). Inhibition by propylthiouracil of the extra thyroidal formation of triiodothyronine from thyroxine. *Acta endocr., Copenh.* **77,** 276–281.

Bernard, C. (1878). "Les Phénomènes de la Vie". Paris.

Bernadis, L. L. and Frohmann, L. A. (1971). Plasma growth hormone responses to electrical stimulation of the hypothalamus in the rat. *Neuroendocrinology,* **7,** 193–201.

Berson, S. A. and Yalow, R. S. (1968). Radioimmunoassay of ACTH in plasma. *J. Clin. Invest.* **47,** 2725–2751.

Berthold, A. A. (1849). Transplantation der Hoden. *Arch. Anat. Physiol. wiss. Med.* **16,** 42–46.

Besser, G. M. (1974). Hypothalamus as an endocrine organ: I. *Br. Med. J.* **3,** 560–564.

Besser, G. M. and Mortimer, C. H. (1974). Hypothalamic regulatory hormones: a review. *J. Clin. Path.* **27,** 173–184.

Besser, G. M., Orth, D. N., Nicholson, W. E., Bynny, R. L., Abe, K. and Woodham, J. P. (1971). Dissociation of the disappearance of bioactive and radioimmunoreactive ACTH from plasma in man. *J. Clin. Endocr. Metab.* **32,** 596–601.

Besser, G. M., Mortimer, C. H., Carr, D., Schally, A. V., Coy, D. H., Evered, D., Kastin, A. J., Turnbridge, W. M. G., Thorner, M. O. and Hall, R. (1974). Growth hormone-release inhibiting hormone in acromegaly. *Br. Med. J.* **1,** 352–355.

Besser, G. M., Paxton, A. M., Johnson, S. A. N., Moody, E. J., Mortimer, C. H., Hall, R., Gomez-Pan, A., Schally, A. V., Kastin, A. J. and Coy, D. H. (1975). Impairment of platelet function by growth hormone-release inhibiting hormone. *Lancet*, **i,** 1166–1168.

Bethune, J. E., Ganong, W. F., Hume, D. M. and Nelson, D. H. (1956). Plasma levels of ACTH in normal subjects and addisonian patients following corticosteroid administration. *J. Clin. Endocr. Metab.* **16,** 913–917.

Bhattacharya, A. N. and Marks, B. H. (1970). Effects of α-methyltyrosine and p-chlorophenylalanine on the regulation of ACTH secretion. *Neuroendocrinology*, **6,** 49–55.

Biedl, A. (1913). "Innere Sekretion; ihre physiologische Grundlagen und ihre Bedeutung für die Pathologie", 2nd edn. Urban u Schwartzenburg, Berlin.

Birge, C. A., Jacobs, L. S., Hammer, C. T. and Daughaday, W. H. (1970). Catecholamine inhibition of prolactin secretion by isolated rat adenohypophysis. *Endocrinology*, **86,** 120–130.

Bishop, W., Fawcett, C. P., Krulich, L. and McCann, S. M. (1972). The effects of hypothalamic lesions on the release of gonadotrophins and prolactin in response to estrogen and progesterone treatment in female rats. *Endocrinology*, **91,** 1404–1410.

Bissonette, T. H. (1932). Modification of mammalian sexual cycles; reactions of ferrets (*Putorius vulgaris*) of both sexes to electric light added after dark in November and December. *Proc. R. Soc. Ser. B.* **110,** 322–336.

Björklund, A. and Falck, B. (1969). Pituitary monoamines of the cat with special reference to the presence of an unidentified monoamine-like substance in the adenohypophysis. *Z. Zellforsch.* **93,** 254–264.

Björklund, A. and Nobin, A. (1973). Fluorescence histochemical and microspectrofluorometric mapping of dopamine and noradrenaline cell groups in the rat diencephalon. *Brain Res.* **51,** 193–205.

Björklund, A., Moore, R. Y., Nobin, A. and Stenevi, U. (1973). The organisation of the tuberohypophyseal and reticulo-infundibular catecholamine neuron systems in the rat brain. *Brain Res.* **51,** 171–191.

Blackard, W. G. and Heidingsfelder, S. A. (1968). Adrenergic receptor control mechanism for growth hormone secretion. *J. Clin. Invest.* **47,** 1400–1414.

Blackard, W. G., Hull, E. W. and Lopez, F. A. (1971). Effect of lipid on growth hormone secretion in humans. *J. Clin. Invest.* **50,** 1439–1443.

Blackwell, R. E. and Guillemin, R. (1973). Hypothalamic control of adenohypophysial secretions. *Annu. Rev. Physiol.* **35,** 357–390.

Blagdon, D. E., Rivier, J. and Goodman, M. (1973). Proposed tertiary structure for hypothalamic thyrotropin-releasing factor. *Proc. Nat. Acad. Sci. USA*, **70,** 1166–1168.

Blake, C. A. (1974). Localisation of the inhibitory actions of ovulation-blocking drugs on release of luteinizing hormone in ovariectomised rats. *Endocrinology*, **95,** 999–1004.

Blake, C. A. and Sawyer, C. H. (1972a). Ovulation blocking actions of urethane in the rat. *Endocrinology*, **91,** 87–94.

Blake, C. A. and Sawyer, C. H. (1972b). Nicotine blocks the suckling-induced rise in circulating prolactin in lactating rats. *Science, N.Y.* **177,** 619–621.

Blake, C. A. and Sawyer, C. H. (1974). Effects of hypothalamic deafferentation on the pulsatile rhythm in plasma concentrations of luteinizing hormone in ovariectomized rats. *Endocrinology*, **94,** 730–736.

Blake, C. A., Scaramuzzi, R. J., Hilliard, J. and Sawyer, C. H. (1973a). Circulating levels of pituitary gonadotrophins and ovarian steroids in rats after hypothalamic deafferentation. *Neuroendocrinology*, **12,** 86–97.

Blake, C. A., Scaramuzzi, R. J., Norman, R. L., Hilliard, J. and Sawyer, C. H. (1973b). Effects of hypothalamic deafferentation on circulating levels of LH, FSH, prolactin and testosterone in the male rat. *Endocrinology*, **92,** 1419–1425.

Bleier, R. (1972). Structural relationships of ependymal cells and their processes within the hypothalamus. *In* "Brain-Endocrine Interaction" (K. M. Knigge, D. E. Scott and A. Weindl, eds) pp. 306–318. Karger, Basel.

Bloom, S. R., Mortimer, C. H., Thorner, M. O., Besser, G. M., Hall, R., Gomez-Pan, A., Roy, V. M., Russell, G. C. G., Coy, D. H., Kastin, A. J. and Schally, A. V. (1974). Inhibition of gastrin and gastric acid secretion by growth hormone release inhibiting hormone. *Lancet*, **ii,** 1106–1109.

Bock, R. (1972). Morphometrische Untersuchungen zum histologischen Nachweis des corticotropin-releasing factor im Infundibulum der Ratte. *Z. Anat. Entwickl. Gesch.* **137**, 1–29.

Bodansky, M. and DuVigneaud, V. (1959). A method of synthesis of long peptide chains using a synthesis of oxytocin as an example. *J. Am. chem. Soc.* **81,** 5688–5691.

Boden, G., Sivitz, M. C., Owen, O., Essa, N. and Landor, J. C. (1975). Somatostatin suppresses secretin release and pancreatic exocrine function. *Diabetes*, **24,** (Suppl. 2) 408.

Bogdanove, E. M. (1963). Direct gonad pituitary feedback: an analysis of effects of intracranial estrogenic depots on gonadotrophin secretion. *Endocrinology*, **73,** 696–712.

Bogdanove, E. M. (1964). The role of the brain in the regulation of pituitary gonadotrophin secretion. *Vitams. & Horm.* **22,** 205–260.

Bøler, J., Enzmann, F., Folkers, K., Bowers, C. Y. and Schally, A. V. (1969). The identity of chemical and hormonal properties of the thyrotropin releasing hormone and pyroglutamyl-histidyl-proline amide. *Biochem. Biophys. Res. Commun.* **37,** 705–710.

Borgeat, P., Labrie, F., Drouin, J., Belanger, A., Immer, H., Sestanj, K., Nelson, V., Gotz, M., Schally, A. V., Coy, D. H. and Coy, A. J. (1974). Inhibition of adenosine 3′, 5′-monophosphate accumulation in anterior pituitary gland *in vitro* by growth hormone-release inhibiting hormone. *Biochem. Biophys. Res. Commun.* **56,** 1052–1059.

Bourguignon, J. P., Burger, H. G. and Franchimont, P. (1974). Radioimmunoassay of serum luteinizing hormone-releasing hormone (LH-RH) after intranasal administration and evaluation of the pituitary gonadotrophic response. *Clin. Endocr.* **3,** 437–440.

Bower, E. A. (1959). Action potentials from uterine sensory nerves. *J. Physiol., Lond.* **148,** 2P–3P.

Bower, A. Sr., Hadley, M. E. and Hruby, V. J. (1971). Comparative MSH release-inhibiting activities of tocinoic acid (the ring of oxytocin), and L-pro-L-leu-gly-NH_2 (the side chain of oxytocin). *Biochem. Biophys. Res. Commun.* **5,** 1185–1191.

Bowers, C. Y., Lee, K. L. and Schally, A. V. (1968). A study on the interaction of the thyrotropin-releasing factor and l-triiodothyronine: Effects of puromycin and cycloheximide. *Endocrinology*, **82,** 75–82.

Bowers, C. Y., Schally, A. V., Enzman, J., Bøler, J. and Folkers, K. (1970). Porcine thyrotropin releasing factor is (pyro)Glu-His-Pro(NH_2). *Endocrinology*, **86,** 1143–1153.

Bowers, C. Y., Friesen, H. G., Hwang, P., Guyda, H. J. and Folkers, K. (1971). Prolactin and thyrotropin release in man by synthetic pyroglutamyl-histidyl-prolinamide. *Biochem. Biophys. Res. Commun.* **45,** 1033–1041.

Bowers, C. Y., Currie, B. L., Johansson, K. N. G. and Folkers, K. (1973a). Biological evidence that separate hypothalamic hormones regulate the follicle stimulating and luteinizing hormone. *Biochem. Biophys. Res. Commun.* **50,** 20–26.

Bowers, C. Y., Friesen, H. G. and Folkers, K. (1973b). Further evidence that TRH is also a physiological regulator of PRL secretion in man. *Biochem. Biophys. Res. Commun.* **51,** 512–521.

Boyd, A. E. III and Jackson, I. M. D. (1975). Prolactin releasing factor in porcine hypothalamic extracts distinct from TRH. *Program 57th Meet. US Endocrine Soc.* p. 300.

Boyd, A. E., Lebovitz, H. E. and Pfeiffer, J. B. (1970). Stimulation of human growth hormone secretion by L-dopa. *N. Engl. J. Med.* **238,** 1425–1429.

Boyd, N. R. H., Jackson, D. B., Hollingsworth, S., Forsling, M. L. and Chard, T. (1972). The development of a radioimmunoassay for oxytocin: the extraction of oxytocin from urine and determination of the excretion rate for exogenous and endogenous oxytocin in human urine. *J. Endocr.* **52,** 59–67.

Bradbury, M. W. B., Burden, J. L., Hillhouse, E. W. and Jones, M. T. (1974). Stimulation electrically and by acetylcholine of the hypophysiotrophic area of the rat hypothalamus *in vitro*. *J. Physiol., Lond.* **239,** 269–283.

Bramble, F. J., Houghton, A. L., Eccles, S. S., Murray, M. A. F. and Jacobs, H. S. (1975). Specific control of follicle stimulating hormone in the male: postulated site of action of inhibin. *Clin. Endocr.* **4,** 443–449.

Brasel, J. A. and Blizzard, R. M. (1974). The influence of the endocrine glands upon growth and development. *In* "Textbook of Endocrinology" (R. H. Williams, ed.) 5th edn, pp. 1030–1058. Saunders, Philadelphia.

Brattin, W. J. and Portanova, R. (1975). Corticosterone modulation of pituitary responsiveness to CRF, effects of actinomycin D and puromycin. *Program 57th Meet. US Endocrine Soc.* p. 145.

Brazeau, P. and Guillemin, R. (1974). Somatostatin: newcomer from the hypothalamus. *N. Engl. J. Med.* **290,** 963–964.

Brazeau, P., Vale, W., Burgus, R., Ling, N., Butcher, M., Rivier, J. and Guillemin, R. (1973). Hypothalamic polypeptide that inhibits the secretion of immunoreactive pituitary growth hormone. *Science, N.Y.* **179,** 77–79.

Breton, B. and Weil, C. (1973). Effects du LH/FSH-RH synthétique et d'extraits hypothalamiques de Carpe sur la sécrétion d'hormone gonadotrophe *in vivo* chez la Carpe (*Cyprinus carpio* L). *C. R. Acad. Sci.* **277,** 2061–2064.

Bridges, P. K. and Jones, M. T. (1973). Relationships between psychological assessments, body-build and physiological stress responses. *J. Neur. Psychiat.* **36,** 839–845.

Brightman, M. W. and Reese, T. S. (1969). Junctions between intimately apposed cell membranes in the vertebrate brain. *J. Cell. Biol.* **40,** 648–677.

Brinck-Johnson, T., Solem, J. H., Brinck-Johnson, K. and Ingvaldsen, P. (1963). The 17-hydroxycorticosteroid response to ACTH, metopiron and bacterial pyrogen. *Acta. Med. Scand.* **173,** 129–138.

British Medical Bulletin, (1974). Radioimmunoassay and saturation analysis. *Br. Med. Bull.* 30, No. 1.

Brodish, A., (1973). Hypothalamic and extrahypothalamic corticotrophin-releasing factors in peripheral blood. *In* "Brain-Pituitary-Adrenal Interrelationships" (A. Brodish and E. S. Redgate, eds) pp. 128–151. Karger, Basel.

Brodish, A. and Long, C. N. H. (1962). ACTH-releasing hypothalamic neurohumor in peripheral blood. *Endocrinology*, **71,** 298–306.

Brown, F. A. (1970). Hypothesis of environmental timing of the clock. *In* "The Biological Clock: Two Views", pp. 15–59. Academic Press, New York and London.

Brown, G. M. and Martin, J. B. (1973). Neuroendocrine relationships. *Prog. Neurol. Psychiat.* **28,** 193–240.

Brown, G. M. and Reichlin, S. (1972). Psychological and neural regulation of growth hormone secretion. *Psychosom. Med.* **34,** 45–61.

Brown, G. M., Krigstein, E., Dankova, J. and Hornykiewicz, O. (1972). Relationship between hypothalamic and median eminence catecholamines and thyroid function. *Neuroendocrinology*, **10,** 207–217.

Brown, M., Rivier, J., Vale, W. and Guillemin, R. (1975). Variability of the duration of inhibition of growth hormone release by N^{α}-acylated-Des-(Ala^1-Gly^2)-H_2-somatostatin analogs. *Biochem. Biophys. Res. Commun.* **65,** 752–756.

Brown, W. A., Van Woert, M. H. and Ambani, L. M. (1973). Effect of apomorphine on growth hormone release in humans. *J. Clin. Endocr. Metab.* **37,** 463–465.

Brown-Grant, K. (1974). Steroid hormone administration and gonadotrophin secretion in the gonadectomized rat. *J. Endocr.* **62,** 319–332.

Brown-Grant, K. and Naftolin, F. (1972). Facilitation of luteinizing hormone secretion in the female rat by progesterone. *J. Endocr.* **53,** 37–46.

Brown-Séquard, C. E. (1889). Des effets produits chez l'homme par des injections sous-cutanées d'un liquide retiré des testicules frais de cobaye et de chien. *C. R. Soc. Biol.* 1, 415–430.

Brownstein, M. J., Palkovits, M., Saavedra, J. M., Bassiri, R. M. and Utiger, R. D. (1974). Thyrotropin-releasing hormone in specific nuclei of rat brain. *Science, N.Y.* **185,** 267–269.

Brownstein, M., Arimura, A., Sato, H., Schally, A. V. and Kizer, J. S. (1975). The regional distribution of somatostatin in the rat brain. *Endocrinology,* **96,** 1456.

Brownstein, M. J., Palkovits, M., Saavedra, J. H. and Kizer, J. S. (1976). Distribution of hypothalamic hormones and neurotransmitters within the diencephalon. *In* "Frontiers in Neuroendocrinology" (L. Martini and W. F. Ganong, eds) pp. 1–23. Raven Press, New York.

Bruce, H. M. (1970). Pheromones. *Br. Med. Bull.* **26,** 10–13.

Brüggemann, J., Adam, A. and Karg, H. (1965). ICSH-Bestimmungen in Hypophysen von Rehböcken (*Capreolus capreolus*) und Hirschen (*Cervus elaphus*) unter Berücksichtigung des Saisoneinflusses. *Acta endocr., Copenh.* **48,** 569–579.

Buckingham, J. L. and Hodges, J. R. (1974a). Feedback control of corticotrophin in the rat. *J. Physiol., Lond.* **239,** 19P–20P.

Buckingham, J. L. and Hodges, J. R. (1974b). Interrelationships of pituitary and plasma corticotrophin and plasma corticosterone in adrenalectomized and stressed adrenalectomized rats. *J. Endocr.* **63,** 213–222.

Buckingham, J. L. and Hodges, J. R. (1977). The use of corticotrophin production by adenohypophyseal tissue *in vitro* for the detection and estimation of potential corticotrophin releasing factors. *J. Endocr.* **72,** 187–193.

Buckler, J. M. H. (1972). Exercise as a screening test for growth hormone release. *Acta endocr., Copenh.,* **69,** 219–229.

Buckman, M. T. and Peake, G. T. (1973). Osmolar control of prolactin secretion in man. *Science, N.Y.* **181,** 755–757.

Bünning, E. (1967). "The Physiological Clock". Springer-Verlag, New York.

Burden, J., Harrison, D. J., Hillhouse, E. W., Ironmonger, M. R. and Jones, M. T. (1975). Effects of chlorpromazine pentabarbitone, vasopressin, angiotensin II, bradykinin and ACTH on the secretion of CRF from the hypothalamus *in vitro*. *J. Endocr.* **67,** 45P.

Burden, J. L., Hillhouse, E. W. and Jones, M. T. (1974a). The inhibitory action of GABA and melatonin on the release of corticotrophin-releasing hormone from the rat hypophysiotrophic area *in vitro*. *J. Physiol., Lond.* **239,** 116P–117P.

Burden, J. L., Hillhouse, E. W. and Jones, M. T. (1974b). A proposed model of the transmitters involved in the control of corticotrophin-releasing hormone. *J. Endocr.* **63,** 20–21.

Burgers, A. C. S., Imai, K. and van Oorth, G. J. (1963). The amount of melanophore-stimulating hormone in single pituitary glands of *Xenopus leavis* kept under various conditions. *Gen. Comp. Endocr.* **3,** 53–57.

Burgus, R. and Guillemin, R. (1970a). Hypothalamic releasing factors. *Annu. Rev. Biochem.* **39,** 499–526.

Burgus, R. and Guillemin, R. (1970b). Chemistry of thyrotropin-releasing factor (TRF). *In* "Hypophysiotropic Hormones of the Hypothalamus" (J. Meites, ed.) pp. 227–241. Williams and Wilkins, Baltimore.

Burgus, R., Dunn, T. F., Desiderio, D., Ward, D. N., Vale, W. and Guillemin, R. (1970). Characterisation of ovine hypothalamic hypophysiotropic TSH releasing factor. *Nature, Lond.,* **226,** 321–325.

Burgus, R., Ling, N., Butcher, M. and Guillemin, R. (1973). Primary structure of somatostatin, a hypothalamic peptide that inhibits the secretion of pituitary growth hormone. *Proc. Natl. Acad. Sci. USA*, **70**, 684–688.

Burgus, R., Nelson, J. and Amoss, M. (1975). Alcohol chloroform extraction of hypothalamic hypophysiotropic peptides. *Methods Enzymol.* **37**, (Pt. B), 402–407.

Burke, C. W. and Roulet, F. (1970). Increased exposure of tissues to cortisol in late pregnancy. *Br. Med. J.* **1**, 657–659.

Burlet, A., Marchetti, J. and Duheille, J. (1973). Etude par immunofluorescence de la répartition de la vasopressine an niveau du système hypothalamo-neurohypophysaire du rat. *C. R. Soc. Biol.* **167**, 924–927.

Burn, J. H. (1975). "The Autonomic Nervous System", 5th edn. Blackwell, Oxford.

Burns, K. (1975). Variation in plasma ACTH levels during the human menstrual cycle. *J. Physiol., Lond.* **49**, 36P.

Burr, W. A., Ramsden, D. B., Griffiths, R. S., Black, E. G., Hoffenberg, R., Meinhold, H. and Wenzel, K. W. (1976). Effect of a single dose of dexamethasone on serum concentration of thyroid hormones. *Lancet* **2**, 58–61.

Burrows, A. W., Cooper, E., Shakespear, R. A., Aickin, C. M., Fraser, S., Hesch, R.-D. and Burke, C. W. (1977). Low serum L-T_3 levels in the elderly sick: protein binding, thyroid and pituitary responsiveness, and reverse T_3 concentrations. *Clin. Endocr.* **7**, 289–300.

Butcher, R. L., Collins, W. E., Fugo, N. W. (1974). Plasma concentrations of LH, FSH, prolactin, progesterone and estradiol-17β throughout the 4-day estrus cycle of the rat. *Endocrinology*, **94**, 1704–1708.

Buttle, H. L. (1974). Seasonal variation of prolactin in plasma of male goats. *J. Reprod. Fert.* **37**, 95–99.

C

Cameron, E. and Foster, C. L. (1971). Some light and electron-microscopical observations on the pars intermedia of the pituitary gland of the rabbit. *J. Endocr.* **49**, 479–485.

Campbell, C. B. G. and Ramaley, J. A. (1974). Retinohypothalamic projections: correlations with onset of the adrenal rhythm in infant rats. *Endocrinology*, **94**, 1201–1204.

Campbell, H. J. (1966). The development of the primary portal plexus in the median eminence of the rabbit. *J. Anat. (London)*, **100**, 381–387.

Camus, J. and Roussy, G. (1914). Polyurie expérimentale par lésions de la base du cerveau; la polyurie hypophysaire. *C.R. Soc. Biol.* **75**, 628–633.

Cannon, W. B. (1929). Organization for physiological homeostasis. *Physiol. Rev.* **9**, 399–431.

Carlson, H. E., Gillin, J. C., Gorden, P. and Snyder, F. (1972). Absence of sleep-related growth hormone peaks in aged normal subjects and in acromegaly. *J. Clin. Endocrinol. Metab.* **34**, 1102–1105.

Carlson, H. E., Jacobs, L. S. and Daughaday, W. H. (1973). Growth hormone, thyrotropin, and prolactin responses to thyrotropin-releasing hormone following diethylstilbestrol pretreatment. *J. Clin. Endocrinol. Metab.* **37**, 488–490.

Carlson, H. E., Mariz, I. K. and Daughaday, W. H. (1974). Thyrotropin releasing hormone stimulation and somatostatin inhibition of growth hormone secretion from perfused rat adenohypophysis. *Endocrinology*, **94**, 1709–1713.

Carrer, H. F. and Taleisnik, S. (1970). Effects of mesencephalic stimulation on the release of gonadotropins. *J. Endocr.* **48**, 527–539.

Carrllo, A. J., Kastin, A. J., Dunn, J. D. and Schally, A. V. (1973). MSH activity in plasma and pituitaries of rats with large hypothalamic lesions. *Neuroendocrinology*, **12**, 120–128.

Castensson, S., Sievertsson, H., Lindeke, B. and Sum, C. Y. (1974). Studies on the inhibition of oxotremorine induced tremor by a melanocyte-stimulating hormone release-inhibiting factor, thyrotropin-releasing hormone and related peptide. *FEBS Letters*, **44,** 101–105.

Céhovic, G. (1967). Effet de MSH sur l'axe hypothalamus-hypophyse-thyroid. *Rev. Europ. Endocrinol.* **4,** 125–138.

Celis, M. E. and Taleisnik, S. (1971a). Formation of a melanocyte-stimulating hormone release-inhibiting factor by hypothalamic extracts from rats. *Intern. J. Neuroscience*, **1,** 223–230.

Celis, M. E. and Taleisnik, S. (1971b). *In vitro* formation of a MSH-releasing agent by hypothalamic extracts. *Experientia*, **27,** 1481–1482.

Celis, M. E. and Taleisnik, S. (1973). Interaction between paraventricular nucleus extracts on the formation of melanocyte stimulating hormone release inhibiting factor. *Life Sci.* **13,** 493–499.

Celis, M. E. and Taleisnik, S. (1974). Estrogen influence on the hypothalamic enzymes involved in the formation of melanocyte stimulating hormone release inhibiting factor (MSH-R-IF). *Proc. Soc. exp. Biol. Med.* **145,** 142–144.

Celis, M. E., Taleisnik, S. and Walter, R. (1971a). Release of pituitary melanocyte stimulating hormone by the oxytocin fragment, H-Cys-Tyr-Ile-Gln-Asn-OH. *Biochem. Biophys. Res. Commun.* **45,** 564–569.

Celis, M. E., Taleisnik, S. and Walter, R. (1971b). Regulation of formation and proposed structure of the factor inhibiting the release of melanocyte-stimulating hormone. *Proc. Nat. Acad. Sci. USA*, **68,** 1428–1433.

Celis, M. E., Hase, S. and Walter, R. (1972). Structure-activity studies of MSH-release-inhibiting hormone. *FEBS Letters*, **27,** 327–330.

Celis, M. E., Macagno, R. and Taleisnik, S. (1973). Inhibitory effect of L-prolyl-L-leucy glycinamide on the secretion of melanocyte-stimulating hormone in rats with median eminence lesions. *Endocrinology*, **93,** 1229–1231.

Ceresa, F., Angeli, A., Buccuzzi, G. and Perotti, L. (1970). Impulsive and basal ACTH secretion phases in normal subjects, in obese subjects with signs of adrenocortical hyperfunction and in hyperthyroid patients. *J. Clin. Endocrinol. Metab.* **31,** 491–501.

Chabraborty, P. K., Reeves, J. J., Arimura, A. and Schally, A. V. (1973). Serum LH-levels in prepubertal female pigs chronically treated with synthetic luteinizing hormone-releasing hormone/follicle stimulating hormone-releasing hormone (LH-RH/FSH-RH). *Endocrinology*, **92,** 55–61.

Chamley, W. A., Findlay, J. K., Cumming, I. A., Buckmaster, J. M. and Goding, J. R. (1974). Effect of pregnancy on the LH response to synthetic gonadotrophin-releasing hormone in the ewe. *Endocrinology*, **94,** 291–293.

Chan, L. T., Schall, S. M. and Saffran, M. (1969). Properties of the corticotrophin-releasing factor of the rat median eminence. *Endocrinology*, **85,** 644–651.

Chan, L. T., Schall, S. M. and Saffran, M. (1970). The rat median eminence as a source of CRF. *In* "Hypophysiotropic Hormones of the Hypothalamus: Assay and Chemistry" (J. Meites, ed.) pp. 253–260. Williams and Wilkins, Baltimore.

Chan, W. Y., Nestor, J. J. Jr., Ferger, M. F. and DuVigneaud, V. (1974). Inhibition of oxytocic responses to oxytocin in pregnant rats by (1-L-Penicillamine) oxytocin and (1-β-mercapto-β,β-diethyl-propinoic acid) oxytocin. *Proc. Soc. exp. Biol. Med.* **146,** 364–366.

Chard, T. (1973). The radioimmunoassay of oxytocin and vasopressin. *J. Endocr.* **58,** 143–160.

Chard, T. and Martin, M. J. (1974). Posterior pituitary peptides. *Br. Med. Bull.* **30,** 74–79.

Charlton, H. M., Gregory, H., Mackinnon, P. C. B., Walpole, A. L. and Worth, R. W. (1973). The effects of injecting synthetic luteinizing hormone-releasing factor in the vole (*Microtus agrestics*). *J. Endocr.* **57,** xxvii.

Chauvet, J., Chauvet, M. T. and Acher, R. (1971). Une éventuelle relation entre le taux d'urée dans le sang et la teneur en vasotocine de la neurohypophyse chez certains poissons. *C.R. Acad. Sci.* **272,** 1669–1671.

Cheesman, D. W. (1970) Structural elucidation of a gonadotropin-inhibitory substance from the bovine pineal gland. *Biochim. Biophys. Acta,* **207,** 247–253.

Cheesman, D. W. and Farris, B. L. (1970). Isolation and characterization of a gonadotrophin-inhibiting substance from the bovine pineal gland. *Proc. Soc. exp. Biol. Med.* **103,** 1254–1257.

Cheifetz, P. N. (1969). Effect of lesions in the globus pallidus on thyroid function. *J. Endocr.* **43,** 36–37.

Chen, C. L., Bixler, E. J., Weber, A. I. and Meites, J. (1968). Hypothalamic stimulation of prolactin release from the pituitary of turkey hens and poults. *Gen. Comp. Endocrinol.* **11,** 489–494.

Cheng, Y-J. and Karavolas, H. J. (1973). Conversion of progesterone to 5α-pregnane-3, 20-dione and 3α-hydroxy-5α-pregnane-20-one by rat medial basal hypothalamus and the effects of estradiol and stage of estrous cycle on this conversion. *Endocrinology,* **93,** 1157–1162.

Chester Jones, I., Henderson, I. W., Wales, N. A. M. and Garland, H. O. (1971). Effects of arginine vasotocin in fish and amphibia. *In* "Neurohypophysial Hormones" (G. E. W. Wolstenholme and J. Birch, eds) pp. 29–44. Churchill Livingstone, London.

Chi, C. C. (1970). Afferent connections to the ventromedial nucleus of the hypothalamus in the rat. *Brain Res.* **17,** 439–445.

Chiang, T. M., Duckworth, W. C., Beachey, E. H. and Kang, A. H. (1975). The effect of somatostatin on platelet aggregation. *Endocrinology,* **97,** 753–756.

Chiodini, P. G., Liuzzi, A., Botalla, L., Oppizzi, G., Müller, E. E. and Silvestrini, F. (1975). Stable reduction of plasma growth hormone (hGH) levels during chronic administration of 2-bromo-alpha-ergocryptine (CB 154) in acromegalic patients. *J. Clin. Endocrinol. Metab.* **40,** 705–708.

Chobsieng, P., Naor, Z. and Koch, Y. (1975). Stimulatory effect of prostoglandin E_2 on LH release in the rat: evidence for hypothalamic site of action. *Neuroendocrinology,* **17,** 12–17.

Chopra, I. J., Chopra, U. and Orgiazzi, J. (1973). Abnormalities of hypothalamo-hypophyseal thyroid axis in patients with Graves' ophthalmopathy. *J. Clin. Endocrinol. Metab.* **37,** 955–967.

Chopra, I. J., Chopra, U., Smith, S. R., Reza, M. and Solomon, D. H. (1975a). Reciprocal changes in serum concentrations of 3,3′5′-triiodothyronine (reverse T_3) and 3,3′5-triiodothyronine (T_3) in systemic illnesses. *J. Clin. Endocrinol. Metab.* **41,** 1043–1049.

Chopra, I. J., Williams, D. E., Orgiazzi, J. and Solomon, D. H. (1975b). Opposite effects of dexamethasone on serum concentrations of 3,3′,5′-triiodothyronine (reverse T_3) and 3,3′5-triiodothyronine (T_3). *J. Clin. Endocrinol. Metab.* **41,** 911–920.

Chowers, I. and McCann, S. M. (1965). Content of luteinizing hormone-releasing factor and luteinizing hormone during the estrus cycle and after changes in gonadal steroid titres. *Endocrinology,* **76,** 700–708.

Chowers, I., Conforti, N. and Feldman, S. (1967). Effect of corticosteroids on hypothalamic corticotrophin. *Neuroendocrinology,* **2,** 219–223.

Christian, J. J. (1971). Population density and reproductive efficiency. *Biol. Reprod.* **4,** 248–294.

Clattenburg, R. E. (1973). Ultrastructure of hypothalamic neurons and of the median eminence. *Canadian J. Neurol. Sci.* **1,** 40–48.

Clayton, G. W., Librik, L., Gardner, R. L. and Guillemin, R. (1963). Studies on the circadian rhythm of pituitary adrenocorticotrophic release in man. *J. Clin. Endocrinol. Metab.* **23,** 975–980.

Clifton, D. K., Steiner, R. A., Resko, J. A. and Spies, H. G. (1975). Estrogen-induced gonadotrophin release in ovariectomized rhesus monkeys and its advancement by progesterone. *Biol. Reprod.* **13,** 190–194.

Cocchi, D., Fraschini, F., Jalanbo, H. and Müller, E. E. (1974). Role of brain catecholamines in the post-castration rise in plasma LH of prepubertal rats. *Endocrinology*, **95,** 1649–1657.

Coceani, F., Puglisi, L. and Lavers, B. (1971). Prostaglandins and neuronal activity in spiral and cuneate nucleus. *Ann. N.Y. Acad. Sci.* **180,** 289–301.

Collin, R. (1928). Etude histophysiologique du complexe tubero-infundibulo-pituitaire. *Arch. Morph. gen. exp.* **28,** 1–100.

Collin, R. (1928). La neurocrinie hypophysaire. Etude histophysiologique du complexe tubero-infundibulo pituitaire. *Arch. d. Morph.* **18,** 1–102.

Collip, J. B., Anderson, E. M. and Thomson, D. L. (1933). The adrenotropic hormone of the anterior pituitary lobe. *Lancet*, **ii,** 347–348.

Collu, R. and Ducharme, J. R. (1975). Role of adrenal steroids in the regulation of gonadotrophin secretion at puberty. *J. Steroid Biochem.* **6,** 869–872.

Colombo, J. A., Baldwin, D. M. and Sawyer, C. H. (1974). Timing of the estrogen-induced release of LH in ovariectomized rats under altered lighting schedules. *Proc. Soc. exp. Biol. Med.* **145,** 1125–1127.

Compher, M. K. Jr. and Dent, J. (1973). Effect of hypothalamic extirpations on thyroidal and intermedinal activity in the red spotted newt. *Gen. Comp. Endocrinol.* **20,** 29–40.

Conroy, R. T. W. L. and Hall, M. C. (1969). Adrenal cortical function and body temperature rhythm after a transatlantic flight. *J. Physiol.* (*London*), **200,** 123P.

Conroy, R. T. W. L. and Mills, J. N. (1970). "Human Circadian Rhythms". Churchill, London.

Constantinidis, J., Geissbuhler, F. and Gaillard, J. M. (1974). Enhancement of cerebral noradrenaline turnover by thyrotrophin-releasing hormone: evidence by fluorescence histochemistry. *Experientia*, **30,** 1182–1183.

Cooper, K. J., Fawcett, C. P. and McCann, S. M. (1974). Inhibitory and facilitatory effects of estradiol-17β on pituitary responsiveness to a luteinizing hormone-follicle stimulating hormone releasing factor (LH-RF/FSH-RF) preparation in the ovariectomized rat. *Proc. Soc. exp. Biol. Med.* **145,** 1422–1426.

Coppen, A., Montgomery, S., Peet, M., Bailey, J., Marks, V. and Woods, P. (1974). Thyrotropin releasing hormone in the treatment of depression. *Lancet*, **ii,** 433–435.

Coppings, R. J. and Malven, P. V. (1975). Rhythmic patterns of endogenous LH release in castrate sheep receiving exogenous LH. *Proc. Soc. exp. Biol. Med.* **148,** 64–68.

Corbin, A. and Beattie, C. W. (1975). Inhibition of the pre-ovulatory proestrous gonadotrophin surge, ovulation and pregnancy with a peptide analogue of luteinizing hormone releasing hormone. *Endoc. Res. Commun.* **2,** 1–23.

Cornblath, M., Parker, M. L., Reisner, S. H., Forbes, A. E. and Daughaday, W. H. (1965). Secretion and metabolism of growth hormone in premature and full-term infants. *J. Clin. Endocrinol. Metab.* **25,** 209–218.

Costa, E., Guidotti, A., Uzunov, P. and Zivkovic, B. (1975). Methods for studying the *in vivo* regulation of cyclic nucleotides in hypothalamic nuclei, pituitary and pineal. *In* "Symposium on Neuroendocrine Regulation and Fertility" (Brain Information Service) pp. 82–84. BRI Publications, Los Angeles.

Cotton, G., Gorman, C. A. and Mayberry, N. E. (1971). Suppression of thyrotropin in myxedema treated with thyroid hormones. *N. Engl. J. Med.* **285,** 529–533.

Cowchock, F. S., Carmel, P. W. and Barrett, R. E. (1974). The distribution of catecholamines in the hypothalamus of the cat. *Neuroendocrinology*, **15,** 209–219.

Cowie, A. T. and Tindal, J. S. (1971). "The Physiology of Lactation". Arnold, London.

Coy, D. H., Coy, E. J., Arimura, A. and Schally, A. V. (1973). Solid phase synthesis of growth hormone-release inhibiting factor. *Biochem. Biophys. Res. Commun.* **54,** 1267–1273.

Coy, D. H., Coy, E. J. and Schally, A. V. (1974a). Stimulatory and inhibitory analogues of luteinizing hormone-releasing hormone. *Biochemistry*, **13,** 323–326.

Coy, D. H., Coy, E. J., Hirotsu, Y. and Schally, A. V. (1974b). Synthesis and biological properties of (2-L-β-(pyrazolyl-3)alanine)-luteinizing hormone-releasing hormone. *J. Med. Chem.* **17,** 140–142.

Coy, D. H., Hirotsu, Y., Redding, T. W., Coy, E. J. and Schally, A. V. (1975). Synthesis and biological properties of the 2-L-β(pyrazolyl-1)alanine analogues of luteinizing hormone-releasing hormone and thyrotropin-releasing hormone. *J. Med. Chem.* **18,** 948–949.

Craig, G. M., Ginsburg, J., Gore, M. and Isaacs, A. J. (1975). Effect of indomethacin on the pituitary response to synthetic luteinising hormone-releasing hormone in women with amenorrhoea or oligomenorrhoea. *J. Endocr.* **66,** 287–288.

Cramer, O. M. and Barraclough, C. A. (1975). Failure to detect LH-RH in third ventricle cerebral spinal fluid under a variety of experimental conditions. *Endocrinology*, **96,** 913–121.

Cramer, O. M. and Porter, J. C. (1973). Input to releasing factor cells. *Progr. Brain Res.* **39,** 73–84.

Crighton, D. B., Foster, J. P., Haresign, W. and Scott, S. A. (1975]. Plasma LH and progesterone levels after single or multiple injections of synthetic LH-RH in anoestrus ewes and comparison with levels during the oestrus cycle. *J. Reprod. Fertil.* **44,** 121–124.

Crim, L. W., Watts, E. G. and Evans, D. M. (1975). The plasma gonadotropin profile during sexual maturation in a variety of salmonid fishes. *Gen. Comp. Endocrinol.* **27,** 62–70.

Crosby, E. C. and Woodburne, R. T. (1940). The comparative anatomy of the preoptic area and the hypothalamus. *Res. Publ. Ass. Res. nerv. ment. Dis.* **20,** 52–169.

Cross, B. A. (1966). Neural control of oxytocin secretion. *In* "Neuroendocrinology" (L. Martini and W. F. Ganong, eds) pp. 217–259. Academic Press, New York and London.

Cross, B. A. and Dyer, R. G. (1971a). Cyclic changes in neurones of the anterior hypothalamus during the oestrous cycle, and the effects of anaesthesia. *In* "Steroid Hormones and Brain Function" (R. Gorski and C. H. Sawyer, eds) pp. 95–102. University of California Press, Los Angeles.

Cross, B. A. and Dyer, R. G. (1971b). Unit activity in diencephalic islands: the effect of anaesthetics. *J. Physiol., Lond.* **212,** 467–481.

Cross, B. A. and Green, J. D. (1959). Activity of single neurones in the hypothalamus: effect of osmotic and other stimuli. *J. Physiol., Lond.* **148,** 554–569.

Cross, B. A. and Kitay, J. I. (1967). Unit activity in diencephalic islands. *Expl. Neurol.* **19,** 316–330.

Cross, B. A. and Silver, I. A. (1965). Effect of luteal hormone on the behaviour of hypothalamic neurones in pseudopregnant rats. *J. Endocr.* **31,** 251–263.

Cross, B. A. and Silver, I. A. (1966). Electrophysiological studies on the hypothalamus. *Br. Med. Bull.* **22,** 254–260.

Cross, B. A., Dyball, R. E. J., Dyer, R. G., Jones, C. W., Lincoln, D. W., Morris, J. F. and Pickering, B. T. (1975). Endocrine neurones. *Recent Prog. Horm. Res.* **31,** 243–294.

Crowe, S. J., Cushing, H. and Homans, J. (1910). Experimental hypophysectomy. *Bull. Johns Hopk. Hosp.* **21,** 127–169.

Crumpton, M. J. (1974). Protein antigens. The molecular bases of antigenicity and immunogenicity. *In* "The Antigens" (M. Sela, ed.) Vol. II, pp. 1–78. Academic Press, London and New York.

Currie, B. L., Sievertsson, H., Bogentoft, C., Chang, J. K., Folkers, K., Bowers, C. Y. and Doolittle, R. F. (1971). Data on structure of bovine LRH by inactivation. *Biochem. Biophys. Res. Commun.* **42,** 1180–1184.

Currie, B. L., Johansson, N. G. K., Griebrock, T. and Folkers, K. (1974). Identification and purification of factor A-GH-RH from hypothalami which releases growth hormone. *Biochem. Biophys. Res. Commun.* **60,** 605–609.

Curry, D. L. and Bennett, L. L. (1974). Reversal of somatostatin inhibition of insulin secretion by calcium. *Biochem. Biophys. Res. Commun.* **60,** 1015–1019.

Curry, D. L., Bennett, L. L. and Li, C. H. (1974). Direct inhibition of insulin secretion by synthetic somatostatin. *Biochem. Biophys. Res. Commun.* **58,** 885–889.

Cushing, H. (1909). The hypophysis cerebri; clinical aspects of hyperpituitarism and hypopituitarism. *J. Amer. Med. Ass.* **53,** 249–255.

Cushing, H. (1912). "The Pituitary Body and its Disorders". Lippincott, Philadelphia.

Cushing, H. and Goetsch, E. (1910). Concerning the secretions of the infundibular lobe of pituitary body and its presence in the cerebrospinal fluid. *Am. J. Physiol.* **27,** 60–86.

Cusick, E. K. and Wilson, F. E. (1972). On control of spontaneous testicular regression in tree sparrows (*Spizella arborea*). *Gen. Comp. Endocrinol.* **19,** 441–456.

D

Dahlberg, B. (1963). Urinary MSH determination as an index of pituitary function. *Ann. N.Y. Acad. Sci.* **100,** 631–638.

Dahlgrén, U. (1914). The electric motor nerve centers in the skates (Rajidae). *Science, N.Y.* **40,** 862.

Dallman, M. F. and Jones, M. T. (1973). Corticosteroid feedback control of stress-induced ACTH secretion. *In* "Brain-Pituitary-Adrenal Interrelationships" (A. Brodish and E. S. Redgate, eds) pp. 176–196. Karger, Basel.

Dallman, M. F. and Yates, F. E. (1968). Anatomical and functional mapping of central neural input and feedback pathways of the adrenocorticoid system. *Mem. Soc. Endocrinol.* **17,** 39–72.

Dallman, M. F. and Yates, F. E. (1969). Dynamic asymmetries in the corticosteroid feedback pathways and distribution, binding and metabolism elements of the adrenocortical system. *Ann. N.Y. Acad. Sci.* **156,** 696–721.

Dallman, M. F., Jones, M. T., Vernikos-Danellis, J. and Ganong, W. F. (1972). Corticosteroid control of ACTH secretion. *Endocrinology*, **91,** 961–968.

Daly, J. R., Myles, A. B., Bacon, P. A., Beardwell, C. G. and Savage, O. (1967). Pituitary adrenal function during corticosteroid withdrawal in rheumatoid arthritis. *Ann. Rheum. Dis.* **26,** 18–25.

Daly, J. R., Fleisher, M. R., Chambers, D. J., Bitensky, L. and Chayen, J. (1974a). Application of the cytochemical bioassay for corticotrophin to clinical and physiological studies in man. *Clin. Endocrinol.* **3,** 335–345.

Daly, J. R., Fletcher, M. R., Glass, D., Chambers, D. J., Bitensky, L. and Chayen, J. (1974b). Comparison of the effects of long-term ACTH and corticosteroid treatment on the responses of plasma GH, ACTH and cortisol to hypoglycaemia. *Br. Med. J.* **2,** 521–524.

Daniel, P. M. (1966). The anatomy of the hypothalamus and pituitary gland. *In* "Neuroendocrinology" (L. Martini and W. F. Ganong, eds) Vol. I, pp. 15–80. Academic Press, New York and London.

Davidson, J. M. (1969). Feedback control of gonadotropin secretion. *In* "Frontiers in Neuroendocrinology" (W. F. Ganong and L. Martini, eds) pp. 343–388. Oxford University Press, New York.

Davidson, J. M., Smith, E. R. and Bowers, C. Y. (1973). Effects of mating on gonadotrophin release in the female rat. *Endocrinology*, **93,** 1185–1192.

Davies, A. G., Duncan, I. F. and Lynch, S. S. (1975a). Autoradiographic localisation of ^{125}I-labelled FSH in the rat hypothalamus. *J. Endocr.* **66,** 301–302.

Davies, I. J., Naftolin, F., Ryan, K. J., Fishman, J. and Siu, J. (1975b). The affinity of catechol estrogens for estrogen receptors in the pituitary and anterior hypothalamus of the rat. *Endocrinology*, **97,** 554–557.

Davis, G. J. and Meyer, R. K. (1973). Seasonal variation in LH and FSH of bilaterally castrated snow shoe hares. *Gen. comp. Endocrinol.* **20,** 61–68.

Dawson, A. B. (1937). The relationships of the epithelial components of the pituitary gland of the rabbit and cat. *Anat. Rec.* **69,** 471–485.

Dawson, A. B. (1948). The relationship of pars tuberalis to pars distalis in the hypophysis of the rhesus monkey. *Anat. Rec.* **102,** 103–121.

Dawson, D. C. and Ralph, C. L. (1974). Light-evoked neural activity in the intermediate lobe of the pituitary in *Rana pipiens*: electrophysiological evidence for a neural pathway linking the eyes with the pars intermedia. *Neuroendocrinology*, **15,** 267–280.

Debeljuk, L., Arimura, A. and Schally, A. V. (1972a). Pituitary responsiveness to LH-releasing hormone in intact female rats of different ages. *Endocrinology*, **90,** 1499–1502.

Debeljuk, L., Arimura, A. and Schally, A. V. (1972b). Studies on the pituitary responsiveness to luteinizing hormone-releasing hormone (LH-RH) in intact male rats of different ages. *Endocrinology*, **90,** 585–588.

Debeljuk, L., Arimura, A. and Schally, A. V. (1973a). Effect of estradiol on the responsiveness to LH-RH in male rats at different times after castration. *Proc. Soc. exp. Biol. Med.* **143,** 1164–1167.

Debeljuk, L., Arimura, A., Shiino, M., Rennels, E. G. and Schally, A. V. (1973b). Effects of chronic treatment with LH/FSH-RH in hypophysectomized pituitary grafted male rats. *Endocrinology*, **92,** 921–930.

de Belleroche, J. S. and Bradford, H. F. (1973). The synaptosomes: an isolated working, neuronal compartment. *In* "Progress in Neurology" (G. A. Kerkut and J. W. Phillips, eds) Vol. 1, pp. 275–298. Pergamon Press, Oxford and New York.

Deery, D. J. (1974). Determination by radioimmunoassay of the luteinizing hormone-releasing hormone (LH-RH) content of the hypothalamus of the rat and some lower vertebrates. *Gen. Comp. Endocrinol.* **24,** 280–285.

Deery, D. J. and Jones, A. C. (1975). Effects of hypothalamic extracts, neurotransmitters and synthetic hypothalamic releasing hormones on adenyl cyclase activity in the lobes of the pituitary of the dogfish (*Scyliorhinus canicula* L). *J. Endocr.* **64,** 49–57.

Deis, R. P. and Orias, R. (1968). The effect of an exteroceptive stimulus on the concentration of melanocyte-stimulating hormone in the pituitary of lactating rats. *J. Physiol., Lond.* **197,** 47–51.

Dekanski, J. (1952). The quantitative assay of vasopressin. *Brit. J. Pharmacol.* **7,** 567–572.

de Kloet, E. R., Van Der Vreis, J. and de Wied, D. (1974). The site of the suppressive action of dexamethasone on pituitary-adrenal activity. *Endocrinology*, **96,** 61–67.

Deller, J. J. Jr., Boulis, M. W., Harriss, W. E., Hutsell, T. C., Garcia, J. F. and Linfoot, J. A. (1970). Growth hormone response patterns to sex hormone administration in growth retardation. *Am. J. Med. Sci.* **259,** 292–297.

del Pozo, E. and Flückiger, E. (1973). Prolactin inhibition: experimental and clinical studies. *In* "Human Prolactin" (J. L. Pasteels and C. Robyn, eds) pp. 291–301. Excerpta Medica, Amsterdam.

Denef, C., Magnus, C. and McEwan, B. A. (1973). Sex differences and hormonal control of testosterone metabolism in rat pituitary and brain. *J. Endocr.* **59,** 605–621.

De Olmos, J. S. and Ingram, W. R. (1972). The projection field of the stria terminalis in the rat brain: an experimental study. *J. Comp. Neurol.* **146,** 303–333.

DePierre, J. W. and Karnovsky, M. L. (1973). Plasma membranes of mammalian cells: a review of methods for their characterisation and isolation. *J. Cell. Biol.* **56,** 275–303.

DeRubertis, F. R. and Kosch, P. C. (1975). Accelerated host metabolism of L-thyroxine during acute injection: role of the leukocyte and peripheral leukocytosis. *J. Clin. Endocrinol. Metab.* **40,** 589–600.

Deuben, R. and Meites, J. (1964). Stimulation of pituitary growth hormone release by a hypothalamic extract *in vitro*. *Endocrinology*, **74,** 408–414.

de Vlaming, V. L. (1972). Environmental control of teleost reproductive cycles: a brief review. *J. Fish. Biol.* **4,** 131–140.

de Wied, D. (1961). The significance of antidiuretic hormone in the release mechanism of corticotrophin. *Endocrinology*, **68,** 956–970.

de Wied, D. and de Jong, W. (1974). Drug effects and hypothalamic-anterior pituitary function. *Annu. Rev. Pharmacol.* **14,** 389–412.

Dhariwal, A. P. S., Krulich, L., Katz, S. H. and McCann, S. M. (1965a). Purification of growth-releasing factor. *Endocrinology*, **77,** 932–936.

Dhariwal, A. P. S., Antunes-Rodrigues, J., Krulich, L. and McCann, S. M. (1965b). Separation of growth hormone-releasing factor (GH-RF) from corticotrophin-releasing factor (CRF). *Neuroendocrinology*, **1,** 341–349.

Dhariwal, A. P. S., Antunes-Rodrigues, J., Reeser, F., Chowers, I. and McCann, S. M. (1966a). Purification of hypothalamic corticotrophin-releasing factor (CRF) of ovine origin (30682). *Proc. Soc. exp. Biol. Med.* **121,** 8–12.

Dhariwal, A. P. S., McCann, S. M., Taleisnik, S. and Tomatis, M. E. (1966b). Purification of hypothalamic melanocyte-stimulating hormone (MSH)-releasing factor with sephadex. *Proc. Soc. Exp. Biol. Med.* **121,** 996–998.

Dhariwal, A. P. S., Russell, S. M., McCann, S. M. and Yates, F. E. (1969). Assay of corticotrophin-releasing factors by injection into the anterior pituitary of intact rats. *Endocrinology*, **84,** 544–556.

Diepen, R. (1962). Nervensystem. Der Hypothalamus. *In* "von Möllendorff's Handbuch der microskopischen Anatomie des Menschen" (W. Bargmann, ed.) Vol. IV, Part 7. Springer-Verlag, Berlin.

Dierickx, K. (1967). The function of the hypophysis without preoptic neurosecretory control. *Z. Zellforsch.* **78,** 114–130.

Dilman, V. M. (1971). Age-associated elevation of hypothalamic threshold to feedback control, and its role in development, aging and disease. *Lancet*, **i,** 1211–1219.

Dixon, J. E. and Acres, S. G. (1975). The inability to demonstrate the non-ribosomal biosynthesis of thyrotropin releasing hormone in hypothalamic tissue. *Fed. Proc. Fed. Am. Soc. Exp. Biol.* **34,** 658.

Döcke, F. and Dörner, G. (1974). Oestrogen and the control of gonadotrophin secretion in the immature rat. *J. Endocr.* **63,** 285–298.

Dodd, J. M. (1975). The hormones of sex and reproduction and their effects in fish and lower chordates: twenty years on. *Am. Zool.* **15,** Suppl., 137–171.

Dodd, J. M., Follet, B. K. and Sharp, P. J. (1971). Hypothalamic control of pituitary function in submammalian vertebrates. *In* "Advances in Comparative Physiology and Biochemistry" (O. E. Lowenstein, ed.) Vol. 4, pp. 113–223. Academic Press, New York and London.

Doerr-Schott, J. and Dubois, M. P. (1975). Localisation et identification d'un centre LRF dans l'encéphae du crapaud, *Bufo vulgaris*, Laur. *C.R. Acad. Sci.* **280,** 1285–1287.

Doerr-Schott, J. and Follenius, E. (1970). Inervation de l'hypophyse intermédiare de *Rana esculenta*, et identification des fibres aminergiques par autoradiographie au microscope électronique. *Z. Zellforsch.* **106,** 99–118.

Döhler, K. D. and Wuttke, W. (1974). Serum LH, FSH, prolactin and progesterone from birth to puberty in female and male rats. *Endocrinology*, **94,** 1003–1008.

Dománski, E. and Polkowska, J. (1973). The hypothalamic areas involved in the control of mammotrophic and lactotrophic processes in sheep. *Endocr. exp.* **7,** 229–246.

Donald, R. A. and Espiner, E. A. (1974). The plasma gonadotrophin response to gonadotrophin-releasing hormone in patients with primary hypogonadism. *J. Clin. Endocrinol. Metab.* **39,** 364–369.

Donoso, A. O., Banzán, A. M. and Barcaglioni, J. C. (1974). Further evidence on the direct action of L-DOPA on prolactin release. *Neuroendocrinology,* **15,** 236–239.

Donovan, B. T. (1970). "Mammalian Neuroendocrinology". McGraw-Hill, London.

Donovan, B. T. (1971). The extra-hypothalamic control of gonadotrophin secretion. *In* "Control of Gonadal Steroid Secretion" (D. T. Baird and J. A. Strong, eds) pp. 1–12. University Press, Edinburgh.

Donovan, B. T. (1972). Neural control of pituitary function-anatomical considerations. *In* "Modern Trends in Endocrinology" (F. T. G. Prunty and H. Gardiner-Hill, eds) pp. 7–14. Butterworths, London.

Donovan, B. T. (1974). The role of the hypothalamus in puberty. *Progr. Brain Res.* **41,** 239–252.

Donovan, B. T. and Peddie, M. J. (1973). The development of the hypophysial portal system in the guinea-pig. *J. Anat.* (*London*), **114,** 292–293.

Donovan, B. T. and ter Haar, M. B. (1974). Luteinizing hormone and follicle stimulating hormone secretion in response to luteinizing hormone-releasing hormone in the adult guinea-pig. *J. Endocr.* **63,** 34P.

Donovan, B. T. and van der Werff ten Bosch, J. J. (1965). "Physiology of Puberty". Williams and Wilkins, Baltimore.

Donovan, B. T., ter Haar, M. B. and Peddie, M. J. (1974). Gonadal-pituitary interaction in the foetal and neonatal guinea-pig. *Les Colloques de l'Institut National de la Santé et de la Recherche Médicale. Endocrinologie Sexuelle de la Période Périnatale,* **32,** 161–176.

Doolittle, R. F. (1969). Pyrrolidonecarboxylyl peptidase: stabilization and purification. *Arch. Biochem. Biophys.* **132,** 80–90.

Dormer, F., Kerbey, A. L., McPherson, M., Manley, S., Ashcroft, S. J. H., Scofield, J. G. and Randle, P. J. (1974). The effect of nickel on secretory systems. Studies on the release of amylase, insulin and growth hormone. *Biochem. J.* **140,** 135–142.

Dörner, G. (1974). Hormone dependent brain differentiation and neuroendocrine functions. *In* "Biological Rhythms in Neuroendocrine Activity" (M. Kawakami, ed.) pp. 220–231. Igaku Shoin, Tokyo.

Dorst, M. and Holm, L. W. (1968). Prolonged gestation in ewes after foetal adrenalectomy. *J. Endocr.* **40,** 293–296.

Douglas, W. W. (1968). Stimulus-secretion coupling: the concept and clues from chromaffin and other cells. *Br. J. Pharmacol.* **34,** 453–474.

Douglas, W. W. (1973). How do neurones secrete peptides? Exocytosis and its consequences, including "synaptic vesicle" formation, in the hypothalamo-neurohypophysial system. *Progr. Brain Res.* **39,** 21–40.

Dreifuss, J. J. and Kelly, J. S. (1972). Recurrent inhibition of antidromically identified rat supraoptic neurones. *J. Physiol., Lond.* **220,** 87–103.

Driver, P. M., El Shahat, A., Boaz, T. G., Forbes, J. M. and Scanes, C. G. (1974). Increase in serum prolactin in sheep associated with long daylength and feeding *ad libitum. J. Endocr.* **63,** 46P.

Drouin, J., DeLéan, A., Ferland, L. and Labrie, F. (1975). Antagonism between estrogens and thyroid hormone in the control of TSH and PRL secretion. *Program. 57th Meet. U.S. Endocrine Soc.* p. 151.

Duchen, L. W. (1962). The effects of ingestion of hypertonic saline on the pituitary gland in the rat: a morphological study of the pars intermedia and posterior lobe. *J. Endocr.* **25,** 161–168.

Ducommon, P., Vale, W., Sakiz, E. and Guillemin, R. (1967). Reversal of the inhibition of TSH secretion due to acute stress. *Endocrinology*, **80,** 953–956.

Dular, R., LaBella, F., Vivian, S. and Eddie, L. (1974). Purification of prolactin-releasing and inhibiting factors from beef. *Endocrinology*, **94,** 563–567.

Dunn, J. D. (1974). Circadian variation in adrenocortical and anterior pituitary hormones. *In* "Biological Rhythms in Neuroendocrine Activity" (M. Kawakami, ed.) pp. 119–139. Igaku Shoin, Tokyo.

Dunn, J. D., Arimura, A. and Scheving, L. E. (1972). Effect of stress on circadian periodicity in serum LH and prolactin concentration. *Endocrinology*, **90,** 29–33.

Dupont, A., Bastarache, E., Endröczi, E. and Fortier, C. (1972). Effect of hippocampal stimulation in the plasma thyrotropin (TSH) and corticosterone responses to acute cold exposure in the rat. *Can. J. Physiol. Pharm.* **50,** 364–367.

Dupont, A., Labrie, F., Pelletier, G., Puriani, R., Coy, E. J. and Schally, A. V. (1974). Organ distribution of radioactivity and disappearance of radioactivity from plasma after administration of (^{3}H) luteinizing hormone-releasing hormone to mice and rats. *Neuroendocrinology*, **16,** 65–73.

Dussault, J. H. and Coulombe, P. (1974). Influence des hormones thyroïdiennes sur la réponse hypophysaire en prolactin (H-Pr) lors d'une épreuve de stimulation à la TRH. *Union Med. Can.* **103,** 417–420.

Duvernoy, H. (1972). The vascular architecture of the median eminence. *In* "Brain-Endocrine Interaction. Median Eminence: Structure and Function" (K. M. Knigge, D. E. Scott and A. Weindl, eds) pp. 79–108. Karger, Basel.

DuVigneaud, V. (1956). Hormones of the posterior pituitary gland: Oxytocin and vasopressin. *In* "The Harvey Lectures", Ser. L, pp. 1–26. Academic Press, New York and London.

DuVigneaud, V., Ressler, C. H. and Trippet, S. (1953a). The sequence of amino acids in oxytocin, with a proposal for the structure of oxytocin. *J. Biol. Chem.* 205, 949–957.

DuVigneaud, V., Lawler, H. C. and Popenoe, E. A. (1953b). Enzymatic cleavage of glycinamide from vasopressin and a proposed structure for this pressor-antidiuretic hormone of the posterior pituitary. *J. Am. Chem. Soc.* **75,** 4880–4881.

Dyball, R. E. J. and Koizumi, K. (1969). Electrical activity in the supraoptic and paraventricular nuclei associated with neurohypophysial hormone release. *J. Physiol., Lond.* **201,** 711–722.

Dyball, R. E. J., Dyer, R. G. and Drewett, R. F. (1974). Chemical sensitivity of preoptic neurones which project to the medial and basal hypothalamus. *Brain Res.* **71,** 140–143.

Dyer, R. G. (1973). An electrophysiological dissection of the hypothalamic regions which regulate the preovulatory secretion of luteinizing hormone in the rat. *J. Physiol., Lond.* **234,** 421–442.

Dyer, R. G. and Dyball, R. E. J. (1974). Evidence for a direct effect of LRF and TRF on single unit activity in the rostral hypothalamus. *Nature, Lond.* **252,** 486–488.

Dyer, R. G., Prichett, C. J. and Cross, B. A. (1972). Unit activity in the diencephalon of female rats during the oestrous cycle. *J. Endocr.* **53,** 151–160.

E

Ebels, I., Cithavel, A., and Moszkowska, A. (1975). Separation of pineal extract by gel-filtration. III. Sheep pineal factors acting either on the hypothalamus or on the anterior hypophysis of mice and rats in *in vitro* experiments. *J. Neural Transm.* **36,** 281–302.

Eccles, J. C. (1964). "The Physiology of Synapses". Springer-Verlag, Berlin.

Eddy, L. J., Hershman, J. M., Taylor, R. E. Jr. and Barker, S. B. (1973). Binding of thyrotropin-releasing hormone by thyrotropin secreting cells. *Biochem. Biophys. Res. Commun.* **54,** 140–146.

Eddy, R. L., Gilliland, P. F., Ibarra, J. D. Jr., McMurry, J. F. Jr. and Thompson, J. Q. (1974). Human growth hormone release. Comparison of provocative test procedures. *Am. J. Med.* **56,** 179–185.

Edwards, C. R. W., Chard, T., Kitau, M. J., Forsling, M. L. and Landon, J. (1972). The development of a radioimmunoassay for arginine vasopressin: production of antisera and labelled hormone; separation techniques; specificity and sensitivity of the assay in aqueous solution. *J. Endocr.* **52,** 279–288.

Edwards, R. G. and Johnson, M. H. (1975). "Physiological Effects of Immunity Against Reproductive Hormones". Cambridge University Press.

Edwardson, J. A. and Bennett, G. W. (1974). Modulation of corticotrophin-releasing factor release from hypothalamic synaptosomes. *Nature* (*London*), **251,** 425–427.

Edwardson, J. A. and Gilbert, D. (1975). Sensitivity of self potentiating effect of luteinising hormone releasing hormone to cycloheximide. *Nature, Lond.* **255,** 71.

Edwardson, J. A. and Gilbert, D. (1976). Application of an *in vitro* perfusion technique to studies of luteinizing hormone release by rat anterior hemipituitaries: self-potentation by luteinizing hormone releasing hormone. *J. Endocr.* **68,** 197–207.

Edwardson, J. A., Bennett, G. W. and Bradford, H. F. (1972). Release of aminoacids and neurosecretory substances after stimulation of nerve-endings (synaptosomes) isolated from the hypothalamus. *Nature, Lond.* **240,** 554–556.

Eféndic, S. and Luft, R. (1975a). Studies on the inhibitory effect of somatostatin on glucose induced insulin release in the isolated perfused rat pancreas. *Acta endocr., Copenh.* **78,** 510–515.

Eféndic, S. and Luft, R. (1975b). Studies on the mechanism of somatostatin action on insulin release in man. I. Effect of blockade of α-adrenergic receptors. *Acta endocr., Copenh.* **78,** 516–523.

Egan, S. M. and Livingstone, A. (1973). Uptake *in vitro* of (^{3}H) oxytocin by the mammary gland and other tissues of the rat. *J. Clin. Endocrinol. Metab.* **58,** 289–303.

Ehrensing, R. H., Kastin, A. J., Schalch, D. S., Friesen, H. G., Vargas, J. R. and Schally, A. V. (1974). Affective state and thyrotropin and prolactin responses after repeated injections of thyrotropin-releasing hormone in depressed patients. *Am. J. Psychiat.* **131,** 714–718.

Ehret, C. F. and Trucco, T. (1967). Molecular models for the circadian clock. 1. The chronon concept. *J. Theoret. Biol.* **15,** 240–262.

Ekins, R. P. (1974). Basic Principles and Methods. *Br. Med. Bull.* **30,** 3–11.

Eldridge, J. C., Dinowski, W. P. and Mahesh, V. B. (1974a). Effects of castration of immature rats on serum FSH and LH, and of various steroid treatments after castration. *Biol. Reprod.* **10,** 438–446.

Eldridge, J. C., McPherson, J. C. and Mahesh, V. B. (1974b). Maturation of the negative feedback control of gonadotropin secretion in the female rat. *Endocrinology,* **34,** 1536–1540.

Eleftheriou, B. E., and Zolovick, A. J. (1968). Effect of amygdaloid lesions on plasma and pituitary thyrotropin levels in the deermouse. *Proc. Soc. Exp. Biol. Med.* **127,** 671–674.

Ellendorf, F., Parvizi, N., Pomerantz, D. K., Hartjen, A., Konig, A., Smidt, D. and Elsaesser, F. (1975). Plasma luteinizing hormone and testosterone in the adult male pig: 24-hour fluctuations and the effect of copulation. *J. Endocr.* **67,** 403–410.

Emerson, C. H., Utiger, R. D. (1975). Thyrotropin Releasing Hormone (TRH) in concentrated extracts of rat plasma. *Program 57th Meet. U.S. Endocrine Soc.* p. 150.

Emerson, C. H., Frohman, L. A., Szabo, M. and Thakkar, I. (1977). TRH immunoreactivity in human urine: evidence for dissociation from TRH. *J. Clin. Endocrinol. Metab.* **45,** 392–399.

Endoczi, E. and Hilliard, J. (1965). Luteinizing hormone releasing activity in different parts of rabbit and dog brain. *Endocrinology,* **77,** 667–674.

Enemar, A., Falck, B. and Iturriza, F. C. (1967). Adrenergic nerves in the pars intermedia of the pituitary in the toad, *Bufo arenarum*. *Z. Zellforsch.* **77,** 325–330.

Engeland, W. C. and Dallman, M. F. (1975). Compensatory adrenal growth is a neurally mediated reflex. *Program 57th Meet. U.S. Endocrine Soc.* p. 147.

Engle, E. T. (1931). The pituitary gonadal relationship and the problem of precocious sexual maturity. *Endocrinology*, **15,** 405–420.

Enzmann, F., Bøler, J., Folkers, K., Bowers, C. Y. and Schally, A. V. (1971). Structure and synthesis of the thyrotropin-releasing hormone. *J. Med. Chem.* **14,** 469–474.

Eskay, R. L. Oliver, C., Grollman, A., and Porter, J. C. (1974). Immunoreactive LRH and TRH in the fetal, neonatal and adult rat brain. *Program 56th Meet. U.S. Endocrine Soc.* p. A83.

Eskay, R. L. Oliver, C., Ben-Jonathan, N. and Porter, J. C. (1975a). Hypothalamic hormones in portal and systemic blood. *In* "Hypothalamic Hormones" (M. Motta, P. Crosignani and L. Martini, eds) pp. 125–137. Academic Press, London and New York.

Eskay, R. L., Warberg, J., Mical, R. S. and Porter, J. C. (1975b). Prostaglandin E_2 induced release of LHRH into hypophysial portal blood. *Endocrinology*, **97,** 816–824.

Espiner, E., Hart, D. S. and Beaven, D. W. (1972). Cortisol secretion during acute stress and response to dexamethasone in sheep with adrenal transplants. *Endocrinology*, **90,** 1510–1514.

Etkin, W. (1967). Relation of the pars intermedia to the hypothalamus. *In* "Neuroendocrinology" (L. Martini and W. F. Ganong, eds) Vol. II, pp. 261–282. Academic Press, New York and London.

Etkin, W. (1970). The endocrine mechanism of amphibian metamorphosis, an evolutionary achievement. *Mem. Soc. Endocrinol.* **18,** 137–153.

Evans, H. M. (1933). Present position of our knowledge of anterior pituitary function. *J. Ann. Med. Ass.* **101,** 425–432.

Evans, H. M. and Long, J. A. (1921). On the association of continued cornification of the vaginal mucosa with the presence of large vesicles in the ovary and the absence of corpus formation. *Anat. Rec.* **21,** 60–61.

Evans, E. S., Rosenberg, L. L., Evans, A. B. and Koneff, A. A. (1964). Relative sensitivity of different biological responses to small quantities of thyroxine and triiodothyronine. *Endocrinology*, **74,** 770–779.

Evered, D. C., Gomez-Pan, A., Tunbridge, W. M. G., Hall, R., Lind, T., Besser, G. M., Mortimer, C. H., Thorner, M. O., Schally, A. V., Kastin, A. J. and Coy, D. H. (1975). Analogues of growth hormone-releasing inhibiting hormone. *Lancet*, **i,** 1250.

Everett, J. W. (1954). Luteotrophic function of autografts of the rat hypophysis. *Endocrinology*, **54,** 685–690.

Everett, J. W. (1956). Functional corpora lutea maintained for months by autographs of rat hypophyses. *Endocrinology*, **58,** 786–796.

Everett, J. W. (1961). The mammalian female reproductive cycle and its controlling mechanisms. *In* "Sex and Internal Secretions" (W. C. Young, ed.) 3rd edn, Vol. 1, pp. 497–555. Williams and Wilkins, Baltimore.

Everett, J. W. (1964). Central neural control of the reproductive functions of the adenohypophysis. *Physiol. Rev.* **44,** 373–431.

Everett, J. W. (1969). Neuroendocrine aspects of mammalian reproduction. *Annu. Rev. Physiol.* **31,** 383–416.

Everett, J. W. (1970). Photoregulation of the ovarian cycle in the rat. *In* "La Photorégulation de la Reproduction chez les Oiseaux et les Mammifères" (J. Benoit and I. Assenmacher, eds) pp. 387–403. C.R.N.S., Paris.

Everett, J. W. and Quinn, D. L. (1966). Differential hypothalamic mechanisms inciting ovulation and pseudopregnancy. *Endocrinology*, **78,** 141–150.

Everitt, A. V. (1976a). The female climacteric. *In* "Hypothalamus, Pituitary and Aging" (A. V. Everitt and J. A. Burgess, eds) pp. 419–430. C. C. Thomas, Springfield.

Everitt, A. V. (1976b). Conclusion: aging and its hypothalamic-pituitary control. *In* "Hypothalamus, Pituitary and Aging" (A. V. Everitt and J. A. Burgess, eds) pp. 676–701. C. C. Thomas, Springfield.

Everitt, A. V. and Burgess, J. A. (1976). "Hypothalamus, Pituitary and Aging". C. C. Thomas, Springfield.

F

Faglia, G., Beck-Peccoz, P., Ferrari, C., Travaglini, P., Ambrosi, B. and Spada, A. (1973). Plasma growth hormone response to thyrotropin releasing hormone in patients with active acromegaly. *J. Clin. Endocrinol. Metab.* **36,** 1259–1262.

Faiman, C. and Winter, J. S. D. (1971). Diurnal cycles in plasma FSH, testosterone and cortisol in men. *J. Clin. Endocrinol. Metab.* **33,** 186–192.

Faiman, C., Reyes, F. I. and Winter, J. S. D. (1974). Serum gonadotropin patterns during the perinatale period in man and the chimpanzee. *Les Colloques de l'Institut National de la Santé et de la Recherche Médicale. Endocrinologie Sexuelle de la Période Périnatale*, **32,** 161–176.

Farner, D. S. (1975). Photoperiodic controls in the secretion of gonadotropins in birds. *Am. Zool.* **15,** 117–135.

Fawcett, C. P., Reed, M., Charlton, H. M. and Harris, G. W. (1968). The purification of luteinizing hormone-releasing factor with some observations on its properties. *Biochem. J.* **106,** 229–236.

Fawcett, C. P., Beezley, A. E. and Wheaton, J. E. (1975). Chromatographic evidence for the existence of another species of luteinizing hormone releasing factor (LRF). *Endocrinology*, **96,** 1311–1314.

Fee, A. R. and Parkes, A. S. (1929). Studies on ovulation. I. The relation of the anterior pituitary body to ovulation in the rabbit. *J. Physiol., Lond.* **67,** 383–388.

Fee, A. R. and Parkes, A. S. (1930). Studies on ovulation. III. Effect of vaginal anesthesia on ovulation in the rabbit. *J. Physiol., Lond.* **70,** 385–388.

Ferguson, D. R. (1969). The genetic distribution of vasopressins in the peccary (*Tyassu angulatus*) and warthog (*Phacochoerus aethiopicus*). *Gen. Comp. Endocrinol.* **12,** 609–613.

Ferguson, D. R. and Heller, H. (1965). Distribution of neurohypophysial hormones in mammals. *J. Physiol.* (*London*), **180,** 846–863.

Ferguson, D. R. and Pickering, B. T. (1969). Arginine and lysine vasopressin in the hippopotamus neurohypophysis. *Gen. Comp. Endocr.* **13,** 425–429.

Ferguson, D. R. and Twite, B. R. (1974). Effect of vasopressin on toad bladder membrane proteins: Relationship to transport of sodium and water. *J. Endocr.* **61,** 501–507.

Ferin, M., Tempone, A., Zimmering, P. E. and Vande Wiele, R. L. (1969). Effects of antibodies to 17β-estradiol and progesterone on the oestrous cycle of the rat. *Endocrinology*, **85,** 1070–1078.

Ferin, M., Halberg, F., Richart, R. M. and Vande Wiele, R. L. (eds) (1974a). "Biorhythms and Human Reproduction". Wiley, New York.

Ferin, M., Tempone, A., Zimmering, P. E. and Van de Wiele, R. L. (1974b). Active immunization to 17β-estradiol and its effects upon the reproductive cycle of the rhesus monkey. *Endocrinology*, **94,** 765–776.

Ferland, L., Labrie, F. Schally, A. G. and Arimura, A. (1975). Antiluteinizing hormone (LH)-releasing activity of several analogues of LH-releasing hormone. *Fertil. Steril.* **26,** 554–559.

Fernstrom, J. D. and Wurtman, R. J. (1972). Brain serotonin content: physiological regulation by plasma neutral amino acids. *Science, N.Y.* **178,** 414–416.

Fichera, G. (1905). Sur l'hypertrophie de la glande pituitaire consecutive à la castration. *Arch. ital. Biol.* **43,** 405–426.

Findlay, A. L. R. (1966). Sensory discharges in lactating mammary glands. *Nature, Lond.* **212,** 1183–1184.

Findlay, A. L. R. (1972). Hypothalamic inputs: methods and five examples. *Progr. Brain Res.* **38,** 163–190.

Fineberg, E. S., Horand, A. A. and Merimee, T. J. (1972). Free fatty acid concentrations and growth hormone secretion in man. *Metabolism,* **21,** 491–498.

Fingerhut, M. and Krieger, D. T. (1974). Plasma growth hormone response to L-dopa in obese subjects. *Metabolism,* **23,** 267–271.

Fink, G. and Harris, G. W. (1970). The luteinizing hormone releasing activity of extracts of blood from the hypophysial portal vessels of rats. *J. Physiol., Lond.* **208,** 221–241.

Fink, G., Gennser, G., Liedholm, P., Thorell, J. and Mulder, J. (1974). Comparison of plasma levels of luteinising hormone releasing hormone in men after intravenous or intranasal administration. *J. Endocr.* **63,** 351–360.

Fink, G., Aiyer, M. S., Jamieson, M. G. and Chiappa, S. A. (1975). Factors modulating the responsiveness of the anterior pituitary gland in the rat with special reference to gonadotrophin releasing hormone (GnRH). *In* "Hypothalamic Hormones" (M. Motta, P. Crosignani and L. Martini, eds) pp. 139–160. Academic Press, London and New York.

Finkelstein, J. W., Boyar, R. M. and Hellman, L. (1974). Growth hormone secretion in hyperthyroidism. *J. Clin. Endocrinol. Metab.* **38,** 634–637.

Fisher, C., Ingram, W. R. and Ranson, S. W. (1938). "Diabetes Insipidus and the Neurohormonal Control of Water Balance: a Contribution to the Structure and Function of the Hypothalamico-Hypophyseal System". Edwards Bros., Ann Arbor.

Fishman, J. and Norton, B. (1975). Catechol estrogen formation in the central nervous system of the rat. *Endocrinology,* **96,** 1054–1059.

Fitzpatrick, R. J. and Bentley, P. J. (1968). The assay of neurohypophysial hormones in blood and other body fluids. *In* "Handbook of Experimental Pharmacology" (B. Berde, ed.) pp. 90–117. Springer-Verlag, Berlin.

Fleischer, N., Donald, R. A. and Butcher, R. W. (1969). Involvement of adenosine 3, '5'-monophosphate in release of ACTH. *Am. J. Physiol.* **217,** 1287–1291.

Fleischer, N., Zimmermann, G., Schindler, W. and Hutchins, M. (1972). Ouabain induced release of ACTH and GH *in vitro*. *Clin. Res.* **20,** 426.

Flerkó, B. (1962). Hypothalamic control of hypophyseal gonadotrophic function. *In* "Hypothalamic Control of the Anterior Pituitary" 1st edn (J. Szentágothai, B. Flerkó, B. Mess and B. Halász, eds) pp. 192–264. Akadémiai Kiadó, Budapest.

Flerkó, B. (1966). Control of gonadotropin secretion in the female. *In* "Neuroendocrinology" (L. Martini and W. F. Ganong, eds) Vol. 1, pp. 613–618. Academic Press, New York and London.

Florsheim, W. H., Rudko, P., Corcorran, N. L. and Bodfish, R. E. (1967). On the mechanism of the inhibition of compensatory ovarian hypertrophy by goitrogens. *Endocrinology,* **81,** 771–778.

Follett, B. K. (1970). Effects of neurohypophysial hormones and their synthetic analogues on lower vertebrates. *In* "International Encyclopedia of Pharmacology and Therapeutics (H. Heller and B. T. Pickering, eds) Section 41, pp. 321–350. Pergamon Press, New York.

Follett, B. K. (1973). Circadian rhythms and photoperiodic time measurement in birds. *J. Reprod. Fertil.* Suppl. **19,** 5–18.

Follett, B. K., Mattocks, P. W. and Farner, D. S. (1974). Circadian function in the photoperiodic induction of gonadotropin secretion in the White-crowned Sparrow, *Zonotrichia leucophrys gambelii. Proc. Nat. Acad. Sci. USA.* **71,** 1666–1669.

Folley, S. J. (1956). "The Physiology and Biochemistry of Lactation". Thomas, Springfield.

Forsling, M. L., Martin, M. J., Sturdy, J. C. and Burton, J. M. (1973). Observations on the release and clearance of neurophysin and the neurohypophysial hormones in the rat. *J. Endocr.* **57,** 307–315.

Fortier, C. (1958). Observations on the so called ACTH releasing activity of neurohypophyseal extract in Saffran and Schally's pituitary incubating system. *Tex. Rep. Biol. Med.* **16,** 68–78.

Fortier, C. (1973). Interrelationships in the control of ACTH and TSH secretion. *In* "Hormones and Brain Function" (K. Lissak, ed.) pp. 93–103. Akadémiai Kiadó, Budapest.

Fortier, C. and DeGroot, J. (1964). Residual synthesis and release of ACTH following electrolytic destruction of the median eminence in the rat. *In* "Major Problems in Neural Endocrinology" (E. Bajusz and G. Jasmin, eds) pp. 203–219. Karger, Basel.

Fortier, C., Delgado, A., Ducommon, P., Ducommon, S., Dupont, A., Jobin, M., Kraicer, J., MacIntosh-Hardt, B., Marceau, H., Miathe, P., Miathe-Voloss, C., Rerup, C. and Van Rees, G. P. (1970). Functional interrelationships between the adenohypophysis, thyroid, adrenal cortex and gonads. *Can. Med. Assoc. J.* **103,** 864–874.

Foster, D. L. (1974). Regulation of gonadotropins during fetal and early postnatal development in the sheep. *Les Colloques de l'Institut National de la Santé et de la Recherche Médicale. Endocrinologie Sexuelle de la Période Périnatale*, **32,** 143–156.

Foster, D. L. and Karsch, F. J. (1975). Development of the mechanism regulating the preovulatory surge of luteinizing hormone in sheep. *Endocrinology*, **97,** 1205–1209.

Foster, J. P., Jeffcoate, S. L., Crighton, D. B. and Holland, D. T. (1976). LH and LH-RH-like immunoreactivity in the jugular blood of sheep at various stages of the oestrous cycle. *J. Endocr.* **68,** 409–417.

Franchimont, P., Becher, H., Ernould, Ch., Thys, Ch., Demoulin, A., Bourgiugron, J. P., Legros, J. J. and Valcke, J. C. (1974). The effect of hypothalamic luteinizing hormone releasing hormone (LH-RH) on plasma gonadotrophin levels in normal subjects. *Clin. Endocrinol.* **3,** 27–39.

Franchimont, P., Chari, S., Hagelstein, M. T. and Duraiswami, S. (1975a). Existence of a follicle-stimulating hormone inhibiting factor 'inhibin' in bull seminal plasma. *Nature, Lond.* **257,** 402–404.

Franchimont, P., Chari, S., Schellen, A. M. and Demoulin, A. (1975b). Relationship between gonadotrophins, spermatogenesis and seminal plasma. *J. Steroid. Biochem.* **6,** 1037–1041.

Francis, M. G. and Peaslee, M. H. (1974). Effect of social stress on pituitary melanocyte-stimulating hormone activity in male mice. *Neuroendocrinology*, **16,** 1–7.

Frantz, A. G., Habif, D. V., Hyman, G. A., Suh, H. K., Sassin, J. F., Zimmerman, E. A., Noel, G. L. and Kleinberg, D. L. (1973). Physiological and pharmacological factors affecting prolactin secretion, including its suppression by L-DOPA in the treatment of breast cancer. *In* "Human Prolactin" (J. L. Pasteels and C. Robyn, eds) pp. 273–290. Excerpta Medica, Amsterdam.

Franz, J., Haselbach, C. H. and Libert, O. (1962). Studies of the effect of hypothalamic extracts on somatotrophic pituitary function. *Acta endocr., Copenh.* **41,** 336–350.

Fraschini, F. and Martini, L. (1970). Rhythmic phenomena and pineal principles. *In* "The Hypothalamus" (L. Martini, M. Motta and F. Fraschini, eds) pp. 529–549. Academic Press, New York and London.

Fraser, H. M. (1976). Physiological effects of antibody to luteinizing hormone-releasing hormone. *In* "Physiological Effects of Immunity Against Reproductive Hormones" (R. G. Edwards and M. Johnson, eds) pp. 137–154. Cambridge University Press.

Fraser, H. M. and Gunn, A. (1973). Effects of antibodies to luteinizing hormone releasing hormone in the male rabbit and on the rat oestrous cycle. *Nature, Lond.* **244,** 160–161.

Fraser, H. M., Gunn, A., Jeffcoate, S. L. and Holland, D. T. (1974). Effect of active immunization to luteinizing hormone releasing hormone on serum and pituitary gonadotrophins, testes and accessory sex organs in the male rat. *J. Endocr.* **63,** 399–406.

Fraser, H. M., Jeffcoate, S. L., Gunn, A. and Holland, D. T. (1975). Effect of active immunisation to luteinizing hormone releasing hormone on gonadotrophin levels in ovariectomized rats. *J. Endocr.* **64,** 191–192.

Freeman, M. E. and Neill, J. D. (1972). The pattern of prolactin secretion during pseudopregnancy in the rat: a daily nocturnal surge. *Endocrinology,* **90,** 1292–1294.

Freeman, M. E., Smith, M. S., Nazian, S. J. and Neill, J. D. (1974). Ovarian and hypothalamic control of the daily surges of prolactin secretion during pseudopregnancy in the rat. *Endocrinology,* **94,** 875–882.

Frieden, E. H. (1951). Experimentally induced changes in the melanophore hormone content of the rat pituitary. *Endocrinology,* **49,** 557–564.

Friedgood, H. B. (1970). The nervous control of the anterior hypophysis. *J. Reprod. Fertil.* Suppl. **10,** 3–14.

Frith, D. A. and Hooper, K. C. (1974). The development of an *in vitro* system for estimating oxytocinase in the rabbit-hypothalamus and the effects of some protein synthesis in-inhibitors and oestradiol-17β on enzyme activity. *Acta endocr., Copenh.* **75,** 443–451.

Frohman, L. A. (1972). Clinical neuropharmacology of hypothalamic releasing factors. *N. Engl. J. Med.* **286,** 1391–1397.

Frohman, L. A. and Stachura, M. (1975). Neuropharmacologic control of neuroendocrine function in man. *Metabolism,* **24,** 211–234.

Frohman, L. A. and Szabo, M. (1975). Evidence for the existence of prolactin releasing activity distinct from TRH in porcine hypothalamic extracts. *Program 57th Meet. U. S. Endocrine Soc.* p. 93A.

Frohman, L. A., Maran, J. W., Yates, F. E. and Dhariwal, A. P. S. (1971). Growth hormone responses to intrapituitary injection of growth hormone-releasing factor in the rat as measured by radioimmunoassay. *Fed. Proc. Fed. Am. Soc. Exp. Biol.* **30,** 198.

Fuji, N. and Yajima, H. (1975). Studies on peptides. III. Application of the trifluoromethanesulphonic acid procedure to the synthesis of a peptide with somatostatin activity. *Chem. Pharm. Bull.* **23,** 1596–1603.

Fujimonto, W. Y., Ensinck, J. W. and Williams, R. H. (1974). Somatostatin inhibits insulin and glucagon release by monolayer cell cultures of rat endocrine pancreas. *Life Sci.* **15,** 1999–2004.

Fujino, M., Kobayashi, S., Obayashi, M., Shinagawa, S., Fukuda, T., Kitada, C. and Nakayama, R. (1972). Structure activity relationships in the C-terminal part of luteinizing hormone-releasing hormone (LH-RH). *Biochem. Biophys. Res. Commun.* **49,** 863–869.

Fujino, M., Fukuda, T., Kobayashi, S. and Obayashi, M. (1973a). Isobenzyloxycarbonyl function, a new convenient amino-protecting group in peptide synthesis. IV. Synthesis of gonadotrophin-releasing hormone (Gn-RH or LH-RH/FSH-RH). *Chem. Pharm. Bull.* **21,** 87–91.

Fujino, M., Shinigawa, S., Obayashi, M., Kobayashi, S., Fukuda, T., Yamazaki, I., White, W. F. and Rippel, R. H. (1973b). Further studies on the structure-activity relationships in the C-terminal part of luteinizing hormone-releasing hormone. *J. Med. Chem.* **16,** 1144–1147.

Furr, B. J. A. and Smith, G. K. (1975). Effects of antisera against gonadal steroids on ovulation in the hen *Gallus domesticus*. *J. Endocr.* **66,** 303–304.

Furr, B. J. A., Onuora, G. I., Bonney, R. C. and Cunningham, F. J. (1973). The effect of synthetic luteinizing hormone-releasing factors on plasma levels of luteinizing hormone in the cockerel. *J. Endocr.* **59,** 495–502.

Furth, J. (1955). Experimental pituitary tumors. *Recent Prog. Horm. Res.* **11,** 221–255.

Fuxe, K. and Hökfelt, T. (1969). Catecholamines in the hypothalamus and pituitary gland. *In* "Frontiers in Neuroendocrinology" (W. Ganong and L. Martini, eds) pp. 47–96. Oxford University Press, New York.

Fuxe, K. and Hökfelt, T. (1970). Central monoaminergic systems and hypothalamic function. *In* "The Hypothalamus" (L. Martini, M. Motta and F. Fraschini, eds) pp. 123–138. Academic Press, New York and London.

Fuxe, K., Hökfelt, T., Jonsson, G., Levine, S., Lidbrink, P. and Lofström, A. (1973). Brain and pituitary-adrenal interaction-studies on central monoamine neurones. *In* "Brain-Pituitary-Adrenal Interrelationships" (A. Brodish and E. S. Redgate, eds) pp. 239–269. Karger, Basel.

Fuxe, K., Hökfelt, T., Lofstrom, A., Johansson, O., Goldstein, M., Jeffcoate, S. L., White, N. and Eneroth, P. (1976). On the role of neurotransmitters and hypothalamic hormones in the control of pituitary function and sexual behaviour. *In* "Subcellular Mechanisms in Reproductive Endocrinology" (F. Naftolin, K. J. Ryan and I. J. Davies, eds), pp. 193–246. Elsevier.

G

Gagel, R. F., Jackson, I. M. D., Deprez, D. P., Papapetrou, P. D. and Reichlin, S. (1976). *In* "Thyroid Research" (J. Robbins and L. E. Braverman, eds). Proceedings of the 7th International Thyroid Conference, Boston, Mass. June 9–13, 1975, pp. 8–10. Excerpta Medica, Amsterdam.

Gaillard, P. J. (1937). An experimental contribution to the origin of the pars intermedia. *Acta neerl. Morph.* **1,** 3–11.

Gale, C. C. (1973). Neuroendocrine aspects of thermoregulation. *Annu. Rev. Physiol.* **35,** 391–430.

Gale, C. C., Toivola, P. and Schonbaum, M. L. D. (1969). Thyroid-sympathetic nervous interactions in thermo-regulation. *Program 51st Meet. U.S. Endocrine Soc.* p. 114.

Gale, C. C., Jobin, M., Proppe, D. W., Notter, D. and Fox, H. (1970). Endocrine thermoregulatory responses to local hypothalamic cooling in unanesthetized baboons. *Am. J. Physiol.* **219,** 193–201.

Gale, C. C., William, R. and Toivola, P. (1971). Thyroid sympathetic nervous system interactions in thermoregulation. *Acta endocr., Copenh.* Suppl. **155,** 34.

Gallo, R. V., Johnson, J. H., Goldman, B. D., Whitmoyer, D. I. and Sawyer, C. H. (1971). Effects of electrochemical stimulation of the ventral hippocampus on hypothalamic electrical activity and pituitary gonadotrophin secretion in female rats. *Endocrinology*, **89,** 704–713.

Galton, V. A. (1975). The metabolism of thyroid hormones in rat pituitary. *7th Internatl. Thyroid Conference, 1975. Excerpta Medica Int. Cong. Ser.* No. 361, p. 76.

Galton, V. A. and Nisula, B. C. (1969). Thyroxine metabolism and thyroid function in the cold-adapted rat. *Endocrinology*, **85,** 79–86.

Gann, D. S. and Cryer, G. L. (1973). Feedback control of ACTH secretion by cortisol. *In* "Brain-Pituitary-Adrenal Inter-relationships" (A. Brodish and E. R. Redgate, eds) pp. 197–233. Karger, Basel.

Ganong, W. F. (1959). Adrenal-hypophysial interrelations. *In* "Comparative Endocrinology" (A. Gorbman, ed.) pp. 187–201. Wiley, New York.

Ganong, W. F. (1963). The central nervous system and the synthesis and release of adrenocorticotrophic hormone. *In* "Advances in Neuroendocrinology" (A. V. Nalbandov, ed.) pp. 92–149. University of Illinois Press, Urbana.

Ganong, W. F. (1970). Control of ACTH and MSH secretion. *In* "The Hypothalamus" (L. Martini, M. Motta and F. Fraschini, eds) pp. 313–333. Academic Press, New York and London.

Ganong, W. F. (1972). Evidence for a central noradrenergic system that inhibits ACTH secretion. *In* "Brain-Endocrine Interaction: Median Eminence: Structure and Function" (K. M. Knigge, D. E. Scott and A. Weindl, eds) pp. 254–266. Karger, Basel.

Ganong, W. F. (1972). Pharmacological aspects of neuroendocrine integration. *Progr. Brain Res.* **38,** 41–54.

Ganong, W. F. (1974). Role of catecholamines and acetyl-choline in the regulation of endocrine function. Review. *Life Sci.* **15,** 1401–1414.

Ganong, W. F. (1975). Brain amines and the control of ACTH and growth hormone secretion. *In* "Hypothalamic Hormones" (M. Motta, P. Crosignani and L. Martini, eds) pp. 237–248. Academic Press, London and New York.

Gautvik, K. M. and Tashjian, A. H. Jr. (1973). Effects of cations and colchicine on the release of prolactin and growth hormone by functional pituitary tumour cells in culture. *Endocrinology*, **93,** 793–799.

Gautvik, K. M., Tashjian, A. H. Jr., Kourides, I. A., Weintraub, B. D., Graeber, C. T., Maloof, F., Suzuki, K. and Zuckerman, J. E. (1974). Thyrotropin-releasing hormone is not the sole physiologic mediator of prolactin release during suckling. *New Engl. J. Med.* **290,** 1162–1165.

Gay, V. L. (1972). The hypothalamus: physiology and clinical use of releasing factors *Fertil. and Steril.* **23,** 50–63.

Gay, V. L. (1974). Decreased metabolism and increased serum concentrations of LH and FSH following nephrectomy of the rat: absence of short-loop regulatory mechanisms. *Endocrinology*, **95,** 1582–1588.

Gay, V. L. and Tomacari, R. L. (1974). Follicle-stimulating hormone secretion in the female rat: cyclic release is dependent on circulating androgen. *Science, N.Y.* **184,** 75–77.

Geffner, D. L., Azukizawa, M. and Hershman, J. M. (1975). Prophythiouracil blocks extrathyroidal conversion of thyroxine to triiodothyronine and augments thyrotropin secretion in man. *J. Clin. Invest.* **55,** 224–229.

Geiger, R., König, W., Sandow, J. and Schally, A. V. (1975). [8-Homoarginin] (Luteinisierendes Hormon freisetzendes Hormon). *Hoppe-Seyler's Z. Physiol. Chem.* **335,** 1526–1534.

Gemzell, C. A., Van Dyke, D. C., Tobias, C. A. and Evans, H. M. (1951). Increase in the formation and secretion of ACTH following adrenalectomy. *Endrocrinology*, **49,** 325–336.

Genazzani, A. R., Lemarchand-Béraud, Th., Aubert, N. L. and Felber, J. P. (1975). Pattern of plasma ACTH, hGH and cortisol during the menstrual cycle. *J. Clin. Endocrinol. Metab.* **41,** 431–437.

Gerich, J. E., Lorenzi, M., Bier, D. M., Schneider, V., Tsalikian, E., Karam, J. H. and Forsham, P. H. (1975a). Prevention of human diabetic ketoacidosis by somatostatin. *New Engl. J. Med.* **292,** 985–989.

Gerich, J. E., Lorenzi, M., Tsalikian, E., Schneider, V., Karam, J. H., Bier, D. M. and Forsham, P. H. (1975b). Role of growth hormone in diabetic ketoacidosis: studies with somatostatin. *Ricerca Scientifica ed Educazione Permanente*, **2** (Suppl. 1), Abstract n. 107. University of Milan, Milan.

Gershengorn, M. C. and Weintraub, B. D. (1975). A new syndrome of hyperthyroidism caused by pituitary hypersection of thyrotropin. (TSH). *Clin. Res.* **23,** 236A.

Geschwind, I. I. (1967). Molecular variation and possible lines of evolution of peptide and protein hormones. *Am. Zool.* **7,** 89–108.

Gibbs, F. P. (1969). Central nervous system lesions that block release of ACTH caused by traumatic stress. *Am. J. Physiol.* **217,** 78–83.

Gilbert, D. M., Bitensky, L., Besser, G. M. and Chayen, J. (1975). Measurable effects of low concentrations (100 fg/ml) of thyrotrophin releasing hormone. *J. Endocr.* **64,** 250–260.

Gillessen, D., Felix, A. M., Lergier, W. and Studer, R. O. (1971). Synthese des "thyrotropin-releasing" hormons (TRH) (Schaf) und verwandter Peptide. *Helv. Chim. Acta,* **53,** 63–72.

Gillham, B., Burden, J. L., Hillhouse, E. W. and Jones, M. T. (1975). Preliminary observations on the nature of corticotrophin releasing hormone from the rat hypothalamus *in vitro. J. Endocr.* **65,** 12P–13P.

Ginsburg, M. and Ireland, M. P. (1966). The role of neurophysins in the transport and release of neurohypophyseal hormones. *J. Endocr.* **35,** 289–298.

Ginsburg, M., MacLusky, N. J., Morris, I. D. and Thomas, P. J. (1975). Physiological variation in abundance of oestrogen specific high affinity binding sites in hypothalamus, pituitary and uterus of the rat. *J. Endocr.* **64,** 443–449.

Ginther, O. J. and Wentworth, B. C. (1974). Effects of synthetic gonadotrophin-releasing hormone on plasma concentrations of luteinizing hormone in ponies. *Am. J. Vet. Res.* **35,** 79–81.

Giroud, A., Kfouri, P. and Martinet, M. (1944). Relations entre la thyroïde et la function mélanophorique hypophysaire. *C.R. Soc. Biol.* **138,** 74–75.

Givens, J. R. and Ney, R. L. (1968). Corticotrophin in blood. *In* "Clinical Endocrinology" (E. B. Astwood and C. E. Cassidy, eds) Vol. 2, pp. 11–24. Grune and Stratton, New York.

Glass, J. D., Schwartz, I. L. and Walter, R. (1972). Bridging of peptides to solid supports through the dinitrophenylene moiety. Bidirectional extension of peptide chains. *J. Amer. Chem. Soc.* **94,** 6209–6211.

Glick, S. M. (1969). The regulation of growth hormone secretion. *In* "Frontiers in Neuroendocrinology" (W. F. Ganong and L. Martini, eds) pp. 141–182. Oxford University Press, New York.

Glowinski, J. (1970). Metabolism of catecholamines in the central nervous system and correlation with hypothalamic function. *In* "The Hypothalamus" (L. Martini, M. Motta and F. Fraschini, eds) pp. 139–152. Academic Press, New York and London.

Goding, J. R., Catt, K. J., Brown, J. M., Kaltenbach, C. C., Cumming, I. A. and Mole, B. J. (1969). Radioimmunoassay for ovine luteinizing hormone. Secretion of luteinizing hormone during estrus and following estrogen administration in sheep. *Endocrinology,* **85,** 133–142.

Golstein-Golaire, J., Vanhoelst, L., Brune, O. D., Leclercq, R. and Capinschi, G. (1970). Acute effects of cold on blood levels of growth hormone, cortisol and thyrotropin in man. *J. Appl. Physiol.* **29,** 622–626.

Gomez-Pan, A., Reed, J. D., Albinus, M., Shaw, B., Hall, R., Besser, G. M., Coy, D. H., Kastin, A. J. and Schally, A. V. (1975a). Direct inhibition of gastric acid and pepsin secretion by growth hormone release-inhibiting hormone in cats. *Lancet,* **i,** 888–890.

Gomez-Pan, A., Tunbridge, W. M. G., Duns, A., Hall, R., Besser, G. M., Coy, D. H., Schally, A. V. and Kastin, A. J. (1975b). Hypothalamic hormone interaction in acromegaly. *Clin. Endocrinol.* **4,** 455–460.

Gonzalez-Barcena, D., Kastin, A. J., Schalch, D. S., Torres-Zamora, M., Perez-Pasten, E., Kato, A. and Schally, A. V. (1973). Responses to thyrotropin-releasing hormone in patients with renal failure and after infusion in normal men. *J. Clin. Endocrinol. Metab.* **36,** 117–120.

Gonzalez-Barcena, A., Kastin, A. J., Miller, M. C., Schalch, D. S., Coy, D. H., Schally, A. V. and Herrara, A. E. (1975). Stimulation of luteinizing hormone (LH) release after oral administration of an analogue of LH releasing hormone. *Lancet*, **ii,** 1126–1128.

Goos, H. J. T. (1969). Hypothalamic control of the pars intermedia in *Xenopus laevis* tadpoles. *Z. Zellforsch.* **97,** 118–124.

Gordin, A., Arimura, A. and Schally, A. V. (1976). Effect of thyroid hormone excess and deficiency on serum thyrotropin in rats immunized passively with antiserum to somatostatin. *Proc. Soc. exp. Biol. Med.* **153,** 319–323.

Gordon, J., Bollinger, J. and Reichlin, S. (1972). Plasma thyrotropin responses to thyrotropin releasing hormone after injection into the third ventricle, systemic circulation, median eminence and anterior pituitary. *Endocrinology*, **91,** 696–701.

Gorski, R. A. (1971). Gonadal hormones and the perinatal development of neuroendocrine function. *In* "Frontiers in Neuroendocrinology" (L. Martini, and W. F. Ganong, eds) pp. 237–290. Oxford University Press, London and New York.

Gorski, R. A. (1973). Perinatal effects of sex steroids on brain development and function. *Progr. Brain Res.* **39,** 149–162.

Gosbee, J. L., Kraicer, J., Kastin, A. J. and Schally, A. V. (1970). Functional relationship between the pars intermedia and ACTH secretion in the rat. *Endocrinology*, **86,** 560–567.

Gourdji, D. and Tixier-Vidal, A. (1966a). Variations du contenu hypophysaire en prolactine chez le canard Pékin mâle au cours du cycle sexuel et de la photostimulation expérimentale du testicule. *C. R. Acad. Sci.* **262,** 1746–1749.

Gourdji, D. and Tixier-Vidal, A. (1966b). Mise en évidence d'un contrôle hypothalamique stimulant de la prolactine hypophysaire chez le canard. *C. R. Acad. Sci.* **263,** 162–165.

Graber, A. L., Ney, R. L., Nicholson, W. E., Island, D. P. and Liddle, G. W. (1965). Natural history of pituitary-adrenal recovery—following long-term suppression with corticosteroids. *J. Clin. Endocrinol. Metab.* **25,** 11–24.

Grandison, L., Gelato, M. and Meites, J. (1974). Inhibition of prolactin secretion by cholinergic drugs. *Proc. Soc. exp. Biol. Med.* **145,** 1236–1239.

Grant, G. and Vale, W. (1975). Pituitary receptor binding assay of hypothalamic releasing factors. *Methods Enzymol.* **37,** (Pt. B) 213–219.

Grant, G., Vale, W., and Guillemin, R. (1972). Interaction of thyrotropin releasing factor with membrane receptors of pituitary cells. *Biochem. Biophys. Res. Commun.* **46,** 28–34.

Grant, G., Vale, W. and Rivier, J. (1973a). Pituitary binding sites for (^{3}H)-labelled luteinizing hormone-releasing factor (LRF). *Biochem. Biophys. Res. Commun.* **50,** 771–778.

Grant, N. H., Clark, D. E. and Rosanoff, E. I. (1973b). Evidence that Pro-Leu-Gly-NH_2, tocinoic acid, and des-Cys-tocinoic acid do not affect secretion of melanocyte stimulating hormone. *Biochem. Biophys. Res. Commun.* **51,** 100–106.

Grant, S. D., Forsham, P. H., and Di Raimondo, V. C. (1965). Suppression of 17-deoxycorticosteroids in plasma and urine by single and divided doses of triamcinolone. *N. Engl. J. Med.* **273,** 1115–1117.

Greeley, G. H. Jr., Muldoon, T. G. and Mahesh, V. B. (1975). Correlative aspects of luteinizing hormone-releasing hormone sensitivity and cytoplasmic oestrogen receptor concentration in the anterior pituitary and hypothalamus of the cycling rat. *Biol. Reprod.* **13,** 505–512.

Green, A. R. and Grahame-Smith, D. G. (1974). TRH potentiates behavioural changes following increased brain 5-hydroxytryptamine accumulation in rats. *Nature, Lond.* **251,** 524–526.

Green, J. D. (1947). Vessels and nerves of the amphibian hypophysis: a study of the living circulation and of the histology of the hypophysial vessels and nerves. *Anat. Rec.* **99,** 21–54.

Green, J. D. (1966). The comparative anatomy of the portal vascular system and of the innervation of the hypophysis. *In* "The Pituitary Gland" (G. W. Harris and B. T. Donovan, eds) Vol. I, pp. 127–146. Butterworths, London.

Green, J. D. (1969). Neural pathway to the hypophysis: anatomical and functional. *In* "The Hypothalamus" (W. Haymaker, E. Anderson and W. J. H. Nauta, eds) pp. 276–310. Thomas, Springfield.

Green, J. D. and Harris, G. W. (1947). The neurovascular link between the neurohypophysis and adenohypophysis. *J. Endocr.* **5,** 136–146.

Green, J. D. and Harris, G. W. (1949). Observations of the hypophysial portal vessels of the living rat. *J. Physiol., Lond.* **108,** 359–361.

Greenspan, F. S., Li, C. H., Simpson, E. and Evans, H. M. (1949). Bioassay of hypophyseal growth hormone: the tibia test. *Endocrinology,* **45,** 455–63.

Greenwald, G. S. (1974). Role of follicle-stimulating hormone and luteinizing hormone in follicular development and ovulation. *In* "Handbook of Physiology", Part IV, Section 7 (E. Knobil and W. H. Sawyer, eds) Part 2, pp. 293–324. American Physiological Society, Washington D.C.

Greenwald, G. S. and Rothchild, I. (1968). Formation and maintenance of corpora lutea in laboratory animals. *J. Anim. Sci.* **27,** Suppl. 1, 139–162.

Greep, R. O. (1936). Functional pituitary grafts in rats. *Proc. Soc. exp. Biol. Med.* **34** 754–755.

Greer, M. A., Fukuda, H., Roberts, L., Allen, C. F., Critchlow, V., and Wilson, M. (1975). Nycthemeral and sex related variations in plasma TSH, T4, and T3 in the rat. *Int. Congr. Ser.* **361,** 4–5. Excerpta Medica, Amsterdam.

Gregerman, R. I. (1971). Intrinsic physiological variables and nonthyroidal illness. *In* "The Thyroid" (S. C. Werner and S. H. Ingbar, eds) pp. 137–152. Harper and Row, New York.

Gregerman, R. I. and Bierman, E. L. (1974). Aging and hormones. *In* "Textbook of Endocrinology" (R. H. Williams, ed.) 5th edn, pp. 1059–1070. Saunders, Baltimore.

Greibrokk, T., Currie, B. S., Johansson, K. N. G., Hansen, J. J. and Folkers, K. (1974). Purification of a prolactin inhibiting hormone and the revealing of hormone D-GHIH which inhibits the release of growth hormone. *Biochem. Biophys. Res. Commun.* **59,** 704–709.

Griffiths, E. C. (1976). Peptidase inactivation of hypothalamic releasing hormone. *Horm. Res.* **7,** 179–191.

Griffiths, E. C. and Hooper, K. C. (1974). Competitive inhibition between oxytocin and luteinising hormone-releasing factor (LRF) for the same enzyme system in the rat hypothalamus. *Acta endocr., Copenh.* **75,** 435–442.

Griffiths, E. C., Hooper, K. C., Jeffcoate, S. L. and Holland, D. T. (1974). The presence of peptidases in rat hypothalamus inactivating luteinising hormone releasing hormone (LH-RH). *Acta endocr., Copenh.* **77,** 435–442.

Griffiths, E. C., Hooper, K. C. and Hopkinson, C. R. N. (1975a). Further studies on enzymatic inactivation of luteinizing hormone-releasing hormone (LH-RH) by peptidases in the rat hypothalamus. *Acta endocr., Copenh.* **79,** 7–15.

Griffiths, E. C., Hooper, K. C., Jeffcoate, S. L. and Holland, D. T. (1975b). Peptidases in different areas of the rat brain inactivating luteinising hormone-releasing hormone (LH-RH). *Brain Res.* **85,** 161–164.

Griffiths, E. C., Hooper, K. C., Jeffcoate, S. L. and Holland, D. T. (1975c). The effects of gonadectomy and gonadal steroids on the activity of the hypothalamic peptidases inactivating luteinising hormone-releasing hormone (LH-RH). *Brain Res.* **88,** 384–38.

Griffiths, E. C., Hooper, K. C., Jeffcoate, S. L. and White, N. (1975d). The presence of peptidases in rat hypothalamus inactivating thyrotrophin releasing hormone (TRH). *Acta endocr., Copenh.* **79,** 209–216.

Griffiths, E. C., Hooper, K. C., Jeffcoate, S. L. and White, N. (1976). Inactivation of thyrotrophin releasing hormone (TRH) by peptidases in different areas of rat brain. *Brain Res.* **105,** 367–380.

Grimm-Jørgensen, Y., and McKelvy, J. F. (1974a). Biosynthesis of thyrotropin releasing factor by newt (*Triturus viridescens*) brain *in vitro*. Isolation and characterization of thyrotropin releasing factor. *J. Neurochem.* **23,** 471–478.

Grimm-Jørgensen, Y., and McKelvy, J. T. (1974b). Control of the *in vitro* biosynthesis of thyrotropin releasing factor (TRF) in the red spotted newt (*Triturus viridescens*). *Program 56th Meet. U.S. Endocrine Soc.* p. A-65.

Grimm-Jørgensen, Y. and Reichlin, S. (1973). Thyrotropin releasing hormone (TRH): neurotransmitter regulation of secretion by mouse hypothalamic tissue *in vitro*. *Endocrinology*, **93,** 626–631.

Grimm-Jørgensen, Y., McKelvy, J. F., and Jackson, I. M. D. (1975). Immunoreactive thyrotropin releasing factor in gastropod circumesophageal ganglia. *Nature, Lond.* **254,** 620.

Grindeland, R. E., Smith, A. T., Evans, E. S. and Ellis, S. (1974). Induction of chronic growth hormone deficiency by anti-GH serum. *Endocrinology*, **95,** 793–798.

Gross, E., Noda, K. and Nisula, B. (1973). Solid phase synthesis of peptides with carboxyl-terminal amides-thyrotropin-releasing factor (TRF). *Angew. Chem. Intern. Ed.* **12,** 664–665.

Grosvenor, C. E. and Mena, F. (1974). Neural and hormonal control of milk secretion and milk ejection. *In* "Lactation" (B. L. Larson and V. R. Smith, eds) Vol. I, pp. 227–276. Academic Press, New York and London.

Grosvenor, C. E. and Whitworth, N. (1974). Evidence for a steady rate of secretion of prolactin following suckling in the rat. *J. Dairy Sci.* **57,** 900–904.

Grosvenor, C. E., McCann, S. M. and Nallar, R. (1965). Inhibition of nursing-induced fall in pituitary prolactin concentration in lactating rats by injection of acid extracts of bovine hypothalamus. *Endocrinology*, **76,** 883–889.

Grosvenor, C. E., Mena, F., Dhariwal, A. P. S. and McCann, S. M. (1967). Reduction of milk secretion by prolactin-inhibiting factor: further evidence that exteroceptive stimuli can release pituitary prolactin in rats. *Endocrinology*, **81,** 1021–1028.

Grumbach, M. M. and Kaplan, S. L. (1973). Ontogenesis of growth hormone, insulin prolactin and gonadotropin secretion in the human foetus. *In* "Foetal and Neonatal Physiology" (K. S. Comline, K. W. Cross, G. S. Dawes and P. W. Nathanielsz, eds) pp. 462–487. Cambridge University Press.

Grumbach, M. M., Roth, J. C., Kaplan, S. L. and Kelch, R. P. (1974). Hypothalamic-pituitary regulation of puberty in man: evidence and concepts derived from clinical research. *In* "The Control of the Onset of Puberty" (M. M. Grumbach, G. D. Grave, and F. E. Mayer, eds) pp. 115–166. J. Wiley, New York.

Gual, C., Scaglia, H. E., Midgley, R. A., Alcocer, J., Echeverria-Rivas, Y. and Lichtenberg, R. (1975). Regulatory effects of steroids on the pituitary response to LH-RH. *J. Steroid. Biochem.* **6,** 1067–1080.

Guansing, A., Leung, Y., Ajlouni, K., and Hagen, T. C. (1975). Effect of hypoglycemia on TSH release in man. *J. Clin. Endocrinol. Metab.* **40,** 755–758.

Guillemin, R. (1967). The adenohypophysis and its hypothalamic control. *Annu. Rev. Physiol.* **29,** 313–348.

Guillemin, R. (1971). Hypothalamic control of the secretion of the adenohypophysial hormones. *Adv. Metab. Disorders*, **5,** 1–51.

Guillemin, R. and Schally, A. V. (1963). Concentrations of CRF in acid extracts of sheep hypothalamus by gel filtration and counter-current distribution. *Texas Rep. Biol. Med.* **21,** 541–545.

Guillemin, R. and Vale, W. (1970). Bioassays of the hypophysiotropic hormones: *in vitro* systems. *In* "Hypophysiotropic Hormones of the Hypothalamus: Assay and Chemistry" (J. Meites, ed.) pp. 21–35. Williams and Wilkins, Baltimore.

Guillemin, R., Hearn, W. R., Cheek, W. R. and Householder, D. E. (1957). Control of corticotrophin release: further studies with *in vitro* methods. *Endocrinology*, **60,** 488–506.

Guillemin, R., Schally, A. V., Anderson, R., Lipscomb, H. S. and Long, J. (1960). Sur l'existence de deux types de substances à activite hypophysiotrope: α-CRF et β-CRF. *C. R. Acad. Sci.* **250,** 4462–4464.

Guillemin, R., Yamazaki, E., Justisz, M. and Sakiz, E. (1962). Présence dans une extrait de tissus hypothalamiques d'une substance stimulant la sécrétion de l'hormone hypophysaire thyréotrope (TSH). Première purification par filtration sur gel sephadex. *C. R. Acad. Sci.* **255,** 1018–1020.

Guillemin, R., Sakiz, E. and Ward, D. N. (1965). Further purification of TSH-releasing factor (TRF) from sheep hypothalamic tissues, with observations on the amino acid composition. *Proc. Soc. exp. Biol. Med.* **118,** 1132–1137.

Guillemin, R., Burgus, R., Salciz, E. and Ward, D. N. (1968). Nouvelles données sur la purification de l'hormone hypothalamique TSH-hypophysiotrope, TRF. *C. R. Acad. Sci.* **262,** 2278–2280.

Gunaga, K. P. and Menon, K. M. J. (1973). Effect of catecholamines and ovarian hormones on cyclic AMP accumulation in rat hypothalamus. *Biochem. Biophys. Res. Commun.* **54,** 440–448.

Gupta, S. K., Ridzon, T. J., Lopez, F. R., Pagliara, A. S., Weldon, V. V. and Daughaday, W. H. (1972). The use of L-dopa in the diagnosis of hyposomatotropism in children—a simplified provocative test. *In* "4th Int. Congr. Endocrinol., Washington", p. 35. Excerpta Medica, Amsterdam.

H

Haase, E., Sharp, P. J. and Paulke, E. (1975). Annual cycle of plasma luteinizing hormone concentrations in wild mallard drakes. *J. exp. Zool.* **194,** 553–558.

Hachmeister, U., Wiegelmann and Solbach, H. G. (1971). Ultrastructure and hormone distribution of the human corticotrophic anterior pituitary cell under normal conditions in corticotrophic adenoma and in exogenous hypercortisolism. *Acta endocr., Copenh.* Suppl. **152,** 90.

Hadley, M. E. and Bagnara, J. T. (1975). Regulation of release and mechanism of action of MSH. *Am. Zool.* **15,** Suppl. 1, 81–104.

Hadley, M. E., Hruby, V. J. and Bower, A. (1975). Cellular mechanisms controlling melanophore stimulating hormone (MSH) release. *Gen. Comp. Endocrinol.* **26,** 24–35.

Hafs, H. D. (1975). Prostaglandins and the control of anterior pituitary hormone secretion. *In* "Hypothalamic Hormones" (M. Motta, P. G. Crosignani and L. Martini, eds) pp. 183–193. Academic Press, New York and London.

Hagen, T. C., Lawrence, A. M. and Kirsteins, L. (1972a). Autoregulation of growth hormone secretion in normal subjects. *Metabolism*, **21,** 603–610.

Hagen, T. C., Lawrence, A. M., Kacherian, E. R. and Kirsteins, L. (1972b). A dialyzable heat stable factor in acromegalic plasma associated with increased *in vitro* growth hormone release. *Proc. Soc. exp. Biol. Med.* **141,** 536–539.

Haighton, J. (1977). An experimental inquiry concerning animal impregnation. *Phil. Trans. Roy. Soc.* (*London*), **87,** 159–196.

Halász, B. (1969). The endocrine effects of isolation of the hypothalamus from the rest of the brain. *In* "Frontiers in Neuroendocrinology" (W. F. Ganong and L. Martini, eds) pp. 307–342. Oxford University Press, London.

Halász, B. (1972). Hypothalamic mechanisms controlling pituitary function. *Progr. Brain Res.* **38,** 97–118.

Halász, B. (1974). Possible role of the hypothalamic hypophysiotrophic area in the control of rhythmic functions of the anterior pituitary. *In* "Biological Rhythms in Neuroendocrine Activity" (M. Kawakami, ed.) pp. 39–53. Igaku Shoin, Tokyo.

Halász, B. and Pupp, L. (1965). Hormone secretion of the anterior pituitary gland after physical interruption of all nervous pathways to the hypophysiotrophic area. *Endocrinology*, **77,** 553–562.

Halász, B., and Szentágothai, J. (1958). Über die unmittelbare Rückwirkung einer vom Hypophysenvorderlappen erzeugten Substanz auf der Hypothalamus. *Acta physiol. Acad. Sci. Hung.* **14,** Suppl. 16.

Halász, B. and Szentágothai, J. (1959). Histologischer Beweis einer nervösen Signalübermittlung von der Nebennierenrinde zum Hypothalamus. *Z. Zellforsch.* **50,** 297–306.

Halász, B., Pupp, L. and Uhlarik, S. (1962). Hypophysiotrophic area in the hypothalamus. *J. Endocr.* **25,** 147–154.

Halász, B., Pupp, L., Uhlarik, S. and Tima, L. (1965). Further studies on the hormone secretion of the anterior pituitary transplanted into the hypophysiotrophic area of the hypothalamus. *Endocrinology*, **77,** 343–355.

Halász, B., Florsheim, W. H., Corcorran, N. I., and Gorski, R. A. (1967a). Thyrotrophic hormone secretion in rats after partial or total interruption of neural afferents to the median basal hypothalamus. *Endocrinology*, **80,** 1075–1082.

Halász, B., Slusher, M. A. and Gorski, R. A. (1967b). Adrenocorticotrophic hormone in rats after partial or total deafferentation of the medial basal hypothalamus. *Neuroendocrinology*, **2,** 43–55.

Halberg, J. (1974). Catfish anyone? *Chronobiologia*, **1,** 127–129.

Hansen, A. P., Christensen, S. E., Iversen, J., Seyer Hansen, K., Orskov, H. and Weeke, J. (1974). Somatostatin in studies of growth hormone, insulin, pancreatic glucagon, and TSH-release in man. *Diabetologica*, **10,** 383.

Hansen, A. P. (1973a). Abnormal serum growth hormone response to exercise in maturity-onset diabetics. *Diabetes*, **22,** 619–628.

Hansen, A. P. (1973b). Serum growth hormone pattern in female juvenile diabetics. *J. Clin. Endocrinol. Metab.* **36,** 638–646.

Hanström, B. (1966). Gross anatomy of the hypophysis of mammals. *In* "The Pituitary Gland" (G. W. Harris and B. T. Donovan, eds) Vol. I, pp. 1–57. Butterworths, London.

Harris, A., Christianson, D., Smith, M. S., Braverman, L. and Vagenakis, A. (1977). The physiological role of TRH in the regulation of TSH and prolactin secretion in the rat. *Clin. Res.* **25,** 463A.

Harris, G. W. (1937). Induction of ovulation in the rabbit by electrical stimulation of the hypothalamo-hypophysial mechanism. *Proc. Roy. Soc.* (*London*) *Ser. B*, **122,** 374–394.

Harris, G. W. (1950). Oestrous rhythm, pseudopregnancy and the pituitary stalk in the rat. *J. Physiol., Lond.* **111,** 347–360.

Harris, G. W. (1955). "Neural Control of the Pituitary Gland". Edward Arnold, London.

Harris, G. W. (1961). The pituitary stalk and ovulation. *In* "Control of Ovulation" (C. Villee, ed.) pp. 56–74. Pergamon Press, London.

Harris, G. W. (1970). Unsolved problems in the portal vessel-chemotransmitter hypothesis. *In* "Hypophysiotropic Hormones of the Hypothalamus: Assay and Chemistry" (J. Meites, ed.) pp. 1–14. Williams and Wilkins, Baltimore.

Harris, G. W. (1972). Humours and hormones. *J. Endocr.* **53,** ii–xxiii.

Harris, G. W. and Jacobsohn, D. (1952). Functional grafts of the anterior pituitary gland. *Proc. Roy. Soc.* (*London*) *Ser. B*, **139,** 263–276.

Harris, G. W. and Johnson, R. T. (1950). Regeneration of the hypophysial portal vessels after section of the hypophysial stalk, in the monkey *Macasus rhesus*. *Nature, Lond.* **165,** 819–820.

Harris, G. W., Michael, R. P. and Scott, P. P. (1958). Neurological site of action of stilboestrol in eliciting sexual behaviour. *In* "Ciba Foundation Symposium on the Neurological Basis of Behaviour" (G. E. W. Wolstenholme and C. M. O'Connor, eds) pp. 236–251. Churchill, London.

Harris, G. W., Manabe, Y. and Ruf, K. A. (1969). A study of the parameters of electrical stimulation of unmyelinated fibres in the pituitary stalk. *J. Physiol., Lond.* **203,** 67–81.

Harris, I. (1966). The chemistry of intermediate lobe hormones. *In* "The Pituitary Gland" (G. W. Harris and B. T. Donovan, eds) Vol. 3, pp. 33–40. University of California Press, Berkeley and Los Angeles.

Harris, M., Dreifuss, J. J. and Debois, J. J. (1975). Excitation of phasically firing supraoptic neurones during vasopressin release. *Nature, Lond.* **258,** 80–82.

Hart, I. C. (1973). Basal levels of prolactin in goat blood measured throughout a 24-hour period by a rapid double antibody-solid phase radioimmunoassay. *J. Dairy Res.* **40,** 235–245.

Hart, I. C. (1974). The relationship between lactation and the release of prolactin and growth hormone in the goat. *J. Reprod. Fertil.* **39,** 485–499.

Hart, I. C. (1975a). Concentrations of prolactin in serial blood samples from goats, before, during and after milking throughout lactation. *J. Endocr.* **64,** 305–312.

Hart, I. C. (1975b). Seasonal factors affecting the release of prolactin in goats in response to milking. *J. Endocr.* **64,** 313–322.

Hastings, J. W. (1970). Cellular-biochemical clock hypothesis. *In* "The Biological Clock: Two Views", pp. 63–91. Academic Press, New York and London.

Hatanaka, C., Obayashi, M., Nishimura, O., Toukai, N. and Fujino, M. (1974). An improved synthesis of thyrotropin-releasing hormone (TRH) and crystallization of the tartrate. *Biochem. Biophys. Res. Commun.* **60,** 1345–1350.

Haterius, H. O. and Derbyshire, A. J. Jr. (1937). Ovulation in the rabbit following upon stimulation of the hypothalamus. *Am. J. Physiol.* **119,** 329–330.

Hawkins, E. F. and Ball, J. N. (1970). Physiological studies on the hypothalamo-hypophysial-interrenal axis in *Poecilia latipinna* (Teleostei). *J. Endocr.* **48,** xxvii.

Hayashi, S., Mennin, S. P. and Gorski, R. A. (1974). Effect of hypothalamic deafferentation on luteinization of intrasplenic ovarian grafts in the ovariectomized rat. *Neuroendocrinology*, **14,** 321–332.

Hayek, A., Driscoll, S. G. and Warshaw, J. B. (1973). Endocrine studies in anecephaly. *J. Clin. Invest.* **52,** 1636–1641.

Hayes, J. R., Johnson, D. G., Koerker, D. and Williams, R. H. (1975). Inhibition of gastrin release by somatostatin *in vitro*. *Endocrinology*, **96,** 1374–1376.

Haymaker, W., Anderson, E. and Nauta, W. J. H. (1969). "The Hypothalamus". C. C. Thomas, Springfield.

Hayward, J. N. (1972). Hypothalamic input to supraoptic neurons. *Progr. Brain Res.* **38,** 145–161.

Hayward, J. N. (1975). Neural control of the posterior pituitary. *Annu. Rev. Physiol.* **37,** 191–210.

Hayward, J. N. and Jennings, D. P. (1973). Activity of magnocellular neuroendocrine cells in the hypothalamus of unanaesthetized monkeys. I. Functional cell types and their anatomical distribution in the supraoptic nucleus and the internuclear zone. *J. Physiol. (London)*, **232,** 515–543.

Heap, R. B., Perry, J. S. and Challis, J. R. G. (1973). Hormonal maintenance in pregnancy. *In* "Handbook of Physiology, Endocrinology" (R. O. Greep and E. B. Astwood, eds) Vol. II Part 2, pp. 217–260. American Physiological Society, Washington, D.C.

Heap, P. F., Jones, C. W., Morris, J. F. and Pickering, B. T. (1975). Movement of neurosecretory product through the anatomical compartments of the neural lobe of the pituitary gland. *Cell Tiss. Res.* **156,** 483–497.

Hedge, G. A. (1971). ACTH secretion due to hypothalamopituitary effects of adenosine 3, ′5′-monophosphate and related substances. *Endocrinology*, **89,** 500–506.

Hedge, G. A. and de Wied, D. (1971). Corticotrophin and vasopressin secretion after hypothalamic implantation of atropine. *Endocrinology*, **88,** 257–259.

Hedge, G. A. and Smelik, P. G. (1968). Corticotrophin-release: inhibition by intrahypothalamic implants of atropine. *Science, N.Y.* **159,** 891–892.

Hedge, G. A., Yates, M. B., Marcus, R. and Yates, F. E. (1966). Site of action of vasopressin in causing corticotropin release. *Endocrinology*, **79,** 328–340.

Heller, H. (1941). Differentiation of an (amphibian) water balance principle from the antidiuretic principle of the posterior pituitary gland. *J. Physiol., Lond.* **100,** 125–141.

Heller, H. (1969). Class and species specific actions of neurohypophysial hormones. *Colloq. Intl. C.N.R.S., Paris*, **177,** 35–44.

Heller, H. (1972). The effect of neurohypophysial hormones on the female reproductive tract of lower vertebrates. *Gen. Comp. Endocrinol. Suppl.* **3,** 703–714.

Heller, H. (1974). Molecular aspects in comparative endocrinology. *Gen. Comp. Endocrinol.* **22,** 315–332.

Heller, H. and Lederis, K. (1960). Posterior pituitary hormones of the hippopotamus. *J. Physiol., Lond.* **151,** 47P–49P.

Heller, H. and Pickering, B. T. (1970). The distribution of vertebrate neurohypophysial hormones and its relation to possible pathways for their evolution. *In* "International Encyclopedia of Pharmacology and Therapeutics" (H. Heller and B. T. Pickering, eds) Section 41, pp. 59–79. Pergamon Press, Oxford.

Heller, H. and Spickett, S. G. (1966). The polymorphisms of the neurohypophysial hormones. *Mem. Soc. Endocrinol.* **15,** 89–109.

Hellman, L., Nakada, F., Curtis, J., Weitman, D., Kream, J., Roffwarg, H., Hellman, S., Fukushima, D. K. and Gallagher, T. F. (1970). Cortisol is secreted episodically by normal man. *J. Clin. Endocrinol. Metab.* **30,** 411–422.

Hershman, J. M. and Pittman, J. A. (1971). Utility of the radioimmunoassay of serum thyrotropin in man. *Ann. Int. Med.* **74,** 481–490.

Hershman, J. M., Read, D. G., Bailey, A. L., Norman, V. D. and Gibson, T. B. (1970). Effect of cold exposure on serum thyrotropin. *J. Clin. Endocrinol. Metab.* **30,** 430–434.

Hild, W. and Zetler, G. (1951). Über das Vorkommen der Hypophysenhinterlappenhormone im Zwischenhirn. *Arch. exp. Path. Pharmak.* **213,** 139–153.

Hillhouse, E. W., Burden, J. L. and Jones, M. T. (1974). Structure-activity relationships of various steroids on the feedback control of corticotrophin releasing hormone. *J. Endocr.* **63,** 22P.

Hillhouse, E. W., Burden, J. L. and Jones, M. T. (1975). The effect of various putative neurotransmitters on the release of corticotrophin-releasing hormone from the hypothalamus of the rat *in vitro*. 1. The effect of acetylcholine and noradrenaline. *Neuroendocrinology*, **17,** 1–11.

Hilliard, J., Penardi, R. and Sawyer, C. H. (1967). A functional role for 20α-hydroxypregn-4-en-3-one in the rabbit. *Endocrinology*, **80,** 901–909.

Hill-Samli, M. and Macleod, R. M. (1974). Interaction of thyrotropin-releasing hormone and dopamine on the release of prolactin from the rat anterior pituitary *in vitro*. *Endocrinology*, **95,** 1189–1192.

Himms-Hagen, J. (1967). Sympathetic regulation of metabolism. *Pharmacol. Rev.* **19,** 367–462.

Hinkle, P. M. and Tashjian, A. H. (1975). Degradation of thyrotropin releasing hormone by the GH_3 strain of pituitary cells in culture. *Endocrinology*, **97,** 324–331.

Hinkle, P. M., Woroch, E. L. and Tashjian, A. H. (1974). Receptor-binding affinities and biological activities of analogues of thyrotropin releasing hormone in prolactin producing pituitary cells in culture. *J. Biol. Chem.* **249,** 3085–3090.

Hinsey, J. C. (1937). The relationship of the nervous system to ovulation and other phenomena of the female reproductive tract. *Cold Spring Harbor Symp. Quant. Biol.* **5,** 269–279.

Hinsey, J. C. and Markee, J. E. (1933). Pregnancy following bilateral section of the cervical sympathetic trunks in the rabbit. *Proc. Soc. exp. Biol. Med.* **31,** 270–271.

Hiroshige, T., Itoh, S. and Sakakura, M. (1969). ACTH release and adenohypophysial glycogenolysis induced by purified hypothalamic CRF. *Endocrinol. Japon.* **1,** 41.

Hiroshige, T., Abe, K., Wade, S. and Kaneko, M. (1973). Sex difference in circadian periodicity of CRF activity in the rat hypothalamus. *Neuroendocrinology,* **11,** 306–320.

Hodges, J. R. (1953). The effect of deoxycorticosterone acetate on the release of adrenocorticotrophic hormone by the pituitary gland. *Br. J. Pharmac. Chemother.* **8,** 242–247.

Hodges, J. R. and Jones, M. T. (1963). The effect of injected corticosterone on the release of adrenocorticotrophic hormone in rats exposed to acute stress. *J. Physiol., Lond.* **167,** 30–37.

Hodges, J. R. and Sadow, J. (1967). Impairment of pituitary adrenocorticotrophic function by corticosterone in the blood. *Br. J. Pharmacol.* **30,** 385–391.

Hodges, J. R. and Vernikos, J. (1958). Influence of circulating ACTH on the pituitary adrenocorticotrophic response to stress in the adrenalectomised rat. *Nature* (*London*), **182,** 725.

Hodges, J. R. and Vernikos, J. (1960). The effects of hydrocortisone on the level of corticotrophin in the blood and pituitary glands of the adrenalectomised and stressed adrenalectomised rats. *J. Endocr.* **55,** 489–497.

Hoffman, J. C. (1973). The influence of photoperiods on reproductive functions in female mammals. *In* "Handbook of Physiology" (R. O. Greep and E. B. Astwood, eds) Vol. II, Part 1, pp. 57–77. American Physiological Society, Washington D.C.

Hofmann, K., Yajima, H., Yanaihara, N., Liu, T. and Lande, S. (1961). Studies on polypeptides. xiii. The synthesis of a tricosapeptide possessing essentially the full biological activity of natural ACTH. *J. Am. Chem. Soc.* **83,** 487–489.

Hohlweg, W. (1934). Veränderungen des Hypophysenvorderlappens und des Ovariums nach Behandlungen mit grossen Dosen von Follikelhormon. *Klin. Wschr.* **13,** 92–95.

Hohlweg, W. and Dohrn, M. (1932). Über die Beziehungen zwischen Hypophysenvorderlappen und Keimdrüsen. *Klin. Wschr.* **11,** 233–235.

Hohlweg, W. and Junkmann, K. (1932). Die hormonal-nervöse Regulierung der Funktion des Hypophysenvorderlappens. *Klin. Wschr.* **11,** 321–323.

Hökfelt, T. and Fuxe, K. (1972). On the morphology and the neuroendocrine role of the hypothalamic catecholamine neurons. *In* "Brain-Endocrine Interaction. Median Eminence: Structure and Function" (K. M. Knigge, D. E. Scott and A. Weindl, eds) pp. 181–223. Karger, Basel.

Hökfelt, T., Fuxe, K., Goldstein, M. and Joh, T. H. (1973). Immunohistochemical localisation of three catecholamine synthesising enzymes: aspects of methodology. *Histochemie,* **33,** 231–254.

Hökfelt, T., Efendic, S., Johansson, O., Luft, R. and Arimura, A. (1974). Immunohistochemical localisation of somatostatin (growth hormone release inhibiting hormone) in the guinea-pig brain. *Brain. Res.* **80,** 165–169.

Hökfelt, T., Efendic, S., Hellerstrom, G., Johansson, O., Luft, R. and Arimura, A. (1975a). Cellular localisation of somatostatin in endocrine-like cells and neurons of the rat with special reference to the A_1 cells of the pancreatic islets and to the hypothalamus. *Acta endocr., Copenh.* **80** (Suppl. 200), 1–41.

Hökfelt, T., Fuxe, K. and Goldstein, M. (1975b). Application of neurohistochemistry to studies on monoamine cell systems with special reference to nervous tissues. *Ann. N.Y. Acad. Sci.* **254,** 407–432.

Hökfelt, T., Fuxe, K., Johansson, O., Jeffcoate, S. L. and White, N. (1975c). Distribution of thyrotrophin releasing hormone (TRH) in the central nervous system as revealed with immunohistochemistry. *Europ. J. Pharmacol.* **34,** 389–392.

Hökfelt, T., Fuxe, K., Johansson, O., Jeffcoate, S. L. and White, N. (1975d). Thyrotrophin releasing hormone (TRH)-containing nerve terminals in certain brain stem nuclei and in the spinal cord. *Neuroscience Letters,* **1,** 133–139.

Hökfelt, T., Hellerstrom, C., Efendie, S., Johansson, O., Luft, R. and Arimura, A. (1975e). Immunohistochemical evidence for a widespread occurrence of somatostatin in neurosecretory cells, primary sensory neurons and endocrine-like cell systems. *Acta endocr., Copenh.* Suppl. **199,** 134.

Hökfelt, T., Johansson, O., Fuxe, K., Goldstein, M., Park, D., Ebstein, R., Fraser, H. M., Jeffcoate, S. L., Efendie, S., Luft, R. and Arimura, A. (1975f). Mapping and relationship of hypothalamic neurotransmitters and hypothalamic hormones. *Proceedings of VIth Int. Congress of Pharmacology,* **3,** 93–110.

Hollister, L. E., Berger, P., Ogle, F. L., Arnold, R. C. and Johnson, A. (1974). Protirelin (TRH) in depression. *Arch. Gen. Psychiat.* **31,** 468–470.

Holmes, R. L. and Ball, J. N. (1974). "The Pituitary Gland: A Comparative Account". Cambridge University Press, London.

Holmes, S. W. and Horton, E. W. (1968). The identification of four prostaglandins in dog brain and their regional distribution in the central nervous system. *J. Physiol., Lond.* **195,** 731–741.

Holton, H. H., Branda, L. A. and Ferrier, B. M. (1973). Synthesis of tritium labelled (8-arginine)-vasopressin. *Can. J. Chem.* **51,** 1910–1914.

Honnebier, W. J. and Swaab, D. F. (1973). The influence of anencephaly upon uterine growth of foetus and placenta and upon gestation length. *J. Obstet. Gynaec. Brit. Cwlth.* **180,** 577–588.

Hooley, R. D., Baxter, R. W., Chamley, W. A., Cumming, I. A., Jonas, H. A. and Findlay, J. K. (1974). FSH and LH response to gonadotrophin releasing hormone during the ovine estrous cycle and following progesterone administration. *Endocrinology,* **95,** 937–942.

Hooper, K. C. (1968). The effects of ovariectomy and injected oestradiol benzoate on polypeptide metabolism in the hypothalamus. *Biochem. J.* **110,** 151–153.

Hopkins, C. R. (1971). Localization of adrenergic fibers in the amphibian pars intermedia by electron microscope autoradiography and their selective removal by 6-hydroxydopamine. *Gen. Comp. Endocrinol.* **16,** 112–120.

Horrobin, D. F., Manku, M. S., Nassar, B. and Evered, D. (1973). Prolactin and fluid and electrolyte balance. *In* "Human Prolactin" (J. L. Pasteels and C. Robyn, eds) pp. 152–155. Excerpta Medica, Amsterdam.

Horstmann, E. (1954). Die Faserglia des Selachiergehirns. *Z. Zellforsch.* **39,** 588–617.

Houssay, B. A., Biassotti, A. and Sammartino, R. (1935). Modifications fonctionelles de l'hypophyse après les lésions infundibulo-tubériennes chez le crapaud. *C.R. Soc. Biol.* **120,** 725–727.

Howe, A. (1973). The mammalian pars intermedia: a review of its structure and function. *J. Endocr.* **59,** 385–409.

Howe, A. and Maxwell, D. S. (1968). Electron microscopy of the pars intermedia of the pituitary gland in the rat. *Gen. Comp. Endocrinol.* **11,** 169–185.

Howe, A. and Thody, A. J. (1969). The effect of hypothalamic lesions on the melanocyte-stimulating hormone content and histology of the pars intermedia of the rat pituitary gland. *J. Physiol., Lond.* **203,** 159–171.

Howe, A. and Thody, A. J. (1970). The effect of injection of hypertonic saline on the melanocyte-stimulating hormone content and histology of the pars intermedia of the rat pituitary. *J. Endocr.* **46,** 201–208.

Howland, B. E. and Skinner, K. R. (1973). Effect of starvation on gonadotrophin secretion in intact and castrated male rats. *Can. J. Physiol. Pharmacol.* **51,** 759–762.

Hruby, V. J., Smith, C. W., Bower, A. Jr. and Hadley, M. E. (1972a). Melanophore-stimulating hormone: release inhibition by ring structures of neurohypophysial hormones. *Science, N.Y.* **176,** 1331–1332.

Hruby, V. J., Smith, W. C., Linn, D. K., Ferger, M. F. and DuVigneaud, V. (1972b). The synthesis and some pharmacological properties of tocinoic acid and deaminotocinoic acid. *J. Am. Chem. Soc.* **94,** 5478–5480.

Humphrey, R. R., Dermody, W. C., Brink, H. O., Bousley, F. G., Schottin, N. H., Sakowsky, R., Vaitkus, J. W., Velosa, H. T. and Reel, J. R. (1973). Induction of luteinizing hormone (LH) release and ovulation in rats, hamsters and rabbits by synthetic luteinizing hormone-releasing factor. (LRF). *Endocrinology,* **92,** 1515–1526.

Hunter, W. M. (1974). Preparation and assessment of radioactive tracers. *Brit. Med. Bull.* **30,** 18–23.

I

Igarashi, M. and McCann, S. M. (1964). A hypothalamic follicle stimulati ng hormone releasing factor. *Endocrinology,* **74,** 446–452.

Igarashi, M., Taya, K., Ishikawa, J. and Yamasaki, M. (1974). Evidences for the existence of FSH-RF distinct from LH-RH/FSH-RH. *In* "Biological Rhythms in Neuroendocrine activity" (M. Kawakami, ed.) pp. 91–101. Igaku Shoin, Tokyo.

Immer, H. U., Sestanj, K., Nelson, V. R. and Goetz, M. P. (1974). Synthesis of somatostatin. *Helv. Chim. Acta,* **57,** 730–734.

Imura, H., Yoshimi, T. and Ikekubo, K. (1971). Growth hormone secretion in a patient with deprivation dwarfism. *Endocr. Jap.* **18,** 301–304.

Imura, H., Nakai, Y. and Yoshimi, T. (1973). Effect of 5-hydroxytryptophan (5-HTP) on growth hormone and ACTH release in man. *J. Clin. Endocr. Metab.* **36,** 204–206.

Imura, H., Kato, Y., Yoshimoto, Y. and Nakai, Y. (1975). The role of biogenic amines in the regulation of pituitary hormone release in man, with special reference to circadian rhythmicity. *In* "Biological Rhythms in Neuroendocrine Activity" (M. Kawakami, ed.) pp. 140–150. Igaku Shoin, Tokyo.

Ingbar, S. H., Braverman, L. E., Dawber, N. A., Lee, G. Y. (1965). A new method for measuring the free thyroid hormone in human serum and an analysis of the factors that influence its concentration. *J. Clin. Invest.* **44,** 1679–1689.

Ingle, D. J. and Kendall, E. C. (1937). Atrophy of the adrenal cortex of the rat produced by the administration of large amounts of cortin. *Science, N.Y.* **86,** 245.

Ingle, D. J., Higgins, G. M. and Kendall, E. C. (1938). Atrophy of the adrenal cortex of the rat. *Proc. Roy. Soc. Biol.* **139,** 263–273.

Ingleton, P. M., Baker, B. I. and Ball, J. N. (1973). Secretion of prolactin and growth hormone by teleost pituitaries *in vitro.* I. Effect of sodium concentration and osmotic pressure during short-term incubations. *J. Comp. Physiol.* **87,** 317–328.

Inoué, S. and Sekaguchi, T. (1973). Neuroendocrine control systems and systems analysis of the brain-hypophyseal gonadal mechanism. *In* "Neuroendocrine Control" (K. Yagi and S. Yoshida, eds) pp. 359–394. Wiley, New York.

Ishida, A. (1970). The oxytocin release and the compound action potential evoked by electrical stimulation of the isolated neurohypophysis of the rat. *Jap. J. Physiol.* **20,** 84–96.

Ishii, S. (1972). Classification and identification of neurosecretory granules in the median eminence. *In* "Brain-Endocrine Interaction. Median Eminence: Structure and Function" (K. M. Knigge, D. E. Scott and A. Weindl, eds) pp. 119–141. Karger, Basel.

Ishikawa, H. and Goto, S. (1973). CRF-stimulated biosynthesis of ACTH in a cell-free system from rat anterior pituitaries. *Biochem. Biophys. Res. Commun.* **52,** 884–889.

Ishikawa, H., Nagayama, T. and Niizuma, K. (1973). The mechanism of the stimulation of LH-production by synthetic LH-releasing hormone (LH-RH) in tissue culture. *Biochem. Biophys. Res. Commun.* **55,** 492–498.

Isidori, A., Spera, G. and Costa, T. (1974). Distribution of radioactivity after administration of I^{125}-labelled luteinizing hormone in rats. *Experientia*, **30,** 1211–1213.

Isles, J. F. and Mulloney, B. (1971). Procion yellow staining of cockroach motorneurons without the use of microelectrodes. *Brain. Res.* **30,** 397–400.

Ito, Y. (1968). Experimental studies on the hypothalamic control of the pars intermedia activity of the frog, *Rana migromaculata. Neuroendocrinology*, **3,** 25–33.

Iturriza, F. C. (1967). The secretion of intermedin in autotransplants of pars intermedia growing in the anterior chamber of intact and sympathectomized eyes of the toad. *Neuroendocrinology*, **2,** 11–18.

Iturriza, F. C. (1973). Autoregulation of the secretion of MSH in incubated toad pars intermedia. *Gen. Comp. Endocrinol.* **20,** 168–171.

IUPAC-IUB. (1975). The nomenclature of peptide hormones. *J. Biol. Chem.* **250,** 3215–3216.

J

Jackson, G. L. (1971). Comparison of rat and chicken luteinizing hormone-releasing factors. *Endocrinology*, **89,** 1460–1463.

Jackson, G. L. and Thurmon, J. C. (1974). Absence of a critical period in estrogen-induced release of LH in the anestrous ewe. *Endocrinology*, **94,** 918–920.

Jackson, I. M. D. (1970). Pituitary enlargement resulting from primary thyroid disease. *Proc. Roy. Soc. Med.* **63,** 578.

Jackson, I. M. D. and Reichlin, S. (1973). TRH radioimmunoassay measurements in normal and altered states of thyroid function in the rat. *Program 49th Meeting American Thyroid Assoc.* p. T4.

Jackson, I. M. D. and Reichlin, S. (1974a). Thyrotropin releasing hormone (TRH): distribution in the brain, blood and urine of the rat. *Life Sci.* **14,** 2259–2266.

Jackson, I. M. D. and Reichlin, S. (1974b). Thyrotropin-releasing hormone (TRH): distribution in hypothalamic and extrahypothalamic brain tissues of mammalian and submammalian chordates. *Endocrinology*, **95,** 854–862.

Jackson, I. M. D. and Reichlin, S. (1975). TRH distribution in brain and urinary excretion following "thyrotrophic area" hypothalamic lesions in the rat. *Program 57th Meeting US Endocrine Soc.* p. 96.

Jackson, I. M. D. and Reichlin, S. (1977a). Brain thyrotrophin-releasing hormone is independent of the hypothalamus. *Nature, Lond.* **267,** 853–854.

Jackson, I. M. D. and Reichlin, S. (1977b). Thyrotropin-releasing hormone: abundance in the skin of the frog, *Rana pipiens. Science, N.Y.* **198,** 414–415.

Jackson, I. M. D., Gagel, R., Papapetrou, P. and Reichlin, S. (1974a). Pituitary, hypothalamic and urinary thyrotropin releasing hormone concentration in altered thyroid states of rat and man. *Clin. Res.* **22,** 342a.

Jackson, I. M. D., Papapetrou, P. and Reichlin, S. (1974b). The metabolism and excretion of TRH in the rat in states of altered thyroid function. *Program 50th Meeting American Thyroid Assoc.* p. T1.

Jackson, I. M. D., Gagel, R., Papapetrou, P., Deprez, D. and Reichlin, S. (1975). TRH excretion and metabolism in man. *Clin. Res.* **23,** 238a.

Jackson, I. M. D., Franco, F. S., Baum, G., Soo-Hoo, F. and Reichlin, S. (1976). Is there, or is there not, thyrotropin-releasing hormone (TRH) in urine? *Program 4th New England Endocrinology Conference*, October 1976, p. 19 (abs).

Jackson, I. M. D., Saperstein, R. and Reichlin, S. (1977). Thyrotropin releasing hormone (TRH) in pineal and hypothalamus of the frog: effect of season and illumination. *Endocrinology*, **100,** 97–100.

Jacobs, L. S., Snyder, P. J., Wilber, J. F., Utiger, R. D. and Daughaday, W. H. (1971). Increased serum prolactin after administration of synthetic thyrotropin releasing hormone (TRH) in man. *J. Clin. Endocrinol. Metab.* **33,** 996–998.

Jacobs, L. S., Snyder, P. J., Utiger, R. D. and Daughaday, W. H. (1973). Prolactin response to thyrotropin releasing hormone in normal subjects. *J. Clin. Endocrinol. Metab.* **36,** 1069–1073.

Jacobsohn, D. (1954). Regeneration of hypophysial portal vessels and grafts of anterior pituitary glands in rabbits. *Acta. endocr., Copenh.* **17,** 187–197.

Jaffe, R. B., Yuen, B. H., Keye, W. R. and Midgley, A. R. (1973). Physiologic and pathologic profiles of circulating human prolactin. *Am. J. Obstet. Gynec.* **117,** 757–770.

Jamieson, M. G. and Fink, G. (1976). Parameters of electrical stimulation of the medial preoptic area for release of gonadotrophins in male rats. *J. Endocr.* **68,** 57–70.

Jard, S. (1966). Etude des effets de la vasotocine sur l'excrétion de l'eau et des électrolytes par le rein de la grenouille *Rana esculenta* L.: Analyse à l'aide d'analogues artificiels de l'hormone naturelle des caractères structuraux requis pour son activité biologique. *J. Physiol.* (*Paris*), **58,** Suppl. 15, 1–124.

Jeffcoate, S. L. (1975a). Problems with the assay of LH-RH in biological fluids. *Biologie Animale*, **16,** 231–240.

Jeffcoate, S. L. (1975b). The control of testicular function in the adult. *Clinics in Endocrinology and Metabolism*, **4,** 521–543.

Jeffcoate, S. L. and Holland, D. T. (1975). Further studies on the nature of the immunoreactive luteinizing hormone-releasing hormone (LH-RH)-like peptide in human urine. *Acta. endocr., Copenh.* **78,** 232–238.

Jeffcoate, S. L. and White, N. (1975a). Is there any thyrotrophin releasing hormone in mammalian extrahypothalamic brain tissue? *J. Endocr.* **67,** 42P.

Jeffcoate, S. L. and White, N. (1975b). Studies on the nature of mammalian hypothalamic thyrotropin releasing hormone using immunochemical, chromatographic and enzymatic techniques. *J. Endocr.* **65,** 83–90.

Jeffcoate, S. L. and White, N. (1975c). The inactivation of thyrotrophin-releasing hormone (TRH) by plasma in thyroid disease. *Clin. Endocr.* **4,** 231–236.

Jeffcoate, S. L. and White, N. (1975d). Clearance and identification of thyrotrophin-releasing hormone in human urine. *Clin. Endocr.* **4,** 421–426.

Jeffcoate, S. L., Greenwood, R. H. and Holland, D. T. (1974a). Blood and urine clearance of luteinizing hormone-releasing hormone in man measured by radioimmunoassay. *J. Endocr.* **60,** 305–314.

Jeffcoate, S. L., Fraser, H. M., Holland, D. T. and Gunn, A. (1974b). Radioimmunoassay of luteinizing hormone-releasing hormone (LH-RH) in serum from man, sheep and rat. *Acta endocr., Copenh.* **75,** 625–635.

Jeffcoate, S. L., Sharp, P. J., Fraser, H. M., Holland, D. T. and Gunn, A. (1974c). Immunochemical and chromatographic similarity of rat, rabbit, chicken, and synthetic luteinizing hormone releasing hormones. *J. Endocr.* **62,** 85–91.

Jeffcoate, S. L., Holland, D. T., White, N., Fraser, H. M., Gunn, A., Crighton, D. B., Foster, J. P., Griffiths, E. C., Hooper, K. C. and Sharp, P. J. (1975). The radioimmunoassay of hypothalamic hormones (TRH and LHRH) and related peptides in biological fluids. *In* "Hypothalamic Hormones" (M. Motta, P. G. Grosignani and L. Martini, eds) pp. 279–298. Academic Press, London and New York.

Jeffcoate, S. L., Holland, D. T. and Fraser, H. M. (1976a). Anti-LH-RH in the study of reproduction. *In* "Physiological Effects of Immunity against Reproductive Hormones" (R. G. Edwards and M. Johnson, eds) pp. 121–136. Cambridge University Press.

Jeffcoate, S. L., White, N., Hökfelt, T., Fuxe, K. and Johansson, O. (1976b). Localisation of thyrotrophin releasing hormone (TRH) in the spinal cord of the rat by immunohistochemistry and by radioimmunoassay. *J. Endocr.* **69,** 9P.

Jeffers, W. A., Livezey, M. M. and Austin, J. H. (1942). A method for demonstrating an antidiuretic action of minute amounts of pitressin: statistical analysis of results. *Proc. Soc. exp. Biol. Med.* **50,** 184–188.

Jeppsson, S., Rannevik, G. and Kullander, S. (1974). Studies on the decreased gonadotrophin response after administration of LH/FSH-releasing hormone during pregnancy and puerperium. *Am. J. Obstet. Gynecol.* **120,** 1029–1034.

Jobin, M., Endröczi, E., Hontela, S. and Fortier, C. (1971). Effect of hypothalamic cooling on the behaviour and the pituitary thyroid activity in the rat. *J. Physiol.* (*Paris*), **63,** 309–311.

Jobin, M., Dulac, S. and Fortier, C. (1973). Effect of cold exposure on thyroxine (T4) and triiodothyronine (T3) secretion in the rat. *Fed. Proc. Fed. Am. Soc. Exp. Biol.* **32,** 110.

Jöchle, W. (1975). Current research in coitus-induced ovulation: a review. *J. Reprod. Fertil.* Suppl. **22,** 165–207.

Johansson, N.-G. (1975). Biosynthesis and secretion of hypothalamic hormones. *In* "Hypothalamic Hormones" (E. S. E. Hafez and J. R. Reel, eds) pp. 17–51. Ann Arbor Science, Michigan.

Johansson, K. N. G., Currie, B. L. and Folkers, K. (1973). Biosynthesis of the luteinizing hormone-releasing hormone in mitochondrial preparations and by a possible pantetheine-template mechanism. *Biochem. Biophys. Res. Commun.* **53,** 502–507.

Johnson, G. E., Flattery, K. V. and Schonbaum, E. (1965). The influence of methimazole on the catecholamine excretion of cold stressed rats. *Can. J. Physiol. Pharmacol.* **45,** 415–422.

Jonas, H. A., Salomonsen, L. A., Burger, H. G., Chamley, W. A., Cumming, I. A., Findlay, J. K. and Goding, J. R. (1973). The role of a gonadotrophin-releasing factor (Gn-RH) in the oestrogen-induced release of FSH and LH in the ewe. *J. Reprod. Fertil.* **32,** 341.

Jones, C. W., Peter, R. E. and Pickering, B. T. (1973). Intraneuronal transport in the hypothalamo-neurohypophysial system of the goldfish (*Carassius auratus*). *Gen. Comp. Endocrinol.* **20,** 99–106.

Jones, E. C. (1975). The post-reproductive phase in mammals. *Frontiers of Hormone Research,* **3,** 1–19.

Jones, M. T., Bridges, P. K. and Leak, D. (1970). Correlation between psychic endocrinological responses to emotional stress. *Progr. Brain. Res.* **32,** 325–335.

Jones, M. T., Brush, F. R. and Neame, R. L. B. (1972). Characteristics of fast feedback control of corticotrophin release by corticosteroids. *J. Endocr.* **55,** 487–489.

Jones, M. T., Tiptaft, E. M., Brush, F. R., Fergusson, D. A. N. and Neame, R. L. B. (1974). Evidence for dual corticosteroid-receptor mechanisms in the feedback control of adrenocorticotrophin secretion. *J. Endocr.* **60,** 223–233.

Jones, M. T., Hillhouse, E. W. and Burden, J. L. (1976a) Effects of various putative neurotransmitters on the secretion of corticotrophin-releasing hormone from the rat hypothalamus *in vitro.* 1. A model of the neurotransmitters involved. *J. Endocr.* **69,** 1–10.

Jones, M. T., Hillhouse, E. W. and Burden, J. L. (1976b). Secretion of corticotrophin releasing factor *in vitro. In* "Frontiers of Neuroendocrinology" (W. F. Ganong and L. Martini, eds) pp. 195–226. Raven Press, New York.

Jones, M. T., Hillhouse, E. W. and Burden, J. L. (1976c). The mechanism of fast and delayed corticosteroid feedback at the hypothalamus. *J. Endocr.* 34P.

Jones, W. C. Jr., Nestor, J. J. Jr. and DuVigneaud, V. (1973). Synthesis and some pharmacological properties of (1-deamino, 9-thioglycine) oxytocin. *J. Am. Chem. Soc.* **95,** 5677–5679.

Jørgensen, C. B. (1970). Hypothalamic control of hypophyseal function in anurans. *In* "The Hypothalamus" (L. Martini, M. Motta and F. Fraschini, eds) pp. 649–661. Academic Press, New York and London.

Jørgensen, C. B. and Larsen, O. (1960). Effect of neurohypophyseal principles on adenohypophyseal activity in toads. *Proc. Soc. exp. Biol. Med.* **103,** 685–688.

Jørgensen, C. B. and Nielsen, L. (1958). Effect of synthetic lysine-vasopressin on adenohypophyseal activity in toads. *Proc. Soc. exp. Biol. Med.* **98,** 393–395.

Jørgensen, C. B. and Vijayakumar, S. (1971). Is the toad pars intermedia under both excitatory and inhibitory nervous control? *Gen. Comp. Endocrinol.* **17,** 575–577.

Joseph, S. A. and Knigge, K. M. (1972). Neural regulation of TSH secretion: thyroxine feedback in the newborn. *Neuroendocrinology*, **10,** 197–206.

Joseph, S. A., Scott, D. E., Vaala, S. S., Knigge, K. M. and Krobisch-Dudley, G. (1973). Localization and content of thyrotropin releasing factor (TRF) in median eminence of the hypothalamus. *Acta. endocr., Copenh.* **74,** 215–225.

Jost, A. (1966). Problems of foetal endocrinology: the adrenal glands. *Recent Prog. Horm. Res.* **22,** 541–574.

Jost, A. (1973). Does the foetal hypophyseal-adrenal system participate in delivery in rats and in rabbits? *In* "Foetal and Neonatal Physiology" (K. S. Comline, K. W. Cross, G. S. Dawes and P. W. Nathanielsz, eds) pp. 589–593. Cambridge University Press.

Jost, A., Dupouy, J. P. and Geloso-Meyer, A. (1970). Hypothalamo-hypophyseal relationships in the foetus. *In* "The Hypothalamus" (L. Martini, M. Motta and F. Fraschini, eds) pp. 605–615. Academic Press, New York and London.

Jost, A., Dopouy, J. P. and Rieutort, M. (1974). The ontogenetic development of hypothalamo-hypophyseal relations. *Progr. Brain. Res.* **41,** 209–219.

K

Kajihara, A. and Kendall, J. W. (1969). Studies on the hypothalamic control of TSH secretion. *Neuroendocrinology*, **5,** 53–63.

Kalra, P. S. and McCann, S. M. (1973). Involvement of catecholamines in feedback mechanisms. *Progr. Brain Res.* **39,** 185–187.

Kamber, B. and Rittel, W. (1968). Eine neue, einfache Methods zur Synthese von Cystinpeptiden. *Helv. Chim. Acta*, **51,** 2061–2064.

Kamberi, I. A. (1973). The role of brain monoamines and pineal indoles in the secretion of gonadotrophins and gonadotrophin-releasing factors. *Progr. Brain Res.* **39,** 261–280.

Kamberi, I. A., Mical, R. S. and Porter, J. C. (1970). Prolactin-inhibiting activity in hypophysial stalk blood and elevation by dopamine. *Experientia*, **26,** 1150–1151.

Kamberi, I. A., Mical, R. S. and Porter, J. C. (1971a). Effect of anterior pituitary perfusion and intraventricular injection of catecholamines on prolactin release. *Endocrinology*, **88,** 1012–1020.

Kamberi, I. A., Mical, R. S. and Porter, J. C. (1971b). Effects of melatonin and serotonin on the release of FSH and prolactin. *Endocrinology*, **88,** 1288–1293.

Kamberi, I. A., Mical, R. S. and Porter, J. C. (1971c). Pituitary portal vessel infusion of hypothalamic extract and release of LH, FSH and Prolactin. *Endocrinology*, **88,** 1294–1299.

Kamm, O., Aldrich, T. B., Grote, I. W., Rowe, L. W. and Bugbee, E. P. (1928). The active principles of the posterior lobe of the pituitary gland. *J. Am. Chem. Soc.* **50,** 573–601.

Kaneko, T., Oka, H., Munemura, M., Suzuki, S., Yasuda, H. and Oda, T. (1974). Stimulation of guanosine 3′5′ cyclic monophosphate accumulation in rat anterior pituitary gland *in vitro* by synthetic somatostatin. *Biochem. Biophys. Res. Commun.* **61,** 53–57.

Kann, G. and Denamur, R. (1974). Possible role of prolactin during the oestrous cycle and gestation in the ewe. *J. Reprod. Fertil.* **39,** 473–483.

Kansal, P. C., Talbert, O. R. and Buse, M. G. (1971). Effect of L-dopa on serum thyroxine and growth hormone. *Program 53rd Meeting U.S. Endocrine Soc.*, p. 211.

Kao, L. W. L. and Weisz, J. (1975). Direct effects of testosterone and its 5α-reduced metabolites on pituitary LH and FSH release *in vitro*: changes in pituitary responsiveness to hypothalamic extract. *Endocrinology*, **96,** 253–260.

Kapen, S., Boyer, R. M. Finkelstein, J. W., Hellman, L. and Weitzmann, E. D. (1974). Effect of sleep-wake cycle reversal on luteinizing hormone secretory pattern in puberty. *J. Clin. Endocrinol. Metab.* **39,** 293–299.

Kaplan, N. M. (1963). Assessment of pituitary ACTH secretory capacity with metryapone: II. Comparison with other tests. *J. Clin. Endocrinol. Metab.* **23,** 953–969.

Kaplan, S. L., Grumbach, M. M., Friesen, H. G. and Costom, B. H. (1972a). Thyrotropin-releasing factor (TRF) effect on secretion of human pituitary prolactin and thyrotropin in children and in idiopathic hypopituitary dwarfism: further evidence for hypophysiotropic hormone deficiencies. *J. Clin. Endocrinol. Metab.* **35,** 825–830.

Kaplan, S. L., Grumbach, M. M. and Shepard, T. H. (1972b). The ontogenesis of human fetal hormones. I. Growth hormone and insulin. *J. Clin. Invest.* **51,** 3080–3093.

Kapłanski, J., Dorst, T. and Smelik, P. G. (1972). Pituitary-adrenal activity and depletion of brain catecholamines after α-methyl-p-tyrosine administration. *Eur. J. Pharmacol.* **20,** 238–240.

Kappeler, H. and Schwyzer, R. (1960). Synthetic peptides related to corticotrophin (ACTH) and the melanophore stimulating hormones (MSH) possessing corticotrophin releasing factor (CRF activity). *Experientia*, **16,** 415–417.

Karavolas, H. J., Hodges, D. and O'Brien, D. (1976). Uptake of [^{3}H] progesterone and [^{3}H] 5-dihydroprogesterone by rat tissue *in vivo* and analysis of accumulated radioactivity: accumulation of 5α-dihydroprogesterone by pituitary and hypothalamic tissues. *Endocrinology*, **98,** 164–175.

Kardon, F. C., Winokur, A. and Utiger, R. D. (1977). Thyrotropin-releasing hormone (TRH) in rat spinal cord. *Brain. Res.* **122,** 578–581.

Karg, H. and Schams, D. (1974). Prolactin release in cattle. *J. Reprod. Fertil.* **39,** 463–472.

Karkun, J. N. and Sen, D. P. (1965). Effect of pregnancy and estrogen treatment on the melanocyte stimulating hormone (MSH) content of the rat pituitary. *India J. Med. Res.* **53,** 226–231.

Karsch, F. J. and Foster, D. L. (1975). Sexual differentiation of the mechanism controlling the preovulatory discharge of luteinizing hormone in sheep. *Endocrinology*, **97,** 373–379.

Karsch, F. J., Weick, R. F., Hotchkiss, J., Dierschke, D. J. and Knobil, E. (1973a). An analysis of the negative feedback control of gonadotrophin secretion utilizing chronic implantation of ovarian steroids in ovariectomized rhesus monkeys. *Endocrinology*, **93,** 478–486.

Karsch, F. J., Weick, R. F., Butler, W. R., Dierschke, D. J., Krey, L. C., Weiss, G., Hotchkiss, J., Yamaji, T. and Knobil, E. (1973b). Induced LH surges in the rhesus monkey: strength-duration characteristics of the estrogen stimulus. *Endocrinology*, **92,** 1740–1747.

Kastin, A. J. and Schally, A. V. (1966). MSH activity in pituitaries of rats treated with hypothalamic extracts. *Gen. Comp. Endocrinol.* **7,** 452–456.

Kastin, A. J. and Schally, A. V. (1967). Auto-regulation of release of melanocyte-stimulating hormone from the rat pituitary. *Nature, Lond.* **213,** 1238–1280.

Kastin, A. J., Schally, A. V., Viosca, S., Barret, L. and Redding, T. W. (1967). MSH activity in the pituitary of rats exposed to constant illumination. *Neuroendocrinology*, **2,** 257–262.

Kastin, A. J., Schally, A. V., Gual, C., Midgley, A. R. Jr., Bowers, C. Y. and Diaz-Infante, A. Jr. (1969a). Stimulation of LH-release in men and women by LH-releasing hormone purified from porcine hypothalami. *J. Clin. Endocrinol. Metab.* **29,** 1046–1050.

Kastin, A. J., Schally, A. V., Viosca, S. and Miller, M. C. (1969b). MSH activity in plasma and pituitaries of rats after various treatments. *Endocrinology*, **84,** 20–27.
Kastin, A. J., Schally, A. V. and Viosca, S. (1971). Inhibition of MSH release in frogs by direct application of L-prolyl-L-leucylglycinamide to the pituitary. *Proc. Soc. exp. Biol. Med*. **137,** 1437–1439.
Kastin, A. J., Ehrensing, R. H., Schalch, D. S. and Anderson, M. S. (1972). Improvement in mental depression with decreased thyrotropin response after administration of thyrotropin-releasing hormone. *Lancet*, **ii,** 740–742.
Kastin, A. J., Nissen, C., Redding, T. W., Nair, R. M. G. and Schally, A. V. (1974a). Delayed disappearance of ^{14}C-labelled Pro-Leu-Gly-NH_2 from the blood of hypophysectomized rats. *Neuroendocrinology*, **16,** 36–42.
Kastin, A. J., Viosca, S. and Schally, A. V. (1974b). Regulation of melanocyte-stimulating hormone release. *In* "Handbook of Physiology" (E. Knobil and W. H. Sawyer, eds) Vol. IV, pp. 551–562. The American Physiological Society, Washington, D.C.
Kastin, A. J., Plotnikoff, N. P., Hall, R. and Schally, A. V. (1975). Hypothalamic hormones and the central nervous system. *In* "Hypothalamic Hormones" (M. Motta, P. Crosignani and L. Martini, eds) pp. 261–268. Academic Press, London and New York.
Katangole, C. B., Naftolin, F. and Short, R. V. (1971). Relationship between blood levels of LH and testosterone in bulls and the effect of sexual stimulation. *J. Endocr*. **60,** 101–106.
Kato, J. (1973). Localization of oestradiol receptors in the rat hypothalamus. *Acta endocr., Copenh*. **72,** 663–670.
Kato, Y., Dupre, J. and Beck, J. C. (1973). Plasma growth hormone in the anesthetized rats: effects of dibutyryl cyclic AMP, prostaglandin E_1, adrenergic agents, vasopressin, chlorpromazine, amphetamine and L-dopa. *Endocrinology*, **93,** 135–146.
Kawakami, M. and Sakuma, Y. (1974). Responses of hypothalamic neurons to the microiontophonesis of LH-RH, LH and FSH under various levels of circulating hormones. *Neuroendocrinology*, **15,** 290–307.
Kawakami, M., Tersawa, E. and Ibuki, T. (1970). Changes in multiunit activity of the brain during the oestrous cycle. *Neuroendocrinology*, **6,** 30–48.
Kawakami, M. and Terasawa, E. (1974). Role of limbic forebrain structures on reproductive cycles. *In* "Biological Rhythms in Neuroendocrine Activity" (M. Kawakami, ed.) pp. 197–219. Igaku Shoin, Tokyo.
Kawakami, M., Negoro, H., Kimura, F., Higuchi, T. and Asai, T. (1975). Neural control of LH release in anterior periventriculo-median eminence-pituitary system. *Neuroendocrinology*, **19,** 137–149.
Keller, H. H., Bartholini, S. and Pletscher, A. (1974). Enhancement of cerebral noradrenaline turnover by thyrotropin-releasing hormone. *Nature, Lond*. **248,** 528–529.
Keller, P. J. (1973). Treatment of anovulation with synthetic luteinizing hormone-releasing hormone. *Am. J. Obstet. Gynecol*. **116,** 698–705.
Kendall, J. W., Rees, L. H. and Kramer, R. (1971). Thyrotropin releasing hormone (TRH) stimulation of thyroidal radioiodine release in the rat, comparison between intravenous and intraventicular administration. *Endocrinology*, **88,** 1503–1506.
Kerdelhué, B., Jutisz, M., Gillesen, D. and Studer, R. O. (1973). Obtention of antisera against a hypothalamic decapeptide (luteinizing hormone/follicle stimulating hormone releasing hormone) which stimulates the release of pituitary gonadotrophins and development of its radioimmunoassay. *Biochem. Biophys. Acta*, **297,** 540–548.
Kikuyama, S., Nagasawa, H., Yanai, R. and Yamanouchi, K. (1974). Effect of perinatal hypothyroidism on pituitary secretion of growth hormone and prolactin in rats. *J. Endocr*. **62,** 213–223.
Kingsley, T. R. and Bogdanove, E. M. (1973). Direct feedback of androgens: localized effects of intrapituitary implants of androgens on gonadotrophic cells and hormone stores. *Endocrinology*, **93,** 1398–1409.

Kirkham, K. E. and Hunter, W. M. (1971). "Radioimmunoassay Methods". Churchill Livingstone, Edinburgh and London.

Kitay, J. I., Holub, N. and Jailer, J. W. (1959). Inhibition of pituitary ACTH release: an extra-adrenal action of exogenous ACTH. *Endocrinology*, **64**, 475–482.

Kizer, J. S., Arimura, A., Schally, A. V. and Brownstein, M. J. (1975). Absence of luteinizing hormone-releasing hormone (LH-RH) from catecholaminergic neurones. *Endocrinology*, **96**, 523–525.

Kluh, J., Sedlakova, E., Barth, T. and Cort, J. H. (1973). Neurophysin binding of vasopressin analogues altered at the NH_2- and COOH-terminal position. *Mol. Pharmacol.* **9**, 414–420.

Knigge, K. M. (1961). Adrenocortical response to immobilisation in rats with lesions in hippocampus and amygdala. *Fed. Proc. Fed. Am. Soc. Exp. Biol.* **20**, 185–188.

Knigge, K. M. and Hays, M. (1963). Evidence of inhibitive role of hippocampus in neural regulation of ACTH release. *Proc. Soc. exp. Biol. Med.* **114**, 67–69.

Knigge, K. M. and Joseph, S. A. (1971). Neural regulation of TSH secretion: sites of thyroxine feedback. *Neuroendocrinology*, **8**, 273–288.

Knigge, K. M. and Joseph, S. A. (1974). Thyrotropin releasing factor (TRF) in cerebrospinal fluid of the 3rd ventricle of rat. *Acta endocr., Copenh.* **76**, 209–213.

Knigge, K. M. and Scott, D. E. (1970). Structure and function of the median eminence. *Amer. J. Anat.* **129**, 223–244.

Knigge, K. M., Scott, D. E. and Weindl, A. (eds). (1972). "Brain-Endocrine Interaction Median Eminence Structure and Function". Karger, Basel.

Knigge, K. M., Joseph, S. A., Silverman, A. J. and Vaula, S. (1973). Further observations on the stimuli and function of median eminence, with reference to the organization of the RF-producing elements in the endocrine hypothalamus. *Progr. Brain Res.* **39**, 7–19.

Knobil, E. (1974). On the control of gonadotrophin secretion in the rhesus monkey. *Recent. Prog. Horm. Res.* **30**, 1–46.

Knowles, F. (1972). Ependyma of the third ventricle in relation to pituitary function. *Progr. Brain Res.* **38**, 255–270.

Kobayashi, T. (1964). Functional morphology of the pars intermedia of the rat hypophysis as revealed with the electron microscope. *Gunma Symp. Endocrinol.* **1**, 173–181.

Kobayashi, F. (1974). Steroidal regulation of hypothalamo-pituitary function in relation to ovulating hormone release in rats. *In* "Psychoneuroendocrinology" (N. Hatotami, ed.) pp. 122–127. Karger, Basel.

Kobayashi, H. and Matsui, T. (1969). Fine structure of the median eminence and its functional significance. *In* "Frontiers in Neuroendocrinology, 1969" (W. F. Ganong and L. Martini, eds) pp. 3–46. Oxford University Press, New York.

Kobayashi, H., Wada, M. and Uemura, H. (1972). Uptake of peroxidase from the third ventricle by ependymal cells of the median eminence. *Z. Zellforsch. mikr. Anat.* **127**, 545–551.

Koch, Y., Chobsieng, P., Zor, U., Fridkin, M. and Lindner, H. R. (1973a). Suppression of gonadotrophin secretion and prevention of ovulation in the rat by antiserum to synthetic gonadotrophin-releasing hormone. *Biochem. Bipohys. Res. Commun.* **55**, 623–629.

Koch, Y., Wilchek, M., Fridkin, M., Chobsieng, P., Zor, U. and Lindner, H. R. (1973b). Production and characterisation of an antiserum to synthetic gonadotrophin releasing hormone. *Biochem. Biophys. Res. Commun.* **55**, 616–622.

Koch, Y., Baram, T., Chobsieng, P. and Fridkin, M. (1974). Enzymatic degradation of luteinizing hormone-releasing hormone (LH-RH) by hypothalamic tissue. *Biochem. Biophys. Res. Commun.* **61**, 95–103.

Kochman, K., Kerdelhue, B., Zor, U. and Jutisz, M. (1975). Studies of enzymatic degradation of luteinizing hormone-releasing hormone by different tissues. *FEBS Letters*, **50**, 190–194.

Koerker, D. J., Harker, L. A. and Goodner, C. J. (1975). Effects of somatostatin on hemostasis in baboons. *N. Engl. J. Med.* **293,** 476–479.

Koikegami, H. (1974). Anatomical considerations of modulatory effects of the limbic and paralimbic structures upon the hypothalamic hormonal or autonomic function. *In* "Biological Rhythms in Neuroendocrine Activity" (M. Kawakami, ed.) pp. 1–38. Igaku Shoin, Tokyo.

König, W. and Geiger, R. (1970). Eine neue Methoda zur Synthese von Peptiden: Aktivierung der Carboxylgruppe mit Dicyclohexylcarbodiimid und 3-Hydroxy-4-oxo-3.4-dihydro-1.2.3-benzotriazin. *Chem. Ber.* **103**, 2034–2040.

König, W. and Geiger, R. (1974). Synthesis of the gonadotrophin-releasing hormone. *In* "Hypothalamic Hormone—Structure, Synthesis and Biological Activity" (D. Gupta and W. Voelter, eds) pp. 93–106. Verlag Chemie, Weinheim.

König, W., Sandow, J. and Geiger, R. (1975). Structure-function relationships of LH-RH/FSH-RH). *In* "Peptides: Chemistry, Structure and Biology" (R. Walter and J. Meienhofer, eds) pp. 883–888. Ann Arbor Science, New York.

Kordon, C., Javoy, F., Vassent, G. and Glowinski, J. (1968). Blockade of superovulation in the immature rat by increased brain serotonin. *Europ. J. Pharmacol.* **4,** 169–174.

Kordon, C., Blake, C. A., Terkel, J. and Sawyer, C. H. (1973/1974). Participation of serotonin-containing neurons in the suckling-induced rise in plasma prolactin levels in lactating rats. *Neuroendocrinology,* **13,** 213–223.

Kourides, I. A., Weintraub, B. D., Levko, M. A. and Maloof, F. (1973). TSH in Graves' disease. *Clin. Res.* **21,** 496.

Kourides, I. A., Weintraub, B. D., Ridgway, E. C. and Maloof, F. (1974). Increase in the beta subunit of human TSH in hypothyroid serum for thyrotropin releasing hormone. *J. Clin. Endocrinol. Metab.* **37,** 836–840.

Kovacs, S., Sandor, A., Vertes, Z. and Vertes, M. (1965). The effect of lesions and stimulation of the amygdala on pituitary thyroid function. *Acta Physiol. Acad. Sci. Hung.* **27,** 221–227.

Köves, K. and Halász, B. (1970). Localization of the neural structures triggering ovulation in the rat. *Neuroendocrinology,* **6,** 180–193.

Kragt, C. L. and Meites, J. (1965). Stimulation of pigeon pituitary prolactin release by pigeon hypothalamic extract *in vitro*. *Endocrinology,* **76,** 1169–1176.

Kraicer, J., Milligan, J. V., Gosbee, J. L., Conrad, R. D. and Branson, C. M. (1969). *In vitro* release of ACTH: effects of potassium, calcium and corticosterone. *Endocrinology,* **85,** 1144–1153.

Krause, R. (1975). The measurement of hypothalamic releasing hormone content and activity in the unanaesthetised rat by means of carotid artery cannulation. *Acta endocr., Copenh.* **78,** 417–427.

Krejči, I., Barth, T., Kupkovy, B., Fuhaufova, L., Flegel, M. and Zaoral, M. (1973). Pharmacology of oxytocin-(1–6)-hexapeptide amide, vasopressin-(1–6)-hexapeptide amide and their deamino analogues. *Eur. J. Pharmacol.* **24,** 179–182.

Krey, L. C., Tyrey, L. and Everett, J. W. (1973a). The estrogen-induced advance in the cyclic LH surge in the rat: dependency on ovarian progesterone secretion. *Endocrinology,* **93,** 385–390.

Krey, L. G., Butler, W. R., Weiss, G., Weick, R. F., Dierschke, D. J. and Knobil, E. (1973b). Influences of endogenous and exogenous gonadal steroids on the actions of synthetic LRF in the rhesus monkey. *In* "Hypothalamic Hypophysiotrophic Hormones (C. Gual and E. Rosemberg, eds) pp. 39–47. Excerpta Medica, Amsterdam.

Krey, L. C., Butler, W. R. and Knobil, E. (1975a). The role of the medial basal hypothalamus in the control of pituitary function in the Rhesus monkey. *In* "Hypothalamic Hormones" (M. Motta, P. G. Crosignani and L. Martini, eds) pp. 255–260. Academic Press, New York and London.

Krey, L. C., Butler, W. R. and Knobil, E. (1975b). Surgical disconnection of the medial basal hypothalamus and pituitary function in the Rhesus monkey. I. Gonadotropin secretion. *Endocrinology*, **96,** 1073–1087.

Krieger, D. T. (1973). Effect of ocular enucleation and altered lighting regimes at various ages on the circadian periodicity of plasma corticosteroid levels in the rat. *Endocrinology*, **93,** 1077–1091.

Krieger, D. T. (1974). Factors influencing the circadian periodicity of plasma corticoid levels. *Chronobiologia*, **1,** 195–216.

Krieger, D. T. and Glick, S. M. (1974). Sleep EEG stages and plasma growth hormone concentration in states of endogenous and exogenous hypercortisolemia or ACTH elevation. *J. Clin. Endocrinol. Metab.* **39,** 986–1000.

Krieger, H. P. L. and Krieger, D. T. (1970). Chemical stimulation of the dopamine of 5–6 di-hydroxytryptamine administration on the circadian periodicity of plasma corticosterone levels in the rat. *Neuroendocrinology*, **17,** 62–74.

Krieger, D. T. and Rizzo, F. (1969). Serotonin mediation of circadian periodicity of plasma 17-hydroxycorticosteroids. *Am. J. Physiol.* **217,** 1703–1707.

Krieger, D. T., Amorosa, L. and Linick, F. (1975). Cyproheptadine-induced remission of Cushing's disease. *New Engl. J. Med.* **293,** 893–896.

Krieger, D. T., Silverberg, A. I., Rizzo, F. and Krieger, H. P. L. (1968). Abolition of the circadian periodicity of plasma 17-OH CS levels in the cat. *Am. J. Physiol.* **215,** 959–967.

Krieger, D. T., Allen, W., Rizzo, F. and Krieger, H. P. L. (1971). Characterisation of the normal temporal pattern of plasma corticosteroid levels. *J. Clin. Endocrinol. Metab.* **32,** 266–284.

Krieger, D. T., Ossowski, R., Fogel, M. and Allen, W. (1972). Lack of circadian periodicity of human serum FSH and LH levels. *J. Clin. Endocrinol. Metab.* **35,** 619–623.

Kronheim, S., Berelowitz, M. and Pimstone, B. L. (1977). The presence of immunoreactive growth hormone release-inhibiting hormone in normal cerebrospinal fluid. *Clin. Endocrinol.* **6,** 411–420.

Krulich, L., Dhariwal, A. P. S. and McCann, S. M. (1968). Stimulatory and inhibitory effects of purified hypothalamic extracts on growth hormone release from rat pituitary *in vitro*. *Endocrinology*, **83,** 783–790.

Krulich, L., Illner, P., Fawcett, C. P., Quijada, M. and McCann, S. M. (1972). Dual hypothalamic regulation of growth hormone secretion. *In* "Growth and Growth Hormone" (A. Pecile and E. E. Miller, eds) pp. 306–316. Excerpta Medica, Amsterdam.

Kuhl, H. and Taubert, H. D. (1975a). Inactivation of luteinizing hormone-releasing hormone by rat hypothalamic L-cystine acrylamidase. *Acta endocr., Copenh.* **78,** 634–648.

Kuhl, H. and Taubert, H. D. (1975b). Short-loop feedback mechanism of luteinizing hormone: LH stimulated hypothalamic L-cystine acrylamidase to inactivate LH-RH in the rat hypothalamus. *Acta endocr., Copenh.* **78,** 649–663.

Kuhl, H., Rosniatowski, C., Bickel, M. and Taubert, H. D. (1974). Stimulation by steroids and age-dependent changes of L-cystine acrylamidase activity in the hypothalamus of the rat. *Acta endocr., Copenh.* **76,** 15–23.

Kuhl, H., Sachs, A., Rosniatowski, C. and Taubert, H. D. (1975). The role of pituitary L-cystine acrylamidase in the feedback control of luteinizing hormone. *Acta endocr., Copenh.* **199,** 195.

Kuhlenbeck, H. (1954). "The Human Diencephalon: A Summary of Development, Structure, Function and Pathology". Karger, Basel.

Kulin, H. E. and Reiter, E. O. (1974). Delayed sexual maturation with special emphasis on the occurrence of the syndrome in the male. *In* "The Control of the Onset of Puberty" (M. M. Grumbach, G. D. Grave and E. E. Mayer, eds) pp. 238–270. Wiley, New York and London.

Kuromizu, K. and Meienhofer, J. (1974). Removal of the N^{α}-benzyloxycarbonyl group from cysteine containing peptides by catalytic hydrogenolysis in liquid ammonia, exemplified by a synthesis of oxytocin. *J. Am. Chem. Soc.* **96,** 4978–4981.

Kuroshima, A., Arimura, A., Bowers, C. Y. and Schally, A. V. (1966). Inhibition of pig hypothalamus extracts of depletion of pituitary prolactin in rats following cervical stimulation. *Endocrinology*, **78,** 216–217.

L

Labhart, A. (1974). "Clinical Endocrinology: Theory and Practice." Springer-Verlag, Berlin.

Labhsetwar, A. P. (1971). Effects of serotonin on spontaneous ovulation: a theory for the dual hypothalamic control of ovulation. *Acta endocr., Copenh.* **68,** 334–344.

Labhsetwar, A. P. (1974). Prostaglandins and catecholamines in gonadotrophin secretion. *Prostaglandins*, **8,** 79–84.

Laborit, H. (1972). Correlations between protein and serotonin synthesis during various activities of the central nervous system. *Res. Commun. Chem. Pathol. Pharmac.* **3,** 51–81.

Labrie, F., Barden, N., Poirier, G., and DeLear, A. (1972). Binding of thyrotropin releasing hormone to plasma membranes of bovine anterior pituitary gland. *Proc. Natl. Acad. Sci. U.S.A.* **69,** 283–287.

Labrie, F., Borgeat, P. and Poirier, G. (1973a). Stimulation of adenosine 3′,5′-monophosphate accumulation in anterior pituitary gland *in vitro* by purified growth hormone-releasing hormone. *Clin. Res.* **21,** 721.

Labrie, F., Gauthier, M., Pelletier, G., Borgeat, P., Lemay, A. and Gouge, J. J. (1973b). Role of microtubules in basal and stimulated release of growth hormone in rat adenohypophysis *in vitro*. *Endocrinology*, **93,** 903–914.

Labrie, F., Borgeat, P., Ferland, L., Lemay, A., Dupont, A., Lemaire, S., Pelletier, G., Barden, N., Drouin, J., de Lean, A., Belanger, A. and Jolicoeur, P. (1975). Mechanism of action and modulation of activity of hypothalamic hypophysiotrophic hormones. *In* "Hypothalamic Hormones" (M. Motta, P. Crosignani and L. Martini, eds) pp. 109–123. Academic Press, London and New York.

Lal, S., de la Vega, C. E., Sourkes, T. L. and Friesen, H. G. (1972). Effect of apomorphine on human-growth-hormone secretion. *Lancet*, **ii,** 661.

Lammers, H. J. and Lohman, A. H. M. (1974). Structure and fiber connections of the hypothalamus in mammals. *Progr. Brain Res.* **41,** 61–78.

Lamming, G. E., Moseley, S. R. and McNeilly, J. R. (1974). Prolactin release in the sheep. *J. Reprod. Fertil.* **40,** 151–168.

Land, R. B., Wheeler, A. G. and Carr, W. R. (1976). Seasonal variation in the oestrogen induced LH discharge of ovariectomized Finnish Landrace and Scottish Blackface ewes. *Annls. Biol. anim. Biochim. Biophys.* **16,** 521–528.

Lande, S. and Lerner, A. B. (1967). The biochemistry of melanotropic agents. *Pharmacol. Rev.* **19,** 1–20.

Landon, J. and Greenwood, F. C. (1968). Homologous radioimmunoassay for plasma levels of corticotrophin in man. *Lancet*, **i,** 273–276.

Landon, J., Wynn, V., and James, V. H. T. (1963). The adrenocortical response to insulin-induced hypoglycaemia. *J. Endocr.* **27,** 183–191.

Landon, J., Rees, L., Bloomfield, G. and Lowry, P. J. (1975). ACTH and its fragments. *Acta endocr., Copenh.* Suppl. **199,** 64.

Landsmeer, L. M. F. (1963). A survey of the analysis of hypophyseal vascularity. *In* "Advances in Neuroendocrinology" (A. V. Nalbandov, ed.) pp. 29–57. University of Illinois Press, Urbana.

Langley, L. L. (1965). "Homeostasis". Chapman and Hall, London.

LaRochelle, F. T. and Freeman, M. E. (1974). Superimposition of thyroid hormone regulation on gonadotrophin secretion. *Endocrinology*, **95,** 379–387.

Lauson, H. D. (1974). Metabolism of the neurohypophysial hormones. *In* "Handbook of Physiology" Vol. IV, Part 1, (E. Knobil and W. H. Sawyer, eds) pp. 287–393. American Physiological Society, Washington, D.C.

Lawson, D. M. and Gala, R. R. (1975). The influence of adrenergic, dopaminergic, cholinergic and serotoninergic drugs on plasma prolactin levels in ovariectomized, estrogen-treated rats. *Endocrinology*, **96,** 313–318.

Lazarus, L. and Eastman, C. J. (1976). Assessment of hypothalamic pituitary function in old age. *In* "Hypothalamus, Pituitary and Aging" (A. V. Everitt and J. A. Burgess, eds) pp. 97–122. C. C. Thomas, Springfield.

Leatham, J. H. (1958). Hormones and protein nutrition. *Recent Prog. Horm. Res.* **14,** 141–176.

Leatherland, J. F. and Ensor, D. M. (1974). Effect of hypothalamic extracts on prolactin secretion in the goldfish, *Carassius auratus* L. *Comp. Biochem. Physiol.* **47**A, 419–426.

Lederis, K. and Jarhysena, K. (1970). Storage of neurohypophysial hormones and the mechanism for their release. *In* "International Encyclopedia of Pharmacology and Therapeutics" (H. Heller and B. T. Pickering, eds) Section 41, Vol. 1, p. 111. Pergamon Press, New York.

Lee, V. W., Cumming, I. A., de Kretser, D. M., Findlay, J. K., Hudson, B. and Keogh, E. J. (1976). Regulation of gonadotrophin secretion in rams from birth to sexual maturity. I. Plasma LH, FSH and testosterone levels. *J. Reprod. Fertil.* **46,** 1–6.

Leeman, S. E. (1974). Substance P. *Life Sci.* **15,** 2033–2044.

Leeman, S. E. and Patton, N. J. (1962). Chemical inactivation of adrenocorticotropin and vasoconstrictor material in sheep hypothalamic extracts. *Biochem. Biophys. Acta*, **59,** 522–524.

Legait, H. and Legait, E. (1962). Relationship between the hypothalamus and pars intermedia in some mammals and amphibians. *In* "Neurosecretion" (H. Heller and R. B. Clark eds) pp. 165–173. Academic Press, London and New York.

Legon, S. J., Coon, G. A. and Karsch, F. J. (1975). Role of estrogen as initiator of daily LH surges in the ovariectomized rat. *Endocrinology*, **96,** 50–56.

Leonardelli, J., Barry, J. and Dubois, M. P. (1973). Mise en évidence par immunofluorescence d'un constituant immunlogiquement apparenté au LRF dans l'hypothalamus et l'éminence médiane chez les mammifères. *C. R. Acad. Sci.* **276,** 2043–2046.

Leppaluoto, J., Virkkunen, P. and Lybeck, H. (1972). Elimination of TRH in man. *J. Clin. Endocrinol. Metab.* **35,** 477–478.

Leppaluoto, J., Ling, N. and Vale, W. (1976). Purification and measurement of urinary TRF. Program V Int. Cong. Endocrinology, Hamburg, July 18–24, 1976, p. 333.

Leung, Y., Guansing, A. R., Ajlouni, K. A., and Hagen, T. C. (1974). Hypoglycemia increases TSH in the rat. *Clin. Res.* **22,** 599A.

Leveque, T. F. (1972). The medial prechiasmatic area in the rat and LH secretion. *In* "Brain-Endocrine Interaction" (K. M. Knigge, D. E. Scott and A. Weindl, eds) pp. 298–305. Karger, Basel.

Leveque, T. F. (1974). Endocrine hypothalamus: an historical review. *Can. J. Neurol. Sci.* 1, 24–28.

L'Hermite, M., Robyn, C., Golstein, J., Rothenbuchner, G., Birk, J., Loos, U., Bonnyns, M. and Vanhaelst, L. (1974). Prolactin and thyrotropin in thyroid diseases: lack of evidence for a physiological role of thyrotropin-releasing hormone in the regulation of prolactin secretion. *Horm. metab. Res.* **6,** 190–195.

Li, C. H., Chung, D., Ramachandran, J. and Gorup, B. (1961). The synthesis of a heptadecapeptide possessing adrenocorticotrophic, melanotropic and lipolytic activities. *J. Am. Chem. Soc.* **84,** 2460–2462.

Libertun, C. and McCann, S. M. (1973). Blockade of the release of gonadotropins and prolactin by subcutaneous or intraventricular injection of atropine in male and female rats. *Endocrinology*, **92,** 1714–1724.

Libertun, C. and McCann, S. M. (1974). Further evidence for cholinergic control of gonadotropin and prolactin secretion. *Proc. Soc. exp. Biol. Med.* **147,** 498–504.

Libertun, C., Orias, R. and McCann, S. M. (1974). Biphasic effect of estrogen on the sensitivity of the pituitary to luteinising hormone-releasing factor (LRF). *Endocrinology*, **94,** 1094–1100.

Licht, P. (1972). Photoperiodic and thermal influence on reproductive cycles in reptiles. *In* "Proceedings of the 4th International Congress on Endocrinology", pp. 185–190. Excerpta Medica, Amsterdam.

Licht, P. (1974). Induction of sperm release in frogs by mammalian gonadotrophin-releasing hormone. *Gen. Comp. Endocrinol.* **23,** 352–354.

Lichtensteiger, W. (1969). Cyclic variations of catecholamine content in hypothalamic nerve cells during the estrous cycle of the rat. *J. Pharmacol. Exp. Ther.* **165,** 204–215.

Lichtensteiger, W. (1970). Effects of endocrine manipulations on the metabolism of hypothalamic monoamines. *In* "Neurochemical Aspects of Hypothalamic Function" (L. Martini and J. Meites, eds) pp. 101–133. Academic Press, New York and London.

Liddle, G. W. (1967). Cushing's syndrome. *In* "The Adrenal Cortex" (A. B. Eisenstein, ed.) pp. 523–552. Little, Brown, Boston.

Liddle, G. W., Island, D. and Meander, C. K. (1961). Normal and abnormal regulation of corticotropin secretion in man. *Recent. Prog. Horm. Res.* **18,** 125–166.

Liggins, G. C. (1968). Premature parturition after infusion of corticotrophin or cortisol into foetal lambs. *J. Endocr.* **42,** 323–329.

Liggins, G. C., Fairclough, R. J., Grieves, S. A., Kendall, J. Z. and Knox, B. S. (1973). The mechanism of initiation of parturition in the ewe. *Recent Prog. Horm. Res.* **29,** 111–150.

Likhite, V. and Sehon, A. (1967). Conjugated and synthetic antigens. *In* "Methods in Immunology and Immunochemistry" (C. A. Williams and M. W. Chase, eds) pp. 120–198. Academic Press, London and New York.

Lincoln, D. W. (1967). Unit activity in the hypothalamus, septum and preoptic area of the rat: characteristics of spontaneous activity and the effect of oestrogen. *J. Endocr.* **37,** 177–189.

Lincoln, D. W. (1969). Correlation of unit activity in the hypothalamus with EEG patterns associated with the sleep cycle. *Expl. Neurol.* **24,** 1–18.

Lincoln, D. W. (1971). Labour in the rabbit: effect of electrical stimulation applied to the infundibulum and median eminence. *J. Endocr.* **50,** 607–618.

Lincoln, D. W. (1974). Maternal hypothalamic control of labor. *Prog. Brain Res.* **41,** 289–306.

Lincoln, D. W. and Wakerley, J. B. (1975). Factors governing the periodic activation of supraoptic and paraventricular neurosecretory cells during suckling in the rat. *J. Physiol., Lond.* **250,** 443–461.

Lincoln, D. W., van der Schoot, P. and Zeilmaker, G. H. (1972). EEG arousal, sympathetic activation and the induction of ovulation by electrochemical stimulation of the rat hypothalamus. *J. Endocr.* **57,** xix–xx.

Lincoln, D. W., Hill, A. and Wakerley, J. B. (1973). The milk-ejection reflex of the rat: an intermittent function not abolished by surgical levels of anaesthesia. *J. Endocr.* **57,** 459–476.

Lincoln, G. A. (1971). The seasonal reproductive changes in the red deer stag (*Cervus elaphus*). *J. Zool.* (*London*), **163,** 105–123.

Lincoln, G. A. (1976). Seasonal variation in the episodic secretion of LH and testosterone in the ram. *J. Endocr.* **69,** 213–226.

Ling, N. and Vale, W. (1975). Analogues of luteinizing hormone-releasing factor (LRF): Synthesis and biological activity of (Nα-Me) Leu-LRF and (D-Ala6, (Nα-Me) Leu7)LRF. *Biochem. Biophys. Res. Commun.* **63,** 801–806.

Ling, N., Burgus, R., Rivier, J., Vale, W. and Brazeau, P. (1973). The use of mass spectrometry in deducing the sequence of somatostatin—a hypothalamic peptide that inhibits the secretion of growth hormone. *Biochem. Biophys. Res. Commun.* **50,** 127–133.

Linzell, J. L. (1973). Innate seasonal oscillations in the rate of milk secretion in goats. *J. Physiol.* (*London*), **230,** 225–233.

Lippa, A. S., Antelman, S. M., Fahringer, E. E. and Redgate, E. S. (1973). Relationship between catecholamines and ACTH: effects of 6-hydroxydopamine. *Nature, Lond.* **241,** 24–25.

Lippe, B. M., Van Herle, A. J., La Franchi, S. H., Uller, R. P., Lavin, N. and Kaplan, S. A. (1975). Reversible hypothyroidism in growth hormone deficient children treated with human growth hormone. *J. Clin. Endocrinol. Metab.* **40,** 612–618.

Lippmann, W. (1975). Stimulation of cyclic AMP accumulation in the rat anterior pituitary *in vitro* by analogues of luteinising hormone-releasing hormone. *Experientia*, **31** 403–404.

Liuzzi, A., Chiodini, P. G., Botalla, L., Cremascoli, G., Müller, E. E. and Silvestrini, F. (1974). Decreased plasma growth hormone (GH) levels in acromegalics following CB 154 (2-Br-alpha-ergocryptine) administration. *J. Clin. Endocrinol. Metab.* **38,** 910–912.

Lloyd, S. J., Josimovich, J. B. and Archer, D. F. (1975). Amenorrhea and galactorrhea: results of therapy with 2-brom-α-ergocryptine (CB-154). *Am. J. Obstet. Gynec.* **122,** 85–89.

Lomax, P., Kokka, N., and George, R. (1970). Thyroid activity following intracerebral injection of morphine in the rat. *Neuroendocrinology*, **6,** 146–152.

London, D. R., Butt, W. R., Lynch, S. S., Marshall, J. C., Owusu, S., Robinson, W. R., and Stephenson, J. M. (1973). Hormonal responses to intranasal luteinizing-hormone releasing hormone. *J. Clin. Endocrinol. Metab.* **37,** 829–831.

Long, J. A. and Evans, H. M. (1922). "The Oestrous Cycle in the Rat and its Associated Phenomena". University of California Press, Berkeley.

Lovinger, R., Boryczka, A. T., Shackelford, R., Kaplan, S. L., Ganong, W. F. and Grumbach, M. M. (1974). Effect of synthetic somatotropin release inhibiting factor on the increase in plasma growth hormone elicited by L-Dopa in the dog. *Endocrinology*, **95,** 943–946.

Lowry, P. J. (1974). A sensitive method for the detection of corticotrophin releasing factor using a perfused pituitary cell column. *J. Endocr.* **62,** 163–164.

Lu, K.-H. and Meites, J. (1972). Effects of L-dopa on serum prolactin and PIF in intact and hypophysectomized, pituitary-grafted rats. *Endocrinology*, **91,** 868–872.

Lu, K.-H. and Meites, J. (1973). Effects of serotonin precursors and melatonin on serum prolactin release in rats. *Endocrinology*, **93,** 152–155.

Lugaro, G., Casellato, M. M., Mazzola, G., Fachini, G. and Carrea, G. (1974). Evidence for the existence in spermatozoa for a factor inhibiting the follicle stimulating hormone releasing hormone synthesis. *Neuroendocrinology*, **15,** 62–68.

Lupulescu, A., Nicolescu, A., Gheorghiescu, B., Merculiev, E. and Lungu, M. (1962). Neural control of the thyroid gland: studies on the role of extrapyramidal and rhinencephalon areas in the development of the goiter. *Endocrinology*, **70,** 517–524.

M

McArthur, J. W. (1972). Gonadotropins in relation to sexual maturity. *In* "Gonadotropins" (B. B. Saxena, C. G. Beling and H. M. Gandy, eds) pp. 487–501. Wiley-Interscience, New York and London.

McCann, S. M. (1974). Regulation of follicle stimulating hormone and luteinizing hormone. *In* "Handbook of Physiology, Endocrinology" (E. Knobil and W. H. Sawyer, eds) Vol. IV Part 2, pp. 489–517. American Physiological Society, Washington, D.C.

McCann, S. M. and Porter, J. C. (1969). Hypothalamic pituitary stimulating and inhibiting hormones. *Physiol. Rev.* **49,** 240–284.

McCann, S. M., Taleisnik, S. and Friedman, H. M. (1960). LH-releasing activity in hypothalamic extracts. *Proc. Soc. exp. Biol. Med.* **104,** 432–434.

McCann, S. M., Antunes-Rodrigues, J., Nallar, R. and Valtin, M. (1966). Pituitary-adrenal function in the absence of vasopressin. *Endocrinology*, **79,** 1058–1064.

McClintock, M. K. (1971). Menstrual synchrony and suppression. *Nature, Lond.* **229,** 244–245.

McCullagh, D. R. (1932). Dual endocrine activity of testis. *Science N.Y.* **76,** 19.

McDonald, R. K., Wagner, H. N. and Weise, V. K. (1957). Relationship between endogenous antidiuretic hormone activity and ACTH release in man. *Proc. Soc. exp. Biol. Med.* **96,** 566–567.

McEwen, B. S. (1973). Glucocorticoid binding sites in rat brain, subcellular and anatomical localizations. *Progr. Brain. Res.* **39,** 87–98.

McEwen, B. S., Weiss, J. M. and Schwartz, L. S. (1969). Uptake of corticosterone by rat brain and its concentration in certain limbic structures. *Brain Res.* **16,** 227–241.

McEwen, B. S., Weiss, J. M. and Schwartz, L. S. (1970). Retention of corticosterone by cell nuclei from brain regions of adrenalectomized rats. *Brain Res.* **17,** 471–482.

McEwen, B. S., Zigmond, E. R. and Gerlach, J. C. (1972). Sites of steroid-binding and action in the brain. *In* "Structure and Function of Neural Tissues" (G. H. Bourne, ed.) Vol. 5, pp. 205–291. Academic Press, New York and London.

Machlin, L. J., Jacobs, L. S., Cirulis, N., Kimes, R. and Miller, R. (1974). As assay for growth hormone and prolactin-releasing activities using a bovine pituitary cell culture system. *Endocrinology*, **95,** 1350–1353.

McIntyre, H. B. and Odell, W. D. (1974). Physiological control of growth hormone in the rabbit. *Neuroendocrinology*, **16,** 8–21.

McKelvy, J. F. (1974). Biochemical Neuroendocrinology. I. Biosynthesis of thyrotropin releasing hormone (TRH) by organ cultures of mammalian hypothalamus. *Brain Res.* **65,** 489–502.

McKelvy, J. F. (1975). A rapid and quantitative procedure for the purification of picomole quantities of salt free synthetic (^{3}H-Pro) thyrotrophin releasing hormone. *Anal. Biochem.* **64,** 609–614.

McKelvy, J. F. and Grimm-Jørgensen, Y. (1975). Studies on the biosynthesis of thyrotrophin releasing hormone *in vitro*. *In* "Hypothalamic Hormones" (M. Motta, P. Crosignani and L. Martini, eds) pp. 13–25. Academic Press, London and New York.

MacLean, P. D. (1972). Cerebral evolution and emotional processes: new findings on the striatal complex. *Ann. N.Y. Acad. Sci.* **193,** 137–149.

MacLeod, R. M. (1969). Influence of norepinephrine and catecholamine-depleting agents on the synthesis and release of prolactin and growth hormone. *Endocrinology*, **85,** 916–923.

MacLeod, R. M. and Lehmeyer, J. E. (1970). Release of pituitary growth hormone by prostaglandins and dibutyryl cyclic AMP in the absence of protein synthesis. *Proc. Natl. Acad. Sci. U.S.A.* **67,** 1172–1179.

MacLeod, R. M. and Lehmeyer, J. E. (1974). Studies on the mechanism of the dopamine-mediated inhibition of prolactin secretion. *Endocrinology*, **94,** 1077–1085.

McNatty, K. P., Sawers, R. S. and McNeilly, A. S. (1974). A possible role for prolactin in control of steroid secretion by the human Graafian follicle. *Nature* (*London*), **250,** 653–655.

McNeilly, A. S. (1974). Prolactin and human reproduction. *Br. J. hosp. Med.* **12,** 57–62.

McNeilly, A. S., Evans, G. E. and Chard, T. (1973). Observations on prolactin levels during the menstrual cycle. *In* "Human Prolactin" (J. L. Pasteels and C. Robyn, eds) pp. 231–232. Excerpta Medica, Amsterdam.

Maetz, J. and Lahlou, B. (1974). Actions of neurohypophysial hormones in fish. *In* "Handbook of Physiology" Section 7, Vol. IV *The Pituitary Gland and its Neuroendocrine Control* (E. Knobil and W. H. Sawyer, eds) Part I, pp. 521–544. American Physiological Society, Washington, D.C.

Makara, G. B. and Stark, E. (1974). Effect of GABA and GABA antagonist drugs on ACTH release. *Neuroendocrinology*, **16,** 178–190.

Makino, T. (1973). Study on the intracellular mechanism of LH release in the anterior pituitary. *Am. J. Obstet. Gynecol.* **115,** 606–614.

Malacara, J. M., Seyler, L. E. and Reichlin, S. (1972). Luteinizing hormone releasing factor activity in peripheral blood from women during the mid-cycle luteinizing hormone ovulatory surge. *J. Clin. Endocrinol. Metab.* **34,** 271–278.

Malkinson, A. M. (1975). "Hormone Action". Chapman and Hall, London.

Malven, P. V. (1975). Prolactin release induced by electrical stimulation of the hypothalamic preoptic area in unanesthetized sheep. *Neuroendocrinology*, **18,** 65–71.

Mann, D. R. and Barraclough, C. A. (1973). Role of estrogen and progesterone in facilitating LH release in 4-day cyclic rats. *Endocrinology*, **93,** 694–699.

Mann, D. R., Korowitz, O. D. and Barraclough, C. A. (1975). Adrenal gland involvement in synchronising the preovulatory release of LH in rats. *Proc. Soc. Exp. Biol. Med.* **150,** 115–120.

Manning, M., Coy, E. and Schally, A. V. (1972). An improved solid phase synthesis of LH-RH/FSH-RH: Progress report. *In* "Chemistry and Biology of Peptides" (J. Meienhofer, ed.) pp. 235–239. Ann Arbor Science Publishers, Michigan.

Manning, M., Blaspiri, L., Acosta, M. and Sawyer, W. H. (1973). Solid phase synthesis of (1-deamino, 4-valine)-8-D-arginine-vasopressin (DVDAVP), a highly potent and specific antidiuretic agent possessing protracted effects. *J. Med. Chem.* **16** 975–978.

Marks, B. H., Hall, M. and Bhattachara, A. N. (1970). Psychopharmacological effects and pituitary adrenal activity. *Progr. Brain Res.* **32,** 57–69.

Marks, N. and Stern, F. (1974). Enzymatic mechanisms for the inactivation of luteinizing hormone-releasing hormone (LH-RH). *Biochem. Biophys. Res. Commun.* **61,** 1458–1463.

Marks, N. and Stern, F. (1975). Inactivation of somatostatin (GH-RIH) and its analogues by crude and partially purified rat brain extracts. *FEBS Letters*, **55,** 220–224.

Marks, N. and Walter, R. (1972). MSH-release-inhibiting factor: inactivation by proteolytic enzymes. *Proc. Soc. exp. Biol. Med.* 140, 673–676.

Marshall, A. J. (1952). The interstitial cycle in relation to autumn and winter sexual behaviour in birds. *Proc. Zool. Soc.* (*London*), **121,** 727–740.

Marshall, F. H. A. (1922). "The Physiology of Reproduction". 2nd edn. Longmans, Green and Co., London.

Marshall, F. H. A. (1942). Enteroceptive factors in sexual periodicity. *Biol. Rev.* **17,** 68–90.

Marshall, F. H. A. and Jolly, W. A. (1907). Results of removal and transplantation of ovaries. *Trans. Roy. Soc. Edinburgh*, **45,** 589–599.

Marshall, F. H. A. and Verney, E. B. (1936). The occurrence of ovulation and pseudopregnancy in the rabbit as a result of central nervous stimulation. *J. Physiol., Lond.* **86,** 327–336.

Marshall, J. C. and Odell, W. D. (1975). Preparation of biologically active ^{125}I-LH-RH suitable for membrane binding studies. *Proc. Soc. Exp. Biol. Med.* **149,** 351–355.

Marshall, J. C., Harsoulis, P., Anderson, D. C., McNeilly, A. S., Besser, G. M. and Hall, R. (1972). Isolated pituitary gonadotrophin deficiency: gonadotrophin secretion after synthetic luteinizing hormone and follicle stimulating hormone releasing hormone. *Br. Med. J.* **4,** 643–645.

Martin, J. B. (1972). Immunosympathectomy as a tool in endocrinologic studies. *In* "Immunosympathectomy" (G. Steiner and E. Schonbaum, eds) pp. 177–194. Elsevier, Amsterdam.

Martin, J. B. (1973). Neural regulation of growth hormone secretion. *N. Engl. J. Med.* **288,** 1384–1393.

Martin, J. B. (1974a). Regulation of the pituitary thyroid system. *In* "Endocrine Physiology, MTP International Review of Science Physiology". (S. M. McCann, ed.) Series 1, Vol. 5, pp. 67–109. Butterworths, London.

Martin, J. B. (1974b). Inhibitory effect of somatostatin (SRIF) on the release of growth hormone (GH) induced in the rat by electrical stimulation. *Endocrinology*, **94,** 497–502.

Martin, J. B. and Jackson, I. M. D. (1975). *In* "Anatomical Neuroendocrinology" (W. E. Stumpf and L. D. Grant, eds). Int. Conf. Neurobiology of CNS-Hormones Interaction, Chapel Hill, pp. 343–353. S. Karger A.G., Basel.

Martin, J. B. and Reichlin, S. (1970). Neural regulation of the pituitary thyroid axis. *In* "Proceedings of 6th Midwest Conference on the Thyroid" (A. D. Kenny and R. R. Anderson, eds) pp. 1–24. University of Columbia Press, Columbia, Mo.

Martin, J. B. and Reichlin, S. (1972). Plasma thyrotropin (TSH) response to hypothalamic electrical stimulation and to injection of synthetic thyrotropin-releasing hormone (TRH). *Endocrinology*, **90,** 1079–1085.

Martin, J. B., Boshans, R. and Reichlin, S. (1970). Feedback regulation of TSH secretion in rats with hypothalamic lesions. *Endocrinology*, **87,** 1032–1040.

Martin, J. B., Kontor, J. and Mead, P. (1973). Plasma GH-responses to hypothalamic, hippocampal and amygdaloid electrical stimulation: effects of variation in stimulus parameters and treatment with α-methyl-p-tyrosine (α-MT). *Endocrinology*, **92,** 1354–1361.

Martin, J. B., Renaud, L. P. and Brazeau, P. Jr. (1974). Pulsatile growth hormone secretion: suppression by hypothalamic ventromedial lesions and by long-acting somatostatin. *Science, N.Y.* **186,** 538–540.

Martin, J. B., Renaud, L. P. and Brazeau, P. (1975). Hypothalamic peptides: new evidence for peptidergic pathways in the CNS. *Lancet*, **ii,** 393–395.

Martini, L. (1966). Neurohypophysis and anterior pituitary activity. *In* "The Pituitary Gland" Vol. 3 (G. W. Harris and B. T. Donovan, eds) pp. 535–577. Butterworths, London.

Martini, L., Franchini, F. and Motta, M. (1968). Neural control of anterior pituitary function. *Recent Prog. Horm. Res.* **24,** 439 485.

Mason, C. A. and Lincoln, D. W. (1976). Visualization of the retinohypothalamic projection in the rat by cobalt precipitation. *Cell Tiss. Res.* **168,** 117–131.

Mason, C. A. and Nishioka, R. S. (1974). The use of the cobalt chloride-ammonium sulphide precipitation technque for the delineation of invertebrate and vertebrate neurosecretory systems. *In* "Neurosecretion: the Final Neuroendocrine Pathway" (F. Knowles and L. Vollrath, eds) pp. 48–58. Springer-Verlag, Heidelberg.

Mason, J. W. (1968). A review of psychoendocrine research on the pituitary-adrenal cortical system. *J. Am. Psychosomat.* **30,** 576–607.

Mason, J. W. and Mougey, E. H. (1972). Thyroid (plasma BEI) response to chair restraint in the monkey. *Psychosom. Med.* **34,** 441–448.

Matsuo, H., Baba, Y., Nair, R. M. G., Arimura, A. and Schally, A. V. (1971). Structure of the porcine LH and FSH-releasing hormone. I. The proposed amino acid sequence. *Biochem. Biophys. Res. Commun.* **43,** 1334–1339.

Mazzi, V. (1970). The hypothalamus as a thermodependent neuroendocrine center in urodales. *In* "The Hypothalamus" (L. Martini, M. Motta and F. Fraschini, eds) pp. 663–673. Academic Press, New York and London.

Mazzi, G., Vellano, C., Colucci, D. and Merlo, A. (1974). Gonadotrophin stimulation by chronic administration of synthetic luteinizing hormone releasing hormone in hypophysectomized pituitary grafted male newts. *Gen. Comp. Endocrinol.* **24,** 1–9.

Meier, A. H. (1973). Daily hormone rhythms in the white-throated sparrow. *Am. Scient.* **61,** 184–187.

Meier, A. H. (1974). Temporal synergism of corticosterone and prolactin controlling seasonal conditions in the White-throated sparrow, *Zonotrichia albicollis*. *In* "Chronobiology" (L. E. Scheving, F. Halberg and J. E. Pauly, eds), pp. 647–651. Igaku Shoin, Tokyo.

Meier, A. H. and Dusseau, J. W. (1973). Daily entrainment of the photoinducible phases for photostimulation of the reproductive system in the sparrows, *Zonotrichia albicollis* and *Passer domesticus*. *Biol. Reprod.* **8,** 400–410.

Meijs-Roelofs, H. M. A., Uilenbroek, J. Th. J., De Jong, F. H. and Welschen, R. (1973). Plasma oestradiol-17β and its relation to serum follicle-stimulating hormone in immature female rats. *J. Endocr.* **59,** 295–312.

Meijs-Roelofs, H. M. A., de Greef, W. J. and Uilenbroek, J. Th. J. (1975) Plasma progesterone and its relationship to serum gonadotrophins in female rats. *J. Endocr.* **64,** 329–336.

Meites, J. (1972). Relation of prolactin to mammary tumourigenesis and growth in rats. *In* "Prolactin and Carcinogenesis" (A. R. Boyns and K. Griffiths, eds) pp. 54–63. Alpha Omega Alpha, Cardiff.

Meites, J. (1973). Control of prolactin secretion in animals. *In* "Human Prolactin" (J. L. Pasteels and C. Robyn, eds) pp. 105–118. Excerpta Medica, Amsterdam.

Meites, J. and Clemens, J. A. (1972). Hypothalamic control of prolactin secretion. *Vitam. & Horm.* **30,** 165–221.

Meites, J. and Nicoll, C. S. (1966). Adenohypophysis: Prolactin. *Annu. Rev. Physiol.* **28,** 57–88.

Meites, J., Kahn, H. and Nicoll, C. S. (1961). Prolactin production by rat pituitary *in vitro*. *Proc. Soc. Exp. Biol. Med.* **108,** 440–443.

Meites, J., Nicoll, C. S. and Talwalker, P. K. (1963). The central nervous system and secretion and release of prolactin. *In* "Advances in Neuroendocrinology" (A. Nalbandov, ed.) pp. 238–277. University of Illinois, Urbana.

Meites, J., Lu, K. H., Wuttke, W., Welsch, C. W., Nagasawa, H. and Quadri, S. K. (1972). Recent studies on functions and control of prolactin secretion in rats. *Recent Prog. Horm. Res.* **28,** 471–516.

Merimee, T. J. and Fineberg, S. E. (1973). Dietary regulation of human growth hormone secretion. *Metabolism,* **22,** 1491–1497.

Merimee, T. J. and Fineberg, S. E. (1974). Growth hormone secretion in starvation: a reassessment. *J. Clin. Endocrinol. Metab.* **39,** 385–386.

Merimee, T. J. and Rabin, D. (1973). A survey of growth hormone secretion and action. *Metabolism,* **22,** 1235–1251.

Mering, J. von and Minkowsky, O., (1889). Diabetes Mellitus nach Pankreas-extirpation. *Arch. exp. Path. Pharmak,* **26,** 371–387.

Mess, B., Zanisi, M. and Tima, L. (1970). Site of production of releasing and inhibiting factors. *In* "The Hypothalamus" (L. Martini, M. Motta and F. Fraschini, eds) pp. 1–18. Academic Press, New York and London.

Metcalf, G. (1974). TRH: A possible mediator of thermoregulation. *Nature, Lond.* **252,** 310–311.

Mey, J. De, Vandesande, F. and Dierickx, K. (1974). Identification of neurophysin producing cells. II. Identification of the Neurophysin I and the Neurophysin II producing neurons in the bovine hypothalamus. *Cell Tiss. Res.* **153,** 531–543.

Mielke, C. H. Jr., Gerich, J. E., Lorenzi, M., Tsalikian, E., Rodvien, R. and Forsham, P. H. (1975). The effect of somatostatin on coagulation and platelet function in man. *New Engl. J. Med.* **293,** 480–483.

Milcu, S. M., Pavel, S., and Neacsu, C. (1963). Biological and chromatographic characterization of a polypeptide with pressor and oxytocic activities isolated from bovine pineal gland. *Endocrinology*, **72,** 563–566.

Millar, R., Aehnett, C., Rossier, G. and Henfricks, S. (1975). Evidence for the existence of higher molecular weight and precursor of luteinizing hormone-releasing hormone (LH-RH). *IRCS Endocrine System*, **3,** 603.

Millar, R. P. and Glover, T. D. (1973). Regulation of seasonal sexual activity in an ascrotal mammal, the rock Hyrax, *Procavia capensis. J. Reprod. Fertil.* Suppl. **19,** 203–220.

Millhouse, O. E. (1971). A Golgi study of third ventricle tanycytes in the adult rodent brain. *Z. Zellforsch.* **121,** 1–13.

Milligan, J. V. and Kraicer, J. (1972). ^{45}Ca uptake during the *in vitro* release of hormones from the rat adenohypophysis. *Endocrinology*, **89,** 766–773.

Milligan, J. V., Kraicer, J., Fawcett, C. P. and Illner, P. (1972). Purified growth hormone-releasing factor increases ^{45}Ca uptake into pituitary cells. *Canad. J. Phys. and Pharmacol.* **50,** 613–617.

Mills, J. N. (1964). Circadian rhythms during and after three months in solitude underground. *J. Physiol.* **174,** 217–231.

Mitchell, J. A. (1975). Functional anatomy of the hypothalamus: possible role of the ependyma. *In* "Hypothalamic Hormones" (E. S. E. Hafez and J. R. Reel, eds) pp. 1–16. Ann Arbor Science, Michigan.

Mitnick, M. and Reichlin, S. (1971). Biosynthesis of TRH by rat hypothalamic tissue *in vitro*. *Science, N.Y.* **172,** 1241–1243.

Mitnick, M., and Reichlin, S. (1972). Enzymatic synthesis of thyrotrophin releasing hormone (TRH) by hypothalamic TRH synthetase. *Endocrinology*, **91,** 1146–1153.

Mitnick, M. A., Valverde, R. C. and Reichlin, S. (1973). Enzymatic synthesis of prolactin-releasing factor (PRF) by rat hypothalamic tissue: Evidence for "PRF synthetase". *Proc. Soc. exp. Biol. Med.* **143,** 418–421.

Miyachi, Y., Chrambach, A., Mecklenburg, R. and Lipsett, M. B. (1973). Preparation and properties of ^{125}I-LH-RH. *Endocrinology*, **92,** 1725–1730.

Moguilevsky, J. A., Enero, M. A. and Szwarcfarb, B. (1973). Biosynthesis *in vitro* of the luteinizing hormone-releasing factor by rat hypothalamic tissue. Influence of castration. *Acta Physiol. Lat.-Amer.* **23,** 492–493.

Moguilevsky, J. A., Enero, M. A., Szwarcfarb, B. and Dosoretz, D. (1975). Effects of castration, and testosterone *in vitro*, on the hypothalamic synthesis of different peptide fractions. *J. Endocr.* **64,** 155–161.

Moldow, R. (1975). Effect of intracranial somatostatin (SRIF) implantation on ACTH release. *Program 57th Meet. U.S. Endocrine Soc.*, p. 146.

Montoya, E., and Wilber, J. F. (1974). Studies of thyrotropin releasing hormone (TRH) physiology by radioimmunoassay. Effects of triiodothyronine (T3) and thyroxine (T4). *Program 56th Meeting U.S. Endocrine Soc.* p. A-65.

Montoya, E., Seibel, M. J., and Wilber, J. F. (1975). Thyrotrophin releasing hormone secretory physiology: studies by radioimmunoassay and affinity chromatography. *Endocrinology*, **96,** 1413–1418.

Monroe, B. G., and Paull, W. K. (1974). Ultrastructural changes in the hypothalamus during development and hypothalamic activity: the median eminence. *Progr. Brain Res.* **41,** 185–208.

Moore, C. R. and Price, D. (1932). Gonad hormone functions, and the reciprocal influence between the gonads and hypophysis with its bearing on sex hormone antagonism. *Am. J. Anat.* **50,** 13–72.

Morel, F. and Jard, S. (1968). Actions and functions of the neurohypophysial hormones and related peptides in lower vertebrates. *In* "Handbook of Experimental Pharmacology" (B. Berde, ed.) Vol. 23, pp. 655–716. Springer-Verlag, Berlin.

Mori, J. and Hafez, E. S. E. (1973). Release of LH by synthetic LH-RH in the monkey, *Macaca fascicularis. J. Reprod. Fertil.* **34,** 155–157.

Moriarty, G. C. and Halmi, N. S. (1972). Adrenocorticotrophin production by the intermediate lobe of the rat pituitary. *Z. Zellforsch.* **132,** 1–14.

Morris, J. F. (1976a). Hormone storage in individual neurosecretory granules: A quantitative ultrastructural approach to hormone storage in the neural lobe. *J. Endocr.* **68,** 209–224.

Morris, J. F. (1976b). Distribution of the neurosecretory granules among the anatomical compartments of the neurosecretory processes. *J. Endocr.* **68,** 225–234.

Mortimer, C. H., Besser, G. M., McNeilly, A. S., Marshall, J. C., Harsoulis, P., Tunbridge, W. M. G., Gomez-Pan, A. and Hall, R. (1973). The LH and FSH releasing hormone test in patients with hypothalamic-pituitary-gonadal dysfunction. *Br. Med. J.* **4,** 73–77.

Mortimer, C. H., Besser, G. M., Hook, J. and McNeilly, A. S. (1974a). Intravenous, intramuscular, subcutaneous and intra nasal administration of LH-FSH-RH: the duration of effect and occurrence of asynchronous pulsatile release of LH and FSH. *Clin. Endocrinol.* **3,** 19–25.

Mortimer, C. H., McNeilly, A. S., Fisher, R. A., Murray, M. A. F. and Besser, G. M. (1974b). Gonadotrophin-releasing hormone therapy in hypogonadal males with hypothalamic or pituitary dysfunction. *Br. Med. J.* **4,** 617–621.

Mortimer, C. H., Besser, G. M. and McNeilly, A. S. (1975). Gonadotrophin releasing hormone therapy in the induction of puberty, potency, spermatogenesis and ovulation in patients with hypothalamic-pituitary-gonadal dysfunction. *In* "Hypothalamic Hormones" (M. Motta, P. Crosignani and L. Martini, eds) pp. 325–335. Academic Press, London and New York.

Moss, R. L. and McCann, S. M. (1973). Induction of mating behaviour in rats by luteinizing hormone releasing factor. *Science N.Y.* **181,** 177–179.

Moss, R. L., Dyball, R. E. J. and Cross, B. A. (1972). Excitation of antidromically identified neurosecretory cells of the paraventricular nucleus by oxytocin applied iontophoretically. *Expl. Neurol.* **34,** 95–102.

Moss, R. L., Dudley, C. A., Foreman, M. M. and McCann, S. M. (1975a). Synthetic LRF: a potentiator of sexual behaviour in the rat. *In* "Hypothalamic Hormones" (M. Motta, P. Crosignani and L. Martini, eds) pp. 269–278. Academic Press, London and New York.

Moss, R. L., Kelly, M., and Riskind, P. (1975b). Tuberoinfundibular neurons: dopaminergic and norepinephrinergic sensitivity. *Brain Res.* **89,** 265–277.

Motokizawa, F. (1974). Olfactory input to the thalamus: electrophysiological evidence. *Brain Res.* **67,** 334–337.

Motta, M. (1964). A short feedback mechanism. *Proc. 2nd Int. Cong. Endocr.* Excerpta Medica Foundation. Series 3, Stop Press Sessions, p. 1276.

Motta, M., Fraschini, F. and Martini, L. (1969). "Short" feedback mechanisms in the control of anterior pituitary function. *In* "Frontiers in Neuroendocrinology" (W. F. Ganong and L. Martini, eds) pp. 211–251. Oxford University Press, New York.

Motta, M., Piva, F., Tima, L., Zanisi, M. and Martini, L. (1971a). Intrahypothalamic localization of the nuclei synthesizing the gonadotropin releasing factors. *J. Neuro-Visc. Relat.* Suppl. X., 32–40.

Motta, M., Schiaffini, O., Piva, F. and Martini, L. (1971b). Pineal principles and the control of adrenocorticotrophin secretion. *In* "The Pineal Gland" (G. E. W. Wolstenholme and J. Knight, eds) pp. 279–291. Churchill Livingstone, London.

Motta, M. Piva, F. and Martini, L. (1973). New findings on the central control of gonadotrophin secretion. *J. Reprod. Fertil.* Suppl. **20,** 27–38.

Mountjoy, C. Q., Weller, M., Hall, R., Price, J. S., Hunter, P. and Dewar, H. H. (1974). A double-blind crossover sequential trial or oral thyrotrophin-releasing hormone in depression. *Lancet*, **i,** 958–960.

Mueller, G. P., Chen, H. J. and Meites, J. (1973). *In vivo* stimulation of prolactin release in the rat by synthetic TRH. *Proc. Soc. Exp. Biol. Med.* **144,** 613–615.

Mueller, G. P., Franco, F. S., Reichlin, S. and Jackson, I. M. D. (1977). Elevated serum thyrotropin (TSH) of Myxedema does not require continuous thyrotropin releasing hormone (TRH) secretion. *Clin. Res.* **25,** 298A.

Muhlemann, M., Titov, M. I., Schwyzer, R. and Rudinger, J. (1972). The use of intermediates with preformed disulphide bridge for the synthesis of oxytocin and deamino-oxytocin. *Helv. Chin. Acta*, **55,** 2854–2860.

Müller, E. E. (1973). Nervous control of growth hormone secretion. *Neuroendocrinology*, **11,** 338–369.

Müller, E. E. and Pecile, A. (1966). Influence of exogenous GH on endogenous growth hormone release. *Proc. Soc. Exp. Biol. Med.* **122,** 1289–1291.

Müller, E. E., Arimura, A., Saito, T. and Schally, A. V. (1967). Growth hormone-releasing activity in plasma of normal and hypophysectomized rats. *Endocrinology*, **80,** 77–81.

Munson, P. L. and Briggs, F. N. (1955). The mechanism of stimulation of ACTH secretion. *Recent Prog. Horm. Res.* **11,** 83–117.

Munson, P. L. and Tai-Chan Reng (1970). Bioassay of the hypophysiotropic hormones: *In vivo* systems. *In* "Hypophysiotrophic Hormones of the Hypothalamus: Assay and Chemistry" (J. Meites, ed.) pp. 36–43. Williams and Wilkins, Baltimore.

Murray, G. R. (1891). Note on the treatment of myxoedema by hypodermic injections of an extract of the thyroid gland of a sheep. *Br. Med. J.* ii, 798.

Murphy, B. E. P. (1973). Does the human foetal adrenal play a role in parturition? *Amer. J. Obstet. Gynaec.* **115,** 521–525.

Murphy, J. T. (1972). The role of the amygdala in controlling hypothalamic output. *In* "The Neurobiology of the Amygdala" (B. E. Eleftheriou, ed.) pp. 371–395. Plenum Press, New York and London.

Murton, R. K. (1975). Ecological adaptation in avian reproductive physiology. *Symp. Zool. Soc.* (*London*), **35,** 149–175.

Myles, A. B., Bacon, P. A. and Daly, J. R. (1971). Single daily dose of corticosteroid treatment. Effects in adrenal function and therapeutic efficiency in various diseases. *Ann. Rheum. Dis.* **30,** 149–153.

N

Naftolin, F., Ryan, K. J. and Petro, Z. (1971). Aromatization of androstenedione by diencephalon. *J. Clin. Endocrinol. Metab.* **33,** 368–370.

Naftolin, F., Ryan, K. J. and Petro, Z. (1972). Aromatization of androstenedione by the anterior hypothalamus of adult male and female rats. *Endocrinology*, **90,** 295–298.

Naik, D. V. (1972). Electron microscopic studies on the pars intermedia in normal and in mice with hereditary nephrogenic diabetes insipidus. *Z. Zellforsch.* **133,** 415–434.

Naik, D. V. (1973). Electron microscopic-immunocytochemical localization of adrenocorticotropin and melanocyte stimulating hormone in the pars intermedia cells of rats and mice. *Z. Zellforsch.* **142,** 305–328.

Naik, D. V. (1975). Immunoreactive LH-RH neurons in the hypothalamus identified by light and fluorescent microscopy. *Cell. Tiss. Res.* **157,** 423–436.

Nair, R. M. G., Barret, J. F., Bowers, C. Y. and Schally, A. V. (1970). Structure of porcine thyrotropin-releasing hormone. *Biochemistry*, **9,** 1103–1106.

Nair, R. M. G., Kastin, A. J. and Schally, A. V. (1971a). Isolation and structure of hypothalamic MSH release-inhibiting hormone. *Biochem. Biophys. Res. Commun.* **43,** 1376–1380.

Nair, R. M. G., Redding, T. W. and Schally, A. V. (1971b). Site of inactivation of thyrotropin-releasing hormone by human plasma. *Biochemistry*, **10,** 3621–3624.

Nair, R. M. G., Kastin, A. J. and Schally, A. V. (1972). Isolation and structure of another hypothalamic peptide possessing MSH-release-inhibiting activity. *Biochem. Biophys. Res. Commun.* **47,** 1420–1425.

Nair, R. M. G., Redding, T. W., Kastin, A. J. and Schally, A. V. (1973). Site of inactivation of melanocyte-stimulating hormone-release-inhibiting hormone by human plasma. *Biochem. Pharmacol.* **22,** 1915–1919.

Nakai, Y. and Gordman, A. (1969). Evidence for a doubly innervated secretory unit in the anuran pars intermedia. *Gen. Comp. Endocrinol.* **13,** 108–116.

Nakamura, M. (1972). Studies on the role of basic amino acid residues of ACTH peptide in steroidogenesis by isolated adrenal cells. *J. Biochem* (*Tokyo*), **71,** 1029–1041.

Nakane, P. K. (1970). Classifications of anterior pituitary cell types with immunoenzyme histochemistry. *J. Histochem. Cytochem.* **18,** 9–20.

Nalbandov, A. V. (1973). Control of luteal function in mammals. *In* "Handbook of Physiology, Endocrinology" (R. O. Greep and E. B. Astwood, eds) Vol. II, Part 1, pp. 153–167. American Physiological Society, Washington D.C.

Naor, Z., Koch, Y., Bauminger, S. and Zor, U. (1975). Action of luteinizing hormone-releasing hormone and synthesis of prostaglandins in the pituitary gland. *Prostaglandins*, **9,** 211–216.

Nathanielsz, R. W., Comline, R. S., Silver, M. and Paisey, R. B. (1972). Cortisol metabolism in the foetal and neonatal sheep. *J. Reprod. Fertil.* Suppl. **16,** 39–59.

Naumenko, E. V. (1968). Hypothalamic chemoreactive structures and pituitary-adrenal function. Effects of local injection of noradrenaline, carbachol and 5-hydroxytryptamine into the brain of guinea-pigs with intact brains and after mesencephalic transection. *Brain. Res.* **11,** 1–10.

Nauta, W. J. H. and Haymaker, W. (1969). Hypothalamic nuclei and fiber connections. *In* "The Hypothalamus" (W. Haymaker, E. Anderson and W. J. H. Nauta, eds) pp. 136–209. C. C. Thomas, Springfield.

Neer, E. J. (1973). Biologically active (^{14}C) vasopressin. A new and simplified synthesis *J. Biol. Chem.* **248,** 1897–1900.

Negro-Vilar, A., Ojeda, S. R. and McCann, S. M. (1973). Evidence for changes in sensitivity to testosterone negative feedback on gonadotrophin release during sexual development in the male rat. *Endocrinology*, **93,** 729–735.

Neill, J. D. (1974). Prolactin: its secretion and control. *In* "Handbook of Physiology" Section 7, Vol. IV (E. Knobil and W. H. Sawyer, eds) Part 2, pp. 469–488. American Physiological Society, Washington, D.C.

Nejad, I. F., Bollinger, J. A., Mitnick, M. and Reichlin, S. (1972). Importance of T3 (triiodothyronine) secretion in altered states of thyroid function in the rat: cold exposure, subtotal thyroidectomy, and hypothysectomy. *Trans. Ass. Am. Phys.* **85,** 295–308.

Nequin, L. G., Alvarez, J. A. and Campbell, C. S. (1975). Alterations in steroid and gonadotrophin release resulting from surgical stress during the morning of proestrus in 5-day cyclic rats. *Endocrinology*, **97,** 718–724.

Newton, B. A. (1970). Compounds affecting DNA structure and function. *Advances in Pharmacology and Chemotherapy*, **8,** 149–184.

Nicholls, T. J. and Storey, C. R. (1976). The effects of castration on plasma LH levels in photosensitive and photorefractory canaries (*Serinus canarius*). *Gen. Comp. Endocrinol.* **30,** 204–208.

Nichols, T., Nugent, C. A. and Tyler, F. H. (1965). Diurnal variation in suppression of adrenal function by glucocorticoids. *J. Clin. Endocrinol. Metab.* **25,** 343–349.

Nicoll, C. S. (1965). Neural regulation of adenohypophysial prolactin secretion in tetrapods: indications from *in vitro* studies. *J. Exp. Zool.* **158,** 203–210.

Nicoll, C. S. (1971). Aspects of the neural control of prolactin secretion. In "Frontiers in Neuroendocrinology" (L. Martini and W. F. Ganong, eds) pp. 291–330. Oxford University Press, New York.

Nicoll, C. S. (1972). Some observations and speculation on the mechanism of "depletion", "repletion" and release of adenohypophyseal hormones. *Gen. Comp. Endocrinol.* (Suppl.) **3,** 86–95.

Nicoll, C. S. (1973). "Spinoff" from comparative studies on prolactin physiology of significance to the clinical endocrinologist. *In* "Human Prolactin" (J. L. Pasteels and C. Robyn, eds) pp. 119–136. Excerpta Medica, Amsterdam.

Nicoll, C. S. (1974). Physiological actions of prolactin. *In* "Handbook of Physiology" Section 7, Vol. IV (E. Knobil and W. H. Sawyer, eds) Part 2, Chapter 32, pp. 253–292. American Physiological Society, Washington D.C.

Nicoll, C. S. and Bern, H. A. (1972). On the actions of prolactin among the vertebrates: is there a common denominator? *In* "Lactogenic Hormones" (G. E. W. Wolstenholme and J. Knight, eds) pp. 299–317. Churchill Livingstone, London.

Nicoll, C. S., Fiorindo, R. P., McKennee, C. J., and Parsons, J. A. (1970). Assay of hypothalamic factors which regulate prolactin secretion. *In* "Hypophysiotrophic Hormones in the Hypothalamus: Assay and Chemistry" (J. Meites, ed.) pp. 115–150. Williams and Wilkins, Baltimore.

Nicoloff, J. S., Fisher, D. A. and Appelman, M. D. Jr. (1970). A new method for the measurement of thyroidal iodine release in man. *J. Clin. Invest.* **49,** 1922–1929.

Nikitovitch-Winer, M. and Everett, J. W. (1958). Functional restitution of pituitary grafts re-transplanted from kidney to median eminence. *Endocrinology*, **63,** 916–930.

Nikitovitch-Winer, M. and Everett, J. W. (1959). Histo-cytologic changes in grafts of rat pituitary on the kidney and upon re-transplantation under the diencephalon. *Endocrinology*, **65,** 357–368.

Nikitovitch-Winer, M. B., Pribble, A. H. and Winer, A. D. (1965). Luteinizing hormone-releasing factor: partial purification. *Am. J. Physiol.* **208,** 1286–1290.

Noel, G. L., Suh, H. K. and Frantz, A. G. (1973). L-dopa suppression of TRH-stimulated prolactin release in man. *J. Clin. Endocrinol. Metab.* **36,** 1255–1258.

Noel, G. L., Suh, H. K. and Frantz, A. G. (1974). Prolactin release during nursing and breast stimulation in postdartum and nonpostpartum subjects. *J. Clin. Endocrinol. Metab.* **38,** 413–423.

Nokin, J., Vekemans, M., L'Hermite, M. and Robyn, C. (1972) Circadian periodicity of serum prolactin concentration in man. *Br. Med. J.* **3,** 561–562.

O

Odell, W. D. and Daughaday, W. H. (1971). "Principles of Competitive Protein Binding Assays." Lippincott, Philadelphia.

Odell, W. D. and Swerdloff, R. S. (1968). Progesterone induced luteinizing and follicle stimulating hormone surge in post-menopausal women: a simulated ovulatory peak. *Proc. Nat. Acad. Sci. U.S.A.* **61,** 529–536.

Odell, W. D. and Swerdloff, R. S. (1974). The role of the gonads in sexual maturation *In* "The Control of the Onset of Puberty" (M. M. Grumbach, G. D. Grave and F. E. Mayer, eds) pp. 313–341. Wiley, New York and London.

Ojeda, S. R., Harms, P. G. and McCann, S. M. (1974a). Effect of blockade of dopaminergic receptors on prolactin and LH release: median eminence and pituitary sites of action. *Endocrinology*, **94**, 1650–1657.

Ojeda, S. R., Harms, P. G. and McCann, S. M. (1974b). Possible role of cyclic AMP and prostaglandin E_1 in the dopaminergic control of prolactin release. *Endocrinology*, **95**, 1694–1703.

Okada, Y., Horikoshi, H. and Baba, Y. (1974). Syntheses and properties of formyl sarkosine[1]-LH-RH and N-Methyl-L-pGly[1]-LH-RH. *Chem. Pharm Bull.* **22**, 721–723.

Oliver, C., Eskay, R. L., Mical, R. S. and Porter, J. C. (1973). Radioimmunoassay for TRH and its determination in hypophysial and portal and peripheral plasma of rats. *Program 49th Meeting American Thyroid Assoc.* p. T4.

Oliver, C., Charvet, J. P., Codaccioni, J. L. and Vague, J. (1974a). Radioimmunoassay of thyrotropin-releasing hormone in human plasma and urine. *J. Clin. Endocrinol. Metab.* **39**, 406–409.

Oliver, C., Charvet, J. P., Codaccioni, J., Vague, J. and Porter, J. C. (1974b). TRH in human CSF. *Lancet*, **i**, 873.

Oliver, C., Eskay, R. L., Ben-Jonathan, N., and Porter, J. C. (1974c). Distribution and concentration of TRH in the rat brain. *Endocrinology*, **95**, 540–546.

Oliver, G. and Schäfer, E. A. (1895). On the physiological actions of extracts of the pituitary body. *J. Physiol. Lond.* **18**, 277–279.

Ondo, J. G., Eskay, R. L., Mical, R. S. and Porter, J. C. (1973). Release of LH by LRH injected into the CSF: A transport role for the median eminence. *Endocrinology*, **93**, 231–237.

Ooki, T., Kotsu, T., Kinutani, M. and Daikoku, S. (1973). Pars intermedia of the hypophysis of rats after early-postnatal lesions of the basal hypothalamus: quantitative and qualitative observations. *Neuroendocrinology*, **11**, 22–45.

Oppenheimer, J. H., Squef, R. S., Surks, M. I. and Hauer, H. (1963). Binding of thyroxine by serum proteins evaluated by equilibrium dialysis and electrophoretic techniques. *J. Clin. Invest.* **42**, 1769–1782.

Orth, D. N. and Island, E. P. (1969). Light synchronisation of the circadian rhythm in plasma cortisol concentration in man. *J. Clin. Endocrinol. Metab.* **29**, 479–486.

Ortowski, H., Richman, P. G. and Master, A. (1969). Isolation and properties of γ-L-glutamylcyclotransferase from human brain. *Biochemistry*, **8**, 1048–1055.

Oshima, K. and Gorbman, A. (1969). Evidence for a double innervated secretory unit in the anuran pars intermedia. *Gen. Comp. Endocrinol.* **13**, 98–107.

Osland, R. B., Gallo, R. V. and Williams, J. A. (1975). *In vitro* release of luteinizing hormone from anterior pituitary fragments superfused with constant or pulsatile amounts of luteinizing hormone-releasing factor. *Endocrinology*, **96**, 1210–1216.

Ôta, K., Ôta, T. and Yokoyama, A. (1974). Plasma corticosterone concentrations and pituitary prolactin content in late pregnancy and their within day fluctuations in the rat. *J. Endocr.* **61**, 21–28.

Oyama, S. N., Kagan, A. and Glick, S. M. (1971). Radioimmunoassay of vasopressin: application to unextracted human urine. *J. Clin. Endocrinol. Metab.* **33**, 739–744.

P

Palay, S. L. (1957). The fine structure of the neurohypophysis. *In* "Ultrastructure and Cellular Chemistry of Neural Tissue" (H. Waelsch, ed) pp. 31–49. Hoeber, New York.

Palka, Y., Coyer, D. and Critchlow, V. (1969). Effects of isolation of medial basal hypothalamus on pituitary adrenal and pituitary ovarian functions. *Neuroendocrinology*, **5**, 333–349.

Palkovits, M., Arimura, A., Brownstein, M., Schally, A. V. and Saavedra, J. M. (1974a). Luteinizing hormone-releasing hormone (LH-RH) content of the hypothalamic nuclei in rat. *Endocrinology*, **95**, 554–558.

Palkovits, M., Brownstein, M., Saavedra, J. M. and Axelrod, J. (1974b). Norepinephrine and dopamine content of hypothalamic nuclei of the rat. *Brain Res.* **77**, 137–149.

Palmer, G. C., Sulser, F. and Robison, G. A. (1973). Effects of neurohumoral and adrenergic agents on cyclic AMP levels in various areas of the rat brain *in vitro*. *Neuropharmacology*, **12**, 327–337.

Papova, N. K., Maslova, L. N. and Naumenko, E. V. (1972). Serotonin and the regulatiou of the pituitary-adrenal axis after deafferentation of the hypothalamus. *Brain. Res.* **11**, 1–10.

Parker, D. C., Rossman, L. G. and Vanderlaan, E. F. (1973). Sleep-related, nyctohemeral and briefly episodic variation in human plasma prolactin concentrations. *J. Clin. Endocrinol. Metab.* **36**, 1119–1124.

Parker, D. C., Rossman, L. G. and Vanderlaan, E. F. (1974). Relation of sleep-entrained human prolactin release to REM-non REM cycles. *J. Clin. Endocrinol. Metab.* **38**, 646–651.

Parry, H. B. and Levett, B. G. (1973). A new hypothalamic pathway to the median eminence containing neurophysin and its hypertrophy in sheep with natural Scrapie. *Nature (London)*, **242**, 63–65.

Passa, P., Gauville, C. and Canivet, J. (1974). Influence of muscular exercise on plasma level of growth hormone in diabetics with and without retinopathy. *Lancet*, **ii**, 72–74.

Pasteels, J. L. (1960). Etude expérimentale des différentes catégories d'éléments chromophiles de l'hypophyse adulte de *Pleurodeles Waltlii*, et de leur controle par l'hypothalamus. *Arch. Biol. (Paris)* **71**, 409–471.

Pasteels, J. L. (1961). Sécrétion de prolactine par l'hypophyse en culture de tissus. *C. R. Acad. Sci.* **253**, 2140–2142.

Pasteels, J. L., Brauman, H. and Brauman, J. (1963). Comparative studies of the secretion of somatotropic hormone by the human pituitary gland *in vitro* and of its lactogenic activity. *C. R. Acad. Sci.* **256**, 2031–2033.

Patel, Y. C., Alford, F. P. and Burger, G. (1972). The 24-hour plasma thyrotropin profile. *Clin. Sci.* **43**, 71–77.

Patel, Y. C., Rao, K. and Reichlin, S. (1977). Somatostatin in human cerebrospinal fluid. *New Engl. J. Med.* **296**, 529.

Patel, Y. C. and Burger, H. B. (1973). Serum thyrotropin (TSH) in pituitary and/or hypothalamic hypothyroidism: normal or elevated basal levels and paradoxical responses to thyrotropin releasing hormone. *J. Clin. Endocrinol. Metab.* **37**, 190–196.

Paton, W. D. M. (1961). A theory of drug action based on the rate of drug-receptor combination. *Proc. Roy. Soc. (Series B)* **154**, 21–69.

Pavel, S., Calb, M. and Georgesen, M. (1975). Reverse of the effects of pinealectomy on the pituitary prolactin content in mice by very low concentrations of vasotocin injected in the third cerebral ventricle. *J. Endocr.* **66**, 289–290.

Pavel, S., Matrescu, L. and Petrescu, M. (1973). Central corticotrophin inhibition by arginine vasotocin in the mouse. *Neuroendocrinology*, **12**, 371–375.

Peake, G. T. (1973). The role of cyclic nucleotides in the secretion of pituitary growth hormone. *In* "Frontiers in Endocrinology" (W. F. Ganong and L. Martini, eds). Oxford University Press.

Peake, G. T., Birge, C. A. and Daughaday, W. H. (1973a). Alterations of radioimmunoassayable growth hormone and prolactin during hypothyroidism. *Endocrinology*, **92**, 487–493.

Peake, G. T., Wilson, M., Steiner, A. L. and Dhariwal, A. P. S. (1973b). Effects of a purified growth hormone-releasing factor on growth hormone secretion and pituitary cyclic nucleotide content. *Metab. Clin. Exp.* **22,** 769–772.

Pearlmutter, F. A., Rapino, E. and Saffran, M. (1974). A semi-automated *in vitro* assay for CRF: Activities of peptides related to vasopressin. *Neuroendocrinology,* **15,** 106–119.

Pecile, A. and Bossa, R. (1976). Age changes in hypothalamic-pituitary function. *In* "Hypothalamus Pituitary and Aging" (A. V. Everitt and J. A. Burgess, eds) pp. 86–96. Thomas, Springfield.

Pecile, A. and Müller, E. (1966). Control of growth hormone secretion. *In* "Neuro-endocrinology" (L. Martini and W. Ganong, eds) Vol. I, pp. 537–556. Academic Press, New York and London.

Pecile, A., Müller, E., Falconi, G. and Martini, L. (1965). Growth hormone-releasing activity of hypothalamic extracts at different ages. *Endocrinology,* **77,** 241–246.

Pecile, A., Müller, E., Felici, M. and Netti, C. (1972). Nervous system participation in growth hormone release from anterior pituitary gland. *In* "Growth and Growth Hormone" (A. Pecile and E. E. Müller, eds) pp. 261–270. Excerpta Medica, Amsterdam.

Peck, W. A. (1975). Comments on the importance of some recent neuroendocrinology advances. *Arch. Intern. Med.* **135,** 1362–1363.

Pelletier, G., Lemay, A., Beraud, G. and Labrie, F. (1972). Ultrastructural changes accompanying the stimulatory effect of N^6-monobutyryl adenosine 3′,5′-monophosphate on the release of growth hormone (GH), prolactin (PRL) and adrenocorticotrophic hormone (ACTH) in rat anterior pituitary gland *in vitro. Endocrinology,* **91,** 1355–1371.

Pelletier, G., Labrie, F., Arimura, A. and Schally, A. V. (1974a). Electron microscopic immunohistochemical localisation of growth hormone release inhibiting hormone (somatostatin) in the rat median eminence. *Am. J. Anat.* **140,** 445–450.

Pelletier, G., Leclerc, R., Labri, F. and Puviani, R. (1974b). Electron microscopic immuno-histochemical localisation of neurophysin in supraoptic and paraventricular nuclei, axons in internal layer of median eminence and in nerve terminals of the posterior pituitary. *Molecular and Cellular Endocrinology,* **1,** 157–166.

Pelletier, G., Leclerc, R. and Dube, D. (1975). Localisation of growth hormone-release-inhibiting hormone (somatostatin) in the rat brain. *Am. J. Anat.* **142,** 397–401.

Pelletier, J. (1973). Evidence for photoperiodic control of prolactin release in rams. *J. Reprod. Fertil.* **35,** 143–147.

Pelletier, J. and Ortavant, R. (1975). Photoperiodic control of LH release in the ram. *Acta endocr., Copenh.* **78,** 442–450.

Pelletier, J. and Signoret, J. P. (1969). Contrôle de la décharge de LH dans le sang par la progésterone et le benzoate d'oestradiol chez la Brebis castrée. *C. R. Acad. Sci.* **269,** 2595–2529.

Pelletier, M. J. (1974). Influence de la lactation sur la décharge préovulatoire de LH induite par l'injection de LRF synthétique chez la brebis durant la période d'anoestrus saisonnier. *C. R. Acad. Sci.* **279,** 179–182.

Peng, M. and Huang, H. (1972). Aging of hypothalamic-pituitary-ovarian function in the rat. *Fert. Steril.* **23,** 535–542.

Pengelly, E. T. (ed.). (1974). "Circannual Clocks: Annual Biological Rhythms". Academic Press, New York and London.

Perks, A. M. (1969). The neurohypophysis. *In* "Fish Physiology" (W. S. Hoar and R. J. Randell, eds) Vol. 2, pp. 111–205. Academic Press, New York and London.

Perks, A. M. and Sawyer, W. H. (1965). A new neurohypophysial principle in the elasmobranch pituitary, *Nature, Lond.* **205,** 154–156.

Perks, A. M. and Vizolyi, E. (1973). Studies on the neurohypophysis in foetal mammals. *In* "Foetal and Neonatal Physiology" (K. S. Comline, K. W. Cross, G. S. Dawes and P. W. Nathanielsz, eds) pp. 430–438. Cambridge University Press.

Perryman, E. K. (1974). Fine structure of the secretory activity of the pars intermedia of *Rana pipiens*. *Gen. Comp. Endocrinol.* **23,** 94–110.

Peter, R. E. (1970). Hypothalamic control of thyroid gland activity and gonadal activity in the goldfish. *Gen. Comp. Endocrinol.* **14,** 334–356.

Peter, R. E. and McKeown, B. A. (1974). Effects of hypothalamic and thalamic lesions on prolactin secretion in goldfish (*Carassius auratus*). *Gen. Comp. Endocrinol.* **23,** 438–452

Pfaff, D. W. (1973). Luteinizing hormone-releasing factor potentiates lordosis behaviour in hypophysectomised ovariectomised female rats. *Science, N.Y.* **182,** 1148–1149.

Pfaff, D. W. and Keiner, M. (1973). Atlas of oestradiol-concentrating cells in the central nervous system of the female rat. *J. Comp. Neurol.* **151,** 121–158.

Pfeiffer, C. A. (1936). Sexual differences of the hypophyses and their determination by the gonads. *Am. J. Anat.* **58,** 195–225.

Phifer, R. F., Spicer, S. S. and Orth, D. N. (1970). Specific demonstration of the human hypophyseal cells which produce adrenocorticotropic hormone. *J. Clin. Endocrinol. Metab.* **31,** 347–361.

Pickering, B. T. (1968). A neurophysin from cod (*Gadus morrhua*) pituitary glands: isolation and properties *J. Endocr.* **42,** 143–152.

Pickering, B. T. (1970). Aspects of the relationships between the chemical structure and biological activity of the neurohypophysial hormones and their synthetic structural analogues. *In* "International Encyclopedia of Pharmacology and Therapeutics" (H. Heller and B. T. Pickering, eds) Section 41, pp. 81–110. Pergamon Press, Oxford.

Pickering, B. T. and Heller, H. (1969). Oxytocin as a neurohypophysial hormone in the holocephalian elasmobranch fish, *Hydrolagus colliei*. *J. Endocr.* **45,** 597–606.

Pickering, B. T., Jones, C. W. and Burford, G. D. (1974). Biochemical aspects of the hypothalamo-neurohypophysial neurone. *In* "Neurosecretion: the Final Neuroendocrine Pathway" (F. G. W. Knowles and J. Vollrath, eds) pp. 72–85. Springer-Verlag, Berlin and New York.

Pietsch, K. (1930). Aufbau und Entwicklung der Pars tuberalis des menschlichen Hirnanhangs in ihren Beziehungen zu den Übrigen Hypophysenteilen. *Z. mikr.-anat. Forsch.* **22,** 227–257.

Pietta, P. G., Cavallo, P. F., Takahashi, K. and Marshall, G. R. (1974). Preparation and use of benzhydrylamine polymers in peptide synthesis. II. Syntheses of thyrotropin-releasing hormone, thyrocalcitonin 26–32, and eledosin. *J. Org. Chem.* **39,** 44–48.

Pimstone, B. L., Becker, D. and Kronheim, S. (1974). Disappearance of plasma growth hormone in acromegaly and protein-calorie malnutrition after somatostatin. *J. Clin. Endocrinol. Metab.* **40,** 168–171.

Pimstone, B. L., Roith, D. L., Epstein, S. and Kronheim, S. (1975). Disappearance rates of plasma growth hormone after intravenous somatostatin in renal and liver disease. *J. Clin. Endocrinol. Metab.* **41,** 392–395.

Pincus, G. (1943). A diurnal rhythm in the secretion of urinary ketosteroids by young men. *J. Clin. Endocrinol. Metab.* **3,** 195–199.

Pitman, R. M., Tweedle, C. D. and Cohen, M. J. (1972). Branching of central neurone: intracellular cobalt injection for light and electron microscopy. *Science, N.Y.* **176,** 412–414.

Plotnikoff, N. P., Prange, A. J. Breese, G. R., Anderson, M. S. and Wilson, I. C. (1972). Thyrotropin releasing hormone: enhancement of DOPA activity by a hypothalamic hormone. *Science, N.Y.* **178,** 417–418.

Plotnikoff, N. P., Prange, A. J. Jr., Breese, G. and Wilson, I. C. (1974). Thyrotropin releasing hormone: enhancement of DOPA activity in thyroidectomised rats. *Life Sci.* **14,** 1271–1278.

Podolsky, S. and Sivaprasad, R. (1972). Assessment of growth hormone reserved: comparison of intravenous arginine and subcutaneous glucagon stimulation tests. *J. Clin. Endocrinol. Metab.* **35,** 580–584.

Polak, J. M., Pearse, A. G. E., Grimelius, L., Bloom, S. R. and Arimura, A. (1975). Growth hormone release-inhibiting hormone in gastrointestinal and pancreatic D cells. *Lancet*, **i**, 1220–1222.

Popa, G. T. and Fielding, U. (1930). A portal circulation from the pituitary to the hypothalamic region. *J. Anat.* (*London*) **65**, 88–91.

Portanova, R. and Sachs, A. (1967). A specific adsorbent for vasopressin: The purification of labelled hormone. *Endocrinology*, **80**, 527–529.

Portanova, R. and Sayer, G. (1973). Isolated pituitary cells: CRF-like activity of neurohypophysial and related polypeptides (37386). *Proc. Soc. exp. Biol. Med.* **143**, 661–666.

Porter, J. C. and Ben-Jonathan, N. (1974). Neuroendocrine mechanisms involved in biorhythms. *In* "Biorhythms and Human Reproduction" (M. Ferin, F. Hallberg, R. M. Richart and R. L. Vande Wiele, eds) pp. 607–619. Wiley, New York.

Porter, J. C. and Rumsfield, H. W. (1956). Effect of lyophilised plasma fractions from hypophyseal portal vessel blood on adrenal ascorbic acid. *Endocrinology*, **58**, 359–364.

Porter, J. C., Dhariwal, A. P. S. and McCann, S. M. (1967). Response of the anterior pituitary-adrenocortical axis to purified CRF. *Endocrinology*, **80**, 679–688.

Porter, J. C., Goldman, B. D. and Wilbur, J. F. (1970). Hypophysiotropic hormones in portal vessel blood. *In* "Hypophysiotropic Hormones of the Hypothalamus: Assay and Chemistry" (J. Meites, ed.) pp. 282–297. Williams and Wilkins, Baltimore.

Porter, J. C., Kamberi, I. A. and Grazia, Y. A. (1971). Pituitary blood flow and portal vessels. *In* "Frontiers in Neuroendocrinology" (L. Martini and W. F. Ganong, eds) pp. 145–175. Oxford University Press, New York.

Porter, J. C., Ben-Jonathan, N., Oliver, C., Eskay, R. L. and Winters, A. J. (1975). The interrelationships of the CSF, hypophyseal portal vessels and the hypothalamus and their role in the regulation of anterior pituitary function. *In* "Symposium on Neuroendocrine Regulation of Fertility", pp. 29–32. Brain Information Service, Los Angeles.

Porter, J. R. (1974). The development for a method for the small scale isolation of ACTH from individual rat pituitary glands and study of factors affecting ACTH biosynthesis *in vitro*. *Dissertation Abstracts Intern.* **B34**, 3448.

Poulain, P. and Carette, B. (1974). Iontophoresis of prostaglandins on hypothalamic neurons. *Brain Res.* **79**, 311–314.

Prange, A. J., Wilson, I. C., Lara, P. P., Alltop, L. B. and Breese, G. R. (1972). Effects of thyrotropin releasing hormone in depression. *Lancet*, **ii**, 999–1002.

Price, H. V., Cone, B. A. and Keogh, M. (1971). The length of gestation in congenital adrenal hyperplasia. *J. Obstet. Gynaec. Brit. Cwlth.* **78**, 430–434.

R

Rabello, M. M., Snyder, P. J. and Utiger, R. D. (1974). Effects on the pituitary thyroid axis and prolactin secretion of single and repetitive oral doses of thyrotrophin releasing hormone (TRH). *J. Clin. Endocrinol. Metab.* **39**, 571–578.

Raghuramulu, N. and Jaya Rao, K. S. (1974). Growth hormone secretion in protein-calorie malnutrition. *J. Clin. Endocrinol. Metab.* **38**, 176–180.

Raisman, G. (1966). Neural connexions of the hypothalamus. *Br. Med. Bull.* **22**, 197–201.

Raisman, G. (1970a). Some aspects of the neural connections of the hypothalamus. *In* "The Hypothalamus" (L. Martini, M. Motta and F. Fraschini, eds) pp. 1–5. Academic Press, New York and London.

Raisman, G. (1970b). An evaluation of the basic patterns of connections between the limbic system and the hypothalamus. *Am. J. Anat.* **129**, 197–202.

Raisman, G. (1972). A second look at the parvicellular neurosecretory system. *In* "Brain-Endocrine Interaction: Median Eminence. Structure and Function" (K. M. Knigge, D. E. Scott and A. Weindl, eds) pp. 109–118. Karger, Basel.

Raisman, G. (1973). Electron microscopic studies of the development of new neurohaemal contacts in the median eminence of the rat after hypophysectomy. *Brain Res.* **55,** 245–261.

Raisman, G. (1974). Anatomical and functional studies on the preoptic area and its involvement in the initiation of ovulation in the female rat. *Advances in the Biosciences,* **15,** 25–34.

Raisman, G. and Field, P. M. (1971a). Anatomical considerations relevant to the interpretation of neuroendocrine experiments. *In* "Frontiers in Neuroendocrinology" (L. Martini and W. F. Ganong, eds) pp. 3–44. Oxford University Press, New York.

Raisman, G. and Field, P. M. (1971b). Sexual dimorphism in the preoptic area of the rat. *Science, N.Y.* **173,** 731–733.

Ramachandran, J., Chung, D. and Li, C. H. (1965). Synthesis of biologically active peptides related to ACTH. *Metabolism,* **13,** Suppl. 1043–1058.

Ramalay, J. A. (1974). Mini review: adrenal-gonadal interactions at puberty. *Life Sci.* **14,** 1623–1633.

Ramirez, V. D. (1973). Endocrinology of puberty. *In* "Handbook of Physiology" (R. O. Greep and E. B. Astwood, eds) Vol. II, Part 1, pp. 1–28. Williams and Wilkins, Baltimore.

Ramirez, V. D. and Kordon, C. (1975). Localisation and subcellular distribution of hypothalamic hormones: studies on luteinizing hormone-releasing hormone (LH-RH). *In* "Hypothalamic Hormones" (M. Motta, P. Crosignani and L. Martini, eds) pp. 57–74. Academic Press, New York and London.

Ramirez, V. D. and McCann, S. M. (1963). A highly sensitive test for LH-releasing activity: the ovariectomized, oestrogen progesterone-blocked rat. *Endocrinology,* **73,** 193–198.

Ramirez, V. D. and McCann, S. M. (1964). Thioglycollate-stable luteinizing hormone and corticotrophin-releasing factors. *Am. J. Physiol.* **207,** 441–445.

Ramirez, V. D. and Sawyer, C. H. (1965). Fluctuations in hypothalamic LH-RF (luteinizing hormone-releasing factor) during the rat estrous cycle. *Endocrinology,* **76,** 282–289.

Ramirez, V. D. and Sawyer, C. H. (1974). Differential dynamic responses of plasma LH and FSH to ovariectomy and to a single injection of estrogen in the rat. *Endocrinology,* **94,** 987–993.

Ramón y Cajal, S. (1894). Algunas contribuciones conociamento de los ganglios del encéfale. *Anal. Soc. espan. Hist. nat.* **23,** 214–215.

Rang, H. P. (1971). Drug receptors and their functions. *Nature* (*London*), **231,** 91–96.

Rannevik, G. and Thorell, J. (1974). Effects of synthetic LH/FSH-releasing hormone (LRH) on plasma LH and FSH in ammenorrhoeic women. Comparison between intravenous and intramuscular administration. *Acta endocr., Copenh.* **75,** 647–652.

Rapoport, B., Refetoff, S., Fang, V. S. and Friesen, H. G. (1973). Suppression of serum thyrotropin (TSH) by L-DOPA in chronic hypothyroidism: interrelationships in the regulation of TSH and prolactin secretion. *J. Clin. Endocrinol. Metab.* **36,** 256–262.

Rasmussen, A. T. (1938). Innervation of the hypophysis. *Endocrinology,* **23,** 263 278.

Rasmussen, H. (1974). Organization and control of endocrine systems. *In* "Textbook of Endocrinology" (R. H. Williams, ed.) pp. 1–30. Saunders, Philadelphia.

Redding, T. W. and Schally, A. V. (1969a). Studies on the inactivation of thyrotrophin releasing hormone (TRH). *Proc. Soc. exp. Biol. Med.* **131,** 415–421.

Redding, T. W. and Schally, A. V. (1969b). Studies on the thyrotropin releasing hormone (TRH) activity in peripheral blood. *Proc. Soc. exp. Biol. Med.* **131,** 421–425.

Redding, T. W. and Schally, A. V. (1970). A study on the mode of administration of thyrotropin-releasing hormone (TRH) in mice. *Neuroendocrinology,* **6,** 329–335.

Redding, T. W. and Schally, A. V. (1971). The distribution of radioactivity following the administration of labelled thyrotropin-releasing hormone (TRH) in rats and mice. *Endocrinology,* **89,** 1075–1081.

Redding, T. W. and Schally, A. V. (1973). The distribution, half-life and excretion of tritiated luteinizing hormone-releasing hormone (LH-RH) in rats. *Life Sci.* **12,** 23–32.

Redding, T. W., Schally, A. V., Arimura, A. and Matsuo, H. (1972). Stimulation of release and synthesis of luteinizing hormone (LH) and follicle stimulating hormone (FSH) in tissue culture of rat pituitaries in response to natural and synthetic LH and FSH releasing hormone. *Endocrinology*, **90,** 764–770.

Redding, T. W., Kastin, A. J., Gonzalez-Barcena, D., Coy, D. H., Coy, E. J., Schalch, D. S. and Schally, A. V. (1973a). The half-life, metabolism and excretion of tritiated luteinizing hormone-releasing hormone (LH-RH) in man. *J. Clin. Endocrinol. Metab.* **37,** 626–631.

Redding, T. W., Kastin, A. J., Nair, R. M. G. and Schally, A. V. (1973b). Distribution, half-life and excretion of ^{14}C- and ^{3}H-labelled L-prolyl-L-leucyl-glycinamide in the rat. *Neuroendocrinology*, **11,** 92–100.

Redding, T. W., Kastin, A. J., Gonzalez-Barcena, D., Coy, D. H., Hirotsu, I., Ruelas, J. and Schally, A. V. (1974). The disappearance, excretion and metabolism of tritiated prolyl-leucyl-glycinamide in man. *Neuroendocrinology*, **16,** 119–136.

Redgate, E. S. (1967). Comparison of plasma corticotropin concentration in intact and adrenalectomised dogs and cats during haemorrhage. *Endocrinology*, **80,** 741–747.

Redgate, E. S. (1970). ACTH release evoked by electrical stimulation of brain stem and limbic sites in the cat: the absence of ACTH release upon infundibular area stimulation. *Endocrinology*, **86,** 806–823.

Redgate, E. S. (1974). Neural control of pituitary adrenal activity in *cyprinus carpio*. *Gen Comp. Endocrinol.* **22,** 35–41.

Redgate, E. S. and Fahringer, E. E. (1973). A comparison of the pituitary adrenal activity elicited by electrical stimulation of preoptic, amygdaloid and hypothalamic sites in the rat brain. *Neuroendocrinology*, **12,** 334–343.

Redgate, E. S., Fahringer, E. E. and Szechtman, L. L. (1973). Effects of the nervous system of pituitary adrenal activity. *In* "Brain Pituitary-Adrenal Interrelationships" (A. Brodish and E. S. Redgate, eds) pp. 152–175. Karger, Basel.

Rees, L. H., Ratcliffe, J. G., Besser, G. M., Kramer, R. and Landon, J. (1973). Comparison of the redox assay for ACTH with previous assays. *Nature, Lond.* **241,** 84–85.

Rees, L. H., Burke, C. W., Chard, R., Evans, S. W. and Letchworth, A. T. (1975). Possible placental origin of ACTH in normal human pregnancy. *Nature, Lond.* **254,** 620–622.

Refetoff, S., Fang, V. S., Rapoport, B. and Friesen, H. G. (1974). Interrelationships in the regulation of TSH and prolactin secretion in man: effects of L-dopa, TRH and thyroid hormone in various combinations. *J. Clin. Endocrinol. Metab.* **38,** 450–457.

Reichlin, S. (1964). Function of the hypothalamus in regulation of pituitary-thyroid activity. *In* "Brain Thyroid Relationships" (M. P. Cameron and M. O'Connor, eds) pp. 17–32. Churchill, London.

Reichlin, S. (1966a). Control of thyrotropin hormone secretion. *In* "Neuroendocrinology" (L. Martini and W. Ganong, eds) Vol. I, pp. 445–536. Academic Press, New York and London.

Reichlin, S. (1966b). Regulation of somatotrophic hormone secretion. *In* "The Pituitary Gland" (G. W. Harris and B. T. Donovan, eds) Vol. II, pp. 270–298. Butterworths, London.

Reichlin, S. (1973a). Hypothalamic-pituitary function. *In* "Proc. 4th Int. Congr. of Endocrinology", Int. Congr. Series 273, pp. 1–16. Excerpta Medica, Amsterdam.

Reichlin, S. (1973b). The physiology of growth hormone regulation: pre-and postimmuno-assay eras. *Metabolism*, **22,** 987–993.

Reichlin, S. (1974). Neuroendocrinology. *In* "Textbook of Endocrinology" (R. H. Williams, ed.) pp. 774–831. Saunders, Philadelphia.

Reichlin, S. (1975). Regulation of the hypophysiotropic secretion of the brain. *Arch. Intern. Med.* **135,** 1350–1361.

Reichlin, S. and Mitnick, M. A. (1973a). Biosynthesis of hypothalamic hypophysiotropic hormones. *In* "Frontiers in Neuroendocrinology" (W. F. Ganong and L. Martini, eds) pp. 61–68. Oxford University Press, New York.

Reichlin, S. and Mitnick, M. A. (1973b). Biosynthesis of thyrotropin releasing hormone and its control by hormones, central monoamines and external environment. *In* "Hypothalamic Hypophysiotropic Hormones" pp. 124–135. Excerpta Medica, ICS 263, Amsterdam.

Reichlin, S. and Mitnick, M. A. (1973c). Enzymatic synthesis of growth hormone-releasing factor (GH-RH) by rat incubates and by extracts of rat and porcine hypothalamic tissue. *Proc. Soc. Exp. Biol. Med.* **142,** 497–501.

Reichlin, S. and Utiger (1967). Regulation of the pituitary thyroid axis in man: relationship of TSH concentration to concentration of free and total thyroxine in plasma. *J. Clin. Endocrinol. Metab.* **27,** 251–255.

Reichlin, S., Volpert, E. M. and Werner, S. C. (1966). Hypothalamic influence on thyroxine monodeiodination by rat anterior pituitary gland. *Endocrinology,* **78,** 302–306.

Reichlin, S., Martin, J. B., Mitnick, M. A., Boshans, R. L., Grimm-Jørgensen, Y., Bollinger, J., Gordon, J. and Malacara, J. (1972). The hypothalamus in pituitary-thyroid regulation. *Recent. Prog. Horm. Res.* **28,** 229–286.

Reichlin, S., Jackson, I. M. D., Boyd, A. E. III, Saperstein, R. and Patel, Y. (1976). Hypophysiotropic hormones. *Annu. Rev. Physiol.* **38,** 389–424.

Reiter, R. J. (1973). Involvement of pineal indoles and polypeptides with the neuroendocrine axis. *Progr. Brain Res.* **39,** 281–287.

Reiter, R. J. (1974). Pineal regulation of hypothalamic pituitary axis: gonadotrophins. *In* "Handbook of Physiology" Section 7, Vol. IV (E. Knobil and W. H. Sawyer, eds) Part 2, pp. 519–562. American Physiological Society, Washington, D.C.

Reiter, R. J. and Vaughan, M. K. (1975). A study of indoles which inhibit pineal antigonadotrophic activity in male hamsters. *Endocrinol. Res. Commun.* **2,** 299–308.

Relkin, R. (1972). Effects of pinealectomy and constant light and darkness on thyrotropin levels in pituitary and plasma or rat. *Neuroendocrinology,* **10,** 46–52.

Relkin, R. (1974). Effects of alterations in serum osmolality on pituitary and plasma prolactin levels in the rat. *Neuroendocrinology,* **14,** 61–64.

Renaud, L. P. and Martin, J. B. (1975). Thyrotropin releasing hormone (TRH): depressant action on central neuronal activity. *Brain Res.* **86,** 150–154.

Renaud, L. P., Martin, J. B. and Brazeau, P. (1975). Depressant action of TRH, LH-RH and somatostatin on activity of central neurons. *Nature, Lond.* **255,** 233–235.

Rennels, E. G., Black, D. E. and Warchol, J. B. (1975). Ultrastructural changes in prolactin cells following *in vitro* incubation of hemi-pituitary glands with dopamine. *Program 57th Meeting U.S. Endocrine Soc.* p. 98.

Resko, J. A., Norman, R. L., Niswender, G. D. and Spies, H. G. (1974). The relationship between progestins and gonadotrophins during the late luteal phase of the menstrual cycle in the rhesus monkey. *Endocrinology,* **94,** 128–135.

Ressler, C. and Popenoe, E. A. (1973). Posterior pituitary hormones. *In* "Methods of Investigative and Diagnostic Endocrinology" (S. A. Berson and R. S. Yalow, eds) Vol. 2, pp. 681–691. North-Holland, Amsterdam.

Rêthelyi, M. and Halász, B. (1970). Origin of the nerve endings in the surface zone of the median eminence. *Exp. Brain Res.* **11,** 145–158.

Retiene, K., Schultz, F. and Marco, J. (1967). Circadian rhythm activity of pituitary adrenal function under resting and stress conditions in rats. *Rass. Neurol. Vegetativa,* **21,** 217–221.

Riegle, G. D. (1976). Aging and adrenocortical function. *In* "Hypothalamus, Pituitary and Aging" (A. V. Everitt and J. A. Burgess, eds) pp. 547–552. C. C. Thomas, Springfield.

Rieutort, M. (1974). Pituitary content and plasma levels of growth hormone in foetal and weanling rats. *J. Endocr.* **60,** 261–268.

Riggs, D. S. (1970). "Control Theory and Physiological Feedback Mechanisms". Williams and Wilkins, Baltimore.

Rioch, D. M., Wislocki, G. B. and O'Leary, L. (1940). A precis of preoptic, hypothalamic and hypophysial terminology with atlas. *Res. Publ. Ass. Res. nerv. ment. Dis.* **20,** 3–30.

Rivier, J. E. F. (1974). Somatostatin: total Solid Phase Synthesis. *J. Am. Chem. Soc.* **96,** 2986–2992.

Rivier, J., Brazeau, P., Vlae, W., Long, N., Burgus, R., Gilon, C., Yardley, J. and Guillemin, R. (1973). Synthèse totale par phase solide d'un tetradecapeptide ayant les propriétés cliniques et biologiques de la somatostatine. *C. R. Acad. Sci.* **276,** 737–740.

Rivier, J., Brown, M. and Vale, W. (1975a). D-Trp^8-somatostatin: an analogue of somatostatin more potent than the native molecule. *Biochem. Biophys. Res. Commun.* **65,** 746–751.

Rivier, J., Ling, N., Monaham, M., River, C., Brown, M. and Vale, W. (1975b). Chemistry and biological properties of somatostatin analogues. *In* "Peptides: Chemistry, Structure and Biology" (R. Walter and J. Meinhofer, eds) pp. 863–870. Ann Arbor Science, Michigan.

Robinson, A. G., Zimmerman, E. A. and Frantz, A. G. (1971). Physiologic investigation of posterior pituitary binding proteins neurophysin I and neurophysin II. *Metabolism,* **20,** 1148–1155.

Robinson, N. and Wells, F. (1973). Distribution and localisation of site of GABA metabolism in the adult rat brain. *J. Anat.* **114,** 65–78.

Rochfort, G. J., Rosenberger, J. and Saffran, M. (1959). Depletion of pituitary corticotrophin by various stresses and by neurohypophysial preparations. *J. Physiol. Lond.* **146,** 105–116.

Rodger, N. W., Beck, J. C., Burgus, R. and Guillemin, R. (1969). Variability of response in the bioassay for a hypothalamic releasing factor based on rat pituitary growth hormone content. *Endocrinology,* **84,** 1373–1383.

Rodgers, C. H. (1975). Neuroendocrine mechanisms responsible for gonadotrophin release. *J. Reprod. Med.* **14,** 1–7.

Rodgers, C. H., Schwartz, N. B. and Nequin, L. G. (1974). Interactions between the ovarian and adrenocortical regulating systems: occurrence of ovulation. *In* "Biological Rhythms in Neuroendocrine Activity" (M. Kawakami, ed) pp. 241–252. Igaku Shoin, Tokyo.

Rodriguez, E. M. and La Pointe, J. (1970). Light and electron microscopic study of the pars intermedia of the lizard, *Klauberina riversiana. Z. Zellforsch.* **104,** 1–13.

Rolland, R., De Jong, F. H., Schellekens, L. S. and Lequin, R. M. (1975a). The role of prolactin in the restoration of ovarian function during the early post-partum period in the human female. II. A study during inhibition of lactation by bromocryptine. *Clin. Endocrinol.* **4,** 27–38.

Rolland, R., Lequin, R. M., Schellekens, L. A. and De Jong, F. H. (1975b). The role of prolactin in the restoration of ovarian function during the early post-partum period in the human female. I. A study during physiological lactation. *Clin. Endocrinol.* **4,** 15–25.

Romeis, B. (1940). Hypophyse. *In* "Handbuch der mikroskopischen Anatomie des Menschen", Vol. 6, Part 3. Springer-Verlag, Berlin.

Rønnekliev, O. K., Krulich, L. and McCann, S. M. (1973). An early morning surge of prolactin in the male rat and its abolition by pinealectomy. *Endocrinology,* **92,** 1339–1342.

Root, A. W., Snyder, P. J., Rezvani, I., DiGeorge, A. M. and Utiger, R. D. (1973). Inhibition of thyrotropin releasing hormone mediated secretion of thyrotropin by human growth hormone. *J. Clin. Endocrinol. Metab.* **36,** 103–107.

Rose, S. and Nelson, J. (1956). Hydrocortisone and ACTH release. *Aust. J. Exp. Biol. Med. Sci.* **34,** 77–82.

Rothenbucher, G., Birk, J., Loos, U., Raptis, S., Fletcher, S. and Pfeiffer, E. F. (1973). Untersuchungen mit ^{125}Jod-markiertem, synthetischem Thyrotropin-Releasing Hormone (TRH). *Klin. Wochenschr.* **51,** 12–14.

Roux, M. (1972). Etude du lobe intermédiaire de l'hypophyse chez la souris normale et throidectomisee: volumétrie, histologie et incorporation de divers éléments radioactifs. *Z. Zellforsch.* **129,** 290–300.

Rowan, W. (1925). Relation of light to bird migration and developmental changes. *Nature, Lond.* **115,** 494–495.

Rowe, J. W., Reichert, J. R., Klein, D. C. and Reichlin, S. (1970). Relation of the pineal gland and environmental lighting to thyroid function in the rat. *Neuroendocrinology,* **6,** 247–254.

Rowe, P. H., Racey, P. A., Lincoln, G. A., Ellwood, M., Lehane, J. and Shenton, J. C. (1975). The temporal relationship between the secretion of luteinizing hormone and testosterone in man. *J. Endocr.* **64,** 17–26.

Royce, P. C. and Sayers, G. (1958a). Blood ACTH: effects of ether, pentobarbital, epirephrine and pain. *Endocrinology,* **63,** 794–800.

Royce, P. C. and Sayers, G. (1958b). Corticotrophin releasing activity of a pepsin-labile factor in the hypothalamus. *Proc. Soc. exp. Biol. Med.* **96,** 677–680.

Rubin, R. J. (1975). Sleep endocrinology studies in man. *Progr. Brain Res.* **42,** 73–80.

Rudinger, J., Pliska, V. and Krejci, I. (1972). Oxytocin analogs in the analysis of some phases of hormone action. *Recent. Prog. Horm. Res.* **28,** 166–172.

Ruf, K. B. (1973). How does the brain control the process of puberty? *Z. Neurol.* **204,** 96–105.

Russell, G. F. (1969). Metabolic, endocrine and psychiatric aspects of anorexia nervosa. *Sci. Basis Med. Ann. Rev.,* 236–255.

Rust, C. C. (1965). Hormonal control of pelage cycles in the short tailed weasel (*Mustela erminea bangsi*). *Gen. Comp. Endocrinol.* **5,** 222–231.

Rust, C. C. and Meyer, R. K. (1969). Hair color, moult and testis size in the male short tailed weasels treated with melatonin. *Science, N.Y.* **165,** 921–922.

S

Saberi, M., Sterling, F. H. and Utiger, R. D. (1975). Reduction in extrathyroidal triiodothyronine production by propylthiouracil in man. *J. Clin. Invest.* **55,** 218–223.

Sachdev, Y., Gomez-Pan, A., Tunbridge, W. M. G., Duns, A., Weightman, D. R., Hale, R. and Goolamili, S. K. (1975). Bromocriptine treatment in acromegaly. *Lancet,* **ii,** 1164–1168.

Sachs, H. and Haller, E. W. (1968). Further studies on the capacity of the neurophysins to release vasopressin. *Endocrinology,* **83,** 251.

Sachs, H., Fawcett, P., Tabatake, Y. and Portanova, R. (1969). Biosynthesis and release of vasopressin and neurophysin. *Recent Prog. Horm. Res.* **25,** 447–464.

Sadow, J. (1973). Release of ACTH from the isolated pituitary gland of the rat: Studies with theophylline. *J. Endocr.* **59,** xxx.

Sadow, J. and Penn, D. (1973). The effect of prostaglandin E_1 on the corticotrophin-releasing activity of vasopressin and hypothalamic extract on the isolated pituitary gland. *J. Endocr.* **57,** xxvii-xxviii.

Saffran, M. (1974). Chemistry of the hypothalamo-hypophysiotrophic factors. *In* "Handbook of Physiology, Endocrinology", Vol. IV, Part 2, pp. 575–586. American Physiological Society, Washington, D.C.

Saffran, M. and Schally, A. V. (1955). *In vitro* bioassay of corticotropin: Modification and statistical treatment. *Endocrinology,* **56,** 523–532.

Saffran, M., Schally, A. V. and Benfry, B. G. (1955). Stimulation of the release of corticotrophin from the adenohypophysis by a neurohypophysial factor. *Endocrinology*, **57**, 439–444.

Saffran, M., Pearlmutter, F. and Rapino, E. (1972). Pressinoic acid: a peptide with potent corticotrophin-releasing activity. *Biochem. Biophys. Res. Commun.* **49**, 748–751.

Salaman, D. F. (1974). The role of DNA, RNA and protein synthesis in sexual differentiation of the brain. *Progr. Brain Res.* **41**, 349–361.

Saland, L. C. (1968). Ultrastructure of the frog pars intermedia in relation to hypothalamic control of hormone release. *Neuroendocrinology*, **3**, 72–88.

Salassa, R. M., Bennett, W. A., Keating, F. R. and Sprague, R. C. (1953). Postoperative adrenal cortical insufficiency. Occurrence in patients previously treated with cortisone. *J. Amer. Med. Ass.* **152**, 1509–1515.

Salem, M. H., Norton, H. W. and Nalbandov, A. V. (1970). A study of ACTH and CRF in chickens. *Gen. Comp. Endocrinol.* **14**, 270–280.

Salmoiraghi, G. C., Costa, E. and Bloom, F. E. (1965). Pharmacology of central synapses. *Ann. Rev. Pharmacology*, **5**, 213–234.

Salmon, M. A. (1972). Russell's diencephalic syndrome of early childhood. *J. Neurol. Neurosurg. Psychiat.* **35**, 196–201.

Samli, M. M., and Geschwind, I. I. (1968). Some effects of energy-transfer inhibitors and of Ca^{++} free or K^{+} enhanced media on the release luteinizing hormone (LH) from the rat pituitary gland *in vitro*. *Endocrinology*, **82**, 225–231.

Samuels, H. H., Tsai, J. S., Casanova, J. and Stanley, F. (1975). *In vitro* characterization of solubilized nuclear receptors from rat liver and cultured GH cells. *J. Clin. Invest.* **54**, 853–865.

Sanchez-Franco, F., Garcia, M. D., Cacicedo, L., Martin-Zurro, A., and Escobar del Rey, F. (1973). Menstrual cycle on thyrotropin (TSH) response to thyrotropin-releasing hormone (TRH). *J. Clin. Endocrinol. Metab.* **37**, 736–740.

Sandow, J. and Babej, M. (1975). Modulating effects of prostaglandins in the hypothalamo-hypophyseal system. *In* "Hypothalamic Hormones: Structure, Synthesis and Biological Activity" (D. Gupta and W. Voelter, eds) pp. 137–156. Verlag Chemie, Weinheim.

Sandow, J. and Heptner, W. (1974). Positive feedback effect of small doses of LH-RH after chronic treatment. *Acta endocr., Copenh.* Suppl. **184**, 28.

Sandow, J., Arimura, A. and Schally, A. G. (1972). Stimulation of growth hormone release by anterior pituitary perfusion in the rat. *Endocrinology*, **90**, 1315–1319.

Sandow, J., Schally, A. V., Uehara, T. and Arimura, A. (1973a). Rat anterior pituitary perfusion: release of RIA-GH purified porcine GH-RH. *Acta endocr., Copenh.* Suppl. **177**, 201.

Sandow, J., Heptner, W. and Vogel, H. G. (1973b). Studies on *in vivo* inactivation of synthetic LH-RH. *In* "Hypothalamic Hypophysiotropic Hormones" (C. Gual and E. Rosemberg, eds) pp. 64–67. Excerpta Medica, Amsterdam.

Sandow, J., Konig, W. and v. Rechenberg, W. (1975). Regulation of MSH-secretion by peptides to oxytocin and vasopressin. *In* "Sixth International Congress of Pharmacology. Abstract No. 199, p. 93.

Sanford, L. M., Winter, J. S. D., Palmer, W. M. and Howland, B. E. (1974). The profile of LH and testosterone in the ram. *Endocrinology*, **96**, 627–631.

Saperstein, R., Mothon, S., and Reichlin, S. (1975). Enzymatic degradation of TRH and LRH by hypothalamic extracts. *Fed. Proc. Fed. Am. Soc. Exp. Biol.* **34**, 239.

Sarantakis, D. and McKinley, W. A. (1973). Total synthesis of hypothalamic somatostatin. *Biochem. Biophys. Res. Commun.* **54**, 234–238.

Sarantakis, W. A., Grant, N. W. and McKinley, W. A. (1975). Structure-activity studies on somatostatin. *In* "Endocrinology" (1975) (I. McIntyre, ed.) pp. 87–94. Academic Press, London and New York.

Sarkar, D. K., Chiappa, S. A., Fink, G. and Sherwood, N. M. (1976). Gonadotrophin-releasing hormone surge in pro-oestrous rats. *Nature, Lond.* **264,** 461–463.

Sassin, J. F., Frantz, A. G., Weitzman, E. D. and Kapen, S. (1972). Human prolactin: 24-hour pattern with increased release during sleep. *Science,* **177,** 1205–1207.

Sassin, J. F., Frantz, A. G., Kapen, S. and Weitzman, E. D. (1973). The nocturnal release of human prolactin is dependent on sleep. *J. Clin. Endocrinol. Metab.* **37,** 436–440.

Sato, T. and George, J. C. (1973). Evidence for the existence of corticotrophin-releasing factor in the pigeon hypothalamus. *Can. J. Physiol. Pharmacol.* **51,** 743–747.

Sato, T., Hirano, T., Tyujo, T., Iesaka, T., Taya, K. and Igarashi, M. (1975). Direct action of prostaglandins on the rat pituitary gland. *Endocrinology,* **96,** 45–49.

Sawano, S., Arimura, A., Schally, A. V., Reddy, T. W. and Shapiro, S. (1969). Neonatal corticoid administration: effects upon adult pituitary growth hormone releasing hormone activity. *Acta endocr., Copenh.* **61,** 57–67.

Sawano, S., Matsuo, I., Yamazaki, M., Baba, Y. and Takahashi, K. (1974) Studies on growth hormone-releasing activity in bovine hypothalami. *In* "Psychoneuroendocrinology" (N. Hatotani, ed.) pp. 232–242. Karger, Basel.

Sawin, C. T., and Hershman, J. M. (1975). Clinical use of thyrotropin releasing hormone (TRH). *In* "Pharmacology of the Thyroid" (G. A. Bray, and J. M. Hershman, eds), pp. 27–32. Pergamon Press, New York.

Sawyer, C. H. (1975). Some recent developments in brain-pituitary-ovarian physiology. *Neuroendocrinology,* **17,** 97–124.

Sawyer, C. H. and Hilliard, J. (1971). Sites of feedback action of estrogen and progesterone. *Excerpta Medica Int. Congr. Ser.* **219,** 716–721.

Sawyer, C. H., Markee, J. E. and Townsend, B. F. (1949). Cholinergic and adrenergic components in the neurohumoral control of the release of LH in the rabbit. *Endocrinology,* **44,** 18–37.

Sawyer, C. H., Hilliard, J., Kanematsu, S., Scaramuzzi, R. and Blake, C. A. (1974). Effects of intraventricular infusions of norepinephrine and dopamine of LH release and ovulation in the rabbit. *Neuroendocrinology,* **15,** 328–337.

Sawyer, W. H. (1968). Phylogenetic aspects of the neurohypophysial hormones. *In* "Handbook of Experimental Pharmacology" (H. Berde, ed.) Vol. 23, pp. 717–747. Springer-Verlag, Berlin.

Sawyer, W. H. (1970). Homologies of structure and function among neurohypophysial peptides. *Miami Winter Symposia.* **1,** 257–269. North-Holland, Amsterdam.

Sawyer, W. H. (1971). Evolution of neurohypophysial peptides among the non-mammalian vertebrates. *In* "Neurohypophysial Hormones" Ciba Study Grp. 39 (G. E. W. Wolstenholme and J. Birch eds) pp. 5–18. Churchill Livingstone, London.

Sawyer, W. H. and Manning, M. (1973). Synthetic analogues of oxytocin and the vasopressins. *Annu. Rev. Pharmacol.* **13,** 5–17.

Sayers, G. and Sayers, M. A. (1947). Regulation of pituitary adrenocorticotrophic activity during the response of the rat to acute stress. *Endocrinology,* **40,** 205–273.

Sayers, G. and Portanova, R. (1974). Secretion of ACTH by isolated anterior pituitary cells kinetics of stimulation by corticotrophin-releasing factor and of inhibition by corticosterone. *Endocrinology,* **94,** 1723–1730.

Sayers, G., Swallow, R. L. and Giordiano, N. D. (1971). An improved technique for the preparation of isolated rat adrenal cells: a sensitive, accurate and specific method for the assay of ACTH. *Endocrinology,* **88,** 1063–1068.

Scapagnini, U. and Preziosi, P. (1972). Role of brain noradrenaline and serotonin in the tonic and phasic regulation of the hypothalamohypophyseal adrenal axis. *Arch. Int. Pharmacodyn. Thér.* **196,** Suppl. 205–220.

Scapagnini, U., van Loon, G. R., Moberg, G. P. and Ganong, W. F. (1970). Effect of α-methyl-p-tyrosine on the circadian variation of plasma corticosterone in rats. *Eur. J. Pharmacol.* **11,** 266–268.

Scapagnini, U., Moberg, G. P., Van Loon, G. R., De Groot, J. and Ganong, W. F. (1971). Relation of brain 5-hydroxytryptamine content to the diurnal variation in plasma corticosterone in the rat. *Neuroendocrinology*, **7,** 90–96.

Schadlow, A. R., Surks, M. I., Schwartz, H. I. and Oppenheimer, J. H. (1972). Specific triiodothyronine binding sites in the anterior pituitary of the rat. *Science, N.Y.* **176,** 1252–1254.

Schalch, D. S. and Reichlin, S. (1968). Stress and growth hormone release. *In* "Growth Hormone" (A. Pecile and E. E. Müller, eds) pp. 211–225. Excerpta Medica, Amsterdam.

Schally, A. V. and Arimura, A. (1972). Growth hormone-releasing hormone (GH-RH) of the hypothalamus; its chemistry and *in vivo* and *in vitro* effects. *In* "Growth and Growth Hormone" (A. Pecile and E. E. Müller, eds) pp. 247–251. Excerpta Medica, Amsterdam.

Schally, A. V. and Bowers, C. Y. (1964). Corticotropin-releasing factor and other hypothalamic peptides. *Metabolism*, **13,** 1190–1205.

Schally, A. V. and Guillemin, R. (1963). Isolation and chemical characterization of a β-CRF from pig posterior pituitary glands. *Proc. Soc. exp. Biol. Med.* **112,** 1014–1017.

Schally, A. V. and Kastin, A. L. (1966). Purification of a bovine hypothalamic factor which elevates pituitary MSH levels in rats. *Endocrinology*, **79,** 768–772.

Schally, A. V., Saffran, M. and Zimmerman, B. (1958). A corticotrophin-releasing factor partial purification and amino acid composition. *Biochem. J.* **70,** 97–103.

Schally, A. V., Anderson, R. N., Lipscomb, H. S., Long, J. M. and Guillemin, R. (1960). Evidence for the existence of two corticotrophin-releasing factors α and β. *Nature, Lond.* **188,** 1192–1193.

Schally, A. V., Kuroshima, A., Ishida, Y., Redding, T. W. and Bowers, C. Y. (1965). The presence of prolactin inhibiting factor (PIF) in extracts of beef, sheep and pig hypothalami. *Proc. Soc. exp. Biol. Med.* **118,** 350–352.

Schally, A. V., Müller, E. E., Arimura, A., Bowers, C. Y., Saito, T., Redding, T. W., Sawano, S. and Pizzolato, P. (1967). Releasing factors in human hypothalamic and neurohypophysial extracts. *J. Clin. Endocrinol. Metab.* **27,** 755–762.

Schally, A. V., Arimura, A., Bowers, C. Y., Kastin, A. J., Sawano, S. and Redding, T. W. (1968a). Hypothalamic neurohormones regulating anterior pituitary function. *Recent Prog. Horm. Res.* **24,** 497–588.

Schally, A. V., Müller, E. E., Arimura, A., Saito, T., Sawano, S., Bowers, C. Y. and Steelman, S. L. (1968b). Growth hormone-releasing factor (GRF): physiological and biochemical studies with GRF preparations of bovine and porcine origin. *Ann. N.Y. Acad. Sci.* **148,** 372–388.

Schally, A. V., Sawano, S., Müller, E. E., Arimura, A., Bowers, C. Y., Redding, T. W. and Steelman, S. L. (1968c). Hypothalamic growth hormone-releasing hormone GRH). Purification and *in vivo* and *in vitro* studies. *In* "Growth Hormone" (A. Pecile and E. E. Muller, eds) pp. 185–203. Excerpta Medica, Amsterdam.

Schally, A. V., Redding, T.W., Bowers, C. Y. and Barrett, J. F. (1969). Isolation and properties of porcine thyrotropin-releasing hormone. *J. Biol. Chem.* **244,** 4077–4088.

Schally, A. V., Arimura, A., Bowers, C. Y., Wakabayashi, I., Kastin, A. J., Redding, T. W., Mittler, J. C., Nair, R. M. G., Pizzolata, P. and Segal, A. J. (1970). Purification of hypothalamic releasing hormones of human origin. *J. Clin. Endocrinol. Metab.* **31,** 291–300.

Schally, A. V., Arimura, A., Baba, Y., Nair, R. M. G., Matsuo, H., Redding, T. W., Debeljuk, L. and White, W. F. (1971a). Isolation and properties of the FSH and LH-releasing hormone. *Biochem. Biophys. Res. Commun.* **43,** 393–399.

Schally, A. V., Arimura, A., Kastin, A. J., Matsuo, H., Baba, Y., Redding, T. W., Nair, R. M. G., Debeljuk, L. and White, W. F. (1971b). Gonadotrophin-releasing hormone: one polypeptide regulates secretion of luteinizing and follicle stimulating hormones. *Science, N.Y.* **173,** 1036–1038.

Schally, A. V., Kastin, A. J. and Arimura, A. (1972). FSH-releasing hormone and LH-releasing hormone. *Vitam. & Horm.* **30,** 83–164.

Schally, A. V., Arimura, A., and Kastin, A. J. (1973a). Hypothalamic regulatory hormones. *Science, N.Y.* **179,** 341–350.

Schally, A. V., Redding, T. W., Takahara, J., Coy, D. H. and Arimura, A. (1973b). Lack of growth hormone-releasing activity of (Pyro)Glu-Ser-Gly-NH_2. *Biochem. Biophys. Res. Commun.* **55,** 556–562.

Schally, A. V., Arimura, A., Takahara, J., Redding, T. W. and Dupont, A. (1974). Inhibition of prolactin release *in vitro* and *in vivo* by catecholamines. *Fed. Proc. Fed. Am. Soc. exp. Biol.* **33,** 237.

Schally, A. V., Kastin, A. J. and Arimura, A. (1975). The hypothalamus and reproduction. *Am. J. Obstet. Gynecol.* **122,** 857–862.

Schally, A. V., Arimura, A., Redding, T. W., Debeljuk, L., Carter, W., Duport, A. and Vilchez-Martinez, J. A. (1976). Re-examination of porcine and bovine hypothalamic fractions for additional luteinizing hormone and follicle-stimulating hormone-releasing activities. *Endocrinology*, **98,** 380–391.

Scharrer, E. (1928). Untersuchungen über das Zwischenhirn der Fische. I. Die Lichtemfindlichkeit blinder Elritzen. *Z. vergleich. Physiol.* **7,** 1–38.

Scharrer, E. and Scharrer, B. (1940). Secretory cells within the hypothalamus. *Res. Publ. Ass. Res. nerv. ment. Dis.* **20,** 170–194.

Scharrer, E. and Scharrer, B. (1954). Hormones produced by neuro-secretory cells. *Recent Prog. Horm. Res.* **10,** 183–240.

Schenkel-Hulliger, H., Koella, H., Hartmann, W. P. and Maitre, L. (1974). Tremorogenic effect of thyrotropin releasing hormone in rats. *Experientia*, **30,** 1168–1170.

Schild, H. O. (1969). The effect of metals on the S-S polypeptide receptor in depolarized rat uterus. *Br. J. Pharmacol.* **36,** 329–349.

Schilling, R. M. and Blobel, M. (1973). Corticotrophin-releasing factor activity in hypothalamic extracts of man. *Acta endocr., Copenh.* **73,** 1–10.

Schleidt, J. (1914). Über die Hypophyse bei feminierten Männchen und maskulierten Weibchen. *Zbl. Physiol.* **27,** 1170–1172.

Schofield, J. G. and Moser, M. F. (1974). Inhibition of brain stimulated release of rat growth hormone by linear somatostatin. *Experientia*, **30,** 708–709.

Schwartz, N. B. (1974). The role of FSH and LH and of their antibodies on follicle growth and ovulation. *Biol. Reprod.* **10,** 236–272.

Schwyzer, R. and Sieber, P. (1963). Total synthesis of adrenocorticotrophic hormone. *Nature, Lond.* **199,** 172–174.

Scott, A. P., Ratcliffe, J. G., Rees, L. H., Landon, J., Bennett, H. P. J., Lowry, P. J. and McMartin, C. (1973). Pituitary peptide. *Nature, Lond.* **244,** 65–67.

Scott, A. P., Lowry, P. J., Ratcliffe, J. G., Rees, L. H. and Landon, J. I. (1974a). Corticotrophin-like peptides in the rat pituitary. *J. Endocr.* **61,** 355–367.

Scott, D. E. (1975). The ultrastructural correlations of circumventricular organ function: the median eminence as a neuroendocrine transducer. *In* "Symposium on Neuroendocrine Regulation of Fertility", pp. 22–28. Brain Information Service, Los Angeles.

Scott, D. E., Dudley, G., Knigge, K. M. and Kozlowski, G. P. (1974b). *In vitro* analysis of the cellular localization of luteinizing hormone releasing factor (LRF) in the basal hypothalamus of the rat. *Cell. Tiss. Res.* **149,** 371–378.

Seltzman, H. H. and Chapman, T. M. (1974). The 1,1,1,3,3,3-hexafluoro-2-(*p*-chlorophenoxymethoxy) propyl (HF-PA) blocking group for the imidazole function of histidine and its use in synthesis of thyrotropin-releasing hormone. *Tetrahedon Letters*, 30, 2637.

Selye, H. (1950). "The Physiology and Pathology of Exposure to Stress". Medical Publishers Inc., Montreal.

Sétáló, G., Vigh, S., Schally, A. V., Arimura, A. and Flerkó, B. (1975) LH-RH-containing neural elements in the rat hypothalamus. *Endocrinology*, **96**, 135–142.

Sétáló, G., Vigh, S., Schally, A. V., Arimura, A. and Flerko, B. (1976). Immunohistological study of the origin of LH-RH-containing nerve fibers of the rat hypothalamus. *Brain Res.* **103**, 597–602.

Setchell, B. P. and Main, S. J. (1974). Bibliography (with review) on inhibin. *Bblphy. Reprod.* **24**, 245–252, 361–367.

Seyler, L. E. and Reichlin, S. (1973). Luteinizing hormone-releasing factor (LRF) in plasma of postmenopausal women. *J. Clin. Endocrinol. Metab.* **37**, 197–203.

Seyler, L. E. and Reichlin, S. (1974a). Episodic secretion of luteinising hormone releasing factor (LRF) in the human. *J. Clin. Endocrinol Metab.* **39**, 471–478.

Seyler, L. E. and Reichlin, S. (1974b). Feedback regulation of circulating LRF concentrations in man. *J. Clin. Endocrinol. Metab.* **39**, 906–912.

Shaar, C. J. and Clemens, J. A. (1974). The role of catecholamines in the release of anterior pituitary prolactin *in vitro*. *Endocrinology*, **95**, 1202–1212.

Shahmanesh, M. and Jeffcoate, S. L. (1976). Absence of an immunologically distinct follicle-stimulating hormone releasing factor in rat stalk median eminence extract. *J. Endocr.* **68**, 89–94.

Shaikh, A. A. and Shaikh, S. A. (1975). Adrenal and ovarian steroid secretion in the rat estrous cycle temporally related to gonadotropins and steroid levels found in peripheral plasma. *Endocrinology*, **96**, 37–43.

Shambaugh, G. E., and Wilber, J. F. (1974). Thyrotropin releasing hormone: measurement in human spinal fluid by radioimmunoassay. *Clin. Res.* **22**, 634a.

Sharp, P. J. (1973). A comparison of circulating levels of luteinizing hormone in intact and gonadectomized growing fowl. *J. Endocr.* **61**, viii.

Sharp, P. J. (1975). A comparison of variations in plasma luteinizing hormone concentrations in male and female domestic chickens (*Gallus domesticus*) from hatch to sexual maturity. *J. Endocr.* **67**, 211–223.

Sharp, P. J. and Moss, R. (1977). The effects of castration on concentrations of luteinizing hormone in the plasma of photorefractory red grouse (*Lagopus lagopus scoticus*) *Gen. comp. Endocrinol.* **32**, 289–293.

Sharp, P. J., Haase, E. and Fraser, H. M. (1975). Immunofluorescent localization of sites binding anti-synthetic LH-RH serum in the median eminence of the greenfinch (*Chloris chloris* L.) *Cell Tiss. Res.* **162**, 83–91.

Shaw, R. W., Butt, W. R. and London, D. R. (1975). The effect of oestrogen pretreatment on subsequent response to LH-RH in normal women. *Clinical Endocrinology*, **4**, 297–304.

Sherwood, N. M. and Timiras, P. S. (1974). Comparison of direct current and radiofrequency lesions in the rostral hypothalamus with respect to sexual maturation in the female rat. *Endocrinology*, **94**, 1275–1285.

Shibusawa, K., Yamomoto, T., Nishi, K., Abe, C. and Tomie, S. (1959). Urinary excretion of TRF in various functional states of the thyroid. *Endocrinol.* (*Japan*), **6**, 131–136.

Shin, S. H. and Kraicer, J. (1974). LH-RH radioimmunoassay and its application: evidence of antigenically distinct FSH-RH and a diurnal study of LH-RH and gonadotrophins. *Life Sci.* **14**, 2491–2496.

Shizume, K. and Lerner, A. B. (1954). Determination of melanocyte stimulating hormone in urine and blood. *J. Clin. Endocrinol. Metab.* **14**, 1491–1510.

Shizume, K., Lerner, A. B. and Fitzpatrick, T. B. (1954). *In vitro* bioassay for the melanocyte stimulating hormone. *Endocrinology*, **54,** 553–560.

Shizumi, K., Matsuzaki, F., Iino, S., Matsuda, K., Nagataki, S., and Okinaka, S. (1962). Effect of electrical stimulation of the limbic system on pituitary thyroid function. *Endocrinology*, **71,** 456–463.

Short, R. V. (1974). Sexual differentiation of the brain of the sheep. *Les Colloques de l'Institut National de la Santé et de la Recherche Médicale. Endocrinologie Sexuelle de la Période Périnatale*, **32,** 121–142.

Shute, C. C. D. (1970). Distribution of cholinesterase and cholinergic pathways. *In* "The Hypothalamus" (L. Martini, M. Motta and F. Fraschini, eds) pp. 167–179. Academic Press, New York and London.

Shute, C. C. D. and Lewis, P. R. (1966). Cholinergic and monoaminergic pathways in the hypothalamus. *Br. Med. Bull.* **22,** 221–226.

Siler, T. M., Yen, S. S. C., Vale, W., and Guillemin, R. (1974). Inhibition by somatostatin on the release of TSH induced in man by thyrotropin releasing factor. *J. Clin. Endocrinol. Metab.* **38,** 742–745.

Silva, J. E. and Larsen, P. R. (1977). Pituitary nuclear 3,5,3′-triiodothyronine and thyrotropin secretion: an explanation for the effect of thyroxine. *Science, N.Y.* **198,** 617–619.

Silverman, A. J., Knigge, K. M. and Peck, W. A. (1972). Transport capacity of median eminence. I. Amino acid transport. *Neuroendocrinology*, **9,** 123–132.

Sinha, D. and Meites, J. (1965). Effects of thyroidectomy and thyroxine on hypothalamic control of "thyrotropic releasing factor" and pituitary content of thyrotropin in rats. *Neuroendocrinology*, **1,** 4–14.

Sinha, Y. N., Selby, F. W. and VanderLaan, W. P. (1974). Relationship of prolactin and growth hormone to mammary function during pregnancy and lactation in the C3H/St mouse. *J. Endocr.* **61,** 219–229.

Siperstein, E. R. (1963). Identification of the adrenocorticotrophin-producing cell in the rat hypophysis by autoradiography. *J. Cell. Biol.* **17,** 521–546.

Siperstein, E. R. and Allison, V. F. (1965). Fine structure of the cells responsible for secretion of adrenocorticotrophin in the adrenalectomised rat. *Endocrinology*, **76,** 70–79.

Siperstein, E. R. and Miller, K. J. (1973). Hypertrophy in the ACTH-producing cells following adrenalectomy. A quantitative electron microscopic study. *Endocrinology*, **93,** 1257–1268.

Sirret, N. E. and Purves, H. D. (1973). The assay of corticotrophin-releasing hormone in ACTH primed grafted rats. *In* "Brain-Pituitary-Adrenal Inter-relationships" (A. Brodish and E. S. Redgate, eds) pp. 79–98. Karger, Basel.

Skadhauge, E. (1973). Renal and cloacal salt and water transport in the fowl (*Gallus domesticus*). *Danish Medical Bulletin*, **20,** Suppl. 1, 1 82.

Sloper, J. C. (1966). The experimental and cytopathological investigation of neurosecretion in the hypothalamus and pituitary. *In* "The Pituitary Gland" (G. W. Harris and B. T. Donovan, eds) Vol. III, pp. 131–239. Butterworths, London.

Sloper, J. C. and Bateson, R. G. (1965). Ultrastructure of neurosecretory cells in the supraoptic nucleus of the dog and rat. *J. Endocr.* **31,** 139–150.

Slusher, M. A. (1964). Effects of chronic hypothalamic lesions on diurnal and stress corticosteroid levels. *Am. J. Physiol.* **206,** 1161–1164.

Smalstig, E. B., Sawyer, B. D. and Clemens, J. A. (1974). Inhibition of rat prolactin release by apomorphine *in vivo* and *in vitro*. *Endocrinology*, **95,** 123–129.

Smelik, P. G. (1963a). Failure to inhibit corticotrophin secretion by experimentally induced increases in corticoid levels. *Acta endocr., Copenh.* **44,** 36–46.

Smelik, P. G. (1963b). Relation between blood level of corticoids and their inhibiting effect of hypophyseal stress response. *Proc. Soc. exp. Biol. Med.* **113,** 616–619.

Smelik, P. G. (1970). Integrated hypothalamic responses to stress. *In* "The Hypothalamus" (L. Martini, M. Motta and F. Fraschini, eds) pp. 491–497. Academic Press, New York and London.

Smelik, P. G. and Sawyer, C. H. (1962). Effects of implantation of cortisol into the brain stem or pituitary gland on the adrenal response to stress in the rabbit. *Acta endocr., Copenh.* **41,** 561–570.

Smith, E. R. and Davidson, J. M. (1974). Location of feedback receptors: effects of intracranially implanted steroids on plasma LH and LRF response. *Endocrinology*, **95,** 1566–1573.

Smith, M. S., Freeman, M. E. and Neill, J. D. (1975). The control of progesterone secretion during the estrous cycle and early pseudopregnancy in the rat: prolactin, gonadotrophin and steroid levels associated with the rescue of the corpus luteum of pseudopregnancy. *Endocrinology*, **96,** 219–226.

Smith, P. E. (1926). Ablation and transplantation of the hypophysis in the rat. *Anat. Rec.* **32,** 221.

Smith, P. E. (1927). The disabilities caused by hypophysectomy and their repair. The tuberal (hypothalamic) syndrome in the rat. *J. Am. Med. Ass.* **88,** 158–161.

Smith, P. E. (1930). Hypophysectomy and a replacement therapy in the rat. *Am. J. Anat.* **45,** 205–274.

Smith, P. E. and Engle, E. T. (1927). Experimental evidence regarding the role of the anterior pituitary in the development and regulation of the genital system. *Am. J. Anat.* **40,** 159–217.

Smith, P. M. and Follett, B. K. (1972). Luteinizing hormone-releasing factor in the quail hypothalamus. *J. Endocr.* **53,** 131–138.

Smythe, G. A. and Lazarus, L. (1974). Suppression of human growth hormone secretion by melatonin and cyproheptadine. *J. Clin. Invest.* **54,** 116–121.

Snyder, P. J. and Utiger, R. D. (1972). Inhibition of thyrotropin response to thyrotropin releasing hormone by small quantities of thyroid hormones. *J. Clin. Invest.* **51,** 2077–2084.

Snyder, P. J. Utiger, R. D. (1973). Repetitive administration of thyrotropin releasing hormone results in small elevations of serum thyroid hormones and in marked inhibition of thyrotropin response. *J. Clin. Invest.* **52,** 2305–2312.

Snyder, S. H., Logan, W. J., Bennett, J. P. and Arregni, A. (1973). Amino acids as central nervous transmitters: biochemical studies. *Neuroscience Research*, **5,** 131–157.

Sollberger, A. (1956). "Biological Rhythm Research". Elsevier, Amsterdam.

Soloff, M. S., Schwartz, T. L. and Steinberg, A. H. (1974). Oxytocin receptors in human uterus. *J. Clin. Endocrinol. Metab.* **38,** 1052–1056.

Sorrentino, S. Jr., and Sundberg, D. K. (1975). Measurement of plasma and hypothalamic luteinizing hormone-releasing hormone in pregnant mare serum-induced ovulating immature rats. *Neuroendocrinology*, **17,** 274–282.

Spatola, A. F., Cornelius, D. A., Hruby, V. J. and Blomquist, A. T. (1974). Synthesis of oxytocin and relatic diastereomers denterated in the half-cystine positions. Comparison of solid-phase and solution methods. *J. Org. Chem.* **39,** 2207–2212.

Spatz, H. (1951). Neues über die Verknüpfung von Hypophyse und Hypothalamus. *Acta neuroveg.* (*Wien*) **3,** 1–49.

Spaulding, S. W., Burrow, G. N., Donabedian, R., and VanWoert, M. (1972) L-Dopa suppression of thyrotropin releasing hormone response in man. *J. Clin. Endocrinol. Metab.* **35,** 182–185.

Speidel, C. C. (1919). Gland-cells of internal secretion in the spinal cord of the skates. *Carnegie Inst. Wash., Publ.* No. 13, 1–31.

Spies, H. G. and Norman, R. L. (1975). Interaction of estradiol and LH-RH on LH in Rhesus females: evidence for a neural site of action. *Endocrinology*, **97,** 685–692.

Spona, J. (1973a). LH-RH interaction with the pituitary plasma membrane. *FEBS Letters*, **34,** 24–26.

Spona, J. (1973b). LH-RH stimulated gonadotrophin release mediated by two distinct pituitary receptors. *FEBS Letters*, **35,** 59–62.

Spona, J. (1974a). LH-RH actions modulated by sex steroids. *Endocr. exper.* **8,** 19–29.

Spona, J. (1974b). LH-RH interaction with the pituitary plasma membrane is affected by sex steroids. *FEBS Letters*, **39,** 221–224.

Spriggs, T. L. B. and Stockham, M. A. (1964). Urethane anaesthesia and pituitary adrenal function in the rat. *J. Pharmacol. Pharmac.* **16,** 603–610.

Stearns, E. L., Winter, J. S. D. and Faiman, C. (1973). Positive feedback effect of progestin upon serum gonadotrophins in oestrogen-primed castrate men. *J. Clin. Endocrinol. Metab.* **37,** 635–638.

Steinach, E. (1894). Untersuchungen zur vergleichenden Physiologie der männlichen Geschlechtsorgane. *Pflüger's Arch. Physiol.* **56,** 304–338.

Steinach, E. (1913). Feminierung von Männchen und Maskulierung von Weibchen. *Zbl. Physiol.* **27,** 717–724.

Steiner, R. J. (1974). Organ distribution of (^{3}H)-TRH in rats. *FEBS Letters*, **38,** 219–221.

Sterling, K. and Hegedus, A. (1962). Measurements of free thyroxine concentration in human serum. *J. Clin. Invest.* **41,** 1031–1040.

Steward, A. D. (1974). Genetic study on the effect of (8-arginine)- and (8-lysine)-vasopressins on adenyl cyclase activity on mouse kidney. *J. Endocr.* **61,** 447–454.

Stewart, A. D. (1971). Genetic variation in the neurohypophysial hormones of the mouse *Mus musculus*. *J. Endocr.* **51,** 191–201.

Stewart-Bentley, M., Odell, W. and Horton, R. (1974). The feedback control of luteinising hormone in normal men. *J. Clin. Endocr. Metab.* **38,** 545–553.

Stibal, S., Bigl, H. and Sterba, G. (1974). Bestimmung der Radioaktivität nach intravenöser Injektion von ^{3}H-markiertem Oxytocin im Hirn der Ratte. *Endokrinologie*, **63,** 33–42.

Stoeckel, M. E., Dellmann, H. D., Porte, A., Klein, M. J. and Stutinsky, F. (1973). Corticotrophic cells on the rostral zone of the pars intermedia and in the adjacent neurohypophysis of the rat and mouse. *Z. Zellforsch.* **136,** 97–110.

Stoner, H. B. and Eslon, P. M. (1971). The effect of injury on monoamine concentrations in the rat hypothalamus. *J. Neurochem.* **18,** 1837–1846.

Strauch, G., Girault, D., Rifai, M. and Bricaire, H. (1973). Alpha-MSH stimulation of growth hormone release. *J. Clin. Endocrinol. Metab.* **37,** 990–993.

Stumpf, W. E. (1968). Estradiol concentrating neurons: topography in the hypothalamus by dry-mount antoradiography. *Science, N.Y.* **162,** 1001–1003.

Sullivan, P. R. C., Bollinger, J. A. and Reichlin, S. (1972). Selective deficiency of tissue triiodothyronine: a proposed mechanism of elevated free thyrozine in the euthyroid sick. *J. Clin. Invest.* **52,** 83a.

Sundberg, D. K., Krulich, L., Fawcett, C. P., Illner, P. and McCann, S. M. (1973). The effect of colchicine on the release of rat anterior pituitary hormone *in vitro*. *Proc. Soc. exp. Biol. Med.* **142,** 1097–1100.

Sundsten, J., Novin, D. and Cross, B. A. (1970). Identification and distribution of paraventricular units excited by stimulation of the neural lobe of the hypophysis. *Expl. Neurol.* **26,** 316–329.

Swaab, D. F. (1972). The hypothalamo-neurohypophysial system and reproduction. *Progr. Brain Res.* **38,** 225–242.

Swaab, D. F. and Honnebier, W. J. (1974). The role of the fetal hypothalamus in development of the feto-placental unit and in parturition. *Progr. Brain Res.* **41,** 255–280.

Swaab, D. F. and Pool, C. W. (1975). Specificity of oxytocin and vasopressin immunofluorescence. *J. Endocr.* **66,** 263–272.

Swaab, D. F., Nijveldt, F. and Pool, C. W. (1975). Distribution of oxytocin and vasopressin within the rat hypothalamus. *Acta. Endocr.* Suppl. **193,** 131.

Swerdloff, R. S. and Walsh, P. C. (1973).Testosterone and oestradiol suppression of LH and FSH in adult male rats: duration of castration, deviation of treatment and combined treatment. *Acta. endocr., Copenh.* **73,** 11–21.

Sydnor, K. L. and Sayers, G. (1954). Blood and pituitary ACTH in intact and adrenalectomised rats after stress. *Endocrinology*, **55,** 621–636.

Syvälahti, E. K. G. (1974). The effect of indomethacin on serum growth hormone, immunoreactive insulin and blood glucose levels of young adult males. *Int. J. Clin. Pharmac.* **10,** 111–116.

Szabo, M. and Frohman, L. A. (1975). Effects of porcine stalk median eminence and prostaglandin E_2 on rat growth hormone secretion *in vitro* and their inhibition by somatostatin. *Endocrinology*, **36,** 955–961.

Szabo, M. and Frohman, L. A. (1976). Suppression of cold stimulated TSH secretion in the rat by anti-TRH serum. Program 58th Annual Meeting, U.S. Endoc. Soc., p. 189.

Szentágothai, J. (1962). *In* "Hypothalamic Control of the Anterior Pituitary" (J. Szentágothai, B. Flerkó, B. Mess and B. Halász, eds) 1st edn p. 95. Akadémiai Kiadó, Budapest.

Szentágothai, J., Flerkó, B., Mess, B. and Halász, B. (1972). "Hypothalamic Control of the Anterior Pituitary", 3rd edn. Akadémiai Kiadó, Budapest.

Szewczuc, A. and Kwiatkowska, J. (1970). Pyrrolidonyl peptidase in animal, plant and human tissues. Occurrence and some properties of the enzyme. *Eur. J. Biochem.* **15,** 92–96.

T

Taber, C. A. and Karavolas, H. C. (1975). Subcellular localization of LH-releasing activity in the rat hypothalamus. *Endocrinology*, **96,** 446–452.

Takahara, J., Arimura, A. and Schally, A. V. (1974a). Suppression of prolactin release by a purified porcine PIF preparation and catecholamines infused into a rat hypophysial portal vessel. *Endocrinology*, **95,** 462–465.

Takahara, J., Arimura, A. and Schally, A. V. (1974b). Stimulation of prolactin and growth hormone release by TRH infused into a hypophysial portal vessel. *Proc. Soc. exp. Biol. Med.* **146,** 831–835.

Takahara, J., Arimura, A. and Schally, A. V. (1975). Assessment of GH-releasing hormone activity in sephadex-separated fractions of porcine hypothalamic extracts by hypophyseal portal vessel infusion in the rat. *Acta endocr., Copenh.* **78,** 428–434.

Takene, K., Setaiski, C., Hirama, M., Yamamoto, M. and Horiuchi, Y. (1966). Effects of a bacterial pyrogena on the pituitary-adrenal axis at various times in the 24-hours. *J. Clin. Endocrinol. Metab.* **26,** 437–442.

Tal, E., Szabo, M. and Burke, G. (1974). TRH and prostaglandin action on rat anterior pituitary: dissociation between cyclic AMP levels and TSH release. *Prostaglandins*, **5,** 175–182.

Taleisnik, S. and Celis, M. E. (1973). Control of melanocyte-stimulating hormone secretion. *In* "Endocrinology. Proc. IV Internat. Cong. Endocrinol." (R. D. Scow, ed.) pp. 100–106. Excerpta Medica, Amsterdam.

Taleisnik, S. and Orias, R. (1965). A melanocyte-stimulating hormone-releasing factor in hypothalamic extracts. *Am. J. Physiol.* **208,** 293–296.

Taleisnik, S. and Orias, R. (1966). Pituitary melanocyte-stimulating hormone (MSH) after suckling stimulus. *Endocrinology*, **78,** 522–526.

Taleisnik, S. and Tomatis, M. E. (1967a). Melanocyte-stimulating hormone releasing and inhibiting factors in two hypothalamic extracts. *Endocrinology*, **81,** 819–825.

Taleisnik, S. and Tomatis, M. E. (1967b). Antagonistic effect on melanocyte-stimulating hormone release of two neural tissue extracts. *Am. J. Physiol.* **212,** 157–163.

Taleisnik, S. and Tomatis, M. E. (1968). Effect of suckling stimuli on the content of hypothalamic factors that control the release of melanocyte stimulating hormone. *Neuroendocrinology*, **3,** 303–308.

Taleisnik, S. and Tomatis, M. E. (1969). Effect of estrogen on pituitary melanocyte-stimulating hormone content. *Neuroendocrinology*, **5,** 24–32.

Taleisnik, S. and Tomatis, M. E. (1970). Mechanisms that determine changes in pituitary MSH activity during pseudopregnancy induced by vaginal stimulation in the rat. *Neuroendocrinology*, **6,** 368–377.

Taleisnik, S., Orias, R. and de Olmos, J. (1966). Topographic distribution of the melanocyte-stimulating hormone-releasing factor in rat hypothalamus. *Proc. Soc. exp. Biol. Med.* **122,** 325–328.

Taleisnik, S., de Olmos, J., Orias, R. and Tomatis, M. E. (1967). Effect of hypothalamic lesions on pituitary melanocyte-stimulating hormone. *J. Endocr.* **39,** 485–492.

Taleisnik, S., Caligaris, L. and Astrada, J. J. (1971). Sex differences in hypothalamo-hypophysial function. *In* "Steroid Hormones and Brain Function" (C. H. Sawyer and R. A. Gorski, eds) pp. 171–184. University of California Press.

Taleisnik, S., Tomatis, M. E. and Celis, M. E. (1972). Role of catecholamines in the control of melanocyte-stimulating hormone secretion in rats. *Neuroendocrinology*, **10,** 235–245.

Taleisnik, S., Celis, M. E. and Tomatis, M. E. (1973). Release of melanocyte-stimulating hormone by several stimuli through the activation of a 5-hydroxytryptamine-mediated inhibitory neuronal mechanism. *Neuroendocrinology*, **13,** 327–338.

Talwalker, P. K., Ratner, A. and Meites, J. (1963). *In vitro* inhibition of pituitary prolactin synthesis and release by hypothalamic extracts. *Am. J. Physiol.* **205,** 213–218.

Tanabe, T., Iino, M., Ooshima, Y. and Takagi, S. F. (1974). An olfactory area in the pre-frontal lobe. *Brain Res.* **80,** 127–130.

Tang, L. L. and Spies, H. G. (1975). Effects of gonadal steroids on the basal and LRF-induced gonadotrophin secretion by cultures of rat pituitaries. *Endocrinology*, **96,** 349–356.

Tashjian, A. H. Jr., Borowsky, N. J. and Jensen, D. K. (1971). Thyrotropin releasing hormone: direct evidence for stimulation of prolactin production by pituitary cells in culture. *Biochem. Biophys. Res. Commun.* **43,** 516–523.

Taurog, A., Oliver, C., Eskay, R. L., Porter, J. C. and McKenzie, J. M. (1974). The role of TRH in the neoteny of the mexican axolotl (*Ambyostoma mexicanum*). *Gen. Comp. Endocrinol.* **24,** 267–279.

Taylor, A. N. (1969). The role of the reticular activating system in the regulation of ACTH secretion. *Brain. Res.* **13,** 234–246.

Terasawa, E. and Sawyer, C. H. (1969). Changes in electrical activity in the rat hypothalamus related to electrochemical stimulation of adenohypophyseal function. *Endocrinology*, **85,** 143–149.

ter Haar, M. B., Mackinnon, P. C. B. and Bulmer, M. G. (1974). Sexual differentiation in the phase of the circadian rhythm of ^{35}S methionine incorporation into cerebral proteins and of serum gonadotrophin levels. *J. Endocr.* **62,** 257–265.

Terkel, J., Blake, C. A. and Sawyer, C. H. (1972). Serum prolactin levels in lactating rats after suckling or exposure to ether. *Endocrinology*, **91,** 49–53.

Thody, A. J. (1974). The lack of effect of MSH-release inhibiting factor (MIF) on the secretion of β-MSH in normal men. *J. Clin. Endocrinol. Metab.* **38,** 491–493.

Thorner, M. O., Besser, G. M., Jones, A., Dacie, J. and Jones, A. E. (1975a). Bromocriptine treatment of female infertility: report of 13 pregnancies. *Br. Med. J.* **4,** 694–697.

Thorner, M. O., Chait, A., Aitken, M., Benker, G., Bloom, S. R., Mortimer, C. H., Sanders, P., Stuart Mason, A. and Besser, G. M. (1975b). Bromocriptine treatment of acromegaly. *Br. med. J.* **1,** 299–303.

Thornton, V. F. (1971). Effect of change of background color on the melanocyte-stimulating hormone content of the pituitary of *Xenopus laevis*. *Gen. Comp. Endocrinol.* **17,** 554–560.

Tilders, F. J. H. and Smelik, P. G. (1975). A diurnal rhythm in melanocyte-stimulatory hormone content of the rat pituitary gland and its independence from the pineal gland. *Neuroendocrinology*, **17,** 296–308.

Tilders, F. J. H., van der Woude, H. A., Mulder, A. H. and Smelik, P. G. (1975). *In vitro* evidence for a MSH-release inhibiting role of dopaminergic innervation of the pars intermedia. *Acta endocr., Copenh.* Suppl. **199,** 135.

Tima, L. and Flerkó, B. (1974). Ovulation induced by norepinephrine in rats made anovulatory by various experimental procedures. *Neuroendocrinology*, **15,** 346–354.

Tindal, J. S. (1974). Hypothalamic control of secretion and release of prolactin. *J. Reprod. Fertil.* **39,** 437–461.

Tindal, J. S. and Knaggs, G. S. (1970). Environmental stimuli and the mammary gland. *Mem. Soc. Endocrinol.* **18,** 239–258.

Tindal, J. S. and Knaggs, G. S. (1972). Pathways in the forebrain of the rabbit concerned with the release of prolactin. *J. Endocr.* **52,** 253–262.

Tindal, J. S. and Knaggs, G. S. (1977). Pathways in the forebrain of the rat concerned with the release of prolactin. *Brain Res.* **119,** 211–221.

Tindal, J. S., Knaggs, G. S. and Turvey, A. (1967). Central nervous control of prolactin secretion in the rabbit: effect of local oestrogen implants in the amygdaloid complex. *J. Endocr.* **37,** 279–287.

Tiptaft, E. M. and Jones, M. T. (1975). Structure-activity relationship of various steroids in the feedback control of ACTH secretion. *J. Endocr.* **64,** 9P.

Tixier-Vidal, A. and Gourdji, D. (1972). Cellular aspects of the control of prolactin secretion in birds. *Gen. Comp. Endocrinol.* Suppl. **3,** 51–64.

Tixier-Vidal, A., Gourdji, D., Pradelles, P., Morget, J. L., Fromageot, P. and Kerdelhue, B. (1975). A cell culture approach to the study of TRH receptors. *In* "Hypothalamic Hormones" (M. Motta, P. Crosignani and L. Martini, eds) pp. 89–107. Academic Press, New York and London.

Tomatis, M. E. and Taleisnik, S. (1968a). Pituitary melanocyte-stimulating hormone and its hypothalamic factors during the estrous cycle. *Acta Physiol. Latinoamer.* **18,** 96–101.

Tomatis, M. E. and Taleisnik, S. (1968b). Influence of reserpine on the pituitary content of melanocyte-stimulating hormone and on hypothalamic factors which effect its release. *J. Endocr.* **42,** 505–512.

Török, B. (1954). Lebendbeobachtung des Hypophysenkreislaufes an Hunden. *Acta morph. Hung.* **4,** 83–89.

Török, B. (1962). Neue Angaben zum Blutkreislauf der Hypophyse. *Anat. Anz.* **109,** (Suppl.) 622–629.

Tsuji, S., Sakoda, M. and Asami, M. (1968). *In* "Integrative Mechanism of Neuroendocrine System" (S. Itoh, ed.) p. 63. Hokkaido University School of Medicine, Japan.

Tuomisto, J., Ranta, T., Männistö, P., Saarinen, A., and Leppaluoto, J. (1975). Neurotransmitter control of thyrotropin secretion in the rat. *Europ. J. Pharmacol.* **30,** 221–229.

Tuppy, H. (1953). The Amino-acid sequence in oxytocin. *Biochem. Biophys. Acta (Amsterdam)*, **11,** 449.

Turkington, M. R. and Everitt, A. V. (1976). The neurohypophysis and aging with special reference to the antidiuretic hormone. *In* "Hypothalamus, Pituitary and Aging" (A. V. Everitt and J. A. Burgess, eds) pp. 123–136. C. C. Thomas, Springfield.

Turkington, R. W. (1972). Inhibition of prolactin secretion and successful therapy of the Forbes-Albright syndrome with L-dopa. *J. Clin. Endocrinol. Metab.* **34,** 306–311.

Tyrrell, J. B., Lorenzi, M., Gerich, J. E. and Forsham, P. H. (1975). Inhibition by somatostatin of ACTH secretion in Nelson's syndrome. *J. Clin. Endocrinol. Metab.* **40,** 1125–1127.

Tyson, J. E., Friesen, H. G. and Anderson, M. S. (1972). Human lactational and ovarian response to endogenous prolactin release. *Science, N.Y.* **177,** 897–900.

U

Ulrich, R. S. and Yuwiler, A. (1973). Failure of 6-hydroxydopamine to abolish the circadian rhythm of serum corticosterone. *Endocrinology*, **92,** 611–614.

Urry, D. W. and Walter, R. (1971). Proposed conformation of oxytocin in solution. *Proc. Natl. Acad. Sci. USA*, **68,** 956–958.

Usunov, P. and Weiss, B. (1972). Psychopharmacological agents and cyclic AMP of rat brain. *Adv. Cyclic Nucleotide Res.* **1,** 435–453.

Uthne, K., Reagan, C. R., Gimpel, L. P. and Kostyo, J. L. (1974). Effects of human somatomedin preparations on membrane transport and protein synthesis in the isolated rat diaphragm. *J. Clin. Endocrinol. Metab.* **39,** 548–554.

Utiger, R. D. and Bassiri, R. M. (1973). Thyrotropin-releasing hormone (TRH) radioimmunoassay. *In* "Hypothalamic Hypophysiotropic Hormones" (C. Gaul and E. Rosemberg, eds) pp. 146–153. Excerpta Medica, Amsterdam.

Uttenthal, L. O. and Hope, D. B. (1970). The isolation of three neurophysins from porcine posterior pituitary lobes. *Biochem. J.* **116,** 899–909.

Uttenthal, L. O. and Hope, D. B. (1972). Neurophysins and posterior pituitary hormones in the suiformes. *Proc. R. Soc. Ser. B.* (*London*), **182,** 73–87.

V

Vaala, S. S. and Knigge, K. M. (1974). Transport capacity of median eminence; *in vitro* uptake of 3H-LRF. *Neuroendocrinology*, **15,** 147–157.

Vagenakis, A. G., Rapoport, B., Azizi, F., Portnoy, G. I., Braverman, L. E. and Ingbar, S. H. (1974). Hyper-response to thyrotropin releasing hormone accompanying small decreases in serum thyroid hormone concentration. *J. Clin. Invest.* **54,** 913–918.

Vagenakis, A. G., Burger, A., Portnay, G. I., Rudolph, M., O'Brian, J. T., Azizi, F., Arky, R. A., Nicod, P., Ingbar, S. H. and Braverman, L. E. (1975). Diversion of peripheral thyroxine metabolism from activating to inactivating pathways during complete fasting. *J. Clin. Endocrinol. Metab.* **41,** 191–194.

Vale, W. and Guillemin, R. (1967). Potassium induced stimulation of thyrotropin release *in vitro*. Requirement for pressure of calcium and inhibition by thyroxine. *Experientia*, **23,** 855–857.

Vale, W., Burgus, R. and Guillemin, R. (1968). On the mechanism of action of TRF: effects of cycloheximide and actinomycin on the release of TSH stimulated *in vitro* by TRF and its inhibition by thyroxine. *Neuroendocrinology*, **3,** 34–36.

Vale, W., Burgus, R., Dunn, T. F. and Guillemin, R. (1971). *In vitro* plasma inactivation of thyrotropin releasing factor (TRF) and related peptides. Its inhibition by various means and by the synthetic decapeptide PCA-His OME. *Hormones* (*Basel*) **2,** 193–203.

Vale, W., Brazeau, P., Grant, G., Nussey, A., Burgus, R., Rivier, J., Ling, N. and Guillemin, R. (1972). Premières observations sur la mode d'action de la somatostatine, un facteur hypothalamique qui inhibe la sécrétion de l'hormone de croissance. *C. R. Acad. Sci.* **275,** 2913–2916.

Vale, W., Grant, G. and Guillemin, R. (1973). Chemistry of the hypothalamic releasing factors—studies on structure—function relationships. *In* "Frontiers in Neuroendocrinology" (W. F. Ganong and L. Martini, eds) pp. 375–413. Oxford University Press.

Valverde, R. C., Chieffo, V. and Reichlin, S. (1972). Prolactin-releasing factor in porcine and rat hypothalamic tissue. *Endocrinology*, **91,** 982–993.

Van Delft, A. M. L., Kaplanski, J. and Smelik, P. G. (1973). Circadian periodicity of pituitary-adrenal function after *p*-chlorophenylalanine administration in the rat. *J. Endocr.* **59,** 465–474.

van Dieten, J. A. M. J., Steijsiger, J., Dullart, J. and van Rees, G. P. (1974). The effect of estradiol benzoate on the pituitary responsiveness to LH-RH in male and female rats. *Neuroendocrinology*, **15,** 182–188.

Vane, J. R. (1971). Inhibition of prostaglandin synthesis or a mechanism of action for aspirin-like drugs. *Nature* (*London*), **231,** 232–235.

Vanhaelst, L., Golstein, J., Van Canter, E., L'Hermite, M. and Robyn, C. (1973). Etude simultanée des variations circadiennes des taux sanguins de la thyréotropine (TSH) et de la prolactine hypophysaires chez l'homme. *C. R. Acad. Sci.* **276,** 1875–1877.

van Loon, G. R. (1973). Brain catecholamines and ACTH secretion. *In* "Frontiers in Neuroendocrinology" (W. F. Ganong and L. Martini, eds) pp. 209–247. Oxford University Press.

van Loon, G. R., Scapagnini, U., Cohen, R. and Ganong, W. F. (1971). Effect of intraventricular administration of adrenergic drugs on the adrenal venous 17-hydroxy-cortico-steroid response to surgical stress in the dog. *Neuroendocrinology*, **8,** 257–272.

van Maanen, J. H. and Smelik, P. G. (1968). Induction of pseudopregnancy in rats following local depletion of monoamines in the median eminence of the hypothalamus. *Neuroendocrinology*, **3,** 177–186.

van Wimersma Greidanus, T. B., Dogterom, J. and de Wied, D. (1975). Intraventricular administration of antivasopressin serum inhibits memory consolidation in rats. *Life Sci.* **16,** 637–644.

Vaughan, M. K., Reiter, R. J. and McKinney, T. (1974). Inhibition of growth of gonadal dependent structures by arginine vasotocin and purified bovine pineal fractions in immature mice and hamsters. *Int. J. Fertil.* **19,** 103–106.

Veber, D. F., Bennett, C. D., Milkowski, J. D., Gal, G., Denkewalter, R. G. and Hirschmann, R. (1971). Synthesis of a proposed growth hormone-releasing factor. *Biochem. Biophys. Res. Commun.* **45,** 235–239.

Velasco, M. E. and Taleisnik, S. (1969). Release of gonadotrophins induced by amygdaloid stimulation in the rat. *Endocrinology*, **85,** 132–139.

Vermes, I., Telegdy, G. and Lissak, K. (1973a). Correlation between hypothalamic serotonin content and adrenal function during acute stress. Effect of adrenal corticosteroids on hypothalamic serotonin content. *Acta. Physiol. Acad. Sci. Hung.* **43,** 33–42.

Vermes, I., Telegdy, G. and Lissak, K. (1973b). Inhibitory action of serotonin on the hypothalamus-induced ACTH release. *Acta. Physiol Acad. Sci. Hung.* **41,** 95–98.

Vermeulen, A. (1976). Leydig-cell function in old age. *In* "Hypothalamus, Pituitary and Aging" (A. V. Everitt and J. A. Burgess, eds) pp. 458–463. C. C. Thomas, Springfield.

Vernikos-Danellis, J. (1964). Estimation of corticotrophin-releasing activity of the rat hypothalamus and neurohypophysis before and after stress. *Endocrinology*, **75,** 514–520.

Vernikos-Danellis, J. (1965a). Effect of stress, adrenalectomy, hypophysectomy and hydrocortisone on the corticotrophin releasing activity of rat median eminence. *Endocrinology*, **76,** 122–126.

Vernikos-Danellis, J. (1965b). Effect of rat median eminence extracts on pituitary ACTH content in normal and adrenalectomised rats. *Endocrinology*, **76,** 240–245.

Vernikos-Danellis, J. and Marks, B. H. (1970). Assay of corticotrophin releasing factor. *In* "Hypophysiotrophic Hormones of the Hypothalamus: Assay and Chemistry" (J. Meites, ed.) pp. 60–68. Williams and Wilkins, Baltimore.

Vernikos-Danellis, J., Berger, P. and Barchas, J. D. (1970). Brain serotonin and pituitary-adrenal function. *Prog. Brain Res.* **39,** 301–311.

Vertes, M., Vertes, T. and Kovacs, S. (1965). Effect of hypothalamic stimulation on pituitary thyroid function. *Acta Physiol. Acad. Sci. Hung.* **27,** 229–235.

Vigh-Teichman, I. (1968). Hydrencephalocriny of neurosecretory material in amphibia. *In* "Zirkumventrikuläre Organe und Liguor" (G. Sterba, ed.) pp. 269–272. Fischer, Jena.

Vigneri, R., Pezzino, V., Squatrito, S., Calandra, A. and Maricchiolo, M. (1974). Sleep-associated growth hormone (GH) release in schizophrenia. *Neuroendocrinology*, **14,** 356–361.

Vilchez-Martinez, J. A., Arimura, A., Debeljuk, L. and Schally, A. V. (1974a) Biphasic effect of estradiol benzoate on the pituitary responsiveness to LH-RH. *Endocrinology*, **94,** 1300–1303.

Vilchez-Martinez, J. A., Coy, D. H. and Arimura, A. (1974b). Synthesis and biological properties of (Leu-6)-LH-RH and (D-Leu-6, Des-Gly-$NH_2{}^{10}$)-LH-RH ethylamide. *Biochem. Biophys. Res. Commun.* **59,** 1226–1232.

Vilchez-Martinez, J. A., Schally, A. V., Coy, D. H., Coy, E., Debeljuk, L. and Arimura, A. (1974c). *In vivo* inhibition of LH release by a synthetic antagonist of LH-releasing hormone (LH-RH). *Endocrinology*, **95,** 213–218.

Vilchez-Martinez, J. A., Coy, D. H., Coy, E. J., Schally, A. V. and Arimura, A. (1975). Anti-luteinizing hormone (LH-releasing activity of several analogues of LH-releasing hor mone. *Fertil. Steril.* **26,** 554–559.

Villee D. B. (1975). "Human Endocrinology". Saunders, Philadelphia.

Vincent, D. A. (1969). Ultrastructural study of the neural relationship of the pars intermedia of the ferret. *J. Endocr.* **43,** xlii-xliii.

Vincent, D. A. and Anand-Kumar, T. C. (1969). Electron microscopic studies on the pars intermedia of the ferret. *Z. Zellforsch.* **99,** 185–197.

Virkunnen, P., Leppaluoto, J. and Lybeck, H. (1973). The inhibition of TSH release from the pituitary by a TRH-related peptide. *Acta endocr., Copenh.* Suppl. **177,** 200.

Virkunnen, P., Lybeck, H., Partanen, J., Ranta, T., Leppaluoto, J. and Seppala, M. (1974). Plasma disappearance of injected synthetic LRH in man as measured by bioassay and radioimmunoassay. *J. Clin. Endocrinol. Metab.* **39,** 702–705.

Visser, T. J., Klootwijk, R. and Hennemann, G. (1975). A radioimmunoassay for the measurement of pyroglutamyl-histidine-proline, a proposed thyrotropin-releasing hormone metabolite. *J. Clin. Endocrinol. Metab.* **40,** 742–745.

Vodian, M. A. and Nicoll, C. S. (1975). Growth hormone releasing factor (GRF) and the bioassay-radioimmunoassay paradox revisited. *Ricerca Scientifica ed Educazione Permanente*, *2* (Suppl. 1), Abstract n. 56, University of Milan, Milan.

Voitkevich, A. A. (1962). Neurosecretory control of the amphibian metamorphosis. *Gen. Comp. Endocrinol.* Suppl. **1,** 133–147.

Vreeburg, J. T. M., Schnetlen, T. J. M. and Braum, H. J. (1975). Specific high affinity binding of 17β estradiol in cytosols from general brain regions and pituitary of intact and castrated adult male rats. *Endocrinology*, **97,** 969–977.

W

Wallace, A. L. C., Stacy, B. D. and Thorburn, G. D. (1973). Regulation of growth hormone secretion in the ovine foetus. *J. Endocr.* **58,** 89–95.

Wakerley, J. B. and Lincoln, D. W. (1971). Phasic discharge of antidromically identified units in the paraventricular nucleus of the hypothalamus. *Brain Res.* **25,** 192–194.

Wakerley, J. B. and Lincoln, D. W. (1973). The milk-ejection reflex of the rat: a 20- to 40-fold acceleration in the firing of paraventricular neurones during oxytocin release. *J. Endocr.* **57,** 477–493.

Wakerley, J. B., Poulain, D. A., Dyball, R. E. J. and Cross, B. A. (1975). Activity of phasic neurosecretory cells during haemorrhage. *Nature, Lond.* **258,** 82–84.

Walter, R. (1971). Conformations of oxytocin and lysine-vasopressin and their relationships to the biology of neurohypophyseal hormones. *In* "Structure-activity Relationships of Protein and Polypeptide Hormones" (M. Margoulies and F. C. Greenwood, eds) I. C. S. No. 241, pp. 181–193. Excerpta Medica Foundation.

Walter, R., Schwartz, J. L., Darnell, J. H. and Urry, D. S. (1971). Relation of the conformation of oxytocin to the biology of neurohypophyseal hormones. *Proc. Natl. Acad. Sci. USA*, **68,** 1355–1359.

Walter, R., Glickson, J. D., Schwartz, I. L. Havran, R. T., Meinhofer, J. and Urry, D. W. (1972a). Conformation of lysine-vasopressin: a comparison with oxytocin. *Proc. Natl. Acad. Sci. USA*, **69,** 1920–1924.

Walter, R., Schwartz, I. L., Hechter, O., Dousa, T. and Hoffmann, P. L. (1972b). Bromoacetyl-oxytocin, an irreversible inhibitor of neurohypophyseal hormone-stimulated adenylate cyclase, and a possible affinity label for hormone receptors. *Endocrinology*, **91,** 39–48.

Walti, M. and Hope, D. B. (1973). Synthesis of oxytocin: suppression of dimer formation. *Experientia*, **29,** 389.

Wang, C. F. and Yen, S. S. C. (1975). Direct evidence of estrogen modulation of pituitary sensitivity to LH-RH during the menstrual cycle. *J. Clin. Invest.* **55,** 201–204.

Watkins, W. B. and Evans, J. J. (1972). Demonstration of neurophysin in the hypothalamo-neurohypophysial system of the normal and dehydrated rat by use of cross-species reactive anti-neurophysins. *Z. Zellforsch.* **131,** 149–170.

Weeke, J. and Hansen, A. P. (1975). Serum TSH and serum T_3 levels during normal menstrual cycles and during cycles on oral contraceptives. *Acta. endocr., Copenh.* **79,** 431–438.

Wei, E., Maran, J. W., Dhariwal, A. P. S. and Yates, F. E. (1973). Regional responses of the pituitary to corticotrophin-releasing factor (CRF) and ammonium ions. *Endocrinology*, **92,** 710–715.

Weil, C., Breton, B. and Reinaud, P. (1975). Etude de la réponse hypophysaire à l'administration de Gr-RH exogène au cours du cycle reproducteur annuel chez la Carpe, *Cyprinus carpio* L. *C. R. Acad. Sci.* **280,** 2469–2472.

Weiner, R. I. (1975). Role of brain catecholamines in the control of LH and prolactin secretion. *In* "Hypothalamic Hormones" (M. Motta, P. Crosignani and L. Martini, eds) pp. 249–253. Academic Press, London and New York.

Weiner, R. I., Pattou, E., Kerdelhué, B. and Kordon, C. (1975). Differential effects of hypothalamic deafferentation upon luteinizing hormone releasing hormone in the median eminence and organum vasculosum of the lamina terminalis. *Endocrinology*, **97,** 1597–1600.

Weiss, P. (1944). Evidence of perpetual proximo-distal growth of nerve fibers. *Biol. Bull.* **87,** 160.

Weiss, P. and Hiscoe, H. B. (1948). Experiments on the mechanism of nerve growth. *J. exp. Zool.* **107,** 315–396.

Weissman, B. A., Daly, J. W. and Skokrich, P. (1975). Diethylstilbestrol-elicited accumulation of cyclic AMP in incubated rat hypothalamus. *Endocrinology*, **97,** 1559–1566.

Weisz, J. and Gibbs, C. (1974). Conversion of testosterone and androstenedione to estrogens *in vitro* by the brain of female rats. *Endocrinology*, **94,** 616–620.

Weitzmann, E. D. (1974). Temporal organization of neuroendocrine function in relation to the sleep-waking cycle in man. *In* "Recent Studies of Hypothalamic Function" (K. Lederis and K. E. Cooper, eds) pp. 26–38. Karger, Basel.

Weitzmann, E. D., Fukushima, D. and Nogeire, C. (1971). Twenty-four hour pattern of the episodic secretion of cortisol in human subjects. *J. Clin. Endocrinol. Metab.* **33,** 14–22.

Weman, B. and Nobin, A. (1973). The pars intermedia of the mink, *Mustela vison*. *Z. Zellforsch*, **143,** 313–327.

Westman, A. and Jacobsohn, D. (1937). Experimentelle Untersuchungen über die Bedeutung des Hypophysen-Zwischenhirnsystems für die Produktion gonadotroper Hormone des Hypophysenvorderlappens. *Acta obst. gynec. scand.* **17,** 235–265.

Westman, A. and Jacobsohn, D. (1938a). Endokrinologische Untersuchungen an Ratten mit durchtrennten Hypophysenstiel. III. Über die luteinizierende Wirkung des Follikelhormons. *Acta obst. gynec. scand.* **18,** 115–123.

Westman, A. and Jacobsohn, D. (1938b). Endokrinologische Untersuchungen an Ratten mit durchtrennten Hypophysenstiel. VI. Produktion und Abgabe der gonadotropen Hormone. *Acta path. microbiol. scand.* **15,** 445–453.

Wettemann, R. P. and Tucker, H. A. (1974). Relationship of ambient temperature to serum prolactin in heifers. *Proc. Soc. exp. Biol. Med.* **146,** 908–911.

Whitaker, S. and La Bella, F. S. (1973). Electron microscopic histochemistry of cholinesterase in the posterior, intermediate and anterior lobes of the rat pituitary. *Z. Zellforsch.* **142,** 69–88.

White, N., Jeffcoate, S. L., Griffiths, E. C. and Hooper, K. C. (1976). The effect of thyroid status on the thyrotrophin releasing hormone (TRH) degrading activity of rat serum. *J. Endocr.* **71,** 13–20.

White, W. F., Chen, A. J., Rippel, R. H., Story, J. C. and Schally, A. V. (1968). Some hypothalamic polyamines that develop pituitary follicle stimulating hormone. *Endocrinology*, **82,** 742–752.

White, W. F., Hedlund, M. T., Rippel, R. H., Arnold, W. and Flouret, G. R. (1973). Chemical and biological properties of gonadotrophin-releasing hormone synthetized by the solid-phase method. *Endocrinology*, **93,** 96–106.

White, W. F., Hedlund, M. T., Weber, G. F., Rippel, R. H., Johnson, E. S. and Wilber, J. F. (1974). The pineal gland: a supplemental source of hypothalamic releasing hormones. *Endocrinology*, **94,** 1422–1426.

Whittaker, V. P. (1959). The isolation and characterisation of acetylcholine-containing particles from brain. *Biochem. J.* **72,** 694–706.

Whitten, W. K. and Champlin, A. K. (1974). The role of olfaction in mammalian reproduction. *In* "Handbook of Physiology" (R. O. Greep and E. B. Astwood, eds) Section 7, Vol. 1, Part 1, pp. 109–123. American Physiological Society, Baltimore.

Wiener, N. (1948). "Cybernetics". Wiley, New York.

Wilber, J. F. (1971). Stimulation of ^{14}C-glucosamine and ^{14}C-alanine incorporation into thyrotropin by synthetic thyrotropin-releasing hormone. *Endocrinology*, **89,** 873–877.

Wilber, J. F. (1973). Thyrotropin releasing hormone: secretion and actions. *Annu. Rev. Medicine*, **24,** 353–364.

Wilber, J. F. and Porter, J. C. (1970). Thyrotropin and growth hormone releasing activity in hypophysial portal blood. *Endocrinology*, **27,** 807–811.

Wilber, J. F. and Seibel, M. J. (1973a). Thyrotrophin-releasing hormone interactions with an anterior pituitary membrane receptor. *In* "Hypothalamic Hypophysiotropic Hormones: Physiological and Clinical Studies" (C. Gual and E. Rosemberg, eds) pp. 182–188. Excerpta Medica, Amsterdam.

Wilber, J. F. and Seibel, M. J. (1973b). Thyrotropin-releasing hormone interaction with an anterior pituitary membrane receptor. *Endocrinology*, **92,** 888–893.

Wilber, J. F. and Utiger, R. D. (1968). *In-vitro* studies on mechanism of action of thyrotropin releasing factor. *Proc. Soc. exp. Biol. Med.* **127,** 488–490.

Wilber, J. F., and Utiger, R. D. (1969). The effect of glucocorticoids on thyrotropin secretion. *J. Clin. Invest.* **48,** 2096–2103.

Wilber, J. F., Nagel, T. and White, W. F. (1971). Hypothalamic growth hormone-releasing activity (GRA): characterization by the *in vitro* rat pituitary and radioimmunoassay. *Endocrinology*, **89,** 1419–1424.

Wilson, J. F. and Dodd, J. M. (1973). Effect of pharmacological agents on the *in vivo* release of melanophore-stimulating hormone in the dogfish, *Scyliorhinus canicula. Gen. Comp. Endocrinol.* **20,** 556–566.

Wilson, S. C. and Sharp, P. J. (1975a). Changes in plasma concentrations of luteinizing hormone after injection of progesterone at various times during the ovulatory cycle of the domestic hen (*Gallus domesticus*). *J. Endocr.* **67,** 59–70.

Wilson, S. C. and Sharp, P. J. (1975b). Episodic release of luteinising hormone in the domestic fowl. *J. Endocr.* **64,** 77–86.

Wilson, S. C. and Sharp, P. J. (1975c). Effects of progesterone and synthetic LH-RH on the release of LH during sexual maturation in the hen (*Gallus domesticus*). *J. Endocr.* **67,** 359–369.

Wilson, S. C. and Sharp, P. J. (1976a). Induction of luteinizing hormone release by gonodal steroids in the ovariectomized domestic hen. *J. Endocr.* **71,** 87–98.

Wilson, S. C. and Sharp, P. J. (1976b). Effects of androgens, oestrogens and deoxycorticosterone acetate on plasma concentrations of luteinizing hormone in laying hens. *J. Endocr.* **69,** 93–102.

Wingstrand, K. G. (1966). Microscopic anatomy, nerve supply and blood supply of the pars intermedia. *In* "The Pituitary Gland" (G. W. Harris and B. T. Donovan, eds) Vol. 3, pp. 1–27. University of California Press, Los Angeles.

Winokur, A. and Utiger, R. D. (1974). Thyrotropin releasing hormone: regional distribution in rat brain. *Science, N.Y.* **185,** 265–267.

Winokur, A., Davis, R., Utiger, R. D. (1977). Subcellular distribution of thyrotropin releasing hormone (TRH) in rat brain and hypothalamus. *Brain Res.* **120,** 423–434.

Winter, J. S. D. and Faiman, C. (1972). Serum gonadotropin concentrations in agonadal children and adults. *J. Clin. Endocrinol. Metab.* **35,** 561–563.

Winters, A. J., Eskay, R. L. and Porter, J. C. (1974). Concentration and distribution of TRH and LRH in the human fetal brain. *J. Clin. Endocrinol. Metab.* **39,** 960–963.

Wislocki, G. B. (1938). The vascular supply of the hypophysis cerebri of the Rhesus monkey and man. *Res. Publ. Ass. nerv. ment. Dis.* **17,** 48–68.

Wislocki, G. B. and King, L. S. (1936). The permeability of the hypophysis and hypothalamus to vital dyes, with a study of the hypophysial vascular supply. *Am. J. Anat.* **58,** 421–472.

Witorsch, R. J. and Brodish, A. (1972). Conditions for the reliable use of lesioned rats for the assay of CRF in tissue extracts. *Endocrinology*, **90,** 552–557.

Wittkowski, W. and Bock, R. (1972). Electron microscopical studies of the median eminence following interference with the feedback system of the anterior pituitary adrenal cortex. *In* "Brain-endocrine Interaction". (K. Knigge, D. E. Scott and A. Weindle, eds) pp. 171–180. Karger, Basel.

Woolf, P. D., Lee, L. A. and Schalch, D. S. (1972). Adrenergic manipulation and thyrotropin-releasing hormone (MRH) induced thyrotropin (TSH) release. *J. Clin. Endocrinol. Metab.* **35,** 616–618.

Worthington, W. C. Jr. (1955). Some observations on the hypophyseal portal system in the living mouse. *Bull. Johns Hopk. Hosp.* **97,** 343–357.

Worthington, W. C. Jr., Folsom, S. E. and Buse, M. G. (1972). Pituitary growth hormone "synthesizing factor" in hypophysial portal blood. *Endocrinology*, **90,** 1664–1667.

Wurtman, R. J. (1967). Effects of light and visual stimuli on endocrine function. *In* "Neuroendocrinology" Vol. II (L. Martini and W. F. Ganong, eds) pp. 19–59. Academic Press, New York and London.

Wurtman, R. J. (1970a). Brain monoamines and endocrine function. *Neurosciences Research Program Bulletin*, **9,** 172–297.

Wurtman, R. J. (1970b). Brain catecholamines and the control of secretion from the anterior pituitary gland. *In* "Hypophysiotropic Hormones of the Hypothalamus: Assay and Chemistry" (J. Meites, ed.) pp. 184–189. Williams and Wilkins, Baltimore.

Wurtman, R. J. and Cardinali, D. P. (1974). The pineal organ. *In* "Textbook of Endocrinology" (R. H. Williams, ed.) pp. 832–840. Saunders, Philadelphia.

Wuu, T. C., Crumm, S. and Saffran, M. (1971) Amino acid sequence of porcine neurophysin I. *J. Biol. Chem.* **246,** 6043–6063.

X

Xuereb, G. P., Prichard, M. M. L. and Daniel, P. M. (1954). The hypophysial portal system of vessels in man. *Quart. J. exp. Physiol.* **39,** 219–230.

Y

Yagi, K. and Sawaki, Y. (1973). Feedback of oestrogen in hypothalamic control of gonadotrophin secretion. *In* "Neuroendocrine Control" (K. Yagi and S. Yoshida, eds) pp. 297–325. University of Tokyo Press, Tokyo.

Yagi, K. and Sawaki, Y. (1975). Recurrent inhibition and facilitation demonstration in the tubero-infundibular system and effects of strychnine and picrotoxin. *Brain Res.* **84,** 155–160.

Yamashiro, D. and Li, C. H. (1973). Synthesis of a peptide with full somatostatin activity. *Biochem. Biophys. Res. Commun.* **54,** 882–888.

Yamamoto, K., Onaya, T., Kotani, M. and Yamada, K. (1972). Stimulation by synthetic thyrotropin-releasing hormone of glucose oxidation in porcine and rat pituitary slices. *Proc. Soc. exp. Biol. Med.* **140,** 677–680.

Yates, F. E. (1974). Modeling periodicities in reproductive, adrenocortical and metabolic systems. *In* "Biorhythms and Human Reproduction" (M. Ferin, F. Halberg, R. M. Richart and R. L. Vande Wiele, eds) pp. 133–142. Wiley, New York.

Yates, F. E. and Maran, J. W. (1975). Stimulation and inhibition of adrenocorticotrophin release. *In* "Handbook of Physiology, Endocrinology", Vol. IV (E. Knobil and W. H. Sawyer, eds) Part 2, pp. 367–404. American Physiological Society, Washington, D.C.

Yates, F. E., Leeman, S. E., Glenister, S. E., Glenister, D. W. and Dallman, M. F. (1961). Interaction between plasma corticosterone concentration and adrenocorticotrophin releasing stimuli in the rat. Evidence for the reset of an endocrine feedback control. *Endocrinology*, **68,** 67–80.

Yates, F. E., Russel, S. M., Dallman, M. F., Hedge, G. A., McCann, S. M. and Dhariwal, A. P. S. (1971). Potentiation by vasopressin of corticotropin release induced by corticotropin-releasing factor. *Endocrinology*, **88,** 3–15.

Yen, S. S. C., van den Berg, G., Rebar, R. and Ehara, Y. (1972). Variation of pituitary responsiveness to synthetic LRF during different phases of the menstrual cycle. *J. Clin. Endocrinol. Metab.* **35,** 931–934.

Yen, S. S. C., Lesley, B. L., Wang, C. R., Leblanc, H. and Siler, T. M. (1975). The operating characteristics of the hypothalamic-pituitary system during the menstrual cycle and observations of biological action of somatostatin. *Recent Prog. Horm. Res.* **31,** 321–363.

Ying, S-Y., Gove, S., Fang, V. S. and Greep, R. O. (1973). Ovulation in postpartum rats. *Endocrinology*, **92,** 108–115.

Yokoyama, A., Tomagane, H. and Ôta, K. (1974). Regulation of the surge of prolactin at proestrus in cycling rats. *In* "Biological Rhythms in Neuroendocrine Activity" (M. Kawakami, ed.) pp. 187–196. Igaku Shoin, Tokyo.

Young, R. T. T. (1973). Effect of propranolol on plasma growth hormone response in insulin-induced hypoglycemia in thyrotoxic patients. *J. Clin. Endocrinol. Metab.* **37** 968–971.

Yudaev, N. A., Utesheva, Z. F., Novikova, T. E., Shvachkin, Y. P. and Smirnova, A. P. (1973). A hypothalamic peptide stimulating the release of growth hormone. *Dokl. Akad. Nauk. USSR B*, **210,** 731–732.

Yuthasastrakosol, P., Palmer, W. M. and Howland, B. E. (1975). Luteinizing hormone, oestrogen and progesterone levels in peripheral serum of anoestrous and cyclic ewes as determined by radioimmunoassay. *J. Reprod. Fertil.* **43,** 57–63.

Z

Zambrano, D. and DeRobertis, E. (1966). The secretory cycle of the supraoptic neurons in the rat. A structural-functional correlation. *Z. Zellforsch.* **73,** 414–431.

Zambrano, D. and DeRobertis, E. (1967). Ultrastructure of the hypothalamic neurosecretory system of the dog. *Z. Zellforsch.* **81,** 264–282.

Zambrano, D., Clarke, W. C., Hawkins, E. F., Sage, M. and Bern, H. A. (1973/74). Influence of 6-hydroxydopamine on hypothalamic control of prolactin and ACTH secretion in the teleost fish, *Tilapia mossambica. Neuroendocrinology*, **13,** 284–298.

Zanartu, J., Dabancens, A., Kastin, J. and Schally, A. V. (1974). Effect of synthetic hypothalamic gonadotrophin-releasing hormone (FSH/LH-RH) in anovulatory sterility. *Fertil. Steril.* **25,** 160–169.

Zanisi, M., Motta, M. and Martini, L. (1973). Feedback activity of testosterone and of its 5α-reduced metabolite. *In* "The Endocrine Function of the Human Testis" (V. H. T. James, M. Serio and L. Martini, eds) Vol. II, pp. 431–438. Academic Press, New York and London.

Zarate, A., Canales, E. S., Sovia, J., Ruiz, F. and MacGregor, C. (1972). Ovarian refractoriness during lactation in women: effect of gonadotrophin stimulation. *Am. J. Obstet. Gynecol.* **112,** 1130–1132.

Zarate, A., Garrido, J., Canales, E. S., Soria, J. and Schally, A. V. (1974). Disparity in the negative gonadal feedback control for LH and FSH secretion in cases of germinal aplasia or sertoli-cell only syndrome. *J. Clin. Endocrinol. Metab.* **38,** 1125–1127.

Ziegler, B. (1963). Licht-und Elektronenmikroskopische untersuchungen an Pars Intermedia und Neurohypophyse der Ratte. *Z. Zellforsch*, **59,** 486–506.

Zimmerman, E. A., Carmel, P. W., Husain, M. K., Ferin, M., Tannenbaum, M., Frantz, A. G. and Robinson, A. G. (1973a). Vasopressin and neurophysin: high concentrations in monkey hypophyseal portal blood. *Science, N.Y.* **182,** 925–7.

Zimmerman, E. A., Hsu, K. C., Robinson, A. G., Carmel, P. W., Frantz, A. G. and Tannenbaum, M. (1973b). Studies of neurophysin secreting neurons with immunoperoxidase techniques employing antibody to bovine neurophysin. I. Light microscopic findings in monkey and bovine tissues. *Endocrinology*, **92,** 931–940.

Zimmerman, E. A., Hsu, K. C., Ferin, M. and Kozlowski, G. P. (1974). Localisation of gonadotrophin-releasing hormone (GnRH) in the hypothalamus of the mouse by immunoperoxidase technique. *Endocrinology*, **95,** 1–8.

Zolovick, A. and Labhsetwar, A. P. (1973). Evidence for the theory of dual hypothalamic control of ovulation. *Nature, Lond.* **245,** 158–159.

Zondek, B. and Tamari, I. (1967). Effects of auditory stimuli on reproduction. *In* "The Effects of External Stimuli on Reproduction" (G. E. W. Wolstenholme and M. O'Connor, eds) pp. 4–16. Churchill, London.

Zor, U., Kaneko, T., Schneider, H. P. G., McCann, S. M., Lowe, I. P., Bloom, G., Borland, B. and Field, J. B. (1969). Stimulation of anterior pituitary adenyl cyclase activity and adenosine 3′5′ cyclic phosphate by hypothalamic extract and prostaglandin E_1. *Proc. Natl. Acad. Sci., USA*, **63,** 918–925.

Zuckerman, S. (1970). The new secretions of the brain. *In* "Beyond the Ivory Tower: the Frontiers of Public and Private Science", pp. 46–60. Weidenfeld and Nicolson, London.

INDEX

(*Numbers in italics refer to tables or figures*)

B

D

E

H

M

N

R